Trace Element Speciation: Analytical Methods and Problems

Editor

Graeme E. Batley
Senior Principal Research Scientist
Centre for Advanced Analytical Chemistry
Division of Fuel Technology
Commonwealth Scientific and Industrial
Research Organization (CSIRO)
Lucas Heights, New South Wales, Australia

CRC Press, Inc.
Boca Raton, Florida

Library of Congress Cataloging-in-Publication Data

Trace element speciation: analytical methods and problems / editor,
Graeme E. Batley
p. cm.
Bibliography: p.
Includes index.
ISBN 0-8493-4712-2
1. Trace elements in water—Speciation. 2. Trace elements in water—Analysis. 3. Trace elements—Bioavailability. I. Batley, Graeme E.
QP534.T722 1989 88-17560
546′.226—dc19 CIP

This book represents information obtained from authentic and highly regarded sources. Reprinted material is quoted with permission, and sources are indicated. A wide variety of references are listed. Every reasonable effort has been made to give reliable data and information, but the author and the publisher cannot assume responsibility for the validity of all materials or for the consequences of their use.

Direct all inquiries to CRC Press, Inc., 2000 Corporate Blvd., N.W., Boca Raton, Florida, 33431.

Second Printing, 1990
Third Printing, 1991

International Standard Book Number 0-8493-4712-2

Library of Congress Card Number 88-17560
Printed in the United States

PREFACE

Biochemists have long been aware that assimilation by the human body of essential trace elements takes place in certain preferred chemical forms or species. In environmental science the importance of chemical speciation mesurement assumed importance in the late 1960s following earlier pollution incidents such as that at Minamata Bay. The toxicity and fate of waterborne metal contaminants was shown to be highly dependent on chemical form, and it was realized that analytical measurement of these forms would be more meaningful than data for total element concentrations. This finding was echoed by monitoring authorities at a time when speciation methods were still evolving.

The development of analytical procedures for measuring specific forms of trace elements often at sub-part-per-billion concentrations represents a major challenge to the analytical chemist. It is especially important that the measured fractions be environmentally significant in terms of bioavailability, toxicity, or transport. The plethora of sophisticated analytical instrumentation available today makes the choice of a technique for speciation analysis a difficult one. The lack of a text that comprehensively reviewed speciation methods led to the genesis of this volume.

With the assistance of my colleagues, this book has been compiled with the analytical chemist specifically in mind, by providing not only a background to speciation studies, but discussing the peculiar problems of each technique to assist in the choice of a technique or method. The chapter authors are all acknowledged leaders in speciation research, and I am grateful for their willing acceptance and cooperation.

My own efforts towards this volume have been facilitated by the help of a number of colleagues and I must particularly thank Mark Florence, Greg Morrison, Bo Hoyer, Gary Low, and Roy Harrison for discussions and comments on manuscripts. The translation of my manuscripts to intelligible typescripts was a challenge expertly accomplished by Lorraine Lloyd, Judith Masson, and Anne Wilson, and for their efforts I am especially grateful. In addition, I must acknowledge the assistance of the library and editorial staff at the Lucas Heights Research Laboratory and my associates in the Division of Fuel Technology.

Graeme Batley

THE EDITOR

Graeme Batley, Ph.D., is a Senior Principal Research Scientist in the Centre for Advanced Analytical Chemistry at the Commonwealth Scientific and Industrial Research Organization (CSIRO) Division of Fuel Technology, at Lucas Heights, New South Wales, Australia.

Dr. Batley received his B.Sc. degree with First Class Honors from the University of New South Wales, Sydney, Australia in 1961. He obtained M.Sc. and Ph.D. degrees in 1964 and 1967, respectively, from the Departments of Analytical Chemistry and Inorganic Chemistry at the same University. After postdoctoral research at the University of Illinois, Urbana-Champaign, he was appointed in 1969 to the Analytical Chemistry Section of the Australian Atomic Energy Commission at the Lucas Heights Research Laboratories. In 1982, this section became part of the CSIRO Division of Energy Chemistry. In 1980 to 1981 he was a visiting scientist at the Canada Centre for Inland Waters, Burlington, Ontario.

Dr. Batley is a Fellow of the Royal Australian Chemical Institute, foundation Chairman of their NSW Environmental Chemistry Group and subeditor for their journal, *Chemistry in Australia.* He is the Australian representative on the IUPAC Commission on Electroanalytical Chemistry and a member of their subcommittee on Electroanalytical Methods of Environmental Trace Analysis.

He has over 100 research publications and has presented numerous invited lectures at international and national meetings and at universities and other research organizations. His current major research interests are the study of the analysis, fate, and transport of trace element species in natural water systems, principally using electroanalytical and chromatographic methods.

CONTRIBUTORS

Graeme E. Batley, Ph.D.
Senior Principal Research Scientist
Centre for Advanced Analytical Chemistry
Division of Fuel Technology
CSIRO
Lucas Heights, New South Wales, Australia

Y. K. Chau, Ph.D., D.Sc.
Research Scientist
National Water Research Institute
Canada Centre for Inland Waters
Burlington, Ontario, Canada

T. Mark Florence, D.Sc.
Senior Principal Research Scientist
Centre for Advanced Analytical Chemistry
Division of Fuel Technology
CSIRO
Lucas Heights, New South Wales, Australia

Ulrich Förstner, Dr.rer.nat.
Professor
Umweltschutztechnik
Technische Universität Hamburg
Hamburg, West Germany

Michael Kersten, Dr. Ing.
Research Scientist
Umweltschutztechik
Technische Universität Hamburg
Hamburg, West Germany

Gary K.-C. Low, Ph.D.
Senior Research Scientist
Centre for Advanced Analytical Chemistry
Division of Feul Technology
CSIRO
Lucas Heights, New South Wales, Australia

Gregory M. P. Morrison, Ph.D.
Research Scientist
Department of Sanitary Engineering
Chalmers University of Technology
Göteborg, Sweden

T. David Waite, Ph.D.
Senior Research Scientist
Environmental Science Division
Australian Nuclear Science and Technology Organization
Lucas Heights, New South Wales, Australia

P. T. S. Wong, Ph.D.
Research Scientist
Great Lakes Laboratory for Fisheries and Aquatic Sciences
Canada Centre for Inland Waters
Burlington, Ontario, Canada

TABLE OF CONTENTS

Chapter 1

COLLECTION, PREPARATION, AND STORAGE OF SAMPLES FOR SPECIATION ANALYSIS

Graeme E. Batley

TABLE OF CONTENTS

I. INTRODUCTION

As has been emphasized many times previously,[1-3] the techniques used in sample collection, preparation, and storage are critical in any analyses being carried out at trace or ultratrace concentrations. Errors introduced in these stages through either sample contamination or adsorptive losses of particular trace elements can negate many hours of more complex analytical measurement. In studies of trace element speciation, potential errors are compounded by the possibility during sample handling and storage of the transformation of chemical species, resulting in an erroneous species distributions being proposed, in addition to errors in total element concentrations. It is essential, therefore, that these operations be performed with an awareness of possible sources of contamination, losses, and species transformations, using a protocol which minimizes such perturbations. This chapter reviews these procedures.

Particular emphasis will be given in this review to natural water samples, where the problems during sampling and storage are more complex than those encountered in solid samples. Natural waters are mixtures containing biological and chemical species in dynamic equilibrium, which can be disturbed by the mere act of sampling. This could be brought about by exposure to oxygen or to container walls, or as a result of changes in pressure or temperature.

II. WATER SAMPLES

A. Water Sampling

A major problem in all sampling exercises is that of obtaining a representative sample. In natural waters, trace element concentrations will vary with depth, salinity, and the proximity to discharge points. The sampler must decide initially upon the information that the samples are required to supply, and design the collection strategy accordingly. If, for example, long-term changes are being examined, a number of defined sampling locations should be selected which can be resampled at regular intervals.

For the collection of discharge waters flowing in pipes or channels, it is important to first establish, by experiment, the homogeneity of the trace element distribution. Sampling from the edges, surfaces, or bottom of discharge channels should be avoided, as these are usually not representative of the bulk sample.

Where sampling is being carried out from a boat, careful consideration must be given to pollution of the sample by the boat itself, whatever its size. Figure 1 illustrates the zones of contamination, A, B, and C associated with a research vessel in open ocean waters. In the wake of the ship, A and B contain waterborne pollutants, while C represents airborne pollutants. Sampling should be performed in zone D from a smaller rubber or aluminium vessel.[4] Similar constraints apply to sampling enclosed waters from smaller vessels, namely sampling from the windward side, preferably from the bow, remote from any outboard motor discharges or antifouling paints.

For surface water sampling, immersion by hand of a polyethylene sampling bottle to well below the surface to avoid any surface film, is generally adequate. Polyethylene gloves should be worn. Surface samples can also be collected using polyethylene buckets attached by a nylon rope. A telescopic bottle holder for contamination-free sampling has been described by Mart[4] (Figure 2). The lower trace metal content of ocean waters necessitates more stringent precautions during sampling than are necessary for estuarine or freshwater sites.

For depth sampling, a range of standard sampling bottles has been used with varied success.[3,5] These are best constructed from polyethylene, polypropylene, polycarbonate, Teflon® or Plexiglas® (Perspex). Samplers should be free of rubber bands, springs or other

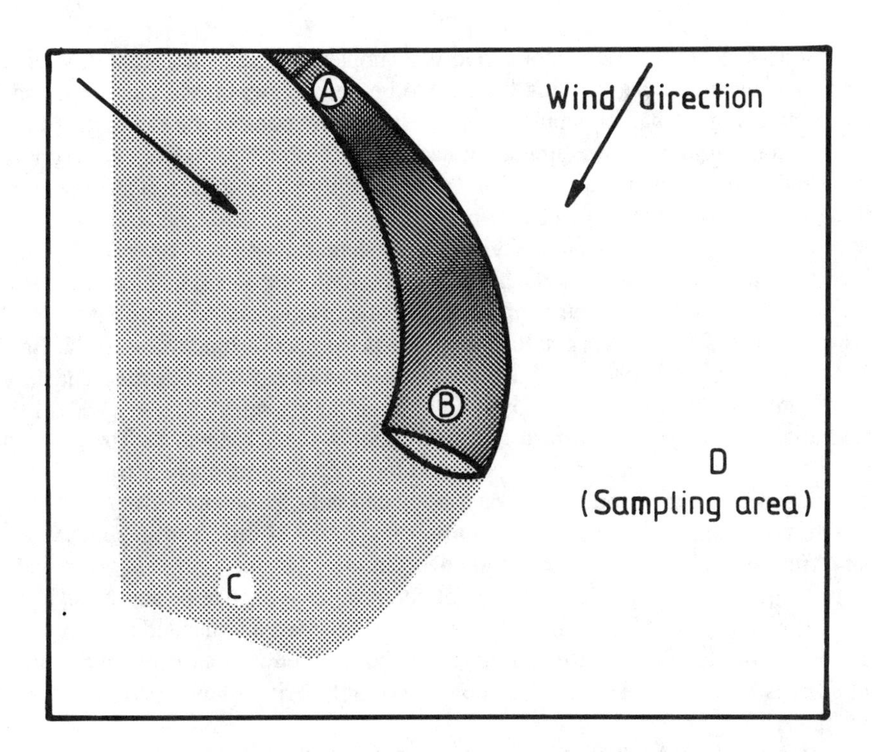

FIGURE 1. Sampling zones for the collection of unpolluted surface seawater. (A) Waters crossed by ship, severely polluted; (B) drifting zone, severely polluted; (C) zone of airborne contamination from ship, moderately polluted; (D) desirable sampling zone. (From Mart, L., *Fresenius Z. Anal. Chem.*, 299, 97, 1979. With permission.)

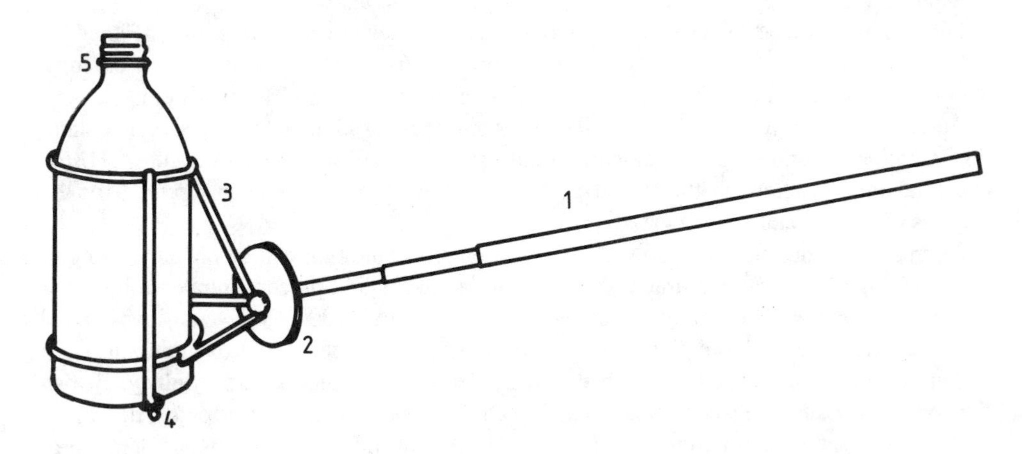

FIGURE 2. Suitable equipment for sampling surface seawater. (1) Telescopic fiberglass bar (3.5 m); (2) polyethylene disk; (3) nylon-coated holder; (4) silicone tubing fastener; (5) polyethylene or Teflon® sampling bottle. (From Mart, L., *Fresenius Z. Anal. Chem.*, 299, 97, 1979. With permission.)

rubber or metallic components. Contamination from rust or grease on the hydrowire should also be avoided. For heavy metal analyses, Van Dorn,[6,7] Niskin,[8-11] Natural Institute of Oceanography (NIO) bottles,[3] and Go-Flo bottles[11] have been used, with the replacement of rubber bands with Teflon®-coated springs and the wrapping rubber of end-closures with thin polyethylene film.

A comparison of surface seawater samples collected in a 1-l FEP Teflon® bottle and in a 30-l Teflon®-coated, polyvinylchloride Go-Flo sampler, showed significant differences in zinc and lead concentrations.[12] Using the Go-Flo bottle, zinc was up to ten times, and lead up to three times higher than in samples collected in Teflon® bottles by a swimmer. Cadmium and copper concentrations were similar in each case. Teflon® samplers, as described by Patterson and Settle,[13] Friemann et al.,[14] and more recently by Brugmann et al.,[15] are without doubt the best for trace metal studies.

Pumping systems offer a valuable alternative for shallow waters ($<$100 m), although the higher surface to volume ratio may enhance the possibility of contamination. The use of a vacuum pump to draw water through carefully cleaned polyethylene tubing to a large Buchner flask is acceptable, although a peristaltic pump also performs this function usefully. Harper[16] described an all-Teflon®, pneumatically driven bellows pump for collecting and filtering water onboard ship from a buoy deployed away from the vessel.

Grab sampling is the procedure most commonly applied to natural waters. The grab sample represents the conditions existing at the point and time of sampling only. A complete picture of the variation in composition of the water body can only be obtained from grab samples taken at different sampling points within a water body, and at different time intervals. The sampling frequency should be sufficient to cover seasonal changes; biweekly or monthly is generally adequate, unless particular events of compositional change or pollutants input are known. The complexity of the sampling profile can be increased as desired by studying both vertical and horizontal transects. Such studies may be required where the mixing of different water bodies is being examined, e.g., the confluence of two rivers having different chemical compositions.

Composite sampling involves the pooling of samples collected at the same site at varying time intervals, or from different sites, or a combination of both site and time variables. This is often the simplest way of achieving a sample truly representative of the main body of water. Usually this procedure is applied to major element analyses, on the combination of a number of grab samples used for specific minor element analyses.

In the sampling of effluent streams, flow-related samples are more important. Either the volume collected or the frequency of collection of a fixed volume is varied to obtain a composite more accurately reflecting the average condition of an effluent system.

Automatic sampling devices are available for continual or high-frequency sample collection. Continual monitoring is often carried out for parameters such as temperature, pH, E_h, and conductivity, although the use of continuous flow systems has been reported for the analysis of trace metals in seawater.[3]

An integrated measure of dissolved trace element concentration can be obtained using *in situ* sampling where water is drawn at a constant rate through a preconcentrator column for a fixed time interval. Filtration and collection on a Chelex®-100 column was investigated by Davey and Soper.[17] A sampler based on the polystyrene-bonded 8-hydroxyquinoline resin system of Willie et al.[18] has been developed by Seastar Instruments, Ltd. (Sidney, British Columbia, Canada). Between 90 and 100% recoveries were obtained for ionic Cd, Cu, Pb and Zn from seawater for sampling flow rates up to 200 ml min^{-1}. Both samplers are of limited value in speciation studies, although it is likely that both collect less than the total metal component. The aluminum hydroxide-coated cation exchange resin column of Zhang and Florence,[19] described in more detail in Chapter 3, is currently being developed as an *in situ* sampler for bioavailable metal species.

B. Water Sample Pretreatment

Unless analyses are to be performed immediately following sample collection, careful consideration must be given to the manner in which the sample is to be treated prior to storage. This treatment will depend on the analyses required. In general, for trace metal analyses, particulate matter is first removed from the sample by filtration or centrifugation.

Preservative reagents may then be added to the sample, and the sample stored in an appropriate container under conditions which minimize contamination or losses of metals from solution.

In unfiltered samples, contact of the dissolved fraction with particulate matter for extended periods of time is likely to lead to changes in the distribution of chemical forms of heavy metals in solution. Several workers[20-22] have shown that the equilibrium times for adsorption and desorption of heavy metals in sediment-water mixtures are rapid and less than 72 h. Maximum adsorption occurs at pH values above 7.5. Concentration factors for metals in sediments may be as high as 50,000. With any change in solution equilibrium after collection, adsorption sites provided by particulate matter will provide a path for removal of metals species,[23] while under some conditions, desorption of adsorbed metal is possible. Studies by Mart,[24] however, showed that the storage of unfiltered seawater samples in polyethylene containers resulted in no losses of total dissolved heavy metals such as Pb, Cu, Cd, and Bi, although in sewage samples, evidence has been presented[25] that particulate matter can adhere to the walls of containers, leading to significant decreases, for example, in total lead.

High bacterial concentrations associated with sedimentary material will also lead to depletion of soluble metal species.[26] McLerran and Holmes[27] have reported the removal of 85% of added ^{65}Zn and 70% of added ^{109}Cd by a bacterial culture in 2 h. The growth of bacteria and algae involves photosynthesis and respiration which will give rise to changes in the carbon dioxide content of the water and therefore in its pH. Changes in pH may result in precipitation, changes in the degree of complexation and adsorptive behavior, and in the rate of redox reactions involving heavy metal species in solution.

Losses of 10 to 30% of dissolved mercury from seawater have been observed when using untreated membrane filters;[26] however, silicone treatment of glass fiber filters was successful in reducing losses to less than 7%. Losses of mercury, copper, and cadmium during filtration were also reported.[22] Approximately 7% of cadmium was lost during the filtration of 25 ml of a sample of tapwater.

Because of the unpredictable nature of bacterial growth in stored samples,[29] it is advisable that filtration be performed as soon as possible after sample collection. If this time is in excess of several hours, the sample should be chilled (not frozen) to near 4°C to retard bacterial growth until filtration can be completed.[30]

It is conventional to filter the sample through a 0.45-μm filter to separate the defined ''dissolved'' and ''particulate'' phases. The former will of course contain colloidal particles (0.1 to 0.001 μm) of both microbiological and bacterial origin as well as soluble species (<0.001 μm). Filters of 0.45 μm pore size retain all phytoplankton and most bacteria. Continued filtration can lead to clogging of pores with a reduction in pore diameter as filtration proceeds. This can be overcome by more frequent replacement of filters in samples with a high content of suspended matter or by use of stirred pressure filtration.

The type of filtration apparatus to be used is also an important consideration. Thought should be given to the materials of construction, Pyrex®, glass, or polycarbonate, and the mode of filtration, vacuum or pressure. Glass filters with rubber bungs are potential sources of contamination. All Pyrex® glass vacuum filter units having a male, standard-taper joint on a 1-l flask are preferred, although in waters buffered to pH 5 and 7, glass filter supports were found to result in greater adsorptive losses of ionic Cu additions than polycarbonate supports.[31] The range of membrane types is discussed in Chapter 3.

Prior to filtration, the apparatus should be cleaned with dilute acid. Soaking overnight in 1 to 3 *M* hydrochloric acid is usual,[3] although for ultratrace analysis, a more rigorous procedure has been described.[24] There are also differences in the procedures recommended for the cleaning of membrane filters. Scarponi et al.[32] reported that seawater leaches metals from uncleaned filters, but after passage of 1-l of seawater, the contamination becomes negligible. Filter surfaces of unconditioned membranes were found to strongly adsorb cadmium and lead from distilled water, but there were no changes in concentrations during filtration of river water.[33] The use of complexing agents as a wash[3] or the soaking of filters in seawater has been recommended.[33] In our laboratory, a 20-ml wash with 2 *M* nitric acid followed by distilled water (50 to 100 ml) is used. The receiving flask should be thoroughly rinsed free of acid with distilled water, then 10 to 20 ml of sample filtered, and the filtrate discarded. Presoaking of filters in acid is only necessary for the analysis of deep ocean waters, reducing total metal concentrations in 47-mm filters to 3 ng for lead and 1 ng for cadmium.[24]

Both pressure and vacuum filtration have been used. Pressure filtration[11,34] offers advantages in terms of filtration speed and is preferable with freshwater samples having a high suspended sediment load, where filtration rates using vacuum, may be as low as 100 ml h^{-1} for 47-mm diameter, 0.45-μm filters. Stirring will prevent the formation of concentration gradients and clogging of filters. Pressure filtration is commonly used with ultrafiltration membranes (Chapter 3).

With either vacuum or pressure filtration, it is essential that only low pressures are used.[24] Rupture of phytoplankton cells will occur at pressures greater than 30 kPa and this may lead to increases in heavy metal concentrations in the filtrate and to changes in heavy metal speciation as a result of the increase in dissolved organic matter. Concentration factors of 3×10^4 for zinc, lead, and Cu in phytoplankton have been reported.[35]

For difficult-to-filter samples, centrifugation can be a useful alternative,[36] but may lead to significant contamination.[37] The effectiveness of centrifugation in separating particulates is a function of the centrifuge speed, time, and particle density, and because of this, centrifuged samples cannot readily be compared with filtered samples. For continuous centrifugation it is necessary to use Teflon® or polyethylene liners which must be rigorously cleaned before use. Similar precautions apply to polyethylene or Teflon® tubes for batch centrifugation. Insignificant mercury losses were observed when using these techniques.[38]

The problems of adsorption of trace metals on container surfaces can be overcome by the use of freeze drying to concentrate samples. Filby et al.[39] used this technique effectively for lake and river water samples. Filtrates were collected directly in polyethylene bags, frozen in dry ice-acetone, then freeze dried. No losses of volatile elements such as antimony and arsenic were reported. This technique is particularly suited to samples for neutron activation analysis.

C. Water Sample Storage

Considerable data are now available on the suitability of sample containers for the storage of filtered water samples. To be able to fully assess these results, it is important to consider the means by which losses of trace metals to container surfaces were estimated. A major problem in storage experiments has been the actual measurement of natural levels of heavy metals, and for this reason, many workers have resorted to radiotracers to follow storage losses. Such studies, however, possibly overestimate these losses as the time of equilibration of added radiotracer will differ for different chemical species of the same metal. While the equilibration time is very rapid for ionic metal and radiotracer spikes, Batley and Florence[40] showed that for lead and copper, for example, a significant percentage of the total metals in an unacidified seawater sample are ''irreversibly'' associated with organic and inorganic matter in solution to the extent that they are not exchanged with added radiotracer after 5 d. While 100% of the tracer may be lost during a test period, this may represent a much smaller percentage of the total metal orginally present. The results of storage experiments based on the addition of ionic metal or radiotracer spikes should therefore be regarded with suspicion.

A second approach to storage studies has been the addition to the natural sample of a measurable ionic metal spike, the losses of which can be readily followed. Experiments of this nature may be misleading since ionic species may only represent a small portion of the naturally present metal. Large ionic metal additions will drastically disturb the solution equilibria in the natural sample and the adsorptive behavior of metal species may be quite different from that of the natural sample. The most reliable experiments are those which attempt to directly measure metal concentrations as a function of storage time.

A range of container materials have been used, including polyethylene, polypropylene, Teflon®, polycarbonate, and borosilicate glass. The choice of container is determined both by its adsorptive properties and the presence of surface impurities.

Glasses have been shown to function as weak ion exhangers.[40] In weakly acidic and slightly alkaline solutions, the negatively-charged silicic acid groups on the surface permit cation exchange.[42,43] Doremus[43] showed that the potential ion exchange capacity of soda glass is significantly higher than a standard polysulfonate resin. The introduction of borosilicate groups alters the adsorptive behavior of the glass, resulting in an order of magnitude decrease in ion exchange ''capacity''. Soda glass containers should therefore be avoided.

The adsorption of metal ions on glasses and oxide surfaces has been the subject of many investigations, and it is now well established that the degree of adsorption is dependent on the ability of the metals to be hydrolyzed. Little adsorption occurs in acid solutions where simple ionic metal species are present, but with increasing pH and the formation of hydrolyzed metal ions having a reduced positive charge, increased adsorption is observed.[44] Significant adsorption occurs at a lower pH for the more readily hydrolyzed metal ions such as Fe(III) and Cr(III).

The adsorptive behavior of hydrophobic organic polymers such as polyethylene or Teflon® is believed to involve ion exchange at a charged double layer on the polymer surface.[45] It has been proposed that on Teflon®, this layer comprises hydroxyl ions sorbed by either Van der Waals forces or by hydrogen bonding.[46] The existence of a negative surface charge has been confirmed by electroosmosis measurements.

In general, adsorptive losses appear to be lower on polyethylene or Teflon® than on Pyrex® glass. The application of a hydrophobic silicone coating to Pyrex® surfaces has been shown to significantly reduce the adsorption of a number of heavy metals.[47-49]

Both polymer materials and glass may contain heavy metal impurities capable of contaminating the sample. These may originate from catalysts, promotors or metal dies used in the manufacturing process. Organic plasticizers may also be released during storage,[50] and these may effect metal speciation either as a result of their redox properties or through metal complexation. Cleaning methods are therefore required to both reduce contamination to acceptable levels and to minimize the potential for heavy metal losses by adsorption onto container walls. These procedures have been comprehensively investigated in a number of studies.[3,51-53]

In general, soaking in dilute acids is sufficient to remove surface metal impurities; however, it is possible that this may activate adsorption sites.[3] Laxen and Harrison[51] investigated a range of procedures for cleaning polyethylene containers for storage of freshwater samples, including soaking of acid-washed containers in solutions of magnesium and calcium sulfates. On the basis of their results, a 48-h soak in 10% HNO_3 was recommended, followed by a water rinse. No changes were observed in the concentrations of Cu, Pb, Cd, and Zn in tapwater (pH 7.4) stored in this manner. These results are consistent with those of Florence[54] who noted no changes in speciation of these elements in feshwater stored in acid-washed polyethylene containers.

Mart[24] recommended washing of polyethylene bottles with 10% HCl heated to 70°C for 4 d to increase the leaching rate, repeated three times, finally using dilute high-purity acid. Other authors[55] reduced the cleaning time by using ultrasonication in 1 *M* HNO_3, then 0.1 *M* HCl for 3 h each, following a 3-*M* HNO_3 soak at 60°C.

Teflon® containers need a stronger cleaning process, as TFE Teflon® is a sintered material which may contain metal contaminants from its manufacture which are difficult to remove. For this purpose, Mart[24] recommended an initial treatment with 50% HNO_3. A similar procedure was recommended for quartzware.

There is now general agreement that polyethylene and Teflon® containers are best for the storage of water samples for trace heavy metal analysis. For preservation of samples, however, a range of treatment options are practiced. Where only a total metal analysis is required, it is usual to minimize adsorptive losses by acidifying the sample after filtration to 0.05 *M* H^+ with either hydrochloric or nitric acids. Acidification before filtration, however, will release metals from particulate matter and is not advisable; however, for some samples, this contribution from particulates to the total dissolved metal content can be shown to be only minor. Note that corrections must be made for acid blanks, and if these are measured by the addition of acid to distilled or resin-purified (Milli-Q) water, it should be borne in mind that often carefully distilled water will be higher in trace metals than the natural water samples being studied. Acid addition is of course detrimental in metal speciation studies.

A number of laboratories practice freezing of water samples after filtration.[56,57] Freezing has been used as a means of concentrating organic materials from water samples[58,59] and intuitively one might expect freezing to affect the distribution of heavy metal species unless a flash freezing in liquid nitrogen was adopted.[60]

Total cadmium, lead, and copper concentrations in unacidified seawater samples stored at -45°C for 3 months, were found to be unchanged, with the only changes in metal speciation being a decrease in labile copper associated with colloidal organic and inorganic matter.[61] Baier[62] reported no differences in lead concentrations between frozen and unfrozen seawater samples; however, freezing of freshwater samples was found to reduce the concentrations of labile copper and lead species, although the total metal concentrations were unchanged.[61]

During storage at room temperature of samples filtered through 0.45-μm membranes, regrowth of particles has been observed[63] to the extent that after several days, large numbers of particles were present in sizes greater than 4 μm in diameter. Evidence was obtained that particle formation occurred by growth and aggregation of bacteria, possibly involving syntrophic growth of more than one bacterial species, although aggregation of colloids could also be occurring. Changes in the bacterial content of samples during storage are unpredictable. Gillbricht[64] observed that in closed systems, the overall concentrations of organic matter decreased during storage, losses most probably occurring as CO_2 during periods of high bacterial growth. Moebus[29] found that the number of colony-forming marine bacteria fluctuated over three orders of magnitude during several months. Such changes can provide pathways for the removal of metals or transformations of metal species. Bacterial activity can be reduced by storage at 4°C, and a decrease in storage losses of heavy metals under these conditions has been reported.[65] With seawater samples, no significant differences with respect to cadmium, lead, and copper speciation were observed between samples stored at room temperature and 4°C over a 3-month period.[61] Similar results were obtained for freshwater samples.[54] Species were measured using anodic stripping voltammetry (ASV) and the speciation scheme described previously.[40] The long-term stability of heavy metal concentrations in unacidified waters stored in properly cleaned containers has since been confirmed in a number of studies.[32,51,55,66]

Although storage experiments have demonstrated the absence of species transformation during long-term storage, the period from the time zero to when the first measurements are made, has not been considered. This may be as long as 8 h, although 2 to 3 h would be more typical. In this regard, the studies of Piotrowitz et al.[67] suggest that perhaps within this period, interruptions to the natural steady-state production and destructive equilibria involving photooxidation and bacterial processes may result in changes in metal species distributions. The time constants for these processes may vary from hours to days. This can only be overcome by either performing *in situ* speciation measurements or greatly reducing the time between sampling and analysis.

Storage conditions applicable to samples for heavy metal analyses do not necessarily apply, for example, to mercury, arsenic, selenium, or organometals. Mercury in particular has been the subject of many storage experiments.[3,68-70] Polyethylene containers are permeable to mercury vapor, representing a potential contamination source.[50,71] Losses of mercury can occur by reduction to the metal or by binding to the container, and trace constituents present in either the water or the plastic container may accelerate these processes.

The high reduction potential of the Hg(II)/Hg(I) couple means that mercury(II) ions present in natural waters will be susceptible to reduction by traces of mild reducing agents in the solution. These may be in the form of plasticizer materials leached from the container, organic detritus, or naturally occuring bacterial reductants. The mercury(I) ions so formed are then capable of disproportionation to metallic mercury and mercury(II). Studies by Shimomura et al.[72] and Toribara et al.[73] suggested this as a likely mechanism for metallic mercury formation. Baier et al.[74] suggested the bacterial conversion of $\mu g\ l^{-1}$ concentrations of inorganic mercury in waters to the organic and/or elemental form as an alternative biochemical route for the production of volatile mercury forms. Other data[68,70] indicate that only a fraction of mercury loss from solution was attributable to the amount adsorbed on the container walls, but because of the nature of bacterial growth in solution, these amounts are subject to some variation. It was verified that mercury loss was less from sterile solutions. Such decomposition can be controlled by refrigeration or freezing of the sample.[70,75]

A range of preservatives has been shown to prevent mercury losses, including oxidants such as permanganate[76] or dichromate,[76,77] or complexing agents like cysteine[78] or humic acids,[79] the oxidants being effective only in acidified solutions. Heiden and Aikens[80] claimed that pretreatment of polyolefin bottles by leaching with chloroform followed by exposure to aqua regia vapor was superior to chemical preservatives. Furthermore, it could be applied to unacidified samples, although the effects on mercury speciation were not investigated.

Methylmercury losses from seawater were found to be significant between pH 4 and 10.[81] In the presence of 20% HCl there were no losses. Acidification was seen to be essential to prevent mercury complexation, possibly by sulfur-containing organics present in seawater.

In studies of arsenic speciation, the stability of inorganic species of arsenic, as arsenic(III) or arsenic(V), in stored samples is of particular concern. Cheam and Agemian[82] found that both valency states were stable in freshwaters, at pH 1.5, 5.2, and 7.2, when stored in polyethylene bottles. Storage in Pyrex® glass bottles was unsatisfactory in all but the low pH samples. The recommended protocol was acidification of the sample with 0.2% sulfuric acid and storage in polyethylene bottles. The same procedure is used to preserve selenium(IV) and (VI).[82]

Arsenic(III) and iron(II) in anoxic interstitial waters are susceptible to oxidation unless air is rigorously excluded.[83] Acidification to pH 2, deoxygenation with nitrogen, and storage below 4°C in polyethylene or stoppered glass flasks has been recommended.

Very little has been reported on the stability of organoarsenic species in solution; however, in the case of organotin compounds, in particular tri- and dibutyltin, polyethylene containers were shown to be unsuitable.[84] Seawater samples containing tributyltin, at 4°C, lost up to 62% after 1 week of storage. Polycarbonate containers were found to be best, with only 3% loss, while Teflon® and Pyrex® glass containers removed 7 and 4%, respectively. Frozen seawater samples containing tributyltin at sub-$\mu g\ l^{-1}$ concentrations showed insignificant losses over 4 to 5 weeks, but only about half of the initial tributyltin concentrations was present after 10 months.[85]

III. SEDIMENT SAMPLES

A. Sediment Sampling

For sediment sampling, grab or core samplers are commonly used depending on the type and quantity of sample required. Grab samplers collect large volumes of samples, but often with considerable disturbance. With samplers of the Ekman and Patterson type, the loss of fines is a particular problem, with noncohesive sediments tending to flow out when the dredge is opened. These are the least desirable of sediment samplers. Large, undisturbed samples are best collected with a box corer. Where depth profiling is of interest, sediment corers should be used. Metal devices can be modified with plastic liners to avoid sample contamination. Perspex or polycarbonate corers (50 cm × 5 cm in diameter) are preferable where the sediments are not too heavily compacted. The leading ends of the corers should be bevelled so that they can readily be immersed by hand by divers or hammered into the sediment. The corers are sealed underwater with polyethylene end caps. In deeper waters, pneumatic or vibrating corers with the ability to be operated from a barge are necessary.

Cores can be extruded using gas pressure, then sectioned and stored in polystyrene containers. Where sections smaller than 5 cm are of interest it is preferable to use a polyvinyl chloride (PVC) corer which can be cut open. In speciation studies it is essential that these operations be performed in an inert atmosphere in a glove box to avoid oxidation of any anoxic material and possible changes in species distribution.

The sampling of suspended sediments is best carried out using sediment traps. Collection efficiencies of traps are highly dependent on their design, in particular the ratio of height to mouth opening, which should exceed five for maximum efficiency. More complex high-efficiency traps (Figure 3) have been reviewed by Hargrave and Burns.[86]

Sediment interstitial waters represent an important phase in the transport of trace elements from the sediment to the overlying water and vice versa. Sampling such waters for speciation studies is difficult, especially in avoiding changes brought about by exposure to oxygen or differences in temperature and pressure. Traditional techniques such as sediment squeezing are of limited value, first because of the difficulty in efficiently removing the water, especially from sandy sediments, and additionally because of the difficulty in avoiding trace metal contamination from the large surface areas exposed to the sample.[87]

Centrifugation, commonly used for interstitial waters in soils, is now finding increased use with sediments. It has the advantage that little sample manipulation is required. The sediment can be placed directly in cleaned polyethylene centrifuge tubes. Water will separate above the sediment for fine-grained compressible sediments; however, for coarser sediments, water collects at the bottom of the tube and a basal collection cup is required (Figure 4). This is more difficult to clean and assemble and requires the use of a filter membrane with the accompanying contamination problems.

Batley and Giles[87] showed that an inert, dense water-immiscible fluorocarbon solvent (Fluorinert® FC-78) would successfully displace interstitial waters from sediments during centrifugation with no loss of trace metals (Figure 4).

In situ samplers avoid many of the problems associated with the above separations and, provided Teflon® or similar noncontaminating components are used, they are suitable for trace metal studies.[88,89] Davison et al.[90] overcame oxidation problems with anoxic sediments by sampling sediment cores by syringe at holes predrilled in the corer and sealed with polyethylene tape. The water was then filtered through a sealed filter unit into a polarographic cell for trace metal analysis. Argon-CO_2, used to degas the sample, brought about changes in both Fe(II) and Mn(II) concentrations, and direct measurement on unstored solutions was recommended.[90]

The use of *in situ* dialyzers (''sediment peepers'') has recently been explored for the study of trace metals in sediment interstitial waters[91-93] following the procedures described by

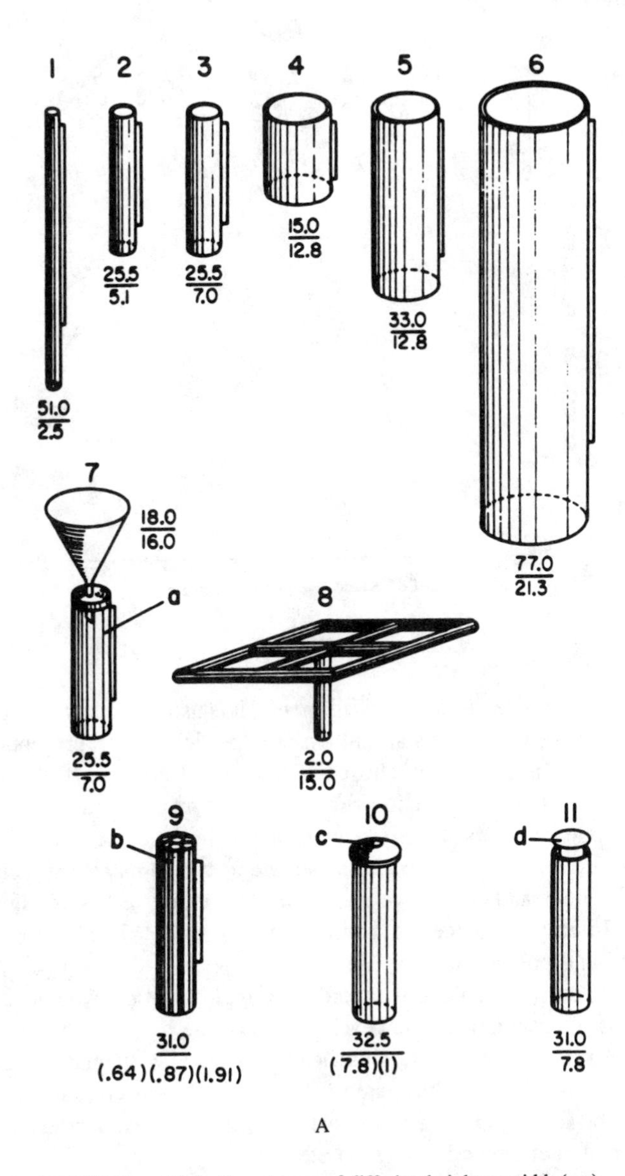

FIGURE 3. (A) Sediment traps of differing height to width (cm) ratios. Trap 9 has three varying-size baffle tubes to fill cylinder; trap 10 has a watch glass cover with a central 1.05-cm hole. (B) Weight of sediment deposited for the above designs. (From Hargrave, B. T. and Burns, N. M., *Limnol. Oceanogr.*, 24, 1124, 1979. With permission.)

Hesslein.[91] The Carignan peeper,[92] described in Chapter 3, gave results similar to those obtained by centrifugation and filtration. It was noted[93] that in some pore water samples which were high in calcium and magnesium carbonates, precipitation of carbonate minerals occurred during centrifugation and filtration due to degassing of dissolved CO_2 and the accompanying pH increase. This change will undoubtedly have serious effects on trace element speciation.

B. Sediment Sample Preparation and Storage

For speciation analyses using selective extraction procedures, sediments should be handled in an inert atmosphere in a glove box to avoid oxidation of anoxic material. Cores can be extruded using gas pressure, then sectioned if necessary for storage in polystyrene containers.

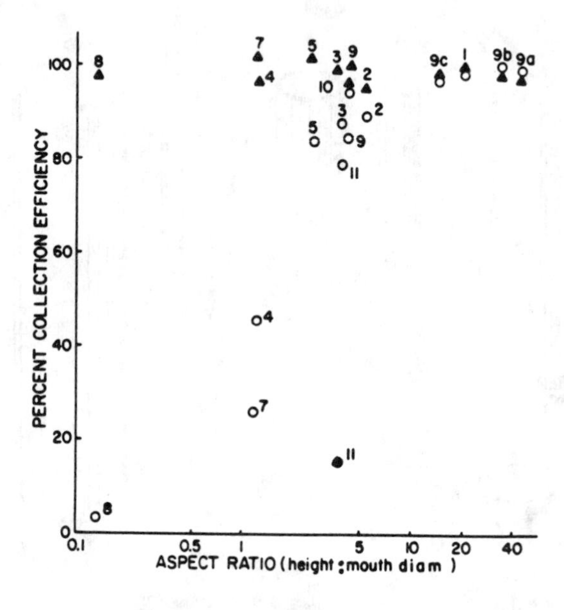

FIGURE 3B

Because of the particle size variation in sediments, it is usual to carry out some preliminary physical separation to remove debris and other large particles not representative of the bulk material. A 20-μm size is commonly chosen as the cut-off point in soil science, and it has been argued[94] that the <20-μm fraction offers the best means of achieving comparability of trace element concentrations. This size will not include coarse silt (20 to 63 μm) and sand particles (>63 μm), and these fractions should also be analyzed if a size fractionation study is required. Other authors[95] have recommended simply a 63-μm sieving to remove sand and gravel. In other instances, a 1-mm screen is used.[96] In all cases metallic sieves should be avoided and polyethylene or nylon used.

Wet sieving is often preferable as it avoids the agglomeration which occurs on drying. Drying at 110°C is often practiced; however, losses of the more volatile elements, such as Hg, can occur. Air drying will also significantly affect Fe speciation, pH, and cation-exchange capacity. Therefore, if these are of interest, or for speciation studies, well-mixed undried sediment or soil samples are best weighed moist or as slurries, and the moisture content of the sample determined on a separate aliquot.

Dried sediments can be stored, without problems, in plastic or glass containers. Moist sediments should be stored at 4°C or preferably frozen to prevent water losses. Drying, even at room temperature, has been found to change soil structure and chemistry, which is likely to be important in speciation studies. For this purpose, samples should be sealed under nitrogen in plastic containers and frozen. This avoids, in particular, the oxidation of iron which will result in changes in the distribution of heavy metals among the different sediment phases.

Additional information on the sampling and storage of sediment samples for speciation studies is presented in Chapter 8.

IV. ATMOSPHERIC SAMPLES

A. Sampling of Aerosols

Ambient aerosols possess particles in the general size range of 0.01 to 10.0 μm and sometimes greater. Their chemical composition is generally size dependent. Batch sampling is usually employed with collection of the aerosol particulates by filtration, impaction, or

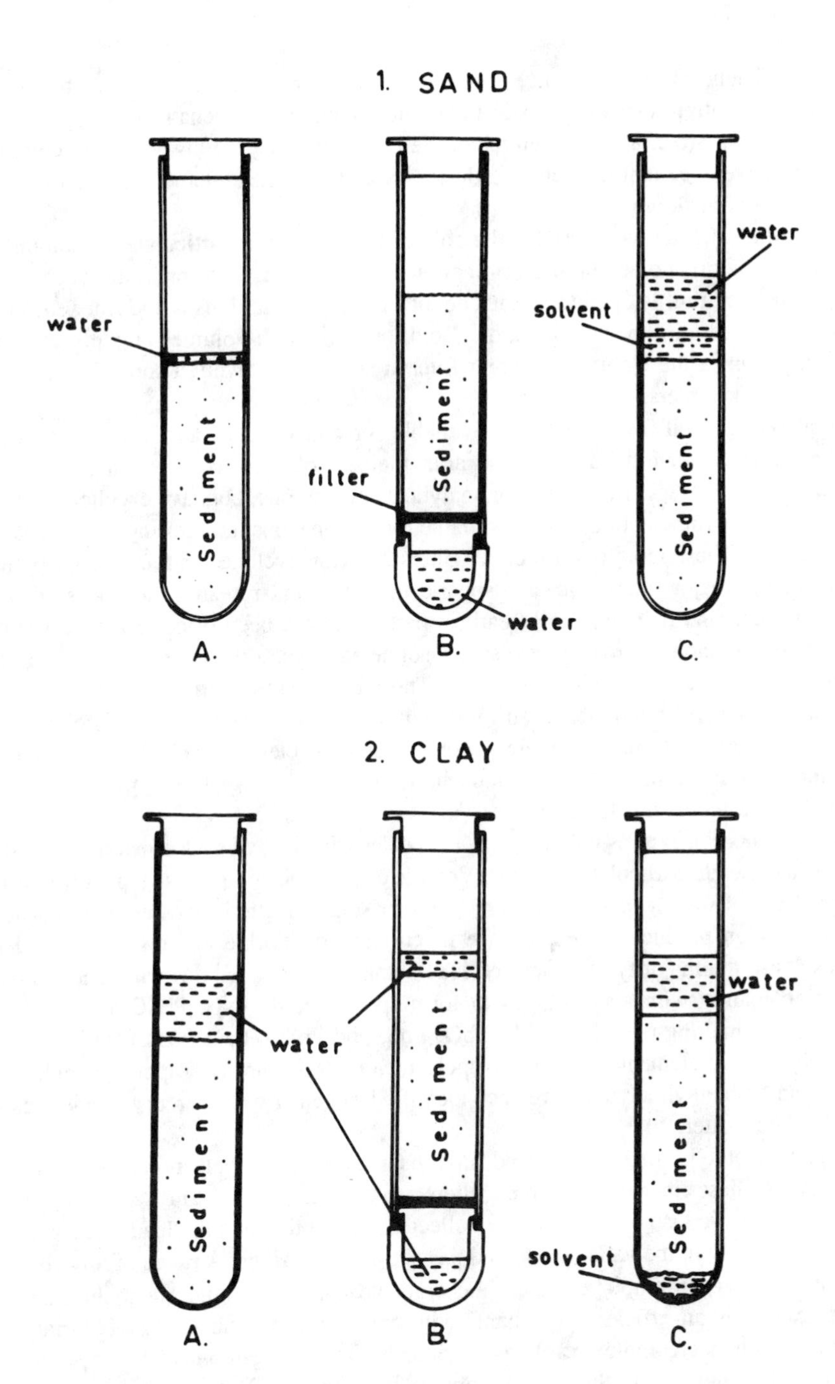

FIGURE 4. Centrifugation methods for the separation of interstitial waters from two different sediment types. (A) Direct centrifugation; (B) basal-cup collection; (C) solvent displacement. (From Batley, G. E. and Giles, M. S., *Water Res.*, 13, 879, 1979. With permission.)

electrostatic precipitation. The method chosen should avoid production of artifact aerosols through gas particle conversion on the filter, avoid loss of material through aerodynamic effects, and avoid contamination of the sample from the filtering materials.

Filtration is the most common procedure. This is usually carried out using either high (10 to 60 $m^3 h^{-1}$)- or low (0.1 to 3 $m^3 h^{-1}$)-volume samplers. Choice of filters is critical, with respect to particle size, collection efficiency, and likely contamination.[97,98] Glass fiber filters have high efficiencies, but are sources of Zn contamination and have too high a background for neutron activation analysis. Plastic fiber filters have poor mechanical strength, and of the fiber filters, quartz filters are preferable. Membrane filters are, however, most commonly used. They have excellent collection efficiencies and low trace element backgrounds, but also have a high airflow resistance.

The viability of filter samplers for the collection of airborne particulates containing elemental mercury, organomercurials, or other elements of high vapor pressure, such as lead, arsenic, and selenium, is open to serious doubt[99] because conditions at the filter surface are conducive to desorption or volatilization. Both low- and high-volume samplers are similar in this regard since the latter use a larger filter area, but effectively a similar face velocity to low-volume samplers.

Several commercial instruments are available for collection of airborne particulates by impaction.[98] These have the ability to separate the aerosols into size fractions. Electrostatic precipitators are used less frequently for analytical applications, but have excellent collection efficiencies.[98] Instrumentation is also available for personal dust monitoring for occupational health surveys. These monitors use either filters or nylon cyclones and are totally portable.

To date, the major interest in atmospheric trace element research has focussed on the behavior of lead and mercury.[99,100] Lead, in particular, has been of concern because of its prominent though now decreasing use as a gasoline additive, although emissions from other sources, e.g., smelters, remain significant. The widespread occurrence and toxicological importance of mercury heightened an interest in atmospheric burdens of this element also from both natural and anthropogenic sources. For both elements, chemical speciation is important in defining not only the atmospheric reaction pathways, but the toxicology of potentially ingested or inhaled forms.

Both elements can exist as volatile gaseous species which may be adsorbed on particulates, or simply as discrete particulates species. Volatile organolead compounds typically constitute between 1 and 4% of lead in urban air,[101] comprising tetraethyl- and tetramethyllead and their degradation products. The presence of ethylene dichloride and dibromide in leaded fuels results in the majority of exhausted lead being present as halides which are converted by reaction mainly with acid sulfates to form species of the type $PbSO_4 \cdot (NH_4)_2SO_4$ and $PbSO_4$. From smelting operations, Pb, PbO, PbS, and $PbSO_4$ have been identified.

In ambient air, elemental mercury vapor, mercury(II) chloride vapor, methylmercuric chloride, and dimethylmercury have been identified, together with mercury species adsorbed on particulate matter.[99,100]

Sampling methods for airborne lead have been comprehensively reviewed by Harrison and Perry.[101] Glass fiber or membrane filters are commonly used to discriminate between involatile inorganic lead salts which are collected efficiently and alkyllead forms which are almost exclusively in the vapor phase and pass through the filter. A range of absorbants and extractants have been employed to isolate organolead forms. Iodine monochloride in hydrochloric acid is an efficient absorbant, with extraction into carbon tetrachloride as the dialkyllead dithizone complex resolving organolead from inorganic lead which, in the presence of EDTA, remains in the aqueous phase.[102] Unambiguous separation from inorganic lead can be achieved using gas chromatography column packing materials.[103] The detection limits will depend both on sampling time and rate. There has been some debate[101,104] as to the nature of vapor phase, nonparticulate lead based on results obtained using both activated charcoal and graphite cup collection subsequent to filtration, with claims that inorganic lead is a significant contributor to what has been assumed to be organolead species.[104] These findings have, however, never been supported by subsequent laboratory studies.

Methods for mercury sampling also include a variety of vapor traps and adsorbants.[100,101] Trujillo and Campbell,[106] for example, developed a two-stage sampler (Figure 5) to collect particulate, organic, and metallic mercury, consisting, respectively of a membrane filter, a

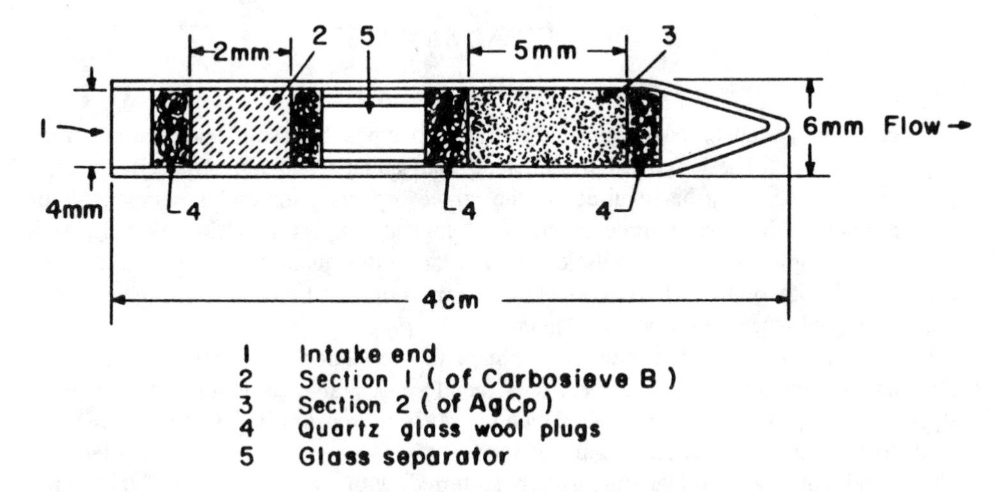

FIGURE 5. Two-stage mercury sampling tube. (From Trujillo, P. E. and Campbell, E. E., *Anal. Chem.*, 47, 1629, 1975. With permission.)

chromatographic packing (Carbosieve® B, [Supelco, Inc.]), and a silvered Chromosorb® P packing, prepared by reducing silver onto the absorbent with dextrose solution. After sampling at flow rates up to 1 l min^{-1}, particulates were dissolved from the filter with nitric acid and thermally desorbed from the other sections for flameless atomic absorption analyses. The specificity of a range of adsorption tubes has been studied by Braman and Johnson.[107] Siliconized Chromosorb® W adsorbed $HgCl_2$; Chromosorb® W treated with 0.5 *M* NaOH removed CH_3HgCl; silvered glass beads removed elemental Hg, with $(CH_3)_2Hg$ being removed by gold-coated glass beads. Absorption by liquids such as acidified permanganate or iodine monochloride solutions has also been practiced.[99] Reliable sampling of particulate mercury, which generally represents a low fraction of the total mercury, usually requires sampling for several hours at flow rates of 0.5 to 1.5 m^3 min^{-1} using high-volume samplers or for several days using low-volume (5 to 50 l min^{-1}) methods. The former, however, can result in some desorption or votatilization of some mercury forms.

Methods for the chemical speciation of metals in dusts and other particulate forms are more fully discussed in Chapter 8. These samples are usually collected by filtration.

B. Atmospheric Deposition

Atmospheric deposition includes both dry deposition (dust particles which settle directly on exposed surfaces) and wet deposition (comprising rainfall and the particulates incorporated therein). Essentially the same collection procedures can be employed for each; however, for the latter, the sampling apparatus is designed to collect only when rain is detected.[91]

Haraldsson and Magnusson[108] recommend the use of a polyethylene funnel mounted on a high-density polyethylene bottle for the collection of total deposition. The main problems in the collection of precipitation are the exclusion of dry deposition, collection of a representative sample, sample contamination, and losses during collection and storage. The exclusion of dry deposition can be achieved using automatic samplers; however, these should also have only polyethylene components contacting the sample. The receiving vessel for both can have acid added to prevent trace metal losses through adsorption on the container surfaces. Acid should, however, be omitted if a separate measure of particulate matter is required. This will necessitate filtration, with the sample being handled similarly to water samples. Dry deposition samples should be sampled and analyzed as described in Chapter 8.

V. BIOLOGICAL SAMPLES

A. Sampling

Biological samples encompass a wide range of matrix types, both solid and liquid, derived from living species. These species will include both aquatic and terrestrial animals and vegetation. Analyses may be required on the whole organism, or on its components or excretion products. In an environmental study, biological samples might include fish, crustacea, seaweed, grasses, fruit, vegetation, leaves, and other plant and animal species. In occupational health studies, the trace element composition of human tissue, hair, sweat, blood, urine, and feces are of potential interest.

As with other matrices, the major concern in the collection of biological samples is contamination. For trace metal studies, it is therefore desirable to avoid metal spatulas, scalpels, scissors, tweezers, or needles which are likely to transmit traces of the element of interest to the sample. Laboratory ware of Pyrex® glass, polyethylene, polypropylene, or Teflon®, and tools made of quartz are usually preferred, with plastic gloves used at all times when handling samples. Powdered gloves should be avoided as the powder is a source of zinc and other trace element contaminants.

In practice, however, stainless steel blades and biopsy needles are commonly used in biological sampling. Stainless steel surgical blades have been found to contribute 3 ng Mn g^{-1}, 15 ng Cr g^{-1}, and 60 ng Ni g^{-1} to liver biopsy samples,[109] but they are more convenient to use than quartz or glass knives, if these levels can be tolerated. Stainless steel biopsy needles can lead to contamination,[110] although Hudnik et al.[111] found similar concentrations of cadmium and manganese in liver tissue, using either glass or Meninghi needles. The latter are used for a long period and the surface layers are therefore effectively cleaned.

Considerable disparity has been noted in the trace element contents of human blood plasma and serum, which can also be attributed to errors introduced during sampling.[110,112] Versieck et al.[110,112] found increases in iron, manganese, nickel, cobalt, molybdenum, and chromium, but not in copper or zinc, particularly in the first 20 ml of blood samples collected with a steel needle. For this reason, Teflon® or polypropylene catheters are preferred.[112] It has also been reported that the use of Vacutainer® or Venoject® tubes for blood collection can also lead to contamination,[112,113] particularly in the latter which has a rubber stopper. Reported discrepancies between lead concentrations in venous and capillary blood samples highlight this problem.[114] For sampling capillary blood, Garnys et al.[115] recommended a finger scrub with a stiff brush and detergent solution, two alcohol swab cleansings, natural drying in air, and spraying the fingertip with a lead-free plastic skin before incision with a microlance.

Dust is an ubiquitous source of zinc, aluminium, and other trace metals and must especially be avoided during the collection of body fluids. Urine is particularly prone to contamination unless careful procedures are employed, namely collection of mid-stream samples in covered acid-washed polyethylene containers away from all dust sources.

In collection of sweat samples for trace metal analysis, precleaning of the sampled area is essential. Stauber and Florence,[116] sampling the inner forearm, used a 1% Extran®-300 (BDH) detergent wash followed by washing with distilled water and drying with a Whatman® 542 filter paper. Sweating was induced by pilocarpine iontophoresis where pilocarpine is forced into the sweat glands by a small electric current (1 mA). The arm is again washed and dried and the sweat then collected on a preweighed, acid-washed Millipore® 25-mm HATF membrane filter placed over the pilocarpine area, covered with a square of acid-washed parafilm, and wrapped in thin polyethylene film. After 30 to 45 min, the filter paper, wet with sweat is placed in an acid-washed plastic vial for weighing then leached with 0.1 *M* $NaHCO_3$, 0.002 *M* HNO_3 for analysis. Trace metal analysis of sweat samples offers an attractive noninvasive alternative to blood analysis for detecting occupational exposure.

B. Sample Preparation and Storage

For many biological samples, a preliminary sample preparation is required to isolate the portion of interest from extraneous, potentially contaminating, material. This should ideally be done as soon as possible after collection. For example, in analyzing shellfish for trace elements, the shell is generally of little concern; nevertheless, it is usually scrubbed to remove excess sediment and debris then opened, and the flesh sampled either totally or in specific organ sections. Seaweeds, sea grasses, and other aquatic biota should also be washed carefully to remove sediment, epiphytes, or similar surface contaminants. Washing is desirable for terrestrial plant material, again, to remove surface contamination. Care must be taken that the sample is representative and it should be remembered that trace elements will concentrate in specific components, e.g., roots, leaves, etc. Furthermore, trace element concentrations may vary greatly with the size of the organism, and this should be taken into consideration if analyzing single samples. The above sample types should then be frozen for storage until, at a later time, they are sectioned or homogenized prior to freeze drying or ashing. Animal and fish tissue or organ samples should also be sectioned if necessary and frozen as soon as possible after collection.

When handling biological samples for trace element analysis, it is desirable to wear polyethylene gloves and to carry out any sectioning of samples on a polyethylene sheet in a laminar-flow work station. Samples should be stored in plastic bags or plastic containers.

For hair samples, a variety of washing procedures have been proposed to remove exogenous elements. For trace metal analyses, ultrasonication with 0.1% Triton® X-100, filtration, washing with methanol, and drying in air or with a hair dryer, appears the most satisfactory procedure.[116]

In analyzing blood samples, an early decision should be made whether whole blood, plasma, or serum analyses are of interest, and necessary separations made. Whole blood or blood fluids can be stored at 4°C for short periods.[117] Freezing is preferable; however, this will lead to hemolysis.[114] Urine samples are also best stored frozen; however, freezing often leads to irreversible precipitation. On thawing, these solids may be solubilized by acid or simply mixed well by shaking before removing a sample aliquot.

Sample homogenizing is often the next step in handling large samples. Contamination from blades should be considered if a laboratory homogenizer is to be used. Portions of the homogenized samples, or smaller whole samples, may be wet ashed by dissolution in concentrated acids[108] or alkalis.[118] Quaternary ammonium hydroxides have also been used as tissue solubilizers.[119] Chau et al.[119] used a 2-h treatment with 20% tetramethylammonium hydroxide to release protein- and lipid-bound alkylleads from fish tissue, algae, and aquatic plants. Dissolution was enhanced by heating to 60°C in a water bath.

Pressurized decomposition with nitric acid in Teflon® bombs is often practiced,[120] and the trace metal blanks have been shown to be extremely low. More often, however, the sample is either freeze dried or oven dried prior to a higher temperature ashing.[121]

Freeze drying is possibly one of the best methods for removing moisture from samples without trace element losses or contamination. Small samples (2 to 5 g) are best handled by this procedure, although some laboratory freeze driers can accommodate larger single samples. Urine and blood samples are often stored in a freeze-dried state. La Fleur[121] found no losses for either inorganic or naturally bound methyl- or phenylmercury in tissue or blood.

Use of an air oven at 100°C is the conventional drying procedure; however, very volatile species, such as elemental mercury, and many organometals may be lost at this stage. Where total analyses are required, samples from this treatment, or after freeze drying, can be wet ashed to bring them into solution, although often the presence of difficult-to-degrade organics makes muffle ashing a useful preliminary stage. Muffle ashing at 500°C will decompose most organic matter, but should be avoided in speciation studies or where Hg, As, Sn, Se, Pb, Ni, and Cr are of interest as losses can occur.[120] Low-temperature plasma ashing is preferable to prevent losses of the less volatile of these elements.

Storage of the freeze-dried or ashed samples poses no problems; however, before extraction or dissolution, the samples must be homogenized unless the whole sample is being used for analysis. An agate mortar and pestle is adequate for crushing and mixing provided the sample is sufficiently dry and brittle.

Liquid samples, such as blood and urine, present particular problems. Acid digestion of raw or freeze-dried samples should be performed with particular care as naturally present surfactants may froth violently leading to loss of sample. Photodecomposition of dilute samples is useful to release organically bound metals or to remove potential organic interferences.[122]

VI. LABORATORY DESIGN AND PRACTICE

A. Clean Laboratories

The extent to which precautions should be taken to provide a dust-free environment for trace element speciation analysis is an area of considerable conjecture. The biggest hurdle in trace analysis is instilling in the minds of both analysts and their associates the need to take extreme precautions to avoid sample contamination.

In the air of an average laboratory, some 3×10^7 particles between 0.5 and 100 $\mu m/m^3$ have been measured.[123] In a class-100 laboratory, a particle count below 100 particles above 0.3 $\mu m/ft^3$ of air is specified. For ultratrace analysis, a class-100 clean laboratory is highly desirable, but highly expensive, and many designs provide a major overkill.[123,124] For example, the expensive exclusion of all-metallic fittings is unwarranted in most instances, and is often quite impractical. Certainly the need for high-efficiency particulate (HEPA) filters is essential to remove dust from the work area (Figure 6). A minimum requirement would be the use of a laminar airflow work station or hood. The use of sticky mats on the floor in front of the hood, and lint-free clothing are also desirable. Overshoes are required if the mats are to function usefully.

Acid digestions cannot of course be performed in conventional laminar-flow cabinets. Experience in our laboratories has shown that fume hoods drawing air from a conventional laboratory may randomly and readily contaminate uncovered samples, e.g., on a hot plate. In our laboratory a standard hood has been modified to include a clean air intake and a laminar air curtain. Commercial hoods are available for use in clean laboratories; however, any exhausting system adds considerably to the size and cost of the cleaned air to the laboratory.

B. Laboratory Apparatus and Reagents

A great deal has been written about sources of laboratory contamination.[3,120,121,123] In particular, metal objects, powder, lint, etc. should be avoided, as already emphasized. For trace metal speciation studies, the use of polyethylene, or Teflon® or other plastic components is preferable. Reagent blanks, however, represent the major sources of error in trace analysis. Ultrapure reagents should be used wherever possible, either commercially obtained or laboratory purified. Electrolysis can be used to remove many metal impurities from salt solutions, while distillation in Teflon® vessels or isothermal distillation are widely practiced for purification of acids.

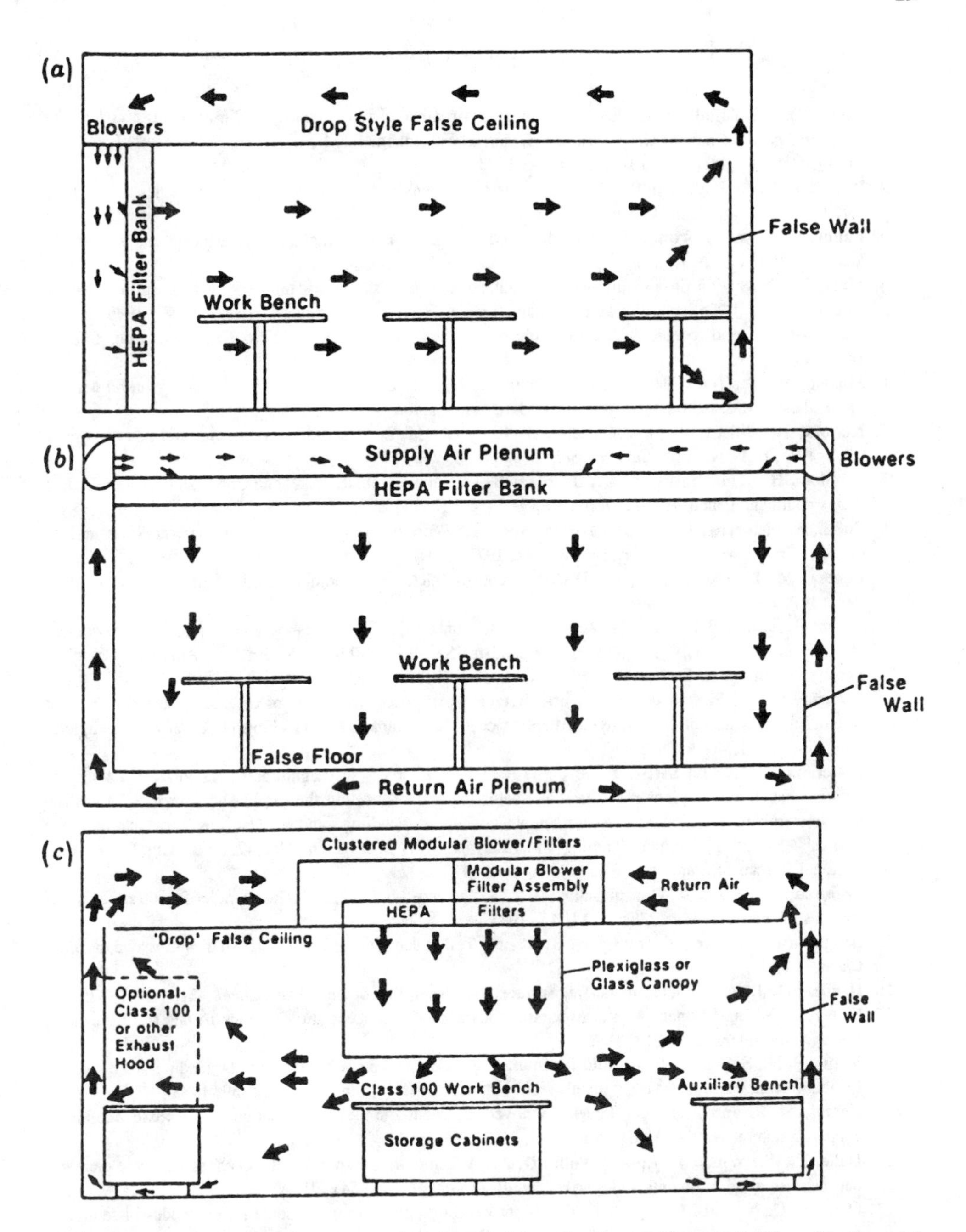

FIGURE 6. Typical clean room designs. (a) Horizontal airflow clean room, (b) vertical airflow clean room (c) N.B.S. clean laboratory. (From Moody, J. R., *Philos. Trans. R. Soc. London Ser.* A, 305, 669, 1982. With permission.)

REFERENCES

1. **Hume, D. N.,** Pitfalls in the determination of environmental trace metals, in *Chemical Analysis of the Environment and Other Modern Techniques,* Ahuja, S., Cohen, E. M., Kniep, T. J., Lambert, J. L., and Zweig, G., Eds., Plenum Press, New York, 1973, 13.
2. **Hamilton, E. I.,** Review of the chemical elements and environmental chemistry — strategies and tactics, *Sci. Total Environ.,* 5, 1, 1976.
3. **Batley, G. E. and Gardner, D.,** Sampling and storage of natural waters for trace metal analysis, *Water Res.,* 11, 745, 1977.
4. **Mart, L.,** Prevention of contamination and other necessary risks in voltammetric trace metal analysis of natural waters. II. Collection of surface water samples, *Fresenius Z. Anal. Chem.,* 299, 97, 1979.
5. **Berman, S. S. and Yeats, P. A.,** Sampling of seawater for trace metals, *CRC Crit. Rev. Anal. Chem.,* 16, 1, 1985.
6. **Matson, W. R.,** Trace Elements, Equilibrium and Kinetics of Trace Metal Complexes in Natural Media, Ph.D. thesis, Massachusetts Institute of Technology, Cambridge, MA, 1968.
7. **Kubota, J., Mills, E. L., and Oglesby, R. T.,** Lead, Cd, Zn, Cu and Co in streams and lake waters of Cayuga Basin, New York, *Environ. Sci. Technol.,* 8, 243, 1974.
8. **Windom, H. L. and Smith, R. G.,** Distribution of cadmium, cobalt, nickel and zinc in southeastern United States continental shelf waters, *Deep-Sea Res.,* 19, 727, 1972.
9. **Fukai, R., Murray, C. N., and Huynh-Ngoc, L.,** Variations of soluble zinc in the Var River and its estuary, *Estuarine Coastal Mar. Sci.,* 3, 177, 1975.
10. **Bender, M. L. and Gagner, C.,** Dissolved copper, nickel and cadmium in the Sargasso Sea, *J. Mar. Res.,* 34, 327, 1976.
11. **Segar, D. A. and Berberian, G. A.,** Trace metal contamination by oceanographic samples, in *Analytical Methods in Oceanography,* Advances in Chemistry Ser. 147, Gibb, T. R. P., Ed., American Chemical Society, Washington, D.C., 1975.
12. **Spencer, M. J., Piotrowicz, S. R., and Betzer, P. R.,** Concentrations of cadmium, copper, lead and zinc in surface waters of the northwest Atlantic Ocean — a comparison of Go-Flo and Teflon water samplers, *Mar. Chem.,* 11, 403, 1982.
13. **Patterson, C. C. and Settle, D. M.,** The reduction of orders of magnitude errors in lead analyses of biological materials and natural waters by evaluating and controlling the extent and sources of industrial lead contamination introduced during sample collecting, handling, and analysis, in Accuracy in Trace Analysis: Sampling, Sample Handling and Analysis, NBS Spec. Publ. No. 422, La Fleur P. D., Ed., National Bureau of Standards, Washington, D.C., 1976, 321.
14. **Freimann, P., Schmidt, D. and Schomaker, K.,** Mercos — a simple Teflon sampler for ultratrace metal analysis in seawater, *Mar. Chem.,* 14, 43, 1983.
15. **Brugmann, L., Geyer, E., and Kay, R.,** A new Teflon sampler for trace metal studies in seawater, *Mar. Chem.,* 21, 99, 1987.
16. **Harper, D. J.,** A new trace metal-free surface water sampling device, *Mar. Chem.,* 21, 183, 1987.
17. **Davey, E. W. and Soper, A. E.,** Apparatus for the in situ concentration of trace metals from seawater, *Limnol. Oceanogr.,* 20, 1019, 1975.
18. **Willie, S. N., Sturgeon, R. E., and Berman, S. S.,** Comparison of 8-quinolinol-bonded polymer supports for the preconcentration of trace metals from seawater, *Anal. Chim. Acta,* 149, 59, 1983.
19. **Zhang, M.-P. and Florence, T. M.,** A novel adsorbent for the determination of the toxic fraction of copper in natural waters, *Anal. Chim. Acta,* 197, 137, 1987.
20. **Duke, T. W., Willis, J. N., and Wolfe, D. A.,** A technique for studying the exchange of trace elements between estuarine sediments and water, *Limnol. Oceanogr.,* 13, 541, 1968.
21. **Murray, C. N. and Murray, L.,** Adsorption-desorption equilibrium of some radionuclides in sediment-freshwater and sediment-seawater systems, in *Radioactive Contamination of the Marine Environment,* IAEA-SM-158, International Atomic Energy Agency, Vienna, 1972, 105.
22. **Gardiner, J.,** The chemistry of cadmium in natural water. II. The adsorption of cadmium in river muds and naturally occurring solids, *Water Res.,* 8, 157, 1974.
23. **Murray, C. N. and Meinke, S.,** Influence of soluble storage material on adsorption and desorption behaviour of cadmium, cobalt, silver and zinc in sediment-freshwater, sediment-seawater systems., *J. Oceanogr. Soc. Jpn.,* 30, 216, 1974.
24. **Mart, L.,** Prevention of contamination and other occurring risks in voltammetric trace metal analysis of natural waters. I. Preparatory steps of filtration and storage of water samples, *Fresenius Z. Anal. Chem.,* 296, 350, 1979.
25. **Patterson, C., Settle, D., and Glover, B.,** Analysis of lead in polluted coastal seawater, *Mar. Chem.,* 4, 305, 1976.

26. **Lee, G. F. and Hoadley, A. W.,** Biological activity in relation to the chemical equilibrium composition of natural waters, in *Equilibrium Concepts in Natural Water Systems,* Advances in Chemistry Ser. 67, American Chemical Society, Washington, D.C., 1967, 319.
27. **McLerran, C. J. and Holmes, C. W.,** Deposition of zinc and cadmium by marine bacteria in estuarine sediments, *Limnol. Oceanogr.,* 19, 998, 1974.
28. **Gardner, D.,** Lipids of Some Invertebrates and Studies of Dissolved Mercury in Seawater, Ph.D. thesis, University of Liverpool, U.K., 1974.
29. **Moebus, K.,** Influence of storage on the antibacterial activity of seawater, *Mar. Biol.,* 12, 346, 1972.
30. **Carpenter, J. H., Bradford, W. L., and Grant, V.,** Processes affecting the composition of estuarine waters, in *Estuarine Research,* Vol. 1, Cronin, L. E., Ed., Academic Press, New York, 1975, 188.
31. **Truitt, R. E. and Weber, J. H.,** Trace metal in filtration losses at pH 5 and 7, *Anal. Chem.,* 51, 2057, 1979.
32. **Scarponi, G., Capodaglio, G., Cescon, P., Cosma, B., and Frache, R.,** Anodic stripping voltammetric determination of the contamination of seawater samples by cadmium, lead and copper during filtration and storage, *Anal. Chim. Acta,* 135, 263, 1982.
33. **Nurnberg, H. W., Valenta, P., Mart, L., Raspor, B., and Sipos, L.,** Application of polarography and voltammetry to marine and aquatic chemistry. II. The polarographic approach to the determination and speciation of toxic trace elements into marine environment, *Fresenius Z. Anal. Chem.,* 282, 357, 1976.
34. **Spencer, D. W. and Brewer, P. K.,** Distribution of copper, zinc and metal in seawater of the Gulf of Maine and the Sargasso Sea, *Geochim. Cosmochim. Acta,* 33, 325, 1969.
35. **Martin, J. H. and Knauer, G. A.,** The elemental composition of plankton, *Geochim. Cosmochim. Acta,* 37, 1639, 1973.
36. **IAEA,** Reference methods for marine radioactivity studies, *IAEA Tech. Rep. Ser.,* 118, 1970.
37. **Abdullah, M. I., El-Rayis, O. A., and Riley, J. P.,** Reassessment of chelating ion-exchange resins for trace metal analysis of seawater, *Anal. Chim. Acta,* 84, 363, 1976.
38. **Gardner, D. and Riley, J. P.,** The distribution of dissolved mercury in the Bristol Channel and Severn Eastuary, *Estuarine Coastal Mar. Sci.,* 1, 191, 1973.
39. **Filby, R. H., Shah, K. R., and Funk, W. H.,** Role of neutron activation analysis in the study of heavy metal pollution of a lake-river system, in *Proc. 2nd Int. Conf. on Nuclear Methods in Environmental Research,* U.S. Energy Research and Development Association, University of Missouri, Columbia, 1974, 10.
40. **Batley, G. E. and Florence, T. M.,** Determination of the chemical forms of dissolved cadmium, lead and copper in seawater, *Mar. Chem.,* 4, 347, 1977.
41. **Helfferich, F.,** *Ion Exchange,* McGraw Hill, New York, 1962, 130.
42. **Adams, P. B.,** Glass containers, in *Ultrapurity, Methods and Techniques,* Zief, M. and Speights, R., Eds., Marcel Dekker, New York, 1972, 297.
43. **Doremus, R. H.,** Ion exchange in glasses, *Ion Exch. Surv.,* 2, 1, 1969.
44. **James, R. O. and Healy, T. W.,** Adsorption of hydrolysable metal ions at the oxide-water interface. III. A thermodynamic model of adsorption, *J. Colloid Interface Sci.,* 40, 65, 1972.
45. **Benes, P. and Smetana, J.,** Radiochemical study of the sorption of trace elements. IV. Adsorption of iron on polythene and its state in aqueous solution, *Coll. Czech. Chem. Commun.,* 34, 1360, 1969.
46. **Starik, I. E., Schebetkovskii, V. N., and Skulskij, I. A.,** The adsorption of radioactive isotopes on polymer adsorbents that are not ion-exchangers III, *Radiokhimiya,* 4, 393, 1063.
47. **Eichholz, G. G., Nagel, A. E., and Hughes, R. B.,** Adsorption of ions in dilute aqueous solutions on glass and plastic surfaces, *Anal. Chem.,* 37, 863, 1965.
48. **Bubic, S., Sipos, L., and Branica, M.,** Comparison of different electro-analytical techniques for the determination of heavy metals in seawater, *Thalassia Jugosl.,* 9, 55, 1973.
49. **West, F. K., West, P. W., and Iddings, P. A.,** Adsorption of traces of silver on container surfaces, *Anal. Chem.,* 38, 1566, 1966.
50. **Robertson, D. E.,** Contamination problems in trace element analysis and ultrapurification, in *Ultrapurity, Methods and Techniques,* Zief, M. and Speights, R., Eds., Marcel Dekker, New York, 1972, 207.
51. **Laxen, D. P. H. and Harrison, A. M.,** Cleaning methods for polythene containers prior to the determination of trace metals in freshwater samples, *Anal. Chem.,* 53, 345, 1981.
52. **Moody, J. R. and Lindstrom, R. M.,** Selection and cleaning of plastic containers for storage of trace element samples, *Anal. Chem.,* 49, 2264, 1977.
53. **Karin, R. W., Buono, J. A., and Fasching, J. L.,** Removal of trace elemental impurities from polyethylene by nitric acid, *Anal. Chem.,* 47, 2296, 1975.
54. **Florence, T. M.,** Trace metal species in freshwaters, *Water Res.,* 11, 681, 1977.
55. **Kinsella, B. and Willis, R. L.,** Ultrasonic bath in container preparation for storage of seawater samples in trace metal analysis, *Anal. Chem.,* 54, 2614, 1982.

56. **Bevan, C. D., Harbison, S. A., Nelson, C. A., and Lakey, J. R. A.,** A trace element study in the Thames Estuary, in *Proc. of a Symp. on the Inputs of Nuclear Releases into the Aquatic Environment,* IAEA-SM-198, International Atomic Energy Agency, Vienna, 1975, 83.
57. **Holliday, L. M. and Liss, P. S.,** The behaviour of dissolved iron, manganese and zinc in the Beaulieu estuary, S. England., *Estuarine Coastal Mar. Sci.,* 4, 349, 1976.
58. **Baker, R. A.,** Trace organic contaminant concentration by freezing. I. Low inorganic aqueous solutions, *Water Res.,* 1, 61, 1967.
59. **Baker, R. A.,** Trace organic contaminant concentration by freezing. II. Inorganic aqueous solutions, *Water Res.,* 1, 97, 1967.
60. **Scheuermann, H. and Hartkamp, H.,** Stabilisierung von wasserproben fur die bestimmung von schwermetallspuren, *Fresenius Z. Anal. Chem.,* 315, 430, 1983.
61. **Batley, G. E. and Gardner, D.,** A study of copper, lead and cadmium speciation in some estuarine and coastal marine waters, *Estuarine Coastal Mar. Sci.,* 7, 59, 1978.
62. **Baier, R.,** Lead Distribution in Coastal Waters, Ph.D. thesis, University of Washington, Seattle, 1971.
63. **Sheldon, R. W., Evelyn, T. P. T., and Parsons, J. R.,** On the occurrence and formation of small particles in seawater, *Limnol. Oceanogr.,* 12, 367, 1967.
64. **Gillbricht, M.,** An oxidative procedure for the determination of organic matter in seawater, *Helgol. Wiss. Meeresunters.,* 6, 76, 1957.
65. **Fukai, R. and Huynh-Ngoc, L.,** Copper, zinc and cadmium in coastal waters in the N.W. Mediterranean, *Mar. Pollut. Bull.,* 7, 9, 1976.
66. **Cescon, P., Scarponi, G., and Maret, I.,** Electrochemical determination of the contamination of seawater samples during storage and filtration, *Sci. Total Environ.,* 37, 95, 1984.
67. **Piotrowitz, S. R., Harvey, G. R., Springer-Young, M., Courant, R. A., and Boren, D. A.,** Studies of cadmium, copper and zinc complexation by marine fulvic and humic materials in seawater using anodic stripping voltammetry, in *Trace Metals in Seawater,* Wong, C. S., Burton, J. D., Boyle, E., Bruland, K., and Goldberg, E. D., Eds., Plenum Press, New York, 1984, 699.
68. **Platell, N. and Webb, J.,** A review of mercury analysis and sampling techniques, in *Examination of Waters; Evaluation of Methods for Selected Characteristics,* Australian Water Resources Council Tech. Paper No. 8., Australian Water Resources Council, Canberra, A. C. T., 1974, 43.
69. **Chilov, S.,** Determination of small amounts of mercury, *Talanta,* 22, 205, 11975.
70. **Avotins, P. and Jenne, E. A.,** The time stability of dissolved mercury in water samples. II. Chemical stabilization, *J. Environ. Qual.,* 4, 515, 1975.
71. **Cragin, J. H.,** Increased mercury contamination of distilled and natural water samples caused by oxidizing preservatives, *Anal. Chim. Acta,* 110, 313, 1979.
72. **Shimomura, S., Nishihara, Y., and Tanase, Y.,** Escape of mercury from diluted mercury(II) solutions, *Jpn. Anal.,* 17, 1148, 1969.
73. **Toribara, T. Y., Shields, C. P., and Koval, L.,** Behaviour of dilute solutions of mercury, *Talanta,* 17, 1025, 1970.
74. **Baier, R. W., Wojnowich, L., and Petrie, L.,** Mercury loss from culture media, *Anal. Chem.,* 47, 2464, 1975.
75. **Gassaway, J. D. and Carr, R. A.,** Problems of storing and analyzing mercury collected in a nearshore environment, in *Proc. Naval Environ. Protection Data Base Instrumentation Workshop,* Naval Civil Engineering Laboratory, Channel Islands Harbor, 1972, 87.
76. **Piccolino, S. P.,** Preparation of mercury in polyethylene containers, *J. Chem. Educ.,* 60, 235, 1983.
77. **Feldman, C.,** Preservation of dilute mercury solutions, *Anal. Chem.,* 46, 99, 1074.
78. **Weiss, H. V., Shipman, W. H., and Guttman, M. A.,** Effective storage of dilute mercury solutions in polyethylene, *Anal. Chim. Acta,* 81, 211, 1976.
79. **Heiden, R. W. and Aikens, D. A.,** Humic acid as a preservative for trace mercury(II) solutions stored in polyolefin containers, *Anal. Chem.,* 55, 2327, 1983.
80. **Heiden, R. W. and Aikens, D. A.,** Pretreatment of polyolefin bottles with chloroform and aqua regia vapor to prevent losses from stored trace mercury(II) solutions, *Anal. Chem.,* 51, 151, 1979.
81. **Ahmed, R. and Stoeppler, M.,** Storage and stability of mercury and methylmercury in seawater, *Anal. Chim. Acta,* 192, 109, 1972.
82. **Cheam, V. and Agemian, H.,** Preservation in organic arsenic species at microgram levels in water samples, *Analyst,* 105, 737, 1980.
83. **Aggett, J. and Kriegman, M. R.,** Preservation of arsenic(III) and arsenic(V) in samples of sediment interstitial water, *Analyst,* 112, 153, 1987.
84. **Dooley, C. A. and Hamer, V.,** Organotin Compounds in its Marine Environment: Uptake Adsorption Behaviour, NOSC Tech. Rep. No. 917, Naval Oceans Systems Center, San Diego, 1983.

85. **Valkirs, A. O., Seligman, P. F., Olson, G. J., Brinckman, F. E., Mathias, C. L., and Bellama, J. M.,** Di-and tributyltin species in marine and estuarine waters. Interlaboratory comparison of the ultratrace analytical methods employing hydride generation and atomic absorption or flame photometric detection, *Analyst,* 112, 17, 1987.
86. **Hargrave, B. T. and Burns, N. M.,** Assessment of sediment trap collection efficiency, *Limnol. Oceanogr.,* 24, 1124, 1979.
87. **Batley, G. E. and Giles, M. S.,** Solvent displacement of sediment interstitial waters before trace metal analysis, *Water Res.,* 13, 879, 1979.
88. **Barnes, R. O.,** An *in situ* interstitial water sampler for use in unconsolidated sediments, *Deep-Sea Res.,* 21, 1125, 1975.
89. **Sayles, F. L., Mangelsdorf, P. C., Wilson, T. R. S., and Hume, D. N.,** A sampler for the *in situ* collection of marine sedimentary pore waters, *Deep-Sea Res.,* 23, 259, 1976.
90. **Davison, W., Woof, C., and Turner, D. R.,** Handling and measurement techniques for anoxic interstitial waters, *Nature (London),* 295, 582, 1982.
91. **Hesslein, R. H.,** An *in situ* sampler for close interval pore water studies, *Limnol. Oceanogr.,* 21, 912, 1976.
92. **Carignan, R.,** Interstitial water sampling by dialysis: methodological notes, *Limnol. Oceanogr.,* 29, 667, 1984.
93. **Carignan, R., Rapin, F., and Tessier, A.,** Sediment pore water sampling for metal analysis: a comparison of techniques, *Geochim. Cosmochim. Acta,* 49, 2493, 1985.
94. **Jenne, E. A., Kennedy, V. C., Burchard, J. M., and Bell, J. W.,** Sediment collection and processing for selective extraction and for total trace element analysis, in *Contaminants and Sediments,* Vol. 2, Baker, R. A., Ed., Ann Arbor Science, Ann Arbor, MI, 1980, 169.
95. **Adams, D. D., Darby, D. A., and Young, R. J.,** Selected analytical techniques for characterizing the metal chemistry and geology of fine-grained sediments and interstitial water, in *Contaminants and Sediments,* Vol. 2, Baker, R. A., Ed., Ann Arbor Science, Ann Arbor, MI, 1980, 3.
96. **Rendell, P. S., Batley, G. E., and Cameron, A. J.,** Adsorption as a control of metal concentrations in sediment extracts, *Environ. Sci. Technol.,* 14, 314, 1980.
97. **Das, H. A., Faanhop, A., and Van der Sloot, H. A.,** Sampling in environmental analysis, in *Environmental Radioanalysis,* Elsevier, Amsterdam, 1983, chap. 7.
98. **WHO,** Evaluation of Exposure to Airborne Particles in the Home Environment, World Health Organization, Geneva, 1984.
99. **Schroeder, W. H.,** Sampling and analysis of mercury and its compounds in the atmosphere, *Environ. Sci. Technol.,* 16, 394A, 1982.
100. **Harrison, R. M.,** Chemical speciation and reaction pathways of metals in the atmosphere, *Adv. Environ. Sci. Technol.,* 17, 319, 1986.
101. **Harrison, R. M. and Perry, R.,** The analysis of tetraalkyl lead compounds and their significance as airborne air pollutant, *Atmos. Environ.,* 11, 847, 1977.
102. **Hancock, S. and Slater, A.,** A specific method for the determination of trace concentrations of tetramethyl and tetraethyl lead vapours in air, *Analyst,* 100, 422, 1975.
103. **Harrison, R. M., Perry, R., and Slater, D. H.,** An adsorption technique for the determination of organic lead in street air, *Atmos. Environ.,* 8, 1187, 1974.
104. **Robinson, J. W., Rhodes, L., and Wolcott, D. K.,** The determination and identification of molecular lead pollutants in the atmosphere, *Anal. Chim. Acta,* 78, 474, 1975.
105. **Hewitt, C. N. and Harrison, R. M.,** Atmospheric concentration and chemistry of alkyllead compounds and environmental alkylation of lead, *Environ. Sci. Technol.,* 21 260, 1987.
106. **Trujillo, P. E. and Campbell, E. E.,** Development of a multistage air sampler for mercury, *Anal. Chem.,* 47, 1629, 1975.
107. **Braman, R. S. and Johnson, D. L.,** Selective adsorption tubes and emission technique for determination of ambient forms of mercury in air, *Environ. Sci. Technol.,* 8, 996, 1974.
108. **Haraldsson, C. and Magnusson, B.,** Heavy metals in rainwater: collection, storage and analysis of samples, In *Proc. 4th Int. Symp. on Heavy Metals in the Environment,* Muller, G. Ed., CEP Consultants, Edinburgh, 1983, 82.
109. **Maienthal, E. J. and Becker, D. A.,** A survey on current literature on sampling, sample handling for environmental materials and long term storage, *Interface,* 5, 49, 1976.
110. **Versieck, J. and Cornelius, R.,** Normal levels of trace elements in human blood plasma or serum, *Anal. Chim. Acta,* 116, 217, 1980.
111. **Hudnik, V., Marolt-Gomiscek, M., and Gomiscek, S.,** The determination of trace metals in human fluids and tissues. III. Contamination in sampling of blood serum and liver tissue and their stability on storage, *Anal. Chim. Acta,* 157, 303, 1984.
112. **Versieck, J., Barbier, F., Cornelius, R., and Hoste, J.,** Sample contamination as a source of error in trace-element analysis of biological samples, *Talanta,* 29, 972, 1982.

113. **Handy, R. W.,** Zinc contamination in vacutainer tubes, *Clin. Chem.*, 25, 197, 1979.
114. **De Silva, P. E. and Donnen, M. B.,** Petrol vendors, capillary blood lead levels and contamination, *Med. J. Aust.*, 1, 344, 1977.
115. **Garnys, V. P., Smythe, L. E., and Freeman, R.,** Petrol vendors, capillary blood lead levels and contamination, *Med. J. Aust.*, 1, 719, 1977.
116. **Stauber, J. S. and Florence, T. M.,** The determination of trace elements in sweat by anodic stripping voltammetry, in Proc. 9th Australian Symp. on Analytical Chemistry, Royal Australian Chemical Institute, Sydney, 1987, 569.
117. **Stoeppler, M.,** Analytical aspects of sample collection, sample storage and sample treatment, in *Trace Element Analytical Chemistry in Medicine and Biology,* Vol. 1, Bratter, P. and Schramel, P., Eds., Walter De Gruyter, Berlin, 1983, 909.
118. **Chapman, J. F. and Dale, L. S.,** The use of alkaline permanganate in the preparation of biological materials for the determination of mercury by atomic absorption spectrometry, *Anal. Chim. Acta,* 134, 379, 1982.
119. **Chau, Y. K., Wong, P. T. S., Bengert, G. A., and Dunn, J. L.,** Determination of dialkyllead, trialkyllead, tetraalkylead and lead(II) compounds in sediment and biological samples, *Anal. Chem.*, 56, 271, 1984.
120. **Iyengar, G. V. and Sansoni, B.,** Sample preparation of biological materials for trace element analysis, in *Elemental Analysis of Biological Materials,* IAEA Tech. Rep. No. 197, International Atomic Energy Agency, Vienna, 1980, 73.
121. **La Fleur, P. D.,** Retention of mercury when freeze-drying biological materials, *Anal. Chem.*, 45, 1534, 1973.
122. **Batley, G. E. and Farrar, Y. J.,** Irradiation techniques for the release of bound heavy metals in natural waters and blood, *Anal. Chim. Acta,* 99, 283, 1978.
123. **Moody, J. R.,** The sampling, handling and storage of materials for trace analysis, *Philos. Trans. R. Soc. London Ser. A,* 305, 669, 1982.
124. **Gardner, D.,** A laminar flow clean-room for trace metal analysis: requirements, design, construction and operation, *Lab. Pract.*, 28, 386, 1979.

Chapter 2

TRACE ELEMENT SPECIATION AND ITS RELATIONSHIP TO BIOAVAILABILITY AND TOXICITY IN NATURAL WATERS

G. M. P. Morrison

TABLE OF CONTENTS

I. INTRODUCTION

In surface waters, trace metals are present in a wide range of chemical forms, in both the particulate and dissolved phases (Table 1). The dissolved phase comprises the hydrated ions, inorganic and organic complexes, together with species associated with heterogeneous colloidal dispersions and organometallic compounds. In some instances these metals are present in more than one valency state. The particulate phase also contains elements in a range of chemical associations, ranging from weak adsorption to binding in the detrital mineral matrix. These species are able to coexist, though not necessarily in thermodynamic, equilibrium with one another.

The aim of speciation studies is to both identify and quantify the many species that together comprise the total trace element concentration. This is usually achieved by physicochemical measurement procedures.[1] The identification of discrete chemical compounds is generally not possible and fractionation into classes of compounds exhibiting similar physical, chemical, or biological behavior has been carried out to provide a useful primary discrimination. The range of physicochemical separations and analytical measurement techniques that have been applied to the study of trace element speciation in surface waters is described in the ensuing chapters. Where the equilibrium data for solution components are known, mathematical modeling provides an alternative predictive method for speciation studies, and this is discussed in Chapter 5.

In formulating an approach to speciation measurements, it is important to consider the aims of the study and whether the planned measurements will provide meaningful data consistent with these aims. Generally the basic reasons for speciation measurements are (1) to study transport and biogeochemical cycling processes, and (2) to predict biological impact.

In the first instance it is important to understand the complex interactions between dissolved, particulate, sedimentary, and biological components (Figure 1). These interactions have not as yet been fully assessed in terms of kinetic considerations; however, some progress has been made in this direction by the measurement of time constants for the interactions of metal species, ligands, and algae in seawater.[2,3]

The second and probably the overriding justification for all speciation measurements is to identify those metal species which are likely to have adverse effects on biota and includes measurements both of bioavailability and toxicity. Ultimately this will involve relating physicochemical speciation measurements to bioassay data for the range of aquatic organisms, including bacteria, unicellular algae, macrophytes, crustacea, and fish.

The fate of accumulated heavy metals is extremely complex and organism dependent. Toxicity occurs when an organism is unable to cope, either by direct usage, storage, or excretion, with additional metal concentration. While normally toxicity is defined by acute effects, less readily detectable chronic effects are important in establishing a permissible concentration.

The interactions of metals with intracellular compartments are highly dependent on chemical form. Some species may be able to chemically bind directly with proteins, enzymes, and other biological molecules, others may adsorb to cell walls, whereas others may diffuse through cell membranes where they may be assimilated, influence enzyme reactions, or exert a toxic effect.

In lower organisms, such as algae, interactions of metal species with individual cells are important. As the primary impact of trace metals will be at this trophic level, it is here that the thrust of speciation toxicity studies should be addressed. Metal species that take part in any biological reactions are presumably bioavailable, although bioavailable is generally taken to mean reaction at the cellular level, i.e., penetration of metals through a biological membrane.

This chapter reviews the present knowledge concerning biochemical processes affecting

Table 1
POSSIBLE FORMS OF TRACE ELEMENTS

Simple ionic species	$Zn(H_2O)_6^{2+}$
Differing valency states	As(III), As(V), Cr(III), Cr(IV)
Weak complexes	Cu-fulvic acid
Adsorbed on colloidal particles	Cu-$Fe(OH)_3$—humic acid
Lipid-soluble complexes	CH_3HgCl
Organometallic species	$CH_3AsO(OH)_2$, Bu_3SnCl
Particulate	Metals adsorbed onto or contained within clay particles

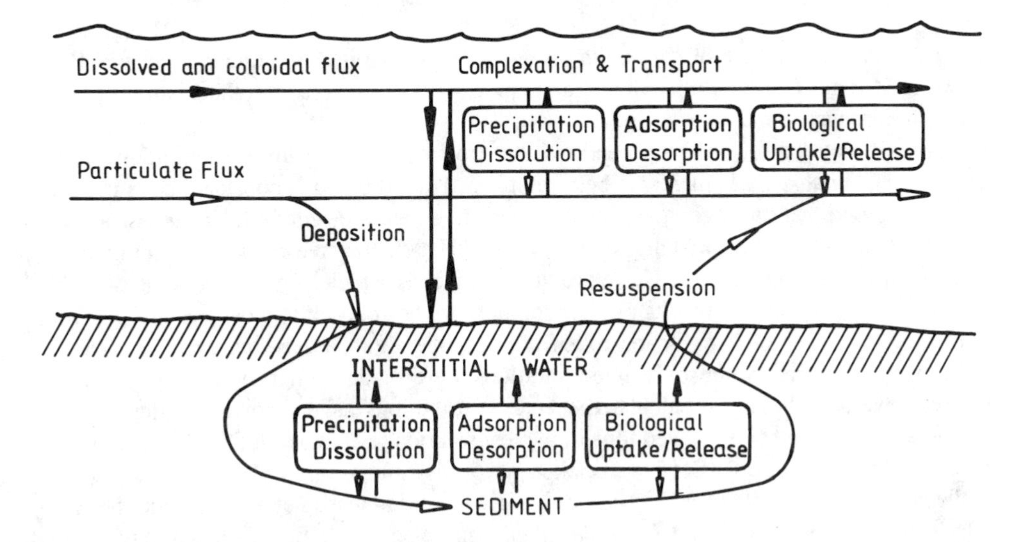

FIGURE 1. Geochemical/biological interactions affecting trace metal speciation. (From Bourg, A. C. M. and Mouvet, C., in *Complexation of Trace Metals in Natural Waters,* Kramer, C. J. M. and Duinker, J. C., Eds., Martinus Nijhoff, The Hague, 1984, 267. With permission.)

metal species uptake, accumulation, and toxicity, in an attempt to provide a background for the selection of an appropriate speciation measurement procedure.

II. BIOAVAILABLE AND TOXIC METAL SPECIES

A. Introduction

Of the species listed in Table 1, three groups have been identified as contributing to the bioavailable fraction. These are simple ionic metal, metal in weak complexes, and lipid-soluble metal. The reasons for this will become apparent when the mechanisms of membrane transport are discussed.

B. Free and Weakly Complexed Metal

Early studies of heavy metal toxicity were carried out using ionic metal additions to laboratory test organisms, with the conclusions based on acute toxicity data. With the realization that speciation was important, the use of synthetic ligands to modify ionic metal concentration was examined.

For example, the toxicity of Cu to the freshwater alga *Scendesmus quadricauda* was demonstrated by Petersen.[5] Using ethylenediaminetetraacetic acid (EDTA) as the complexing ligand, free Cu ion was found to be toxic in the range 10^{-10} to 10^{-12} *M*. Zevenhuisen et

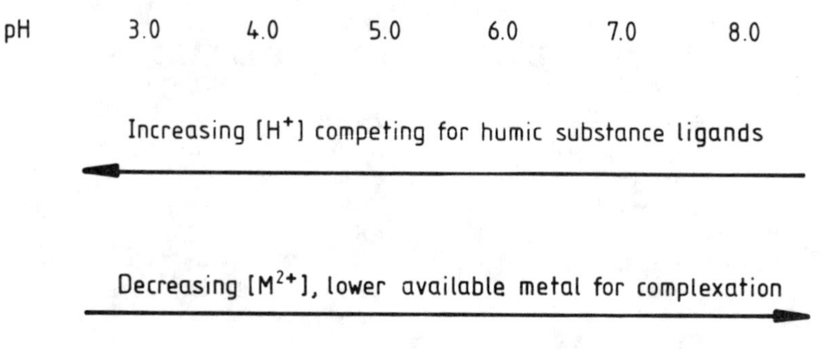

FIGURE 2. Dependence of metal complexation by humic substances on pH.

al.[6] found that the bacterium *Klebsiella aerogenes* was inhibited by concentrations of 10^{-6} to 10^{-10} *M* ionic Cu. Mortality of the amphipod *Hyallela azteca* correlates strongly with ionic Pb concentrations.[7]

However, Petersen et al.[8] have reported that Cu toxicity (as P uptake inhibition) to the green alga *S. quadricauda* increases between pH 5.0 and 6.5 and is constant above pH 6.5. It was proposed that hydrogen ion competes with free metal ion for cellular binding sites at low pH, thereby reducing metal toxicity. Above pH 6.5, the free metal concentration decreases due to the presence of Cu hydroxide complexes which may not be as toxic. This theory is supported by the results of Harrison et al.[9] who found that a decrease of pH from 7 to 5 led to a 50 to 60% reduction of Zn flux into the alga *Chlamydomonas variabilis*.

In many instances, when only ionic additions were made, conclusions as to the toxic species have been based on calculated species distributions using known equilibrium data for metals and the other dissolved ions in the test solution. For example, Theis and Dodge[10] reported that Cu^{2+} and $CuOH^{+}$ were taken up by the midge larva *Chironomous tentans*, while no uptake was noted for Cu-NTA or Cu-glycine. Magnusson et al.[11] came to the same conclusions for the crustacean *Daphnia magna* and also found that carbonate complexes were not toxic. Guy and Kean[12] showed that five organic ligands with relatively high Cu stability constants were toxic at a concentration of 10^{-8} *M* Cu. The toxicity was higher for citric acid (10^{-10} *M* Cu) and ethylenediamine (10^{-9} *M* Cu) which have low stability constants with Cu. These small organic ligands may transport Cu into the cell via the biological membrane.[13] Copper toxicity towards fish was found to be less at higher pH and water hardness.[14] Using EDTA as the complexing ligand, Allen et al.[15] found that Zn^{2+} and $ZnOH^{+}$ were the toxic species to the alga *Microcystis aeruginosa*. Both the carbonate and free ionic forms of Pb have been reported as toxic to rainbow trout.[16]

The metal complexing ligands present in natural waters include humic, fulvic, and tannic acids; complex aromatic polymer molecules having carboxylic acid, hydroxyl, and amine functionality. In addition, extracellular polymers derived from excretion products, exudates, and munchates of aquatic organisms have been shown to have metal-binding ability.[17]

Organic ligands of high molecular weight in pond water were found to reduce Cd toxicity to the crustacean *Simocephalus serrulatus,* but not to the mosquito fish *Gambusia affinis*.[18]

Saar and Weber[19] have shown that the complexation of metals by humic substances is highly dependent on pH (Figure 2). At low pH values hydrogen ion saturates the metal binding sites on the humic substance, whereas at high pH values free metal ion is not an important species. A relatively neutral pH value is therefore favored for metal complexation by humic substances.

It is believed that complexation of metals by humic substances reduces metal bioavailability and, in turn, potential toxicity. Winner[20] demonstrated that humic substances reduced both the acute (3-d) and chronic (42-d) toxicity of Cu to the crustacean *D. pulex*. The LC_{50} value provided a linear relationship with increasing humic substance concentration. Similar conclusions can be derived from the results of Sedlacek et al.[21] for the effect of Cd on the alga *Selenastrum capricornutum* Printz.

Humic substances clearly represent an important complexation and transport mechanism for heavy metal ions in natural water, although in acidic waters metals may be released to more biologically available free and weakly complexed forms.

C. Lipid-Soluble Metal

Lipid-solubility can be estimated by solvent extraction using a solvent mixture with similar dielectric properties to the lipid bilayer. A comparison of the toxicity to algae of Cu complexes with a range of synthetic ligands has shown the greater toxicity of lipid-soluble Cu complexes.[22] It was demonstrated that as little as 2 $\mu g\ l^{-1}$ Cu, as 8-hydroxyquinolinol complex, is highly toxic to the marine diatom *Nitzschia closterium*.[22] It has also been shown that certain uncharged organic ligands facilitate the diffusion of metals both through an artificial lipid bilayer[23,24] and into olive oil.[25]

D. Sediment-Bound Metal

The bioavailability and toxicity of sediment-bound metal species has been comprehensively reviewed by Luoma[13] and Langston and Bryan.[26] Uptake of sediment-bound metal may occur either by phagocytosis (enclosure of external particulate material in the cell) or by direct ingestion.[26] However, field studies have generally shown a poor correlation between metal concentrations in detritus feeders and surrounding sediment. Acid-extractable (1 *M* HCl) metal seemed to provide the best, although weak, correlation.[13]

III. MECHANISMS OF METAL TRANSPORT ACROSS BIOMEMBRANES

A. Introduction

Mechanisms for the transport of ionic and molecular species across biological membranes have been the subject of detailed investigations in recent years. It is now generally accepted that three major transport routes exist (Figure 3) including passive diffusion, carrier-mediated pathways involving interactions with membrane components, and active transport driven by the potential gradient across the membrane.

B. Passive Diffusion

Some metal species can passively diffuse through aqueous pores in the cell membrane. The rates of diffusion will be a function of molecular size; therefore, colloidal species will be virtually excluded with free metal able to diffuse at a faster rate. Nevertheless, transport of free metal ion is more likely to be mediated by interactions with membrane-bound proteins. In the case of lipid-soluble species, however, passive diffusion provides a *rapid* pathway across the cell membrane and rapid penetration by this route may explain their high toxicity.

C. Facilitated (Carrier-Mediated) Transport

Ionophores (viz., compounds such as protein which increase membrane permeability to ions) may facilitate the transport of metals across biomembranes by forming complexes which are lipid soluble. Metals are weakly bound to the receptor ionophore on one side of the membrane, and the hydrophobic complex then diffuses to the other side where the metal may be released into the cytosol. Indeed, it has been stated that the available evidence indicates facilitated transport of metals across biomembranes.[13,27] However, no study has definitely proven this to be the case.

An example of an ionophore is lasolocid A which contains lipophilic and lipophobic conformations[28] (Figure 4A). The carboxylic acid group on the lipophobic end can complex metals from the water body. The ionophore transports the complexed metal ion through the

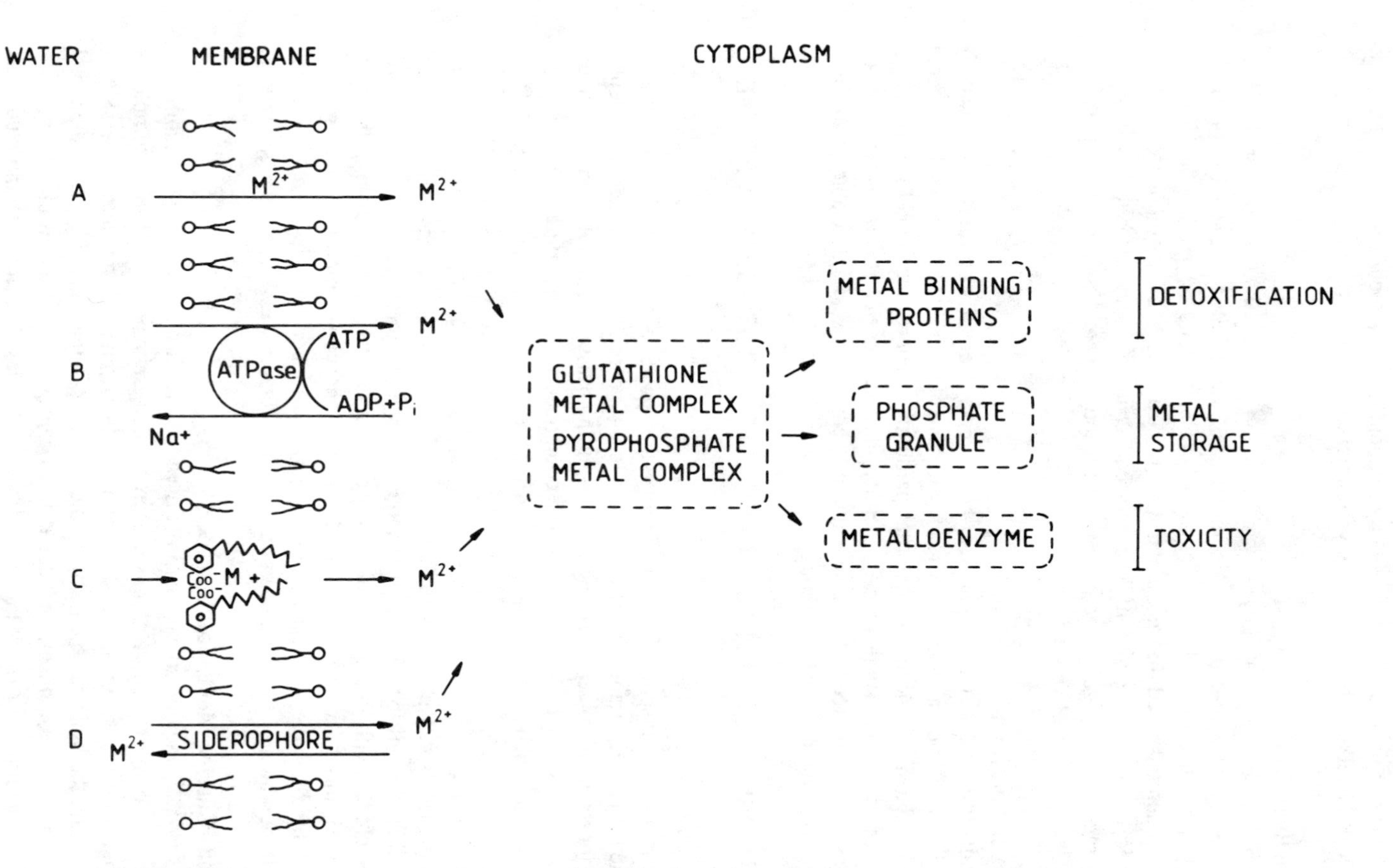

FIGURE 3. Metal uptake, transport, and cell interactions in a biological cell. Transport mechanisms across the biomembrane include passive diffusion (A), active transport (B), facilitated transport (C), and extracellular ionophore diffusion (D).

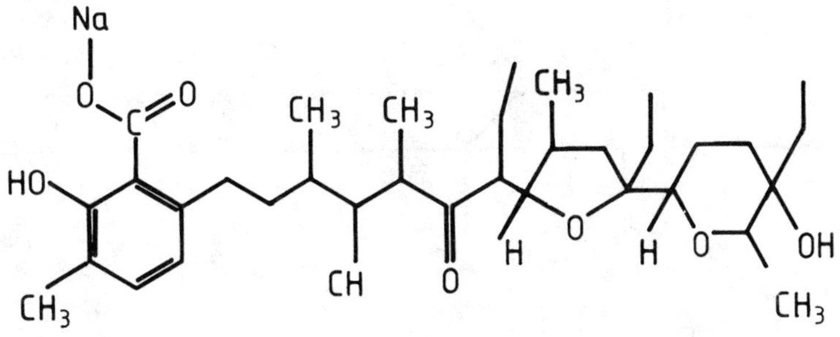

FIGURE 4. (A) Lasolocid A. (B) Sequence of lasolocid-Na^+ decomplexation at a lipid-water interface. (From Brasseur, R. and Ryusschaert, J., *Biochem. J.*, 238, 9, 1986. With permission.)

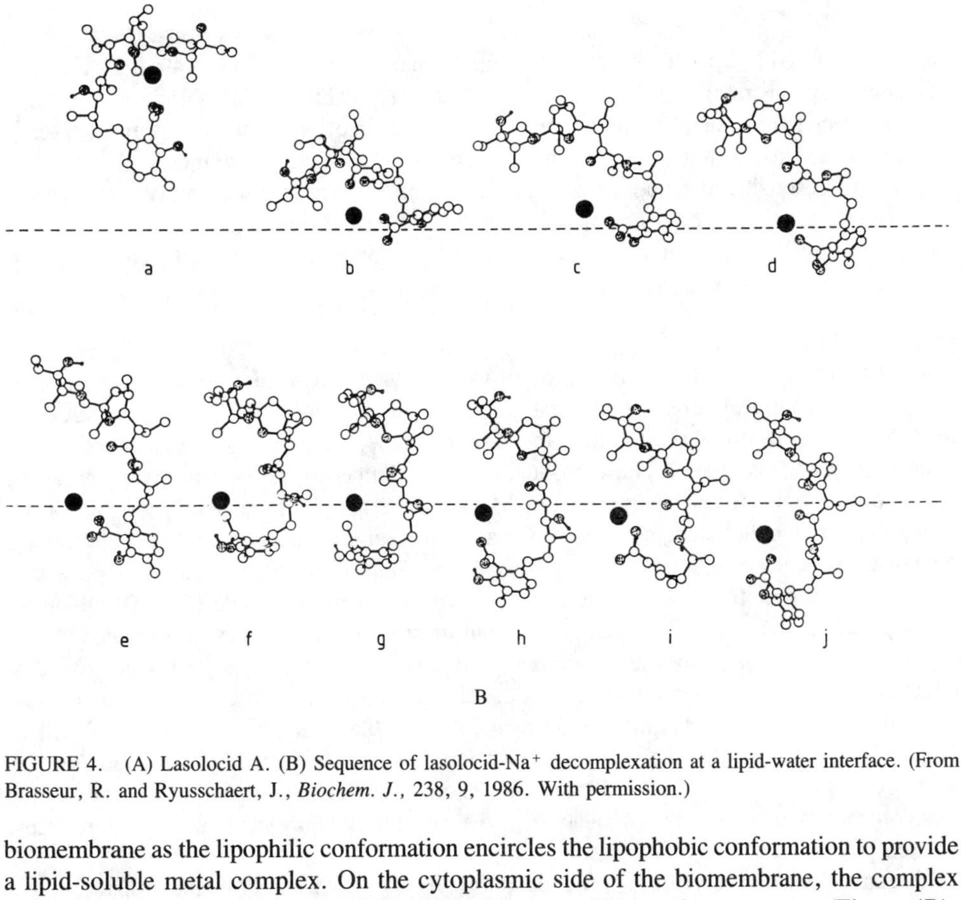

biomembrane as the lipophilic conformation encircles the lipophobic conformation to provide a lipid-soluble metal complex. On the cytoplasmic side of the biomembrane, the complex unwinds and the metal may be released to strongly binding cell components (Figure 4B). The ionophore can then diffuse back to the outer surface of the membrane again. The major role of the ionophoric carrier for Ni transport in the cyanobacterium *Anabaena cylindrica* may be the transport of Mg ions into the cell.[29] In this case the metal ion is a low-affinity alternative substrate.

Binding of metal ion to membrane-bound ionophores depends on interactions of metal with complexing ligands in the bulk solution. In addition, uptake of metal by ionophores is dependent on the kinetics of metal complex interactions.

Extracellular ionophores might also facilitate the transport of metal ions across the biomembrane of bacteria and blue-green algae. Siderophores are avid ferric ion-binding, low-

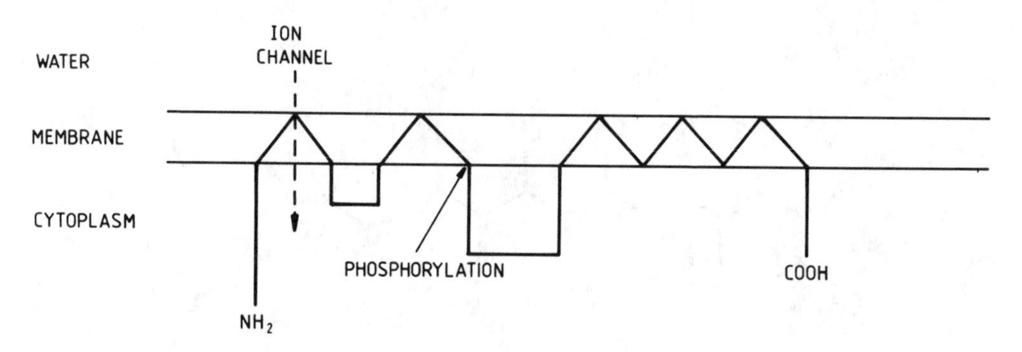

FIGURE 5. Model of ATPase spanning the biomembrane. (From Serrano, R., Kielland-Brandt, M. C., and Fink, G. R., *Nature (London)*, 319, 691, 1986. With permission.)

molecular-weight compounds[30] excreted by nearly all known microorganisms.[31] Siderophores transport required Fe and therefore have high K_m (maximum uptake concentration) and V_{max} (maximum uptake rate) values. The low K_m and V_{max} values of membrane carriers may protect the cell from metal toxicity[30] and at the same time, allow sufficient metal to enter for the metal nutrient (e.g., copper, zinc) requirements of metalloenzymes.

It seems reasonable to propose that facilitated transport processes can prevent toxic quantities of metal from entering the cell. Carriers should only be specific to metals required in small quantities, such as Cu and Zn, and may prevent overloading of the latter or the entry of toxic quantities of nonessential metals.

D. Active Transport

The driving force for the accumulation of many solutes in organisms is a transmembrane proton electrochemical potential gradient ($\Delta_{\mu}^{\sim H+}$). At steady state, a $\Delta_{\mu}^{\sim H+}$ value of approximately 70 mV is established across the biomembrane.[32] Adenosine triphosphate (ATP) hydrolyzed by the enzyme ATPase produces the $\Delta_{\mu}^{\sim H+}$ that drives nutrient uptake either by cotransport or countertransport processes. Metal uptake may be regulated in a similar way, especially for essential elements such as Cu and Zn. Campbell and Smith[30] have shown that Ni transport and accumulation in the cyanobacterium *Anabaena cylindrica* may be moderated by an active carrier transport process. *N,N'*-dicyclohexyl carbodiimide (DCCD) which inhibits proton translocating enzymes, was found to prevent Ni transport into the cell.

ATPase may be an important component for active metal uptake by organisms. This enzyme, with a molecular mass of approximately 170,000 Da, consists of over 1000 amino acids and has about 8 transmembrane domains, most of the protein being cytoplasmic. Shull et al.[33] have demonstrated the existence of an ion channel where two hydrophobic domains span the membrane (Figure 5). Passage of required metal ions through the hydrophilic junction may be powered by connection to the phosphorylation site, via an energy transduction pathway.

Evidence for the involvement of metals in the nutrient/ATPase membrane transfer system is provided by the reduction of Na influx in the waterbug *Corixa punctata* as a response to either pH 3.0 or high Al concentration (10 mg l^{-1}).[35] The ATPase exchanges, Na (out) and K (in) ions across the membrane against concentration gradients.[36] Involvement of metal in the transfer system may explain why exposure of the cyanobacterium *A. variabilis*[37] to Cu and Pb results in a loss of K, whereas Ca, Mg, and Na protect, but K enhances, the toxicity of Cu to the alga *Chlorella saccarophila*.[38]

In addition, 0.5 to 1.0 mg l^{-1} of Cu resulted in altered hemolymph osmolality in the crab *Carcinus maenus*.[39] This suggests that Cu has an effect on ion regulatory processes and explains why Cu toxicity to crabs increases at low salinity. Similar ideas have been forwarded by Viarengo[27] to explain the disruption of the ionic balance of fish by Cd and Hg.

However, metals may also have a more direct toxic effect due to inhibition of ATPase activity. This inhibition may be indicated by changes in adenylate energy due to metal inhibition of the phosphorylation site. A considerable decrease in adenylate energy[40] has been found in the freshwater clam *Corbicula fluminea*, when exposed to 0.6 mg of Cd per liter.

Metals may inhibit ATPase in a similar way to DCCD inactivating an aspartic acid group which is involved in proton translocation.[41] The log equilibrium constant[42] of aspartic acid with Cu is 8.6, which may explain the measured equilibrium constant of Cu with a marine alga *Nitzschia closterium* (log K = 9.11).[43]

Stoecker et al.[44] demonstrated chronic and acute toxicity, 10^{-10} and $10^{-12.8}$ *M*, respectively, of Cu to marine planktonic ciliates, by measuring abnormal motility. At such low concentrations, the toxicity of Cu is probably a membrane effect, as intracellular detoxification and slow metal flux across the membrane would be expected to provide sufficient protection for the cell. Although Stoecker et al.[44] suggested that Cu antagonism to Zn was a result of competition by Cu for Zn activated enzymes, an alternative explanation might be the inhibition of ATPase transport of Zn by binding at the aspartate site.

Competition of toxic free and weakly complexed metal for aspartate binding sites on the membrane-bound ATPase molecule may provide a potent toxic effect to aquatic organisms. With inhibition of ATPase, the cell may literally starve to death due to the prevention of nutrient uptake.

IV. METAL ACCUMULATION BY ORGANISMS

Organisms may regulate metal concentration either by prevention of uptake or by excretion. However, in polluted waters metals show a tendency to bioconcentrate. Lipophilic metal species, such as methylmercury, bioconcentrate to a factor of 10^6 to 10^8 in fish,[45] undoubtedly as a result of the rapid diffusion of lipid-soluble metal species across the biomembrane. Free and weakly complexed metals are transported across the cell membrane at a slower rate and provide bioconcentration factors of between 10^2 and 10^5.[9,46-49]

The surface area of an organism is critical to the passive metal diffusion process into the cell. Evidence for this is provided by an experimental decrease in Cr and Cu concentration with increasing organism size which has been found for aquatic insects.[50,51] It is therefore not surprising that bacterial and algal communities, which have the highest surface area/body weight ratio, frequently have the highest metal concentration in the food web.

Fisher[52] has demonstrated that marine picoplankton also have a high metal bioconcentration rate due to their high surface area/volume ratio. Figure 6 demonstrates the relationship between surface area/volume ratio and metal bioconcentration. Clearly the lower trophic levels, such as bacteria and pico/nanoplankton, can contain considerable metal concentrations, and this may be particularly important for metals such as Hg which magnify through the food web.

Harvey et al.[53] have shown that the enrichment of Pb in the surface layer of a salt marsh sediment strongly correlated with the numbers of surface layer bacteria. Bacteria have the highest surface area per unit weight available for metal diffusion, and it is probable that they represent an essential component in the flow of metals through the food web.

Algae are frequently sampled as periphyton (aufwuchs) on artificial substrates. Selby et al.[49] analyzed metal bioconcentration in a microcosm containing a regulated Cd concentration of 22 μg l^{-1}. Periphyton contained 1000 μg of Cd per gram, and the macrophyte population (*Potemogeton* sp.) contained 232 μg of Cd per gram. Collector-feeders (*Arctopsyche* sp., *Ephemerella* sp.) and grazers (*Brachycentrus* sp., *Physa* sp.) had relatively high Cd con-

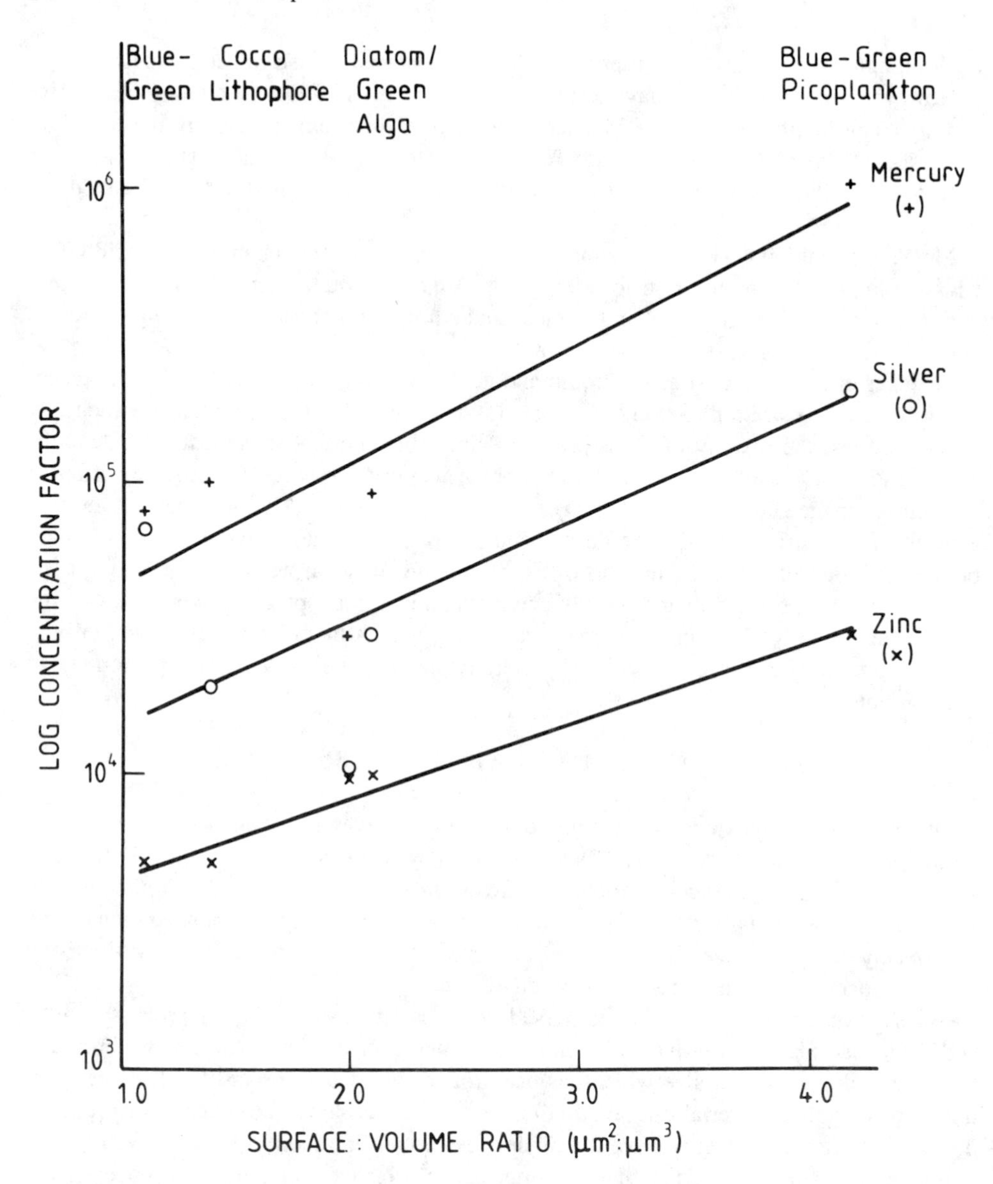

FIGURE 6. Concentration factors for metals in marine plankton over a range of surface to volume ratios. (From Fisher, N. S., *Mar. Biol.*, 87, 137, 1985. With permission.)

centrations of 71 to 109 μg of Cd per gram due to their consumption of the algal, and probably also bacterial, communities. The lowest Cd concentrations of 1.7 to 10.3 $\mu g\ g^{-1}$ were found in the predators (*Hesperoperla* sp., *Isoperla* sp., *Skwala* sp., and *Antherix* sp.). Thus no magnification of inorganic metal species was found to occur in the food web.

Periphyton communities are effective at concentrating lipid-soluble metal species. Stokes et al.[45] showed that the filamentous algae, *Mougeotia* sp. and *Spirogyra* sp., concentrated methylmercury in lake water to 40 to 330 $\mu g\ g^{-1}$, despite a probable lake concentration of 1 $ng\ l^{-1}$. Unlike inorganic metal species, lipid-soluble organic metal species are retained in the food web, and fish in the studied lake contained 40 to 300 $ng\ g^{-1}$ methylmercury.

Periphyton are clearly useful indicators of metal accumulation in the phytoplankton community. However, there is a need to identify the effect of individual metal species on phytoplankton structure both in terms of metal accumulation and toxicity.

Aquatic bryophytes are abundant and widespread water plants found in both acid-[54] and metal-polluted[55,56] waters. Mouvet[57] has shown that *Fontinalis antipyretica* L. ex Hedw can be transplanted from an unpolluted to a polluted river. Exposure to Cu and Cr pollution achieved equilibrium in several weeks.

Indigenous bryophyte populations in rivers draining old metal mines have been shown to contain up to 16 mg g^{-1} Pb and 7 mg g^{-1} Zn.[55] Everard and Denny[56] analyzed *Elodea canadensis* shoots in a Pb mine-polluted lake and uptake was only shown to be significant when the lake sediment containing the Pb reservoir was disturbed. However, Wehr and Whitton[58] found no evidence of seasonal changes for Zn, Cd, and Pb tips of *Rhynchostegium riporioides* in relatively unpolluted streams.

The use of aquatic bryophytes can only provide a relative degree of metal pollution and has not been calibrated in terms of a dose-response relationship. Sampling indigenous populations may roughly reflect integrated long-term metal loadings; transplanted bryophytes are preferable as they indicate short-term metal exposure. However, bryophytes may not reflect sporadic metal discharges which may be damaging to sensitive ecosystem diversity.

Foster[46] has shown that both macroalgal diversity and abundance are reduced by acute metal pollution. Algal species present in metal-contaminated communities usually produce slime and mucilage and are indicated by the presence of *Microspora* sp. and *Ulotrichales* sp. The production of slime may provide a barrier to metal adsorption onto the plant surface or act as a extracellular metal detoxification mechanism by complexing free and weakly bound metals.

Metal concentrations in mussels and oysters in seawater have been extensively studied in the U.S.[59,60] ^{210}Pb was shown to have a half-life of 2 years in mussels, which indicates that the range of Pb values obtained (0 to 18μg g^{-1}) is representative of metal pollution during the previous 2 years. Although *Mytilus* sp. provide an indication of metal pollution, Coleman et al.[61] have shown that there is not necessarily a direct correlation between metal accumulation by mussels and metal concentration in the surrounding water.

Lobel[62] found that up to 53% of the Zn body burden of *M. edulis* accumulates in the kidney. For the same mussel, Elliot et al.[63] found interactions between Zn, Cd, and Cu, although no mechanism for uptake interference was proposed.

Generally, aquatic organisms accumulate metals, although a direct dose-response relationship has not been demonstrated. Certainly lipid-soluble metal complexes are bioconcentrated to a greater extent than free and weakly complexed metal, and only lipid-soluble metal complexes magnify through the food web. The accumulation of metals in aquatic organisms provides only a gross estimate of integrated metal exposure. Biochemical response methods have to be invoked to provide a satisfactory assessment of toxic metal species exposure.

V. METAL TOXICITY AND DETOXIFICATION

When the capacity of a cell to detoxify accumulated metal is exceeded, damage to cytoplasmic constituents will occur. Signs of intracellular toxicity may include ultrastructural deformities, as well as reductions in cell division rate, respiration, photosynthesis, motility, electron transport activity, and ATP production.

Besides speciation, metal toxicity may also depend on the physiological pH. The depolymerization of RNA by Pb(II) in yeast has been shown to proceed at a faster rate at pH 7.4 than at pH 5.0, possibly because 7.4 is close to the value for the pK_a of a water molecule bound to Pb(II).[64] Brown et al.[64] suggested that a hydroxyl group bound to the Pb(II) ion is involved in the catalytic nucleophilic attack.

Free metal ion may also exert toxic effects as a result of the competitive inhibition of enzyme systems.[65] Metals may replace Mg at sulfhydryl binding sites, and certainly a strong correlation has been observed between sulfhydryl binding and toxicity in the marine phytoplankton *Asterionella japonica*.[66] Stauber and Florence[43] have demonstrated the possibility of the intracellular reaction between Cu and reduced glutathione in *Nitzschia closterium*. Glutathiones defend the cell against peroxide damage, and their inactivation by metals leaves the cell open to lipid peroxidation.[27] Evidence has also been presented for the inhibition of the enzyme catalase by Cu; catalase provides a cell defense mechanism against oxygen free radicals.[43] Phillips and Depledge[67] have proposed that the presence of arsenolipids and arsenosugars in the gastropod *Kemifusus* sp., as well as in kelp and giant clams, is accounted for by the interference of As with phospholipid synthesis.

Another possible effect of metals on organisms is a loss of lysosomal membrane stability.[27,68] This may lead to a leakage of hydrolytic enzymes into the cytosol and catabolic breakdown of the cell.

Lipid-soluble complexes are usually very stable and are therefore unlikely to dissociate and provide free ion toxicity.[22] It was proposed that the extreme toxicity of the lipid-soluble Cu xanthate complex to the seawater alga *N. closterium* was due to the catalysis of the intracellular formation of highly destructive hydroxyl free radicals from molecular oxygen. Florence et al.[69] demonstrated that the Cu(I) complex of 2,9-dimethyl-1,10-phenanthroline can react with H_2O_2 to produce oxygen free radicals. Certainly the prevalent toxic effect of metal on cell contents is not fully understood. However, the available evidence conclusively shows that redox balance, enzyme systems, replication, and lysosome stability may be affected by the presence of toxic levels of metal ions.

Cells have the capability of preventing toxic metal action by inducing the formation of metal-binding granules or proteins (Figure 3). Many organisms contain phosphate granules which have a high concentration of metals.[70] However, although Thomas and Ritz[71] have demonstrated the initiation of granule formation by metals in the barnacle *Eliminius modestus*, phosphate granules do not necessarily represent a metal detoxification system. Langston and Zhou[72] suggested that phosphate granules are a regulation (homeostasis) mechanism for Zn and Cu. Excess Zn and Cu are then utilized in metalloenzymes and hemocyanin. Phosphate granules contain a significant amount of pyrophosphate which forms strong complexes with metals under the normal physiological conditions of the cytosol. It is therefore tempting to propose that pyrophosphate acts as a receiver for nutrient metal from either facilitated diffusion or active transport (Figure 3).

In many organisms metal detoxification may occur in cytoplasmic compartments such as lipofuscin granules,[73] lysosomes, and vesicles.[74] In some cases these compartments may respond to elevated metal concentrations by the production of metal-binding proteins. For example, certain cells in the periwinkle *(Littorina littorea)* produce metal-binding proteins in response to elevated metal concentrations in the respiratory current.[74]

The most documented metal detoxification protein is metallothionein (Figure 3) which was first isolated from vertebrates. Metallothioneins are low-molecular-weight (6000 to 10,000 Da) cytoplasmic proteins which contain 28 to 30% cysteine and, therefore, bind metals (between 6 and 11% of total dry weight) as mercaptide complexes. Only ferritin has a higher metal content.[75] Metallothioneins are produced, often in the kidneys and liver,[76] in response to high metal concentrations.[77] Metals seem to trigger a specific messenger RNA for thionein synthesis, and although metallothionein has a half-life of only 4 to 5 d, continued synthesis allows long-term metal retention.[76] Invertebrates produce a similar metallothionein-type (metal-binding) protein which is lower in cysteine (6 to 20%) and, probably as a direct consequence, metal content.[77] A similar Cu-binding protein has been reported for the fungus *Neurospora crassa*.[78]

Although it is not proven that a dose-response relationship exists between metals and metal-binding protein production, Cd has been shown to induce the formation of a high-sulfhydryl-content protein in the winkle *L. littorea*.[72] It was suggested that glutathione and cysteine are synthesised in the cytosol as a response to incoming Cd, acting as a thiol reservoir for metal-binding protein induction. This is supported by the findings of Thomas et al.,[79] which demonstrate an increase of glutathione in the liver of the mullet *Mugil cephalus* as a response to Cd. Viarengo[27] suggests that binding of glutathione by Cd stimulates γ-glutamylcysteine synthetase (the enzyme responsible for glutathione synthesis) activity and therefore increases cell glutathione concentration.

Metal-binding proteins may therefore serve as detoxification agents due to stable intracellular metal binding, preventing free metal ion diffusion which would certainly exert damage on various enzymes and proteins which are essential for biological reactions. Metal-binding proteins have been shown to prevent lysosome destabilization by Cd.[68] In addition, metal-binding proteins may store and regulate essential elements[76] such as Zn and Cu, and it has been proposed that metal-binding proteins may act as a Cu donor to active sites of the Cu-containing enzyme tyrosinase.[78]

Metal toxicity may occur as a result of metal-binding protein spillover (Figure 3). Brown et al.[80] carried out a gel filtration separation of low-molecular-weight binding proteins (3000 to 20,000 Da) from high-molecular-weight metalloenzymes (>20,000 Da), extracted from the organs of seawater organisms. Although metal spillover into enzymes was not found in coastal water organisms, Cd spillover occurred in the kidney of scorpion fish after a 4-d exposure to 20 to 50 mg l^{-1} Cd. In addition, exposure of the American oyster to 160 μg of Cd per liter in seawater for 2 weeks was found to induce metal-binding proteins.[81] Metal spillover was observed for bluegills in a power station effluent,[82] Cu displacing Zn both in the metalloenzyme and metal-binding protein fractions. Bluegills in the effluent were found to have increased deformities and reduced reproductive capacity.

The interpretation of metal spillover results should be treated with caution, as shrimps and gastropods have been shown to produce metal-tolerant enzymes as an adaption to polluted environments,[65] and therefore the presence of metal in the enzyme fraction may not positively identify toxic effects. Rainbow[70] has shown that oxidation of the metal-binding protein during separation may induce apparent metal spillover. In addition, Langston and Zhou[72] suggest that Cd spillover may be an apparent phenomenon due to the transport of this metal between tissues.

Metal-binding proteins are synthesized in aquatic organisms as a response to toxic intracellular levels of Cd and possibly other metals, and it is possible that glutathione acts as a source of thiol for the synthesis. Metal-binding proteins undoubtedly provide a metal detoxification mechanism for cells and may also be involved in the homeostatic regulation of Zn and Cu.

VI. CONCLUSIONS

Both lipid-soluble metal complexes and metals in free or weakly complexed forms have been shown to be the species most likely to penetrate a cell membrane and hence be considered bioavailable. Their transport processes are, however, quite different, the former utilizing passive diffusion and the latter an active or facilitated transport mechanism.

The detoxification response of the cell to intracellular metal involves accumulation in metal-binding proteins, phosphate granules, and lysosomes. The inducement of the formation of metal detoxification proteins has been used as an indicator of bioavailable metal species; however, this approach is suspect since a clear dose-response relationship has not been established. A more promising approach is likely to be one which attempts to differentiate those species causing intracellular toxic reactions.

Bioassays provide the most definitive measure of bioavailable metal species. They are, however, lengthy and expensive and the results can be misleading because of interferences from algal exudates or other complexing agents present in the culture medium. An alternative is to use chemical procedures to quantify the bioavailable species. This may involve a physiochemical separation procedure or the use of an analytical technique that is species selective. The development of such procedures represents an important challenge for the analytical chemist. Approaches towards the solution of this priority problem are described in subsequent chapters.

ACKNOWLEDGMENTS

The author wishes to acknowledge an overseas fellowship award from the Natural Environment Research Council, U.K. Graeme Batley, Mark Florence, and Jenny Stauber made valuable comments on the final manuscript.

REFERENCES

1. **Florence, T. M.,** Electrochemical approaches to trace element speciation in waters: a review, *Analyst,* 111, 489, 1986.
2. **Piotrowicz, S. R., Harvey, G. R., Springer-Young, M., Courant, R. A., and Boren, D. A.,** Studies of cadmium, copper and zinc complexation by marine fulvic and humic materials in seawater, in *Trace Metals in Seawater,* Wong, C. S., Burton, J. D., Bruland, K., and Goldberg, E. D., Eds., Plenum Press, New York, 1984, 699.
3. **Piotrowicz, S. R., Springer-Young, M., Puig, J. A., and Spencer, M. J.,** Anodic stripping voltammetry for evaluation of organic-metal interactions in seawater, *Anal. Chem.,* 54, 1367, 1982.
4. **Bourg, A. C. M. and Mouvet, C.,** A heterogeneous complexation model of the adsorption of trace metals on natural particulate matter, in *Complexation of Trace Metals in Natural Waters,* Kramer, C. J. M. and Duinker, J. C., Eds., Martinus, Nijhoff, The Hague, 1984, 267.
5. **Petersen, R.,** Influence of copper and zinc on the growth of a freshwater alga, *Scendesmus quadricauda:* the significance of chemical speciation, *Environ. Sci. Technol.,* 16, 443, 1982.
6. **Zevenhuisen, L. P. T. M., Dolfing, J., Eshuis, E. J., and Scholten-Koersehman, I. J.,** Inhibitory effects of copper on bacteria related to the free ion concentration, *Microbiol. Ecol.,* 5, 139, 1979.
7. **Freedman, M. E., Cunningham, P. M., Schindler, J. E., and Zimmerman, M. J.,** Effect of lead speciation on toxicity, *Bull. Environ. Contam. Toxicol.,* 25, 389, 1980.
8. **Petersen, H. G., Healey, F. P., and Wagemann, R.,** Metal toxicity to algae: a highly pH dependent phenomenon, *Can. J. Fish. Aquat. Sci.,* 41, 974, 1984.
9. **Harrison, G. I., Campbell, P. G. C., and Tessier, A.,** Effects of pH changes on Zn uptake by *Chlamydomonas variabilis* grown in batch culture, *Can. J. Fish. Aquat. Sci.,* 43, 687, 1986.
10. **Theis, T. L. and Dodge, E. E.,** Effect of chemical speciation on the uptake of copper by *Chironomous tentans, Environ. Sci. Technol.,* 13, 1287, 1979.
11. **Magnusson, V. R., Harris, D. K., Sun, M. S., Taylor, D. K., and Glass, G. E.,** Relationships of activities of metal-ligand species to aquatic toxicity, in *Chemical Modelling in Aqueous Systems,* Jenne, E. A., Ed., ACS Symp. Ser. No. 93, American Chemical Society, Washington, D. C., 1979, 633.
12. **Guy, R. D. and Kean, A. R.,** Algae as a chemical speciation monitor, I. A comparison of algal growth and computer calculated speciation, *Water Res.,* 14, 891, 1980.
13. **Luoma, S. N.,** Bioavailability of trace metals to aquatic organisms — a review, *Sci. Total Environ.,* 28, 1, 1983.
14. **Borgmann, U.,** Metal speciation and toxicity of free metal ions to aquatic biota, *Adv. Environ. Sci. Technol.,* 13, 47, 1983.
15. **Allen, H. E., Hall, R. H., and Brisbin, T. D.,** Metal speciation effects on aquatic toxicity, *Environ. Sci. Technol.,* 14, 441, 1980.
16. **Davies, P. H., Goettl, J. P., Sinley, J. R., and Smith, N. F.,** Acute and chronic toxicity of lead to rainbow trout, *Salmo Gairneri,* in hard and soft water, *Water Res.,* 10, 199, 1976.
17. **Rudd, T., Sterritt, R. M., and Lester, J. N.,** Complexation of heavy metals by extracellular polymer in the activated sludge process, *J. Water Pollut. Control Fed.,* 56, 1260, 1984.
18. **Geisy, J. P., Leversee, G. J., and Williams, D. R.,** Effects of naturally occurring aquatic organic fractions on cadmium toxicity to *Simocephalus serrulatus (Daphnidae)* and *Gambusia affinis (Poeciliidae), Water Res.,* 11, 1013, 1977.

19. **Saar, R. A. and Weber, J. H.,** Fulvic acid: modifier and metal-ion chemistry, *Environ. Sci. Technol.*, 16, 510A, 1982.
20. **Winner, R. W.,** Bioaccumulation and toxicity of copper as affected by interactions between humic acid and water hardness, *Water Res.*, 19, 449, 1985.
21. **Sedlacek, J., Källquist, T., and Gjessing, E.,** Effect of aquatic humus on uptake and toxicity of cadmium to *Selenastrum capricornutum* Printz, in *Aquatic and Terrestrial Humic Materials*, Christman, R. F. and Gjessing, E. T., Eds., Ann Arbor Science, Ann Arbor, MI, 1983, 495.
22. **Florence, T. M., Lumsden, B. G., and Fardy, J. J.,** Evaluation of some physico-chemical techniques for the determination of the fraction of dissolved copper toxic to the marine diatom, *Nitzschia closterium, Anal. Chim. Acta*, 151, 281, 1983.
23. **Gutknecht, J.,** Inorganic mercury (Hg^{2+}) transport through lipid bilayer membranes, *J. Membr. Biol.*, 61, 61, 1981.
24. **Bienvenue, E., Boudou, A., Desmazes, J. P., Gavach, C., Georgescauld, D., Sandeaux, J., and Seta, P.,** Transport of mercury compounds across bimolecular lipid membranes: effect of lipid composition, pH and chloride concentration, *Chem. Biol. Interact.*, 48, 91, 1984.
25. **Simkiss, K.,** Lipid solubility of heavy metals in saline solutions, *J. Mar. Biol. Assoc. U.K.*, 63, 1, 1983.
26. **Langston, W. J. and Bryan, G. W.,** The relationships between metal speciation in the environment and bioaccumulation in aquatic organisms, in *Complexation of Trace Metals in Natural Waters*, Kramer, C. J. M. and Duinker, J. C., Eds., Martinus Nijhoff, The Hague, 375, 1984.
27. **Viarengo, A.,** Biochemical effects of trace metals, *Mar. Pollut. Bull.*, 16, 153, 1985.
28. **Brasseur, R. and Ruysschaert, J.,** Conformation and mode of organisation of amphiphilic membrane components: a conformational analysis, *Biochem. J.*, 238, 1, 1986.
29. **Jarrell, K. F. and Sprott, D.,** Nickel transport in *Methanobacterium bryantii, J. Bacteriol.*, 151, 1195, 1982.
30. **Campbell, P. M. and Smith, G. D.,** Transport and accumulation of nickel ions in the cyanobacterium *Anabaena cylindrica, Arch. Biochem. Biophys.*, 244, 470, 1986.
31. **Smarrelli, J. and Castignetti, D.,** Iron aquisition by plants: the reduction of ferrisiderophores by higher plant NADH:nitrate reductase, *Biochim. Biophys. Acta*, 882, 337, 1986.
32. **Boudou, A., Georgescauld, D., and Desmazes, J. P.,** Ecotoxicological role of the membrane barriers in transport and bioaccumulation of mercury compounds, in *Aquatic Toxicology*, Nriagu, J. O., Ed., John Wiley & Sons, New York, 1983, 117.
33. **Shull, G. E., Schwartz, A., and Lingrel, J. B.,** Amino-acid sequence of the catalytic subunit of the (Na^+, K^+) ATPase deduced from a complementary DNA, *Nature (London)*, 316, 691, 1985.
34. **Serrano, R., Kielland-Brandt, M. C., and Fink, G. R.,** Yeast plasma membrane ATPase is essential for growth and has homology with ($Na^+ + K^+$), K^+ and Ca^{2+}-ATPases, *Nature (London)*, 319, 689, 1986.
35. **Witters, H., Vangenechten, J. H. D., Puymbroeck, S. V., and Vanderborght, O. L. J.,** Interference of aluminium and pH on the Na-influx in aquatic insect *Corixa punctata* (Illig.), *Bull. Environ. Contam. Toxicol.*, 32, 575, 1984.
36. **Cantley, L. C.,** Structure and mechanism of the (Na,K)-ATPase, *Curr. Top. Bioenerg.*, 11, 201, 1981.
37. **Jensen, T. E., Rachlin, J. W., Jani, V., and Warkentine, B. E.,** Heavy metal uptake in relation to phosphorus nutrition in *Anabaena variabilis (Cyanophyceae), Environ. Pollut. Ser. A*, 42, 261, 1986.
38. **Folsom, B. R., Popescu, W. A., and Wood, J. M.,** Comparative study of aluminium and copper transport and toxicity in an acid-tolerant freshwater green alga, *Environ. Sci. Technol.*, 20, 616, 1986.
39. **Bjerregaard, P. and Vislie, T.,** Effect of copper on ion- and osmoregulation in the shore crab, *Carcinus maenus, Mar. Biol.*, 91, 69, 1986.
40. **Giesy, J. P., Duke, C. S., Bingham, R. D., and Dickson, G. W.,** Changes in phosphoadenylate concentrations and adenylate energy charge as an integrated biochemical measure of stress in invertebrates: the effects of cadmium on the freshwater clam *Corbicula fluminea, Toxicol. Environ. Chem.*, 6, 259, 1983.
41. **Senior, A. E. and Wise, J. G.,** The proton-ATPase of bacteria and mitochondria, *J. Membr. Biol.*, 73, 105, 1983.
42. **Martell, A. E. and Smith, R. M., Eds.,** *Critical Stability Constants*, Vol. 1, Plenum Press, New York, 1974, 25.
43. **Stauber, J. L. and Florence, T. M.,** Reversibility of copper-thiol binding in *Nitzschia closterium* and *Chlorella pyrenoidosa, Aquat. Toxicol.*, 8, 223, 1986.
44. **Stoecker, D. K., Sunda, W. G., and Davis, L. H.,** Effects of copper and zinc on two planktonic ciliates, *Mar. Biol.*, 92, 21, 1986.
45. **Stokes, P. M., Dreier, S. I., Farkas, M. O., and McLean, R. A. N.,** Mercury accumulation by filamentous algae: a promising biological monitoring system for methyl mercury in acid-stressed lakes, *Environ. Pollut., Ser. B*, 5, 255, 1983.
46. **Foster, P. L.,** Species associations and metal contents of algae from rivers polluted by heavy metals, *Freshwater Biol.*, 12, 17, 1982.

47. **Friant, S. L. and Koerner, H.,** Use of an *in situ* artificial substrate for biological accumulation and monitoring of aqueous trace metals. A preliminary field investigation, *Water Res.*, 15, 161, 1981.
48. **Ramelow, G. J., Maples, R. S., Thompson, R. L., Mueller, C. S., Webre, C., and Beck, J. N.,** Periphyton as monitors for heavy metal pollution in the Calcasieu river estuary, *Environ. Pollut.*, 43, 247, 1987.
49. **Selby, D. A., Ihnat, J. M., and Messer, J. J.,** Effects of subacute cadmium exposure on a hardwater mountain stream microcosm, *Water Res.*, 19, 645, 1985.
50. **Smock, L. A.,** Relationships between metal concentrations and organism size in aquatic insects, *Freshwater Biol.*, 13, 313, 1983.
51. **Darlington, S. T., Gower, A. M., and Ebdon, L.,** The measurement of copper in individual aquatic insect larvae, *Environ. Technol. Lett.*, 7, 141, 1986.
52. **Fisher, N. S.,** Accumulation of metals by marine picoplankton, *Mar. Biol.*, 87, 137, 1985.
53. **Harvey, R. W., Lion, L. W., Young, L. Y., and Leckie, L. O.,** Enrichment and association of lead and bacteria at particulate surfaces in a salt-marsh surface layers, *J. Mar. Res.*, 40, 1201, 1982.
54. **Caines, L. A., Watt, A. W., and Wells, D. W.,** The uptake and release of some trace metals by aquatic bryophytes in acidified waters in Scotland, *Environ. Pollut., Ser. B*, 10, 1, 1985.
55. **Burton, M. A. S. and Peterson, P. J.,** Metal accumulation by aquatic bryophytes from polluted mine streams, *Environ. Pollut.*, 19, 39, 1979.
56. **Everard, M. and Denny, P.,** Flux of lead in submerged plants and its relevance to a freshwater system, *Aquat. Bot.*, 21, 181, 1985.
57. **Mouvet, C.,** Accumulation of chromium and copper by the aquatic moss *Fontinalis antipyretica* L. ex Hedw. transplanted in a metal-contaminated river, *Environ. Technol. Lett.*, 5, 541, 1984.
58. **Wehr, J. D. and Whitton, B. A.,** Accumulation of heavy metals by aquatic mosses. III. Seasonal changes, *Hydrobiologia*, 100, 285, 1983.
59. **Goldberg, E. D. and Martin, J. H.,** Metals in seawater as recorded by mussels, in *Trace Metals in Seawater*, Vol. 9, NATO Conf. Ser. IV, Wong, C. S., Bruland, K. W., and Burton, J. D., Eds., Plenum Press, New York, 1983, 811.
60. **Goldberg, E. D.,** The mussel watch concept, *Environ. Monit. Assessment*, 7, 91, 1986.
61. **Coleman, N., Mann, T. F., Mobley, M., and Hickman, N.,** *Mytilus edulis planulatus:* an ''integrator'' of cadmium pollution?, *Mar. Biol.*, 92, 1, 1986.
62. **Lobel, P. B.,** Role of the kidney in determining the whole soft tissue zinc concentration of individual mussels *(Mytilus edulis), Mar. Biol.*, 92, 355, 1986.
63. **Elliot, N. G., Swain, R., and Ritz, D. A.,** Metal interaction during accumulation by the mussel *Mytilus edulis planulatus, Mar. Biol.*, 93, 395, 1986.
64. **Brown, R. S., Hingerty, B. E., Dewan, J. C., and Klug, A.,** Pb(II)-catalysed cleavage of the sugar-phosphate backbone of yeast $tRNA^{Phe}$ — implications for lead toxicity and self-splicing RNA, *Nature (London)*, 303, 543, 1983.
65. **Nevo, E., Ben-Shlomo, R., and Lavie, B.,** Mercury selection of allozymes in marine organisms: reduction and verification in nature, *Proc. Natl. Acad. Sci. U.S.A.*, 81, 1258, 1984.
66. **Fisher, N. S. and Jones, G. J.,** Heavy metals and marine phytoplankton: correlation of toxicity and sulfhydryl-binding, *J. Phycol.*, 17, 108, 1981.
67. **Phillips, D. J. H. and Depledge, M. H.,** Chemical forms of arsenic in marine organisms, with emphasis on *Hermifuscus* species, *Water Sci. Technol.*, 18, 213, 1986.
68. **Viarengo, A., Moore, M. N., Mancinelli, G., Mazzucotelli, A., Pipe, R. K., and Farrar, S. V.,** Metallothioneins and lysosomes in metal toxicity and accumulation in marine mussels: the effect of cadmium in the presence and absence of phenanthrene, *Mar. Biol*, 94, 251, 1987.
69. **Florence, T. M., Stauber, J. L., and Mann, K. J.,** The reaction of copper-2,9-dimethyl-1,10-phenanthroline with hydrogen peroxide, *J. Inorg. Biochem.*, 24, 243, 1985.
70. **Rainbow, P. S.,** The biology of heavy metals in the sea, *Int. J. Environ. Stud.*, 25, 195, 1985.
71. **Thomas, P. G. and Ritz, D. A.,** Growth of zinc granules in the barnacle *Eliminius modestus, Mar. Biol.*, 90, 255, 1986.
72. **Langston, W. J. and Zhou, M.,** Evaluation of the significance of metal-binding proteins in the gastropod *Littorina littorea, Mar. Biol.*, 92, 505
73. **George, S. G.,** Heavy metal detoxification in *Mytilus* kidney — an *in vitro* study of Cd and Zd binding to isolated tertiary lysosomes, *Comp. Biochem. Physiol.*, 76C, 59, 1983.
74. **Mason, A. Z., Simkiss, K., and Ryan, K. P.,** The ultrastructural localization of metals in specimens of *Littorina littorea* collected from clean and polluted sites, *J. Mar. Biol. Assoc. U.K.*, 64, 699, 1984.
75. **Kojima, T. and Kagi, J. H. R.,** Metallothionein, *Trends Biochem. Sci.*, 90, 1978.
76. **Cherian, M. G. and Goyer, R. A.,** Metallothioneins and their role in the metabolism and toxicity of metals, *Life Sci.*, 23, 1, 1978.
77. **Roesijaldi, G.,** The significance of low molecular weight, metallothionein-like proteins in marine invertebrates: current status, *Mar. Environ. Res.*, 4, 167, 1980.

78. **Lerch, K.,** Copper metallothionein, a copper-binding protein from *Neurospora crassa, Nature (London),* 284, 368, 1980.
79. **Thomas, P., Woffard, H. W., and Neff, J.,** Effects of cadmium on glutathione content of mullet *(Mugil cephalus)* tissues, in *Physiological Mechanisms of Marine Pollutant Toxicity,* Vernberg, W. B., Calabrese, A., Thurberg, F. P., and Vernberg, F. J., Eds., Academic Press, New York, 1982, 109.
80. **Brown, D. A., Bay, S. M., and Gossett, R. W.,** Using the natural detoxification capacities of marine organisms to assess assimilative capacity, in *7th Symp. on Aquatic Toxicology and Hazard Assessment,* Cardwell R. D., Purdy, R., and Bahner, R. C., Eds., American Society for Testing and Materials, Philadelphia, 1985, 364.
81. **Frazier, J. M.,** Bioaccumulation of cadmium in marine organisms, *Environ. Health Perspect.,* 28, 75, 1979.
82. **Harrison, F. L.,** Effect of physiochemical form on copper availability to aquatic organisms, in *7th Symp. on Aquatic Toxicology and Hazard Assessment,* Cardwell, R. D., Purdy, R., and Bahner, R. C., Eds., American Society for Testing and Materials, Philadelphia, 1985, 469.

Chapter 3

PHYSICOCHEMICAL SEPARATION METHODS FOR TRACE ELEMENT SPECIATION IN AQUATIC SAMPLES

Graeme E. Batley

TABLE OF CONTENTS

I. INTRODUCTION

The separations of metal species based on differences in some physicochemical property, e.g., ionic charge or molecular size, were among the first techniques to be applied in metal speciation.[1-3] Combinations of these procedures with highly sensitive analytical techniques, such as anodic stripping voltammetry (ASV), form the basis of a number of more complex, schemes for resolving heavy metal species in natural waters.[4]

A major assumption in the application of separation techniques is that the selective removal of one or more species does not appreciably disturb the solution equilibria. This is difficult to accept, especially in systems as dynamic as natural waters. Indeed, if the act of sampling itself does not disturb the equilibria, there is evidence[5] that the classical 0.45-μm filtration to remove particulates may indeed do so, with regrowth of filtrable particles occurring with time in the filtrate. It is important therefore to acknowledge the limitations of such procedures. Operationally defined separations do, however, provide a useful means of comparing different samples and currently afford the best way, given the limitations of metal speciation modeling.[2,4] The problems associated with specific separation methods, their advantages, and limitations, will be reviewed in this chapter.

II. SEPARATION TECHNIQUES

A. Filtration

The use of a 0.45-μm membrane filter to separate particulate from dissolved species in a water sample is a well-established practice. This distinction is, of course, arbitrarily chosen and it is well-known that the "dissolved" fraction will contain filtrable matter, including colloids, which can be separated by conventional screen filters extending over the range 0.45 to 0.01 μm nominal pore diameter.

Filtration techniques have been reviewed elsewhere,[6] and the need to avoid sample contamination or losses due to adsorption on filter surfaces cannot be too highly stressed. Vacuum or pressure filtration are both acceptable, although the former is more widely practiced. Filtration assemblies are usually made of polypropylene or Pyrex® glass. Stirring during vacuum filtration was found to offer little advantage for filtrations above 0.01 μm.[7]

The choice of membrane filter is critical and has been the subject of several studies.[7,8] The conventional cellulose acetate or cellulose nitrate filters (Millipore® or Sartorius®) are effective depth filters, having thicknesses in the range 100 to 150 μm. They are thus able to trap particles during filtration and present a much greater surface for adsorption. The thinner (5 to 10 μm) polycarbonate filters (Nuclepore®) are more uniform in their size distribution and can more closely be considered as absolute screen filters. Not surprisingly, therefore, they reveal quite different size distributions, for example, for the filtration of colloidal iron in a freshwater sample (Figure 1).[7] Nuclepore® filters were found to be fairly accurate in their size distribution, generally allowing the passage of particles smaller than the nominal pore size. The deeper filters retained more than their nominal rating for pore sizes 0.45 μm and above, while the smallest pore size filters appeared to be passing particles greater than the nominal rating.

Clogging was a major problem with both membrane types.[7-9] Permeability of a 0.45 μm filter to colloidal particles was found to decrease dramatically beyond a certain filter load[9] (Figure 2), with the polycarbonate filters showing a greater susceptibility to clogging than cellulose acetate membranes. For groundwater samples, only the first 100 to 200 ml filtered through 0.45 μm polycarbonate filters could be considered to be well defined with a filter pore size equivalent to a 0.1-μm membrane after filtration of 500 ml. If a depth filter is used as a prefilter, clogging of both filters is dramatically reduced;[8], however, the risk of adsorptive losses of metal species is increased.

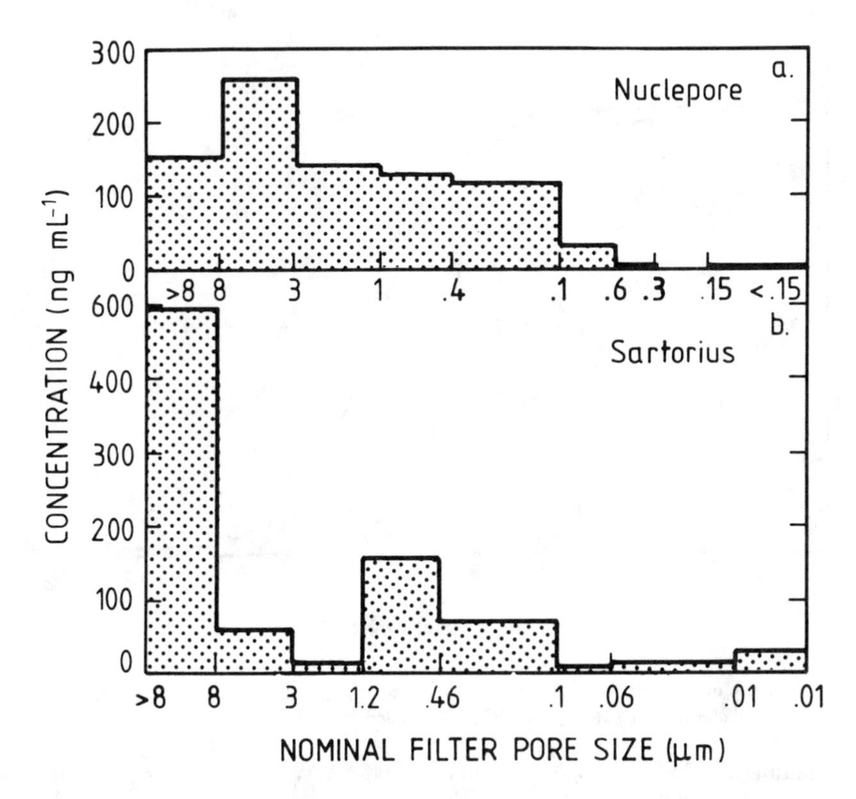

FIGURE 1. Size distribution of iron in a freshwater sample for two filter types. (From Laxen, D. P. H. and Chandler, I. M., *Anal. Chem.* 54, 1350, 1982. With permission.)

It has been recommended that filters be changed once the filtration pressure builds up to 30 kPa.[7] It should be noted, however, that as with pressure filtration, extreme suction pressures will lead to rupture of phytoplankton cells which could contribute both dissolved organic matter and trace metals to the filtrate. The latter are concentrated in some marine algae by four orders of magnitude in excess of their seawater concentrations.[2]

Adsorption on membrane filter surfaces can represent a significant source of trace metal losses.[8] These can be minimized by preconditioning the filter with sample; however, particles retained by the membrane may enhance adsorptive losses. The recommended washing procedure of Salbu et al.[8] involves 24 h soaking in 0.1*M* nitric acid, a 3-d soak in distilled water, and before use, conditioning by passage of 250 ml of distilled water and 50 ml of sample. In our laboratory, washing with 2 *M* nitric acid (20 ml) followed by distilled water ($\simeq$ 50 ml) and sample ($\simeq$ 20 ml) has been found adequate for most applications.

B. Ultrafiltration

The term ultrafiltration is commonly used to describe the separation, on the basis of size at the molecular level (molecular filtration), of dissolved species by passage through molecular filters. The filters generally consist of a thin organic polymeric hydrous gel film supported on porous polyethylene or cellulose ester bases. The pore sizes of the membranes are normally specified to retain most molecules above a nominal molecular weight limit (NMWL). Typically these are 210, 500, 1000, 10,000, 50,000, 100,000 and 300,000.[9] In fact, they do not retain *all* molecules larger than an absolute molecular weight cut-off, as molecules can deform such that, for example, a linear polymer may pass through the membrane while a globular molecule of the same molecular weight will be retained. Ogura[10] showed that cytochrome c (M_r 12,600) was 95% retained by a 100,000-NMWL membrane,

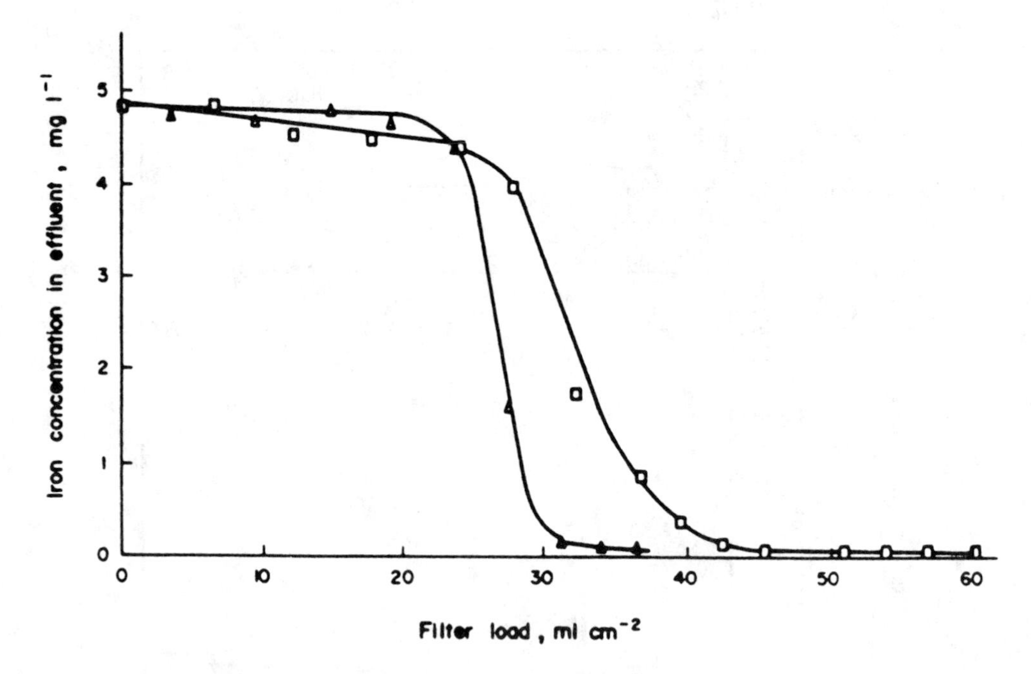

FIGURE 2. Effect of filter load on the retention of iron colloids. (□) Millipore®; (△) Nuclepore®. (From Danielsson, L.G., *Water Res.*, 16, 179, 1982. With permission.)

and cobalamin (M_r 1357) was 76% retained by a 500-NMWL filter. While the larger NMWL ultrafilters are essentially nonionic, those of 10,000 NMWL and below have ionic sites. In particular, 500-NMWL membranes possess a net negative charge which undoubtedly will influence the separation process.[11]

Prewashing of filters with dilute acid is essential to remove both heavy metal contaminants[6] and organic wetting agents[12] which may interfere with the subsequent analysis. Wheeler[13] recommended stirring the membrane for 4 h with distilled water, with frequent changes of water, to remove added glycerine or ethanol. Ultrasonication would enhance this process. The membrane is then placed in the filtration cell and distilled water is filtered until the organic carbon content of the filtrate falls below 1 $\mu g\ l^{-1}$. Other authors[14] have used only the latter step, with UV absorbance monitoring at 200 nm to detect a carbon-free filtrate. Laxen and Harrison[15] preconditioned ultrafilters by successive soaking in dilute NaOH (pH <11), 0.5% high-purity HNO_3 and finally distilled water. Filters were stored in a 10% ethanol in water solution in acid-washed plastic petri dishes at 4°C before use. Stirred filtration cells were precleaned by soaking in a 5% solution of Decon® 90 surface active agent for 24 h between each experiment.

When filtering a sample, it is desirable to discard the first 10% and the last 50% of the ultrafiltrate, collecting only the middle fraction. The initial volume can contain a reduced concentration of species in the ultrafiltrate because of dilution by distilled water trapped in the filter, while, in some cases, this may be masked by slow clogging of the filter, particularly if a solution containing an abundance of high-molecular-weight species is being filtered through a low-porosity membrane. The last 50% of the ultrafiltrate often contains higher concentrations of organic carbon because of leakage of high-molecular-weight substances occurring as the concentration of filtered species above the membrane increases[13] (Figure 3). Adsorption, molecular association, and coagulation are also likely occurrences with this concentration increase. As a general rule, it is preferable not to filter solutions containing more than a few $g\ l^{-1}$ on 10,000-NMWL filters or a few hundred $mg\ l^{-1}$ on 1000-NMWL. At least the last 10% of the ultrafiltrate should be rejected.

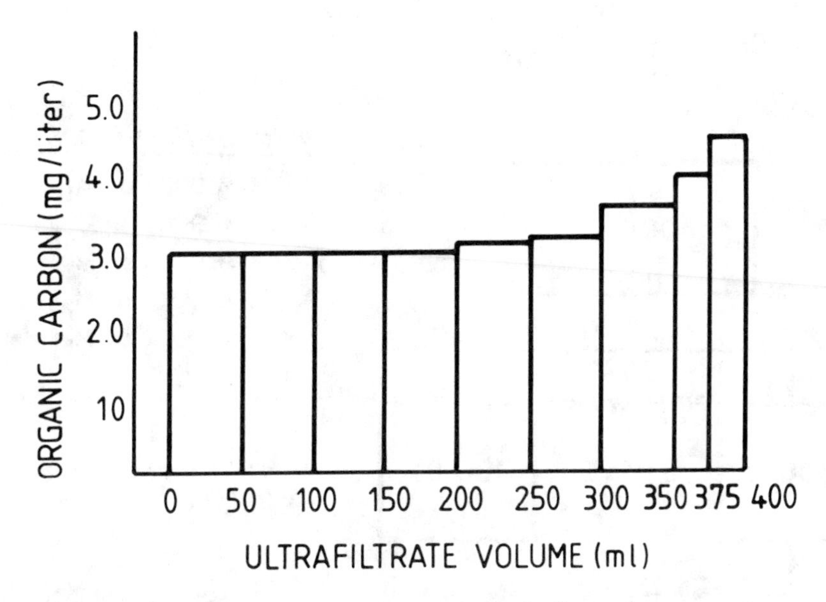

FIGURE 3. Concentration of organic substances in the ultrafiltrate of a water sample as a function of volume collected (From Wheeler, J. R., *Limnol. Oceanogr.*, 21, 846, 1976. With permission.)

To characterize metals associated with high-molecular-weight organics, the most satisfactory procedure is direct cascade filtration, through increasingly finer filters, but rejecting the last 10 to 20% of each filtrate as described above. This procedure has been applied by Smith[14] and Giesy et al.[16] For comparison, the "washing" procedure, where the cell volume is kept constant by the addition of distilled water adjusted to the same pH as the sample, has been investigated by Buffle et al.[17] This technique is useful for evaluating the retention factor R for a filtered compound which is related by the equation

$$\ln (0.01\ F_c) = -R\ (V/Vo)$$

where F_c is the fraction of the compound left in the filtration cell, Vo is the initial volume of sample in the cell, and V is the washing volume. Ideally, R values approach 300 for fully retained compounds, and unity for those not retained. The reproducibility of this procedure is good, and for the separation of humic and fulvic acids from lower molecular weight compounds in natural waters, it has been recommended that the last step in the cascade filtration use the washing technique with a 210-NMWL filter and V/Vo near five.[16]

The washing procedure may, however, result in irreversible changes in metal-organic associations. To minimize this and other interferences in metal species distribution, Hoffmann et al.[18] employed a mass balance approach, based on both organic carbon and metals, to sequential ultrafiltration. Only half of the retentate volume is collected at each cascade stage. The remaining retentate at each stage is returned for metal and carbon analysis, with the collected ultrafiltrate being fractionated by the next lower pore size filter. Thus, for a four-stage filtration using 100,000-, 25,000-, 10,000-, and 1000-NMWL filters, if DOC_1, DOC_2, DOC_3, DOC_4, and DOC_5 are, respectively, the carbon contents in the ranges >100,000, 25,000 to 100,000, 10,000 to 25,000, 1000 to 10,000, and <1000 NMWL, and if L_1, L_2, L_3, and L_4 are the successive volume reduction factors, the measurements shown in Figure 4 will permit calculation of the DOC fractions:

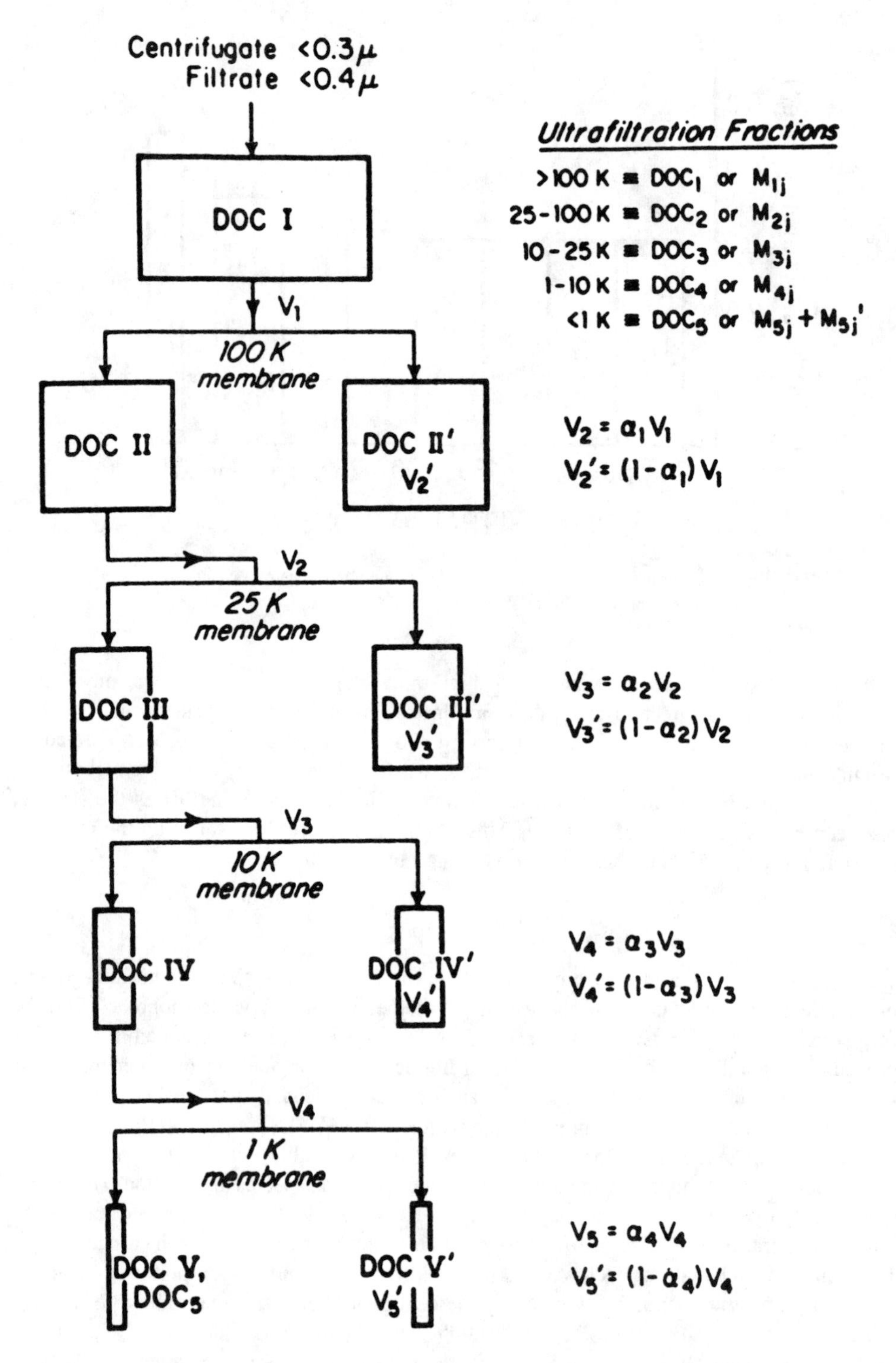

FIGURE 4. Flow diagram for sequential ultrafiltration. (From Hoffman, M. R., Yost, E. C., Eisenreich, S. J., and Maier, W. J., *Environ. Sci. Technol.*, 15, 655, 1981. With permission.)

$$DOC_1 = (DOC\ I) - (DOC\ II)/\alpha_1$$

$$DOC_2 = (DOC\ II)/\alpha_1 - (DOC\ III)/(\alpha_1\alpha_2)$$

$$DOC_3 = (DOC\ III)/(\alpha_1\alpha_2) - (DOC\ IV)/(\alpha_1\alpha_2\alpha_3)$$

$$DOC_4 = (DOC\ IV)/(\alpha_1\alpha_2\alpha_3) - (DOC\ V)/(\alpha_1\alpha_2\alpha_3\alpha_4)$$

$$DOC_5 = (DOC\ V)/(\alpha_1\alpha_2\alpha_3\alpha_4)$$

Similarly for metals:

$$M_{1j} = [M(I)] - M(II)/\alpha_1$$

$$M_{2j} = [M(II)/\alpha_1] - M(III)/(\alpha_1\alpha_2)$$

$$M_{3j} = [M(III)/(\alpha_1\alpha_2)] - M(IV)/(\alpha_1\alpha_2\alpha_3)$$

$$M_{4j} = [M(IV)/(\alpha_1\alpha_2\alpha_3)] - M(V)/(\alpha_1\alpha_2\alpha_3\alpha_4)$$

$$M_{5j} = [M(V)/(\alpha_1\alpha_2\alpha_3\alpha_4)] - M_{5j}'$$

where M_{ij} is the mass of metal M_j in the ith molecular-weight fraction, which is nominally associated with dissolved organic carbon or colloidal material, and M_{5j}' is the summation of free aquated metal and monomeric or polymeric inorganic complexes which, because of their size, should appear in the smallest size fraction.

Ultrafilters are reusable, but it should be noted that surfaces can easily be scratched. After use they should be stored in water or ethanol at 4°C or with the addition of sodium azide (0.1%) to prevent bacterial growth.

Ultrafiltration cells consist of a membrane support, a source of inert gas pressure to enhance transport through the membrane, and some form of agitation. Typical pressures of <30 kPa and 2-ml min^{-1} flow rates are obtained with >10,000-NMWL filters, with 300 kPa giving 2-ml min^{-1} flows with the finer filters. During filtration, concentration polarization occurs at the membrane surface with the formation of a boundary layer which may accumulate as a semisolid gel. Increased pressure will simply compact this gel; however, flow rates can be enhanced by agitation using either a stirrer bar or a vibrating forced-flow agitator which will redistribute the boundary layer into the bulk solution. The desirable membrane-limited flow is pressure dependent, whereas gel-limited flow is not. The latest ultrafiltration equipment utilizes tangential flow to avoid damage to shear-sensitive molecules. As an alternative to membrane ultrafilters, conical filters have been developed by the Amicon Corporation and others for centrifugal filtration. With a fixed angle rotor, concentration polarization during centrifugation is avoided, as the retained material accumulates at the outer edge of the membrane. Filters of 10,000 and 30,000 NMWL are available.

For trace metal applications, the absence of sources of contamination should be fully explored. Metal screen filter supports and rubber O-rings, for example, should be absent, otherwise meaningless fractionations will be obtained. Blank determinations should be carried out by filtration of distilled water. The contamination problems cannot be too highly stressed, especially where small volumes of solution are being filtered through large-surface-area membranes. The problem of adsorptive losses in these samples has not been fully explored, but might be expected to be substantial, especially on the finer, charged membranes. Laxen and Harrison[15] observed the adsorption was especially significant with 500-NMWL Amicon® UM05 filters and only marginally less for Berghof BM5 filters of the same size. This led them to recommend that a 0.015-μm ultrafilter be the smallest size used in any fractionation scheme.

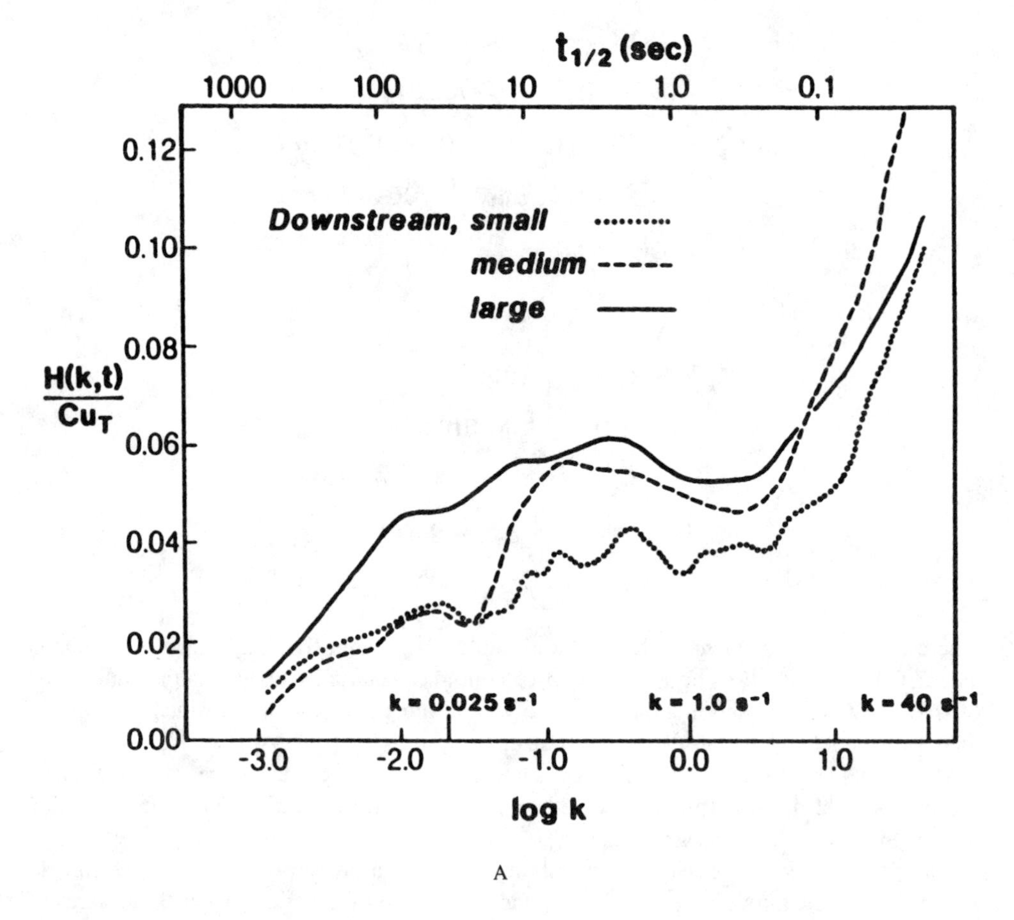

A

FIGURE 5. Kinetic distribution of bound copper in seawater samples. (From Olson, D. L. and Shuman, M. S., *Geochim. Cosmochim. Acta,* 49, 1371, 1985. With permission.)

The relevance of separations based on size with respect to biological availability is questionable.[19] Certainly biomembranes would be impermeable to very large molecules, while this may also be so for some smaller species. Of more significance is the lability of the metals in these larger molecules. Thus, Olson and Shuman[20] examined the kinetics of copper dissociation from their ultrafiltered size fractions by following the rates of formation of the copper complex with 4-(2-pyridylazo)resorcinol, where the limiting step was the dissolution of the humic material or similar copper complexes. The kinetic stabilities of these species are a function of pH and salinity. In keeping with isotope exchange data,[21] a significant fraction of copper was found to be nonlabile. The concept of a kinetic spectrum for bound copper (Figure 5A) represents a new approach to speciation measurements. These are obtained by numerical differentiation of the pseudo-first-order kinetic data. The area under the entire spectrum is equal to the sum of the initial concentrations of the reacting components. It is convenient to subdivide these spectra into four regions based on k values, $k < 0.025$ s^{-1}, $0.025 < k < 1.0$ s^{-1}, $1.0 < k < 40$ s^{-1}, and $k > 40$ s^{-1}, to obtain the kinetic distribution shown in Figure 5B.

Few speciation techniques are indeed equilibrium measurements. Physicochemical speciation methods disturb the equilibrium through removal of selected species, generally with a change in bulk solution composition. The time scale of the measurement and any kinetic contribution to the separation process will therefore be important,[18] as will be shown later with respect to the use of chelating resin extractions.

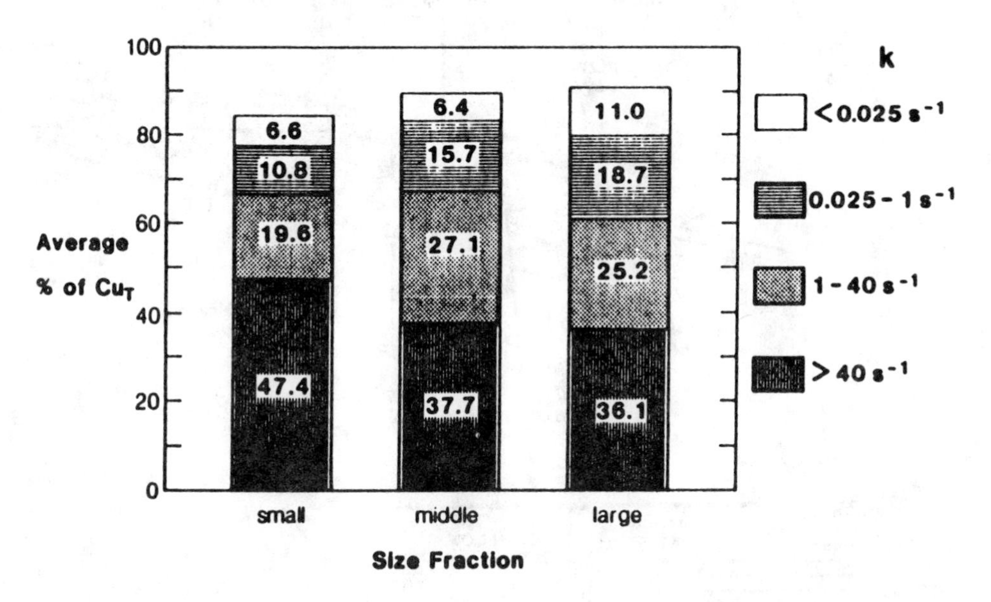

FIGURE 5B

C. Centrifugation

This technique is a rarely used size fractionation method based on particle density. By varying the rotation speed and centrifugation times, good separations of suspended and colloidial particles can be achieved. Continuous-flow centrifugation is useful for separating suspended particles; however, colloids remain suspended unless high-speed ultracentrifugation or density gradient techniques are employed.[1,22] Aggregation and sorption effects may bias results while contamination is likely to be significant, especially during ultracentrifugation.[8]

Few applications to metal speciation have been reported. Salbu et al.[8] separated 0.4- and 0.1-μm fractions; however, nonspecific size separation were apparent for both aluminium and manganese due to aggregation effects.

More recently the technique of field flow fractionation[22-25] has displayed promise for particulate fractionation. In the earlier studies[22] this technique used a horizontal cell with sample injected from one end. A cross-flow of a suitable carrier solvent resulted in a flow profile which deposited particulates in zones of decreasing particle density along the longitudinal flow axis (Figure 6). A variant of the technique uses a similar cell wrapped in an annular shape for use in a centrifuge basket. Particles are forced to the outside wall to form clouds, the thickness of which depends on their mass and on the centrifugal field.[25] It is possible both to collect the separated particles and to record, using a suitable detector, the particle size distribution (fractogram) of colloidal matter in a water sample. No attempts have yet been made to measure metal species in the separated colloids.

D. Dialysis

Dialysis can be used to fractionate metal species by their differential rates of diffusion through porous membranes. Whereas in ultrafiltration the membrane functions as a mechanical screen to fully separate species, dialysis is a rate process, with diffusion of membrane-permeable species being driven by their concentration gradient. Semipermeable membranes separate species primarily on the basis of size. Cellulose nitrate, collodion, and gelatin are typical membrane materials. The former are negatively charged due to acidic

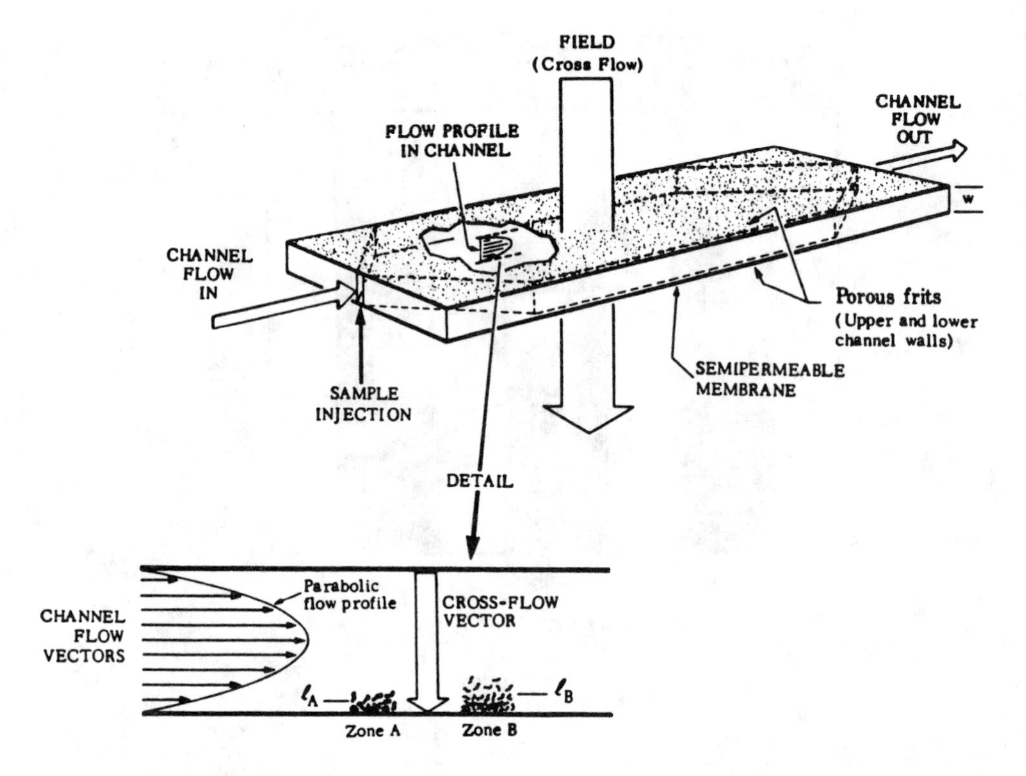

FIGURE 6. Schematic representation of a flow field-flow fractionation obtained showing the porous frits that allow the cross-flow to pass. Detail is the cross section of the channel. (From Beckett, R., *Environ. Technol. Lett.*, 8, 339, 1987. With permission.)

groups on the molecular skeleton and, because of this, are also good adsorbents for ionic metal species. Diffusion is not limited only by pore size, but also by interaction between solute, solvent, and membrane, including membrane swelling, hydration, and adsorption effects. In classical dialysis, the solvent (diffusate) and the solvent plus solute (retentate) are separated by a membrane. Both solutions are stirred. This technique has been applied to seawater samples by Hood and co-workers,[26-27] using a 4.8-nm-pore-size cellulose dialysis tubing containing 500 ml seawater aliquots. These were dialyzed for 24 h against distilled water in a rotating dialyzer. The distilled water was changed frequently, and the test concluded when a negative test for chloride was obtained. At equilibrium, free or dialyzable metal is measured in the diffusate while the nondialyzed portion is analyzed for trace (free plus complexed) metals.

Guy and Chakrabarti[28] used a 14,000-NMWL dialysis bag to separate metal complexes of humic and tannic acids from simple carbonate and EDTA complexes. The bag contained 20 ml of distilled water equilibrated with a 200-ml solution containing 0.01 M KNO_3 and the appropriate metal species. Again, a 24-h equilibrium was necessary (Figure 7). It was possible to estimate the stability constants for metal-humic acid complexes from a study of dialyzable metal vs. total metal for a fixed humic acid concentration.

The value of dialysis titrations for measurements of the complexing capacities of large organic molecules, such as humic and fulvic acids, was further explored by Truitt and Weber,[29] using 1000- NMWL Spectrapor® dialysis bags. Each addition required a 48-h equilibration. The method was also applied to natural freshwaters[30] dialyzed against 0.001 M KNO_3 having a conductance similar to the samples. The method requires that no free ligands permeate the membrane; however, up to 10% of the copper binding capacity of soil fulvic acids could permeate the membrane, while at pH values of 7 or above, results were

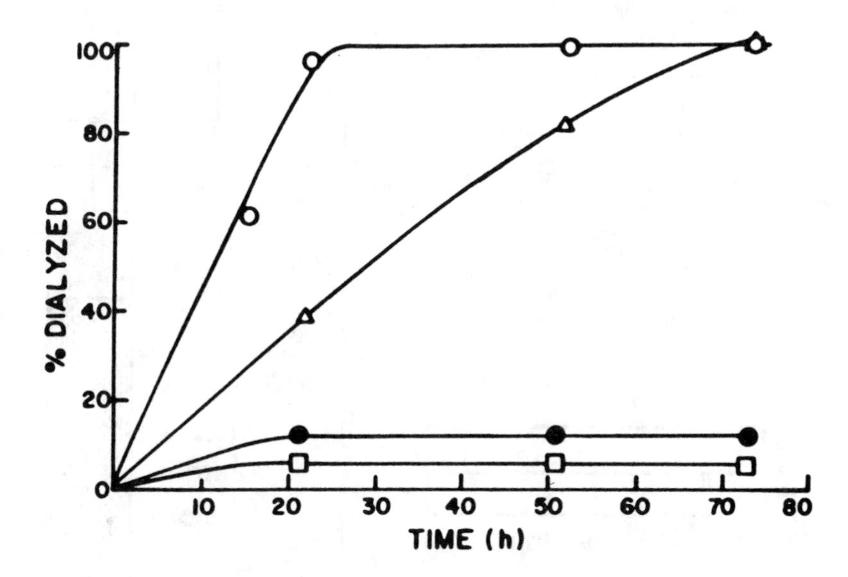

FIGURE 7. Dialysis as a function of time. (○) Cu EDTA^{-2} (Spectrapor® 2); (△) Cu EDTA^{-2} (Spectrapor® 3); (●) copper/humic acid, (□) copper/tannic acid (Spectrapor® 3). (From Guy, R. D. and Chakrabarti, C. L., *Can. J. Chem.*, 54, 2600, 1976. With permission.)

complicated by the formation of nondialyzable copper inorganic species such as $Cu_2(OH)_2^{2+}$. While the technique is potentially attractive since chemical equilibria are not greatly disturbed by the addition of electrolytes, it is a time-consuming exercise and similar results could be obtained by ion-selective electrode measurements. A greater number of metal ions can, however, be studied using the dialysis technique.

As a practical note, the need for cleaning of commercial dialysis bags prior to use should be stressed. A recommended procedure[30] is soaking in warm water to remove preservatives, then in 0.1% Na_2S solution at 60°C for 15 min, before rinsing with warm distilled water. This is followed by soaking in 3% H_2SO_4 at 60°C for 5 min, rinsing in distilled water, and storing until needed in distilled water.

Benes and co-workers[31-34] have extensively studied the applications of dialysis to metal speciation in natural waters. In particular, they investigated *in situ* dialysis, where an acid-washed dialysis bag, made from 4.8-nm-pore-size Visking tubing (NMWL = 10,000), and filled with 100 ml of distilled water, is immersed in a river. This system has the advantage that the supply of dialysate is effectively unlimited. The bags were left in place for a number of weeks, but dialytic equilibrium is established during the first week if not within 24 h. The method should be useful, therefore, for the analysis of seasonal variations of metal species. Unlike ultrafiltration separations, no significant adsorptive losses were observed. In studies of groundwater, coatings of iron compounds on the surface of dialysis bags were often noted after continuous operation.[8] These coatings impeded dialysis, prompting the conclusions that dialysis is better suited to rivers and lakes than slow-moving groundwaters

In situ dialysis is suitable for semicontinuous monitoring only if (1) the diffusion coefficients of the dialyzing species do not change, e.g., with large temperature or ionic strength effects, (2) the properties of the diffusion layers are not altered by flow rate, microbial attack, or plugging with slime, and (3) the concentrations of dialyzed ions are maintained below 1% of their minimum concentrations outside the cell.[33] This latter condition can be achieved by removal of the species by adsorption or ion exchange. This can be done in a batch process,[35] or continuously where diffusate is circulated in a closed loop through a column of adsorbent.[36] In this procedure possible losses of ionic species can occur through adsorption on the tubing and other parts of the system. In the experiments of Hart and

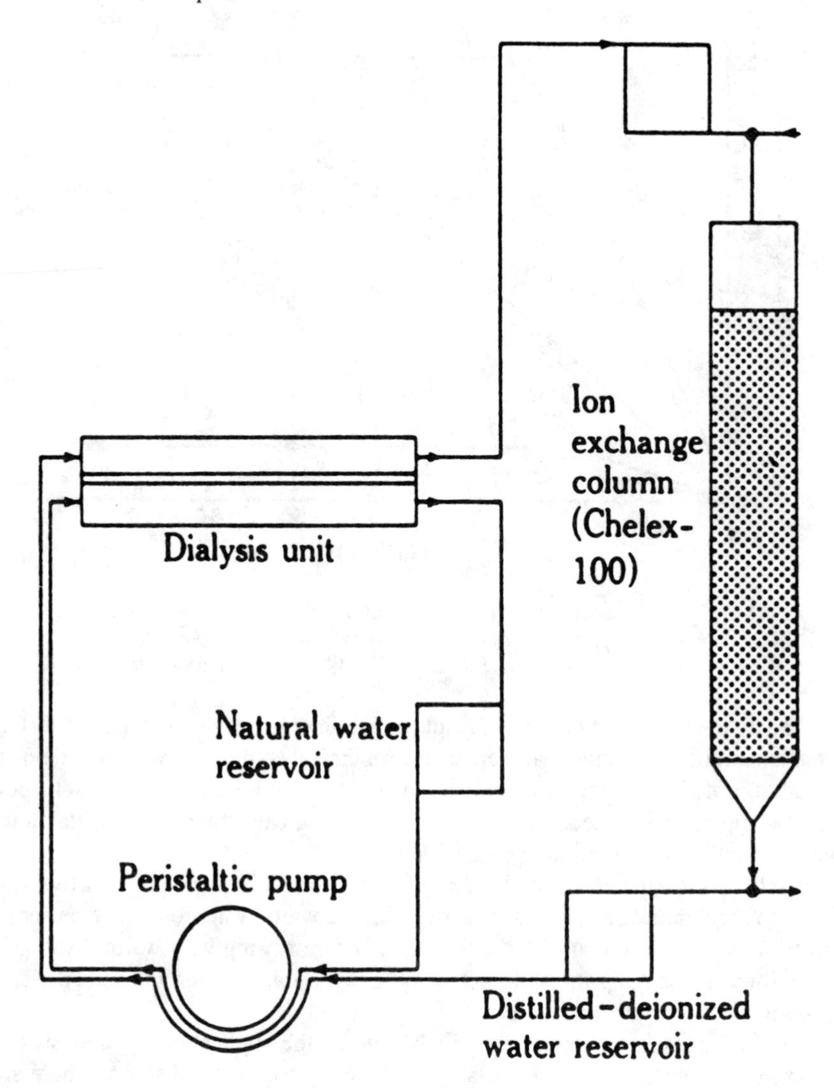

FIGURE 8. Dialysis-ion exchange apparatus. (From Hart, B. T. and Davies, S. H. R., *Aust. J. Mar. Freshwater Res.*, 28, 105, 1977. With permission.)

Davies,[36] water was passed through an autoanalyzer dialysis unit at 5 ml min^{-1} with the diffusate passing through a Chelex®-100 column (Figure 8). Equilibrium was established in 5 to 6 h for heavy metal standards near 200 μg l^{-1}, but the same may not necessarily be true for a natural water. Complexes of EDTA with cadmium, lead, copper, and zinc were not dialyzed within the 5-h period. This is contrary to the results of Guy and Chakrabarti[28] who found some 10% of Cu $EDTA^{2-}$ dialyzed in 5 h, but 40 to 100% after 24 h, depending on the membrane used (Figure 7). In general, the diffusion of negatively charged species is slower because of the charge on the membrane.

Morrison[35] used a 38-mm-diameter dialysis bag (Spectrapor®, NMWL 100) containing Chelex®-100 in an attempt to measure bioavailable metals in polluted waters. The bag was precleaned with 0.01 *M* EDTA, pH 8, and after adding resin, was immersed in the sample solution for between 1 and 4 d. At the end of this period the resin was released to a separating column for elution with 2 *M* nitric acid prior to metal analysis. As expected, the dialysis membrane, because of its large surface area, also can also adsorb equivalent concentrations of metals to those dialyzed. These concentrations were taken as an indication of metal species which might not readily penetrate a biological membrane. The method was not examined for model compounds. This system has also been used for *in situ* monitoring of stormwater. The results are expressed as a metal flux, pg mm^{-2} h^{-1} across the membrane. An integrated flux is obtained over the sampling period.

In situ dialysis is also being increasingly used for the sampling of dissolved metal species in sediment interstitial waters.[28-41] A sampler utilizing dialysis bags mounted at intervals within an acrylic plastic (Plexiglas®) cylinder has been used for the sampling of dissolved silica.[38] For close interval studies, Hesslein[39] designed a sampler consisting of small chambers, ≈2.5 ml volume, machined in an acrylic plastic sheet and covered with a dialysis membrane. Although these were initially applied to phosphate analysis, Carignan et al.[40,41] recognized their potential for trace metal studies. The low trace metal content of acrylic plastic, coupled with its rigidity, make it ideally suited to this application. The dialyzers, or sediment peepers, consisted of a 30 × 15 × 1-cm Plexiglas® plastic in which 6.5 × 3.6 × 0.6-cm chambers spaced 1 cm center to center were machined. The plate was covered with the appropriate dialysis membrane kept in place by a 0.2-cm-thick Plexiglas® sheet with apertures matching the chambers and fastened to the main body with nylon screws spaced 15 cm apart. Prior to use, the chambers must be filled with distilled water and, if being used in anoxic sediments, should be deoxygenated by placing the chambers in acrylic cylinders filled with distilled water and bubbling with nitrogen for 24 h. For sampling interstitial waters, the peepers are inserted into the sediment by hand and left in place for 1 week before removal. The technique minimizes changes in speciation that might occur as a result of oxygen exposure or temperature changes during conventional centrifugation or filtration of some waters.

The Donnan equilibrium effect, in which electrical neturality is maintained on both sides of a membrane, can affect dialysis speciation results where, for example, high concentrations of a charged colloidal particle are being dialyzed against water, since to maintain neutrality, dissociation of water might occur altering the pH of the solution. This is less likely to be a problem with natural freshwaters.

The possibility of slow dissolution of complexes or dissolution of colloids is also possible if the ionic concentration of trace metal falls to a fraction of its original value during dialysis, and this can also cause errors. *In situ* dialysis minimizes these errors.

If an ion-exchange membrane is used in dialysis, Donnan equilibrium effects can be used to advantage in both speciation and preconcentration. Ions of the same charge as the membrane will be excluded and will diffuse more slowly through the membrane phase.[42] For two cations, A and B, of charge y and z, respectively, in solutions 1 and 2, separated by a cation exchange membrane,

$$\left[\frac{[A]_1}{[A]_2}\right]^{y/z} = \frac{[B]_1}{[B]_2}$$

When B is in trace concentrations compared to A, for speciation studies one requires $[B]_1 = [B]_2$, i.e., $[A]_1 = [A]_2$, whereas for preconcentration, $[A]_1 \gg [A]_2$.

A number of papers have exploited dialysis at ion-exchange membranes for trace metal preconcentration[43-45] (Figure 9a). Maximum enrichment can be obtained for small volumes (4 ml) of high ionic strength receiver electrolytes circulated through a 200-ml water sample for as little as 1 h (Figure 9b).

Several workers[46,47] have proposed metal speciation procedures using cation-exchange membranes. Guy and co-workers[46] used a Nafion® 811X perfluorosulfonate membrane and a 0.3-*M* sodium nitrate receiver electrolyte buffered to the sample pH. The separation was based on dialysis kinetics. Preliminary studies suggest a difference in rates for anionic, neutral, and cationic species. A high salt content is, however, needed to prevent metal

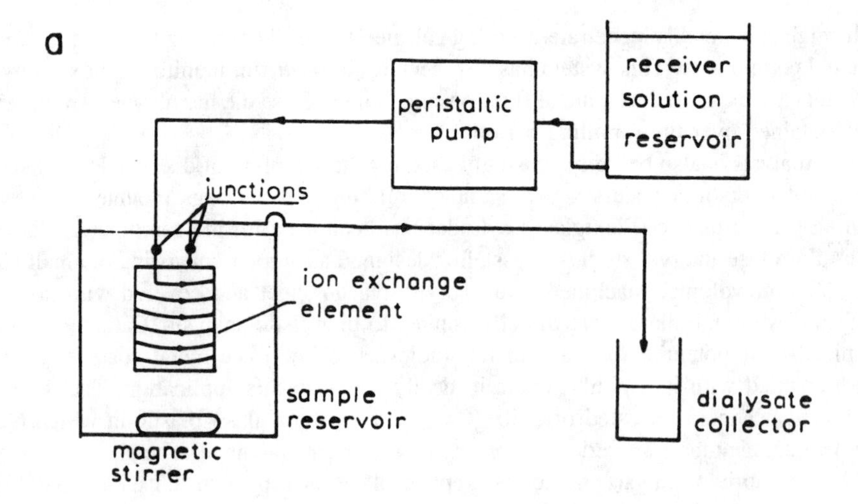

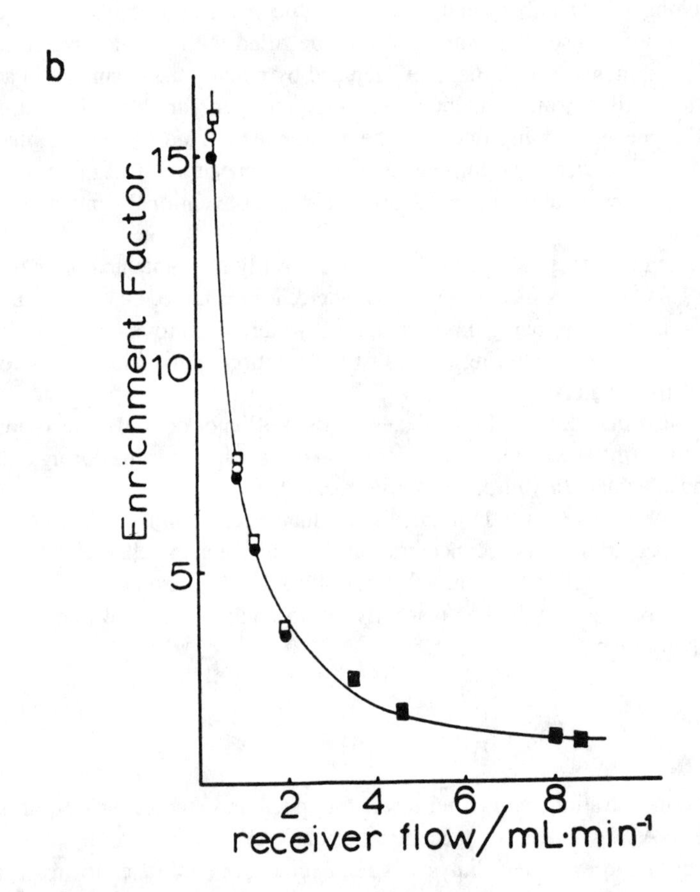

FIGURE 9. Tubular flow Donnan dialysis (a) experimental system (b) effect of flow rate on enrichment efficiency. Receiver electrolyte at 0.2 *M* Mg^{2+}, 0.5 *M* Al^{3+}; sample 2×10^{-4} *M* Cu (□), Cd (○), and Zn (●). (From Cox, J. A. and Twardowski, Z., *Anal. Chem.*, 52, 1503, 1980. With permission.)

adsorption and preconcentration on the membrane, and this will release metal associated with some colloidal species.

Cox et al.[47] measured a dialysis-labile metal fraction by comparing the enrichment factor determined for transport of a given metal from a 500-ml sample free of complexing agents, across a cation exchange membrane into 5 ml of 0.6 *M* Na_2SO_4, with that for the same metal in a natural water sample. The ratio after a 30-min dialysis they called the labile fraction. Bound metals were determined either as the difference between this and total metals or by dialyzing into acid solution, which leads to dissociation of most bound complexes, and comparing enrichment factors as before. These results compared well with labile species delineated on the basis of resin-exchange kinetics.[48,49] Such procedures are not applicable to seawater samples. Although the rate of attainment of cation exchange equilibrium is much faster than that for chemical equilibrium, there is significant back-diffusion into the sample solution which could well perturb the preexisting chemical equilibrium. For example, a pH change of one unit was observed on dialysis into 0.1 *M* HCl, 0.6 *M* Na_2SO_4 to determine bound species. Although overcoming the need for long equilibration times of conventional dialysis, ion-exchange dialysis requires considerable further study before being acceptable for speciation studies.

Dialysis membranes are available commercially in tube form or in sheets. Dialysis beakers or fiber cartiridges are available (Amicon Corporation, U.S.A.) using clusters of hollow fibers of specified NMWL from 3000 to 100,000. The high surface areas of the latter (0.01 to 5 m^2), while suited to rapid organic compound separation and purification, could prove deleterious as adsorptive sites in trace metal studies. Such losses are greatest for anionic species.[8] It has also been noted[8] that the small applied pressure (30 kPa) necessary in some system permits deformable molecules to penetrate the fibers.

E. Gel Filtration

Gel filtration is a liquid chromatographic technique where the separation is on the basis of molecular size. The stationary phase consists of an uncharged, cross-linked macromolecular gel which is swollen in the presence of the solvent. The partitioning of solute molecules is governed exclusively be steric effects. The largest molecules are least retarded, while smaller molecules are able to enter the spaces between the chains of the gel matrix and are readily retarded.[50] Gels can be obtained with specific exclusion limits or fractionation ranges. Dextran gels (Sephadex®) are available covering the ranges of M_r <700, <1500, 100 to 5000, 500 to 10,000, 1000 to 50,000, and higher. Polyacrylamide gels (e.g., Bio-Gel® P) cover similar ranges, although their swelling characteristics differ. Both types are suitable for use with aqueous solutions. In addition, porous glass beads, silica, and polystyrene substrates have been used.[50]

For use in open columns, the gels must first be hydrated by allowing them to swell in the eluant solvent. This typically takes 4 d at room temperature, or 1 h at 90°C. The gel suspension should be degassed after gentle stirring and settling, then the slurry is added to the column. The gel columns may be calibrated for molecular weight determination from the elution volumes for standards of known molecular weights.

The first applications to the study of natural waters were to determine the nature of dissolved organic matter.[47,51] Steinberg[52] extended this work to metal-organic associations by simultaneously measuring organic carbon and metal concentrations in the eluted portions (Figure 10). A limitation of the technique is that preconcentration is generally required to attain concentrations detectable using graphite furnace atoms absorption analysis of collected fractions. The 250-fold concentration achieved by sample evaporation is likely to produce changes in metal speciation. Samples were fractionated on 40 × 1.2-cm-diameter Sephadex® G-15 gel and the polymeric fraction rechromatographed on Sephadex® G-50 after reconcentration. The dissolved material was eluted with water only. Prior to the initial precon-

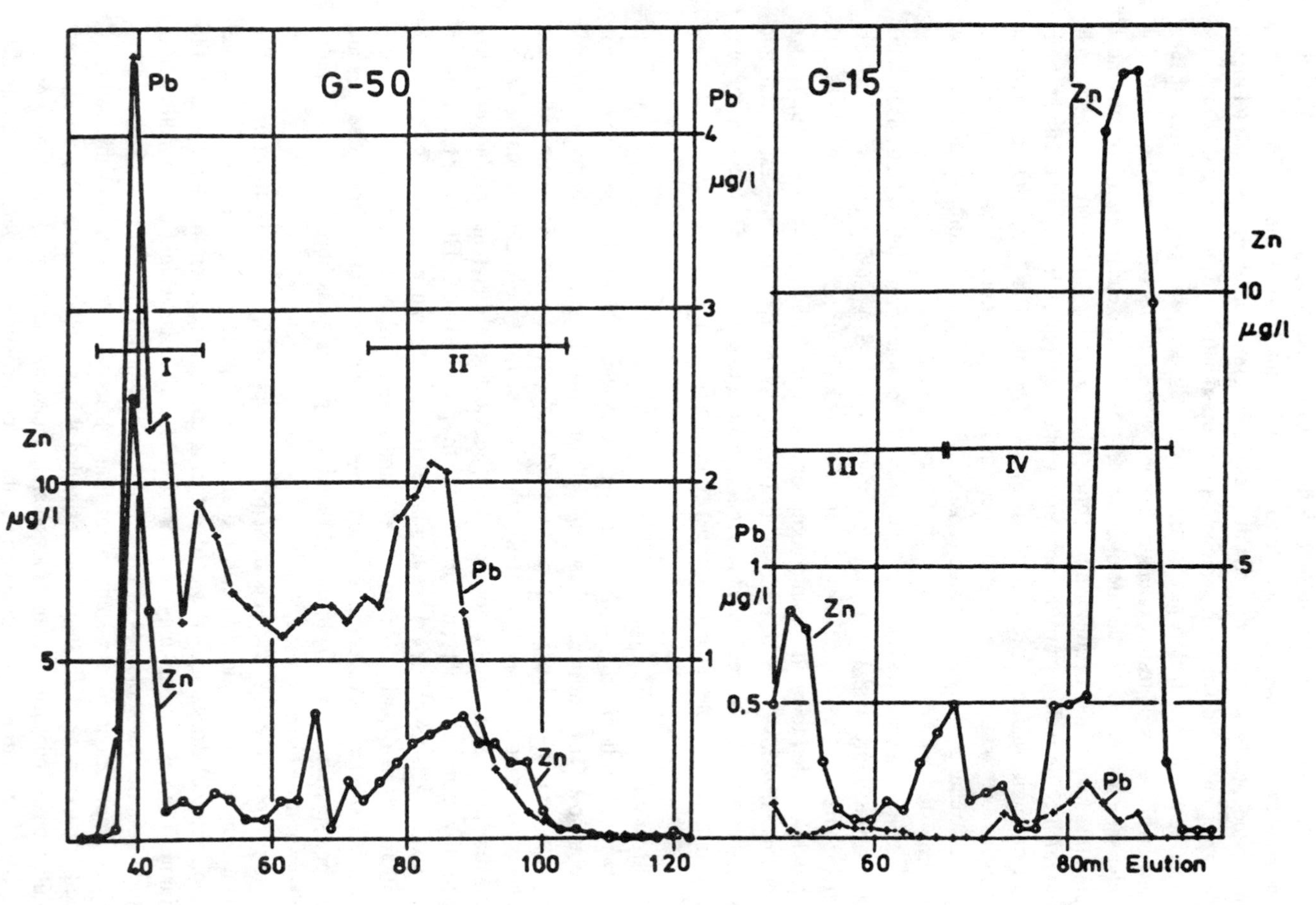

FIGURE 10. Distribution of Zn and Pb in fractions of a hard water separated by two-step gel chromatography. Molecular weight fractions: I and II, polymeric; III, oligomeric; IV, monomeric. (From Steinberg, C., *Water Res.*, 14, 1239, 1980. With permission.)

centration, carbonates were removed by acidification and degassing, and neutralized to pH 7. This process is also likely to alter species. For a brook water, three major fractions: a polymeric >1500 Da, an oligomeric, containing both 750- and 580-Da size classes, and a monomeric fraction were identified. The polymeric fraction separated further into fractions of ≃5300 and >10,000 Da, respectively. Steinberg[52] concluded that the fulvic acid fraction contributed more to chelation than other organic species.

The majority of applications of gel filtration to metal speciation have been to the study of the metal complexing by macromolecular ligands. Mantoura and Rile[53] used a Sephadex® G-15 column and a buffered eluent containing a known concentration of the metal ion of interest. A similar solution of the complexing agent was injected into the flow system where it continuously binds with metal until coming to equilibrium with the free metal ion in solution. The resulting metal deficiency moves more slowly down the column than the metal complex. The complexed metal at equilibrium with free metal ions and the overall stability constant can then be examined. This technique has been applied to humic and fulvic acids in sediments.[54] Additons of heavy metal spikes of increasing concentrations to sample solutions before separation allows determination of the metal binding capacity in gel-separated fractions.[55-57] Total metal concentrations in the eluted samples can be monitored together with analysis for total organic carbon.

Gerritse and Salomons[58] compared the inorganic speciation of cadmium, lead, and zinc in synthetic salt solutions (0.005 *M* KCl, 0.025 *M* $CaCl_2$, and 0.0025 $Ca(NO_3)_2$) with their behavior in natural waters by passage through Sephadex® G-15, G-25, and G-100 gels. These cover the molecular weight ranges up to 1500, 5000, and 100,000 Da, respectively. Samples were eluted in their solution matrix. In the pH range 6 to 8, the main difference between surface waters and synthetic salt solutions is in the high-molecular-weight range, >50,000. It would appear that the formation of hydrated and/or carbonate species can make a significant contribution to the elution profile between 1500 and 5000 Da. This study used radiotracer spikes of ^{65}Zn, ^{210}Pb, and ^{109}Cd, preequilibrated with the samples and detected in the effluent. Equilibration of radiotracers with some bound metal species has been shown to be very slow,[21] thus introducing major errors in this type of separation study, although it would not be a limitation in the study of metal binding capacities.[53]

While gel filtration has considerable potential for the separation of metals bound to macromolecular species, the major limitation currently is the detectability of both metals and carbon in the eluted fractions. Continuous monitoring of the effluent is out of the question unless gross preconcentration of the sample can be achieved. With minor preconcentration, batch analysis of the eluted fractions must be undertaken. Evaporation is an unsatisfactory procedure for sample concentration because of likely changes in speciation. Adsorptive preconcentration is a more acceptable procedure.

F. Adsorption and Ion Exchange

A range of adsorbants have been investigated for the separation of metal species from aqueous solutions. These include porous organic polymers, ion-exchange materials, and silica-based adsorbents used in reversed-phase high-performance liquid chromatography (HPLC). These materials can be highly selective towards particular species and can thus be used to advantage in metal speciation studies.

Cross-linked hydrophobic styrene-divinylbenzene copolymers (Amberlite® XAD-1, XAD-2, and XAD-4; Johns-Manville's Chromosorb® 102 and Bio-Rad® SM-2, SM-4) have been extensively studied for the adsorption of trace organics from waters.[59,60] The more hydrophobic methyl methacrylate-based copolymers (Amberlite® XAD-7, XAD-8, and Bio-Rad® SM-7) are suitable for polar organics.

More than 92% of humic acids and 75% of fulvic acids were removed from seawater using XAD-2 columns at flow rates of 40 bed volumes per hour, provided the pH was below

2.2.[61] At the natural pH, mainly nonpolar hydrocarbons are adsorbed.[62] Mackey[63,64] found XAD resins to be unsuitable for studying metal speciation because they adsorb significant amounts of uncomplexed metal cations. The adsorption of ionic zinc and copper was independent of pH and believed to occur at impurity sites on the resin. The adsorbed species could not be eluted by organic solvents such as methanol or acetonitrile.

Sugimura and co-workers[65,66] found that inorganic iron was not adsorbed onto XAD-2 from seawater acidified to pH 3, whereas organic iron compounds were fully retained. In agreement with this, Florence and Batley,[67] using the equivalent resin Bio-Rad® SM-2, showed that removal of ionic copper, lead, cadmium, and zinc from distilled water was significant only at pH values above 4.

Although Mackey's experiments used consistently higher concentrations of metals than did the other authors, the most likely explanation for the difference in behavior is the nature of the resin. Indeed, other authors[68,69] have stressed the problems with impurities in this type of resin. The cleaning procedure might well be critical in either removing these, or in avoiding activating of impurity sites to ionic metal adsorption. Mackey[64] did find that methylation reduced inorganic metal species adsorption.

Typical clean-up procedures for XAD resin first involve Soxhlet extractions for 8 to 24 h each with methanol, acetonitrile, and diethyl ether[64] or water, methanol, and methylene chloride[64,68] or acetone, methanol, and distilled water.[70] The resins are normally stored under methanol and washed with distilled water before use. Mackey[64] pretreated columns by washing with 10% Suprapur® HCl in methanol, 5% Suprapur® NH_3 in methanol, and 10^{-3} *M* sodium EDTA, before use, to remove metal impurities. Bio-Rad® SM-2 appears to contain less leachable metal, and washing with methanol, 0.1 *M* HCl, and water is adequate.[67]

A range of bonded reversed-phase HPLC packing materials have also been used to separate dissolved organics and their metal complexes from waters.[68-79] Both preconcentrator columns and analytical columns have been used. The former are commercially available, prepacked columns containing C_{18} (octadecylsilane), C_2, or phenyl moieties bonded to a silica support (Sep-pak cartridges, Waters® Associates, MA, or Bond Elut® columns, Analytichem, CA). Of these columns, the phenyl and C_{18} Sep-paks were most efficient,[77] but still were only able to isolate less than 40% of the toal dissolved organic matter from estuarine or marine waters.[76-79] It has been claimed, however, that the nonpolar fraction is the most geochemically significant.[78] The organically bound copper collected in this manner accounted for between 10 and 60% of the total disolved copper in Atlantic Ocean waters.[78] Again, the phenyl and C_{18} columns gave the best results.[77]

This separation technique first requires activation and cleaning of the cartridges using successive 10-ml rinses of methanol, 0.3 *M* HCl, methanol, and distilled water. Acid washing does hydrolize some of the bonded groups which are removed by the second methanol wash; however, the low acid concentration used minimizes hydrolysis while giving low-metal blanks.[77] Filtered seawater (1 l) was passed through the precleaned cartridges at 12 to 15 ml min^{-1}.[71] The preconcentrated species could be measured by elution with methanol; however, successive elutions with 50% methanol in water, and 100% methanol, fractionated the sample, with the majority of the recovered copper-organics being measured in the first extract.

Sep-pak C_{18} cartridges used to study copper-organic species in seawater exhibited a linear retention of organics for seawater for volumes up to 900 ml for the 50% methanol-soluble fraction, and to 1300 ml for the 100% methanol-soluble fractions. Similar behavior was obtained for copper recoveries as determined in the elutriates by graphite furnace atomic absorption spectrophotometry.

Although it has been claimed that the measured copper did not originate from inorganic complexes nor was associated with colloidal iron(III) oxide,[71] Mills et al.[72] found that some 15% of the total copper isolated from seawater by Sep-paks required 0.6 *M* HCl for its elution and was most probably present as ion-exchanged inorganic forms. It is well known that free silanol groups have an ion exchange capacity for uncomplexed metal ions.[73,74] In conventional C_{18} reversed-phase columns, less than 50% of the available silanol groups are unreacted and only 60 to 70% of these can be eliminated by silanization or endcapping.[80] Retention of inorganic copper and zinc was also confirmed by tracer studies. The possibility that this occurs through organic matter preconcentrated on the adsorbent seemed unlikely. The inorganic traces were not eluted by methanol or methanol/water and, at best, 40% was removed by 0.6 *M* HCl. Inorganic metal species can, however, be eluted with EDTA.[126]

It seems generally agreed[74-76] that C_{18} solid-phase extraction does not retrieve all dissolved copper and zinc from surface seawater. Less than 40% of the total dissolved copper is isolated, although voltammetric studies suggest that both of these metals are present principally as organic complexes. Recoveries from deep waters are more efficient. Furthermore, the technique does not substantially remove copper-complexing organic ligands from seawater.[76,79,81]

Charcoal has also been investigated as an adsorbent for both inorganic and complexed metals in seawater.[82] Blank metal concentrations and the possibility of the denaturing of metal complexes make it unsuitable for most applications.

Direct ion exchange separation has been applied by several authors,[83-86] but is is of limited value in speciation studies[2] principally because separations based on charge have little significance in terms of either transport or biological behavior. Filby et al.,[84] for example, used anion- and cation-exchange columns sequentially to resolve anionic, cationic, and neutral species in a river water. A similar procedure has been applied to chromium speciation.[85] Although the more toxic hexavalent chromium might be assumed to be the only species removed by an anion-exchange column, speciation calculations[87,88] predict the anionic trivalent species $Cr(OH)_4^-$ to the major component of seawater. The cation-exchangeable species, however, should all be trivalent chromium.[2]

It has been proposed that cation-exchange resins be used to measure a labile metal component in water samples.[86] Indeed, separate studies[89] showed cation-exchangeable copper correlated well with that fraction toxic to the marine diatom *Thalassiosira pseudonona*. In these experiments a fixed volume of seawater (250 ml) was pumped through a column of prewashed cation-exchange resin (1 g of Bio-Rad® AG 50W-X12, 200-400 mesh hydrogen form) to adsorb cationic forms of copper. Speciation calculations show these to be $CuCl^+$ and $Cu(OH)^+$; other neutral species $CuCl_2$ and $CuCO_3$ are not adsorbed. Based on the distribution coefficient measured in artificial seawater (λ^i), inorganic copper ($\overline{Cu}_i$) in a natural sample was related to sorbed copper ($\overline{Cu}_x$) using the empirical equation

$$[Cu_i] = [Cu]_x/\lambda_i$$

Experiments verified a linear dependence of adsorbed copper on solution concentration. As expected, however, other cationic species in higher concentrations than in the artificial seawater will bias the results. These include iron or organic complexes with certain amino acids (e.g., histidine). On the basis that both instances are unlikely, i.e., iron is fairly predictable and humic and fulvic acids constitute the major organic species, the method appeared to give a reasonable measure of bioavailable copper.

By far, the greatest application of ion exchange to speciation has been in the use of chelating resins. Although originally proposed for the total preconcentration of trace metals in seawater,[90] their use in speciation studies expanded rapidly following the findings of Florence and Batley[91,92] that the resin was selective in the species it adsorbed. Chelex®-100 is a styrene-divinylbenzene copolymer resin incorporating iminodiacetate chelating groups. The resin has a pore size near 1.5 nm, resulting in its exclusion of large molecules and colloidal particles. Solutions of copper and lead adsorbed on colloidal ferric hydroxide[127] or of large organic dyes[93] are rejected by the resin, while the external surface of the resin beads was found to contribute little to their total adsorptive capacity.[2,93,94] Chelex®-100 therefore provides a rapid method for the separation of ionic metal from colloidally associated metal species and metal present in strong complexes.

Preparation of the resin is critical. In the H^+-form, the adsorptive capacity of Chelex®-100 is poor; however, the passage of seawater will gradually replace H^+ by Na^+, Mg^{2+}, and Ca^{2+}, with the effluent pH rising to near 6.[90] The resin should be prepared free of metal by acid washing followed by copious rinses with distilled water. The resin has been variously used in the sodium, calcium, and ammonium forms; however, the common feature in these procedures is the washing of the resin bed after conversion to the appropriate form, with a buffer of the same pH as the sample to be analyzed, so that the resin will not alter the pH of the sample solution.[48,49,95] There appears to be no particular advantage in using a particular one of these forms. The method is virtually contamination free, as the usual procedure is to measure the metal uptake from a sample by difference. If required, however, metal can be effectively eluted from the resin using dilute acid.

Both batch and column separations have been used. Hart and Davies[36] used a Chelex® column to concentrate dialyzable heavy metals, but advocated a 24-h batch equilibrium with 0.5 g of Chelex® in 200 ml of sample to determine an ion-exchangeable fraction.[96,97] A similar batch resin procedure was incorporated in a scheme by Harrison and Laxen[98,99] (Figure 11). This method is convenient for field studies.

As stated earlier, the timescale of an analytical method is important in defining the reactivity of metal species. Indeed, the ratio of uptake of metals by a biological system will be defined at least in part by the kinetics of dissociation of metal species at membrane surfaces.[18] Figura and McDuffie[48,49] have developed a speciation scheme (Figure 12, top) based on a scale of lability (Figure 12, bottom), using anodic stripping voltammetry (ASV) and both column and batch exchange with Chelex®-100. The Chelex® column separation is considered as a transient technique with a time scale, dependent on flow rate and column length, of typically 6 to 9 s. This is considerably longer than the dissociation time for labile species measured by ASV, which is of the order of milliseconds. Zinc complexes of humic acid were defined as labile by both ASV and Chelex® column, whereas the copper complex was 5.6% Chelex® column labile, but 0% ASV labile (Figure 12, bottom). A second category of lability, "slowly labile", was obtained from batch measurements with 250-ml samples equilibrated for 72 h with 2 g of resin. In the batch method, exchange is governed both by complex dissociation kinetics and the mass transfer rate of free metal to resin. Hence, a fairly long equilibration time is required. The batch method gave Cd-$EDTA^{2-}$ as 100% labile, whereas it was 14% labile by column exchange.

Chelex®-100 has also found application in complexing capacity measurements, as it provides a means of removing free metal ion.[100-102] It will also dissociate copper from weak complexes, hence, the significance of the results is questionable.

The ultimate aim of speciation techniques is surely to equate results to the bioavailable fraction. The toxicities to salmon of cadmium in river water and sewage treatment plant effluents were found to correlate with Chelex®-labile fractions determined by batch equilibration.[103] In this instance, this results may simply reflect a large ionic metal component ($\approx 5 \times 10^{-6}$ *M*). Other immobilized chelating agents have, however, been examined to better reflect a bioavailable fraction. Both 8-hydroxyquinoline and the bis(dithiocarbamate) of *N*-propylethylenediamine(diamine), immoblized on glass, were found to be more effective than Chelex®-100 in removing copper from water samples containing fulvic acids.[104] Similar results were obtained using thiol resins,[67] chosen on the basis of the greater affinity of copper for sulfur ligands and the assumption that transport of copper across biological membranes was principally mediated by sulfur proteins. Recent studies[105] have shown that binding to protein carboxylic or amino groups is more likely, with intracellular thiols being important in reversing copper toxicity to algae.

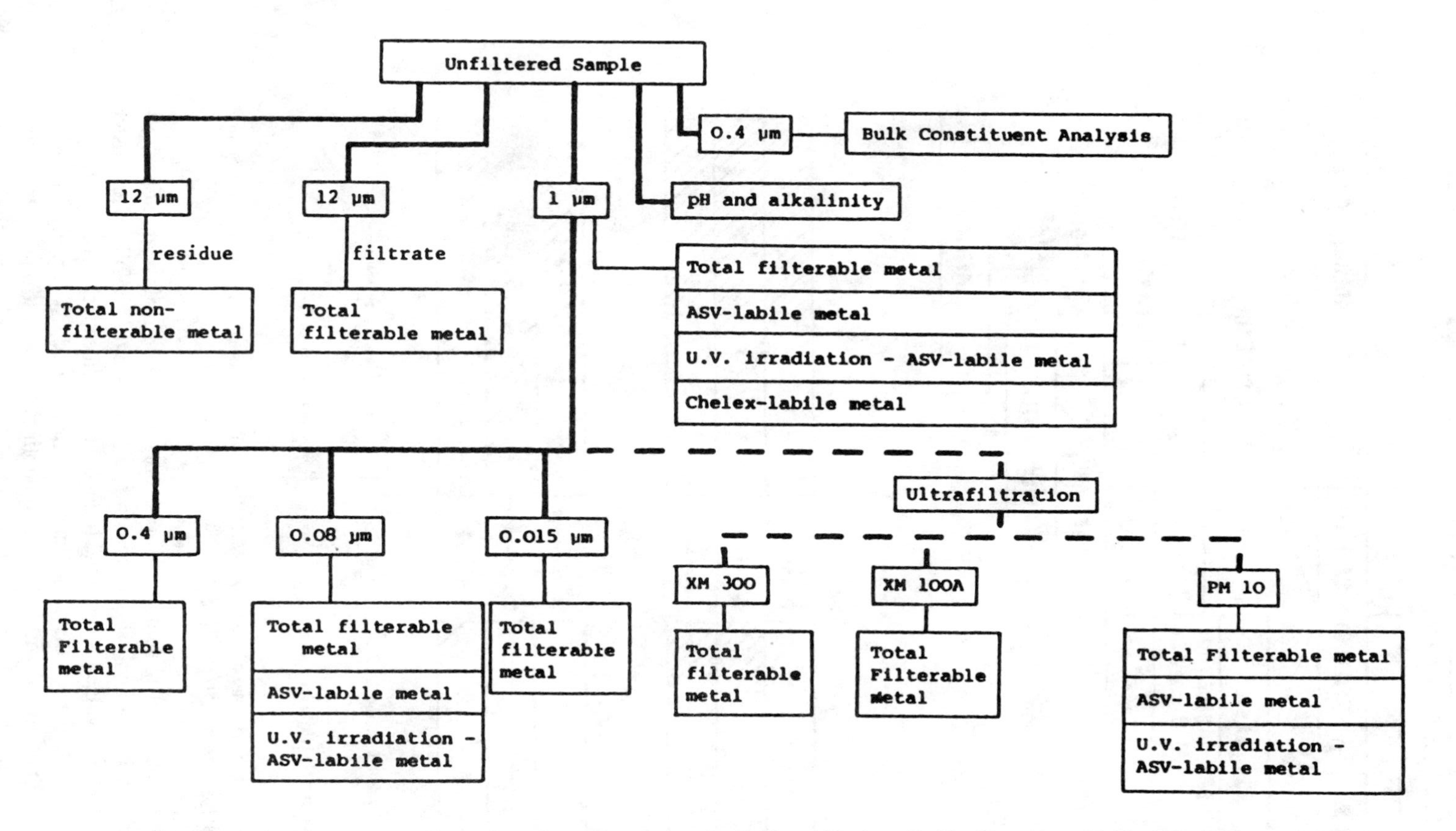

FIGURE 11. Speciation scheme of Laxen and Harrison. (From Laxen, D. P. H. and Harrison, R. M., *Water Res.*, 17, 71, 1981. With permission.)

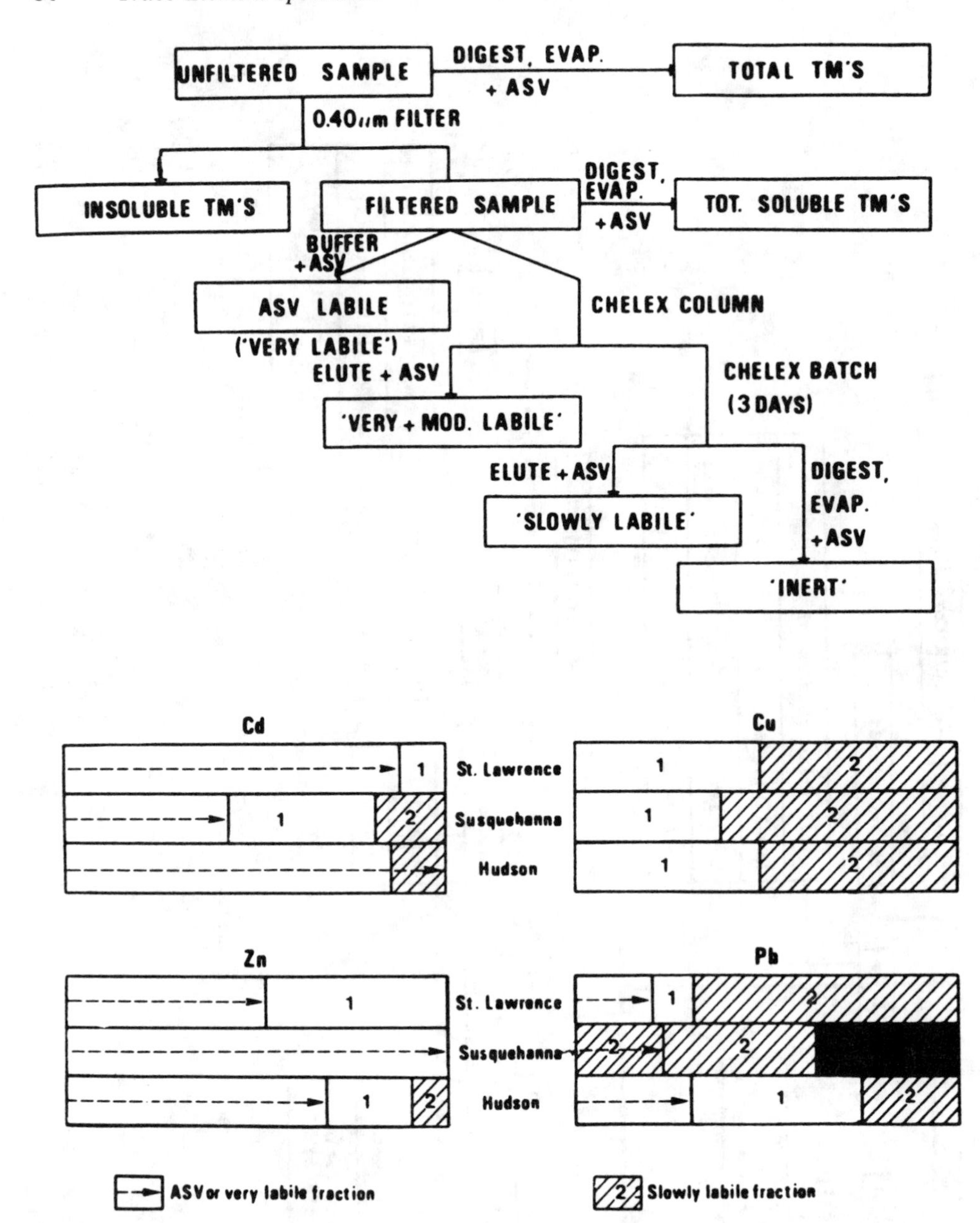

FIGURE 12. Kinetic speciation scheme of Figura and McDuffie (top) with selected results (bottom) . (From Figura, P. and McDuffie, B., *Anal. Chem.*, 52, 1433, 1980. With permission.)

C_{18} cartridges have been used in immobilize chelators such as 8-hydroxyquinoline, 8-aminoquinoline, pyridylazoresorcinol, salicylaldoxime, and bipyridyl.[106] While silica-based cartridges containing the immoblized ligands had limited adsorption capacity for metals such as copper, lead, cadmium, and zinc, the same ligands on resin-based C_{18} cartridges showed better metal retention. When coated with 8-hydroxyquinoline, the resin strongly retained all ionic metals and metals bound to natural chelators such as humic, tannic, and fulvic acids, although retention was quantitative only if the ligands were first premixed with the metal

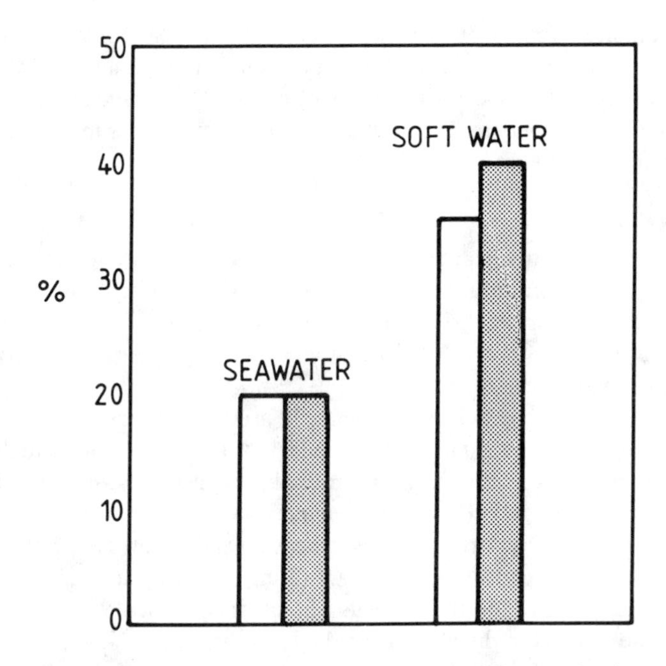

FIGURE 13. Comparison of metal adsorbed by $Al(OH)_3$-coated cation-exchange resin and fraction toxic to algae. (Open areas) Adsorbed fraction; (shaded areas) toxic fraction. (From Zhang, M.-P. and Florence, T. M., *Anal. Chim. Acta,* 197, 137, 1987. With permission.)

species. Weaker ligands such as 8-aminoquinoline were examined in an attempt to more closely model bioavailable metals, as it has been reported[105] that the equilibrium constants for reaction between Cu^{2+} and marine algae are in the range log k = 9 to 11. Comparisons of the fractions of copper extracted from a range of natural complexes showed that coatings of these ligands on resin-based C_{18} cartridges overestimated the toxic copper fraction as determined by algal bioassays. Copper removal by an aluminum hydroxide-coated cation-exchange resin did, however, correlate well with the fraction toxic to algae (Figure 13).[107]

The aluminum hydroxide cation-exchange columns were prepared using Baker aromatic sulfonic acid disposable extraction columns, H^+ form, containing about 1 ml of cation-exchange resin (J.T. Baker Chemical Co., Cat. No. 7090-3). These were converted to Na^+ form by passing 10 ml of 1 *M* NaCl solution through the column until the pH of the effluent was close to neutral. The resin was then converted to the Al^{3+} form by passing 10 ml of 0.1 *M* $AlCl_3$. The ion exchange capacity of a column was 1.1 mmol Na^+. Determination of aluminum in the effluent showed that 0.16 mmol of Al^{3+} was retained in the column. The column was washed with water to remove excess Al^{3+}, then 5 ml of 0.5 *M* NH_3 was passed through the column to convert the Al^{3+} to aluminium hydroxide, and excess NH_3 removed by water washing. When not in use, the column was filled with water to prevent the aluminum hydroxide coating from drying out and losing its adsorptive properties.

Adsorbed metal could be removed using dilute acids; however, this method also removed the aluminum coating. Quantitative removal of adsorbed copper, lead, cadmium, and zinc was achieved using 10 ml of 0.02 *M* diethylenetriaminepentaacetic acid (DTPA) (pH 8). With DTPA as leachate, the aluminum hydroxide coating was unaffected and, after washing with water, the column could be reused many times before the resin needed to be recoated. Flow rate in the range 3 to 50 ml/min had no affect on the adsorption efficiency.

Some lipid-soluble organometallic complexes are also extremely toxic to aquatic organisms.[108] In seawater, the toxic complexes copper oxine and copper ethylxanthate were quantitatively adsorbed on the top of the aluminum hydroxide column as a thin, colored layer. These complexes could be selectively leached from the column by methanol, and ionic copper removed with DTPA. With this procedure, the aluminum hydroxide column is able to determine the two most toxic forms of copper, i.e., ionic and lipid soluble. It is planned to include these columns as part of an *in situ* instrument for monitoring bioavailable metals in natural water bodies.

G. Solvent Extraction

The earliest attempts to separate trace metal species in seawater relied upon solvent extraction, using chloroform to remove originically bound copper.[109] Between 10 and 60% of the total copper was extractable. Chloroform is a poor solvent choice because it is usually badly contaminated with copper. Stiff[110] used a hexane extraction to quantify copper associated with humic substances in seawater.

Octanol-water partition coefficients are typically used as a measure of lipid solubility and bioaccumulation of selected organics.[110] *N*-hexane-10% butanol, having a dielectric constant similar to membrane lipid mixtures, has been used to measure the lipid-soluble fraction of copper in natural waters.[110,112] Freshwater samples were buffered to pH 5.7 with acetate before extraction to prevent adsorptive losses of heavy metals even onto siliconized glassware; seawater was extracted directly. The extractant (5 ml) was shaken for 2 min with the water sample (25 ml), separated, and the aqueous phase evaporated to dryness in a silica dish. The residue was dissolved in dilute nitric acid and analyzed after UV irradiation for metals by ASV. Extractable metals were determined by difference between these and the total metal. In general, solvent extraction procedures are accompanied by higher metal blanks than many of the solid adsorbents. There is also a problem with colloidal species in extractions because these may accumulate at the interface, producing errors in the measured dissolved portions. Lipid solubility as measured by solvent extraction does not, however, necessarily correlate with toxicity.[113]

H. Foam Fractionation

The association of trace metals species in seawater with surface-active organic metals can be investigated using the technique of foam fractionation.[114] Prefiltered air, passed at ≃18 $cm^{-3}s^{1-}$ through seawater in a 1.5-m × 6-cm-diameter column having a sintered glass disc, produces a foam which can be removed through an inclined side arm and collected. Samples of up to 100 l can be processed by repeated partial draining of the column and refilling.

The speciation of copper, lead, and mercury was found to be significantly influenced by surface-active organic matter.[114] Increases in surface-active forms of copper occurred after sediment resuspension and also following the decline of a large phytoplankton bloom, with the copper presumably being released from decaying phytoplankton.

The surface-active species separated by foam fractionation will include metals present as truly dissolved neutral or ionic complexes, those complexed or adsorbed to macromolecules or colloidal particles, and those adsorbed or incorporated into micro- and macroscopic particulates. It is noteworthy that the fraction of surface-active copper in seawater measured by Wallace[114] was similar to the copper fraction separated by reversed-phase HPLC in separate studies.[73,77]

I. Photooxidation

Ultraviolet photolysis, although not strictly a separation procedure, has formed on important component step of many speciation studies for bringing about transformations of species into readily detectable forms. The early studies of Armstrong and Tibbitts[115] demonstrated that many dissolved organics in waters could be readily photooxidized, and their irradiation apparatus (Figure 14) has formed the basis of a number of commercial photolysis

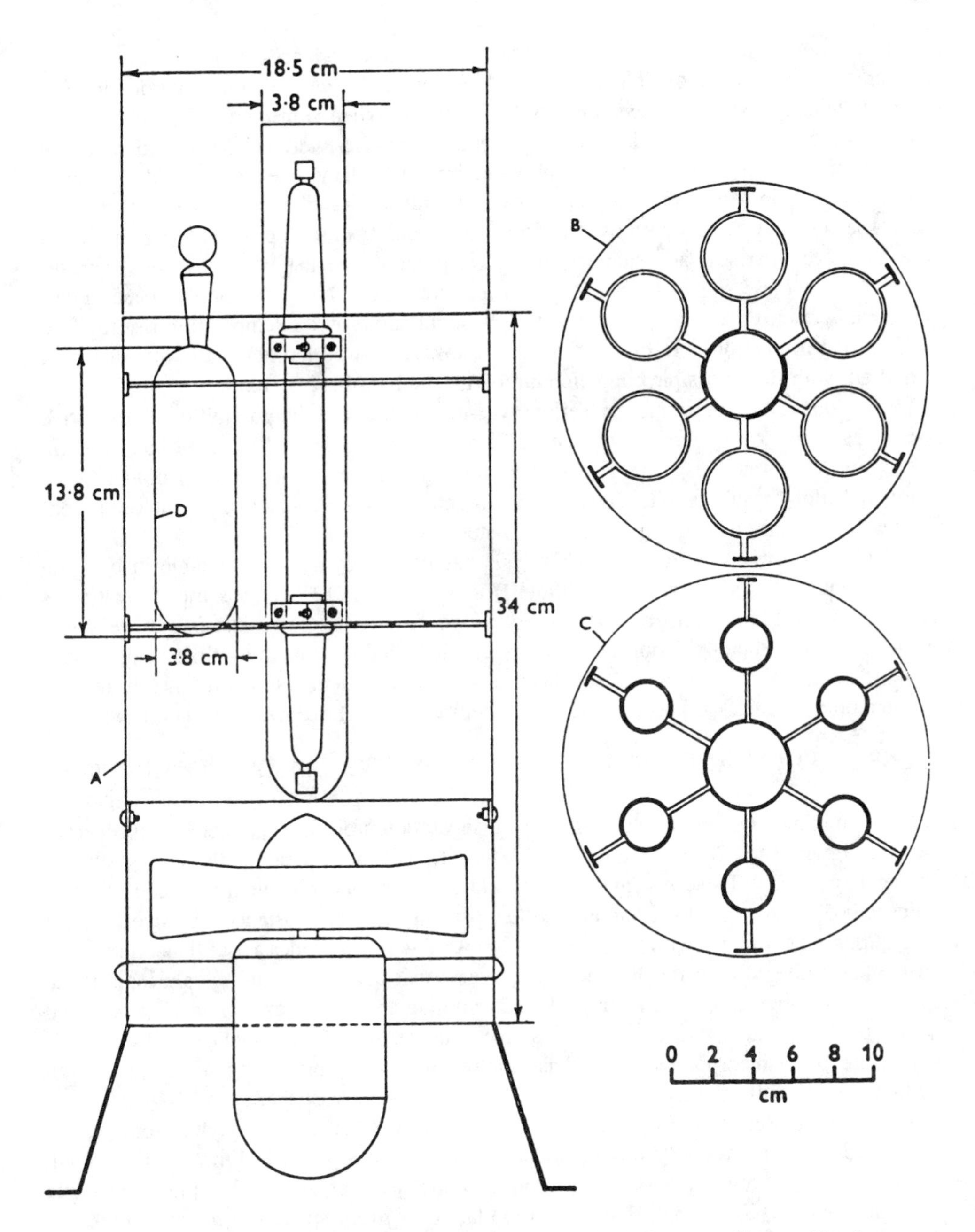

FIGURE 14. UV irradiation apparatus. (A) Section of the reactor; (B and C) upper and lower spiders for support of lamp and sample tubes; (D) 100 ml sample tube. (From Armstrong, F. A. J. and Tibbitts, S., *J. Mar. Biol. Assoc.*, 48, 143, 1968. With permission.)

units. In trace metal studies, a 550-W lamp has been shown to supply sufficient incident radiation over 4 to 5 h to natural waters, in 2- to 5-cm-diameter silica tubes, to bring about release of metals bound to organic chelators.[115] Irradiations in acidified solutions will therefore transform all heavy metal species into simple ionic forms. At the natural pH, release of metals from inorganic colloidal species will not occur; however, metal released from organic complexes could be lost from solution due to adsorption on the silica reaction tubes, adsorption onto particulate matter, or precipitation as metal hydroxides. For example, losses of aluminum after UV irradiation have been reported by Campbell et al.[117]

The apparent heterogeneity of colloidal species in natural waters can also be a source of problems during UV irradiation. Humic acid, for example, will stabilize hydrous iron(II) oxide colloid in seawater, preventing its flocculation. When subjected to UV irradiation, decomposition of the humic acid from a synthetic humic acid-hydrous iron(III) oxide colloid has been shown to result in precipitation of the iron(III). Heavy metals originally bound to the humic acid in a natural water sample will be released and, together with other ionic metal species, may be adsorbed by the precipitated iron[118] No such problems were observed in estaurine or marine water samples; however, precipitation has been observed in freshwater.[15,119] In these samples, an increase in pH from near 7 to 8.5 has also been noted, presumably due to the loss of dissolved CO_2 brought about by the temperature rise to 50 to 60°C during irradiation.[15] In the same instance, losses of ASV-labile metals were observed consistent with their possible coprecipitation with the hydrous iron oxide.

Irradiations can be performed in Teflon® vessels to minimize the possibility of adsorptive losses. Here the low UV transmittance of Telfon® is supplemented by good internal light reflection. Recent studies[110] have, however, highlighted the possibility of photodecomposition of Teflon® resulting in the release of hydrogen and fluoride ions to the sample solution. This will be highly deleterious in speciation studies.

In valency state speciation, UV photolysis has been shown to be valuable in bringing about quantitative reduction of selenium(VI) to selenium(IV).[121] Such a transformation is particularly useful for electrochemical detection where only the tetravalent form of selenium is electroactive. Photooxidation of arsenic species, including dimethylarsinic acid, methylarsonic acid, arsenobetaine, and arsenic(III), in natural waters resulted ultimately in their conversion to arsenic(V). This process is considerably faster in freshwater than in seawater.[122]

III. SPECIATION SCHEMES BASED ON PHYSICOCHEMICAL SEPARATIONS

By combining a number of physicochemical separation procedures, a plethora of schemes has been devised to divide the total metal concentration into operationally defined classes of metal species.[2,4] These schemes vary considerably in their complexity, and in the significance of the analytical speciation that they provide. In many instances physicochemical separations are combined with measurements by ASV which provides a useful discrimination on the basis of electrochemical lability, between weak complexes and simple ionic metal species dissociable under the measurement and solution conditions, and a nonlabile or bound metal species fraction which is electroinactive (Chapter 4). Measurements at different pH values are often incorporated on the assumption that, with increasing acidity, colloidal hydrous oxides will be dissolved and metal complexes will dissociate.

The first and still the most comprehensive scheme was that of Batley and Florence[59,123] (Figure 15). It combines ASV measurements before and after passage of the sample through a Chelex®-100 column, after UV irradiation, and after passage of the UV-irradiated sample through Chelex®-100. It enables up to seven classes of metal species to be quantified.

M + ML1 + MA1 — In seawater at natural pH, these groups will consist principally of simple inorganic complexes such as chloride, sulfate, carbonate, and hydroxide. In freshwater, citrate and amino acid complexes may be present

MA2, MA4, ML2, and ML4 — As significant concentrations of metal complexes, stronger then the corresponding metal-Chelex®-100 complexes, are unlikely to be present in seawater, these species would be mainly metal adsorbed on, or occluded in, organic and inorganic colloidal particles. The situation may be different in freshwater, which generally has a higher organic content, and hence possibly higher concentrations of strongly chelating organic ligands such as humic and fulvic acids.

MA3 and ML3 — These species may include some metal dissociated from colloids and retained by the resin. Also included are some of the humic and fulvic acid complexes, which should be at least partially dissociated by Chelex®-100 resin.

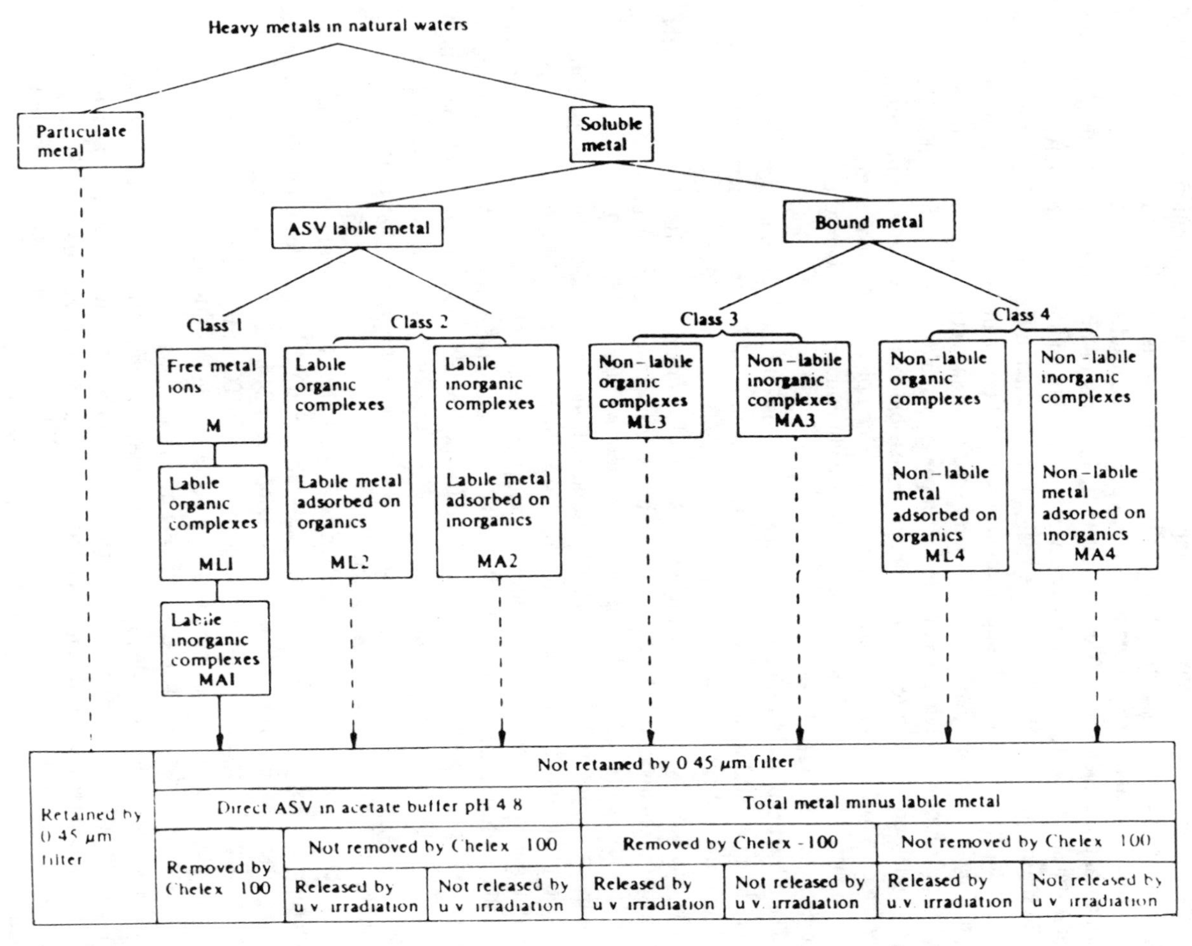

FIGURE 15. Speciation scheme of Batley and Florence. (From Batley, G. E. and Florence, T. M., *Mar. Chem.*, 4, 347, 1976. With permission.)

The scheme is based on the findings, discussed earlier, that Chelex®-100 will differentiate free or weakly complexed metal ions from metals connected with collodial particles which are unable to penetrate the pore network of the resin, and metals in strong complexes unable to be dissociated during the time scale of the column separation. The shortcomings of UV irradiation procedures have been discussed in the preceeding section. Despite the possibility of unwanted species transformations during this operation, as observed with synthetic colloids, good metal balances were obtained in studies in natural river, ocean, and estuarine waters.[116] However, in freshwater, having high colloidal iron concentrations, the use of UV irradiations to release nonlabile organically bound metals is clearly not feasible.

A criticism of this type of speciation scheme is that the results it generates are operationally defined and are therefore difficult to interpret in terms of bioavailability. With respect to metal transport and for comparing one water body with another, it is certainly possible to make meaningful observations on the basis of the speciation data concerning differences in trace metal associations.[94,118] Indeed, applications of this scheme were important in revealing a previously undocumented aspect of metal speciation in waters,[2] namely that despite predicted distributions, high percentages of copper, lead, and cadmium are naturally present in colloidal forms. Even though it may not be possible to unequivocally define organic and inorganic colloidal forms, in coastal seawater this combined fraction accounted for 56% of total copper, 67% of total lead, and 33% of total cadmium, while in river water, the respective percentages were 52, 24, and 5.

The classifications afforded by this scheme are possibly more complex than are required to define biological availability, while the analysis time of 8 h required per sample would preclude its application to routine analysis.

The basis of the Figura and McDuffie scheme[49] using Chelex®-100 has already been discussed. Combined with ASV measurements, it permits the measurement of "very labile", "moderately labile", "slowly labile", and inert fractions (Figure 12). The ASV-labile measurements are carried out at pH 6.3 and at optimum plating potentials for each metal, to avoid reduction of some complexed species at the high cathodic potentials necessary for the simultaneous deposition of cadmium, lead, and copper. The scheme does not provide mutually exclusive separations. The "very labile fraction measured by ASV on the filtered sample will not necessarily be fully retained by the Chelex® column, and could be as high as 75% for some metals giving erroneous "moderately labile" numbers by difference. Nevertheless, the scheme is simple and it is likely that this type of operational approach could be useful in relating to bioavailability.

The scheme proposed by Hart and Davies[96,97] (Figure 16) does not attempt to separate exclusive classes, but defines filtrable, dialyzable, and exchangeable metal by procedures already discussed. Exchangeable metal is obtained by a 24-h metal equilibration with Chelex®-100. Dialyzable metals are concentrated on Chelex®-100 during a continuous 5-h dialysis. The operations are relatively simple and the data can be usefully evaluated.

The scheme of Laxen and Harrison[15] (Figure 11) attempts to consider the whole spectrum of metal species in the size continuum between dissolved and particulate species. It measures ASV-labile before and after UV irradiation and a Chelex®-labile (batch) fraction of the fraction passing a 1-μm filter and also undertakes separations using 0.4-, 0.08-, and 0.015-μm filters as well as a range of three ultrafilters. The filtration separations were seen as important in more closely defining the colloidal size fractions.

Anion and cation exchange separations with ASV measurements on acidified samples have been incorporated in a scheme by Lecomte et al.[124] Hasle and Abdullah[125] proposed a scheme on the basis of ultrafiltration through a 1000-NMWL filter and ASV measurements at pH 5.5, pH 2.8, and after acid digestion. The combination of XAD resin and chelating resins has been explored by Sugimura et al.[66] The major limitations of all these schemes lie in their complexity and in their inability to provide data which can be readily interpreted in terms of bioavailability.

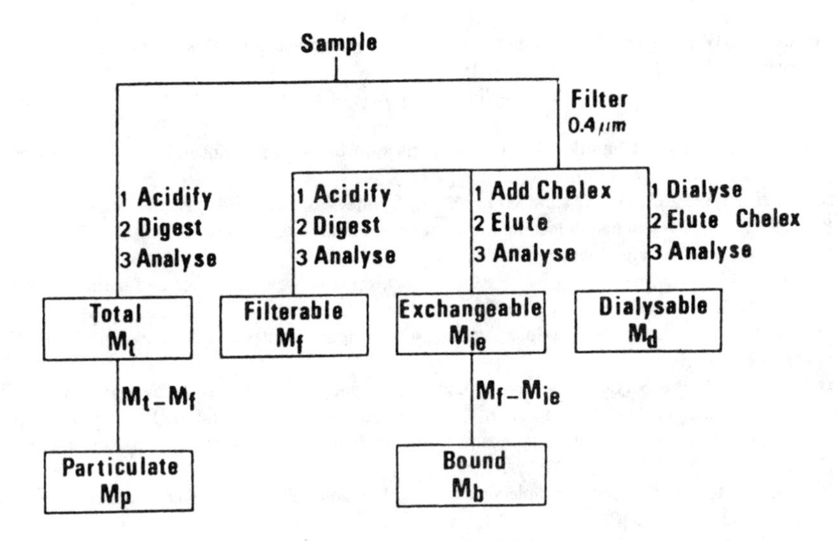

FIGURE 16. Speciation scheme of Hart and Davies. (From Hart, B. T. and Davies, S. H. R., *Aust. J. Mar. Freshwater Res.*, 32, 175, 1981. With permission.)

IV. CONCLUSIONS

As long as we are unable to reliably model equilibrium trace element speciation in waters because of deficiencies in our knowledge of equilibrium constants for the full range of solution components, physicochemical separations will continue to be important in any speciation measurement procedure. As has been shown, however, there are limitations to these procedures, not the least of which are the effects that such separations have on solution equilibria which may occur even from the moment of sampling. Indeed, physicochemical separations are only truly applicable where the kinetics of solution interactions are slow, relative to the time scale of the analytical measurement.

Without doubt, future research will contine to be directed towards physicochemical separations which are able to be related to the bioavailable or toxic species fractions. This will be limited by the fact that this fraction may differ for different organisms, but one can rationalize that the common, important step is the transport across the biological membrane. Although kinetically inert species may be readily measured and quantified, the difficulties with more labile species may require measurements on the basis of their kinetic spectrum.

Apart from the development of equilibrium techniques to detect ultratrace species concentrations, there should be increasing emphasis on *in situ* analysis, where the disturbances brought about by the separation processes are nullified by the apparently infinite sample volume.

REFERENCES

1. **Steinnes, E.,** Physical separation techniques in trace element speciation studies, in *Trace Element Speciation in Surface Waters,* Leppard, G. G., Ed., Plenum Press, New York, 1981, 37.
2. **Florence, T. M. and Batley, G. E.,** Chemical speciation in natural waters, *CRC Crit. Rev. Anal. Chem.*, 9, 219, 1980.
3. **De Mora, S. J. and Harrison, R. M.,** The use of physical separation techniques in trace metal speciation studies, *Water Res.*, 17, 723, 1983.
4. **Batley, G. E.,** The current status of trace element speciation studies in natural waters, in *Trace Element Speciation in Surface Waters,* Leppard, G. G., Ed., Plenum Press, New York, 1981, 17.

5. **Sheldon, R. W., Evelyn, T. P. T., and Parsons, J. R.,** On the occurrence and formation of small particles in seawater, *Limol. Oceanogr.*, 12, 367, 1967.
6. **Batley, G. E. and Gardner, D.,** Sampling and storage of natural waters for trace metal analysis, *Water Res.*, 11, 745, 1977.
7. **Laxen, D. P. H. and Chandler, I. M.,** Comparison of filtration techniques for size distribution in fresh water, *Anal. Chem.*, 54, 1350, 1982.
8. **Salbu, B., Bjornstad, H. E., Lindstrom, N. S., Lyndersen, E., Bresik, E. M., Ramback, J. P., and Paus, P. E.,** Size fractionation techniques in the determination of elements associated with particulates of colloidal material in natural waters, *Talanta*, 32, 907, 1985.
9. **Danielsson, L. G.,** On the use of filters for distinguishing between dissolved and particulate fractions in natural waters, *Water Res.*, 16, 179, 1982.
10. **Ogura, N.,** Molecular weight fractionation of dissolved organic matter in coastal seawater by ultrafiltration, *Mar. Biol.*, 34, 305, 1974.
11. **Oda, H. and Okobe, S.,** Some speciations of zinc and copper in coastal and oceanic waters by ultrafiltration, *J. Coll. Mar. Sci. Technol.*, Tokai Univ. (Tokai Daigaku Kiyo Kaiyogakubu), 10, 81, 1977.
12. **Cooney, D. O.,** Interference of contaminants from membrane filters in ultraviolet spectrophotometry, *Anal. Chem.*, 52, 1068, 1980.
13. **Wheeler, J. R.,** Fractionation by molecular weight of organic substances in Georgia coastal water, *Limnol. Oceanogr.*, 21, 846, 1976.
14. **Smith, R.,** Evaluation of combined application of ultrafiltration and complexation capacity techniques of natural waters, *Anal. Chem.*, 48, 74, 1976.
15. **Laxen, D. P. and Harrison, R. M.,** A Scheme for the physico-chemical speciation of trace metals in freshwater samples, *Sci. Total Environ.*, 19, 59, 1981.
16. **Giesy, J. P., Leversee, G. J., and Williams, D. R.,** Effects of naturally occurring aquatic organic fractions on cadmium toxicity to Simocephalus Serrulatus *(Daphnidae)* and Gambusia Affinis *(Poeciliidae)*, *Water Res.*, 11, 1013, 1977.
17. **Buffle, J., Deladoey, P., and Haerdi, W.,** The use of ultrafiltration for the separation and fractionation of organic ligands in fresh waters, *Anal. Chim. Acta*, 101, 329, 1978.
18. **Hoffman, M. R., Yost, E. C., Eisenreich, S. J., and Maier, W. J.,** Characterization of soluble and colloidal-phase metal complexes in river water by ultrafiltration. A mass-balance approach, *Environ. Sci. Technol.*, 15, 655, 1981.
19. **Turner, D.,** Relationships between biological availability in chemical measurements, in *Metal Ions in Biological Systems*, Vol. 18, Siegel, H., Ed., Marcel Dekker, New York, 1984, 137.
20. **Olson, D. L. and Shuman, M. S.,** Copper dissociation from estuarine humic materials, *Geochim. Cosmochim. Acta*, 49, 1371, 1985.
21. **Batley, G. E. and Florence, T. M.,** Determination of chemical forms of dissolved cadmium, lead and copper in seawater, *Mar. Chem.*, 4, 347, 1976.
22. **Lammers, W. T.,** Speciation of suspended and colloidal particles from natural water, *Environ. Sci. Technol.*, 1, 52, 1967.
23. **Beckett, R., Zhang, J., and Giddings, J. C.,** Determination of molecular weight distributions of fulvic and humic acids using flow field-flow fractionation, *Environ. Sci. Technol.*, 21, 249, 1987.
24. **Giddings, J. C., Williams, P. S., and Beckett, R.,** Fractionating power in programmed field-flow fractionation: experimental sedimentation field decay, *Anal. Chem.*, 59, 28, 1987.
25. **Beckett, R.,** The application of field-flow fractionation techniques to the characterization of complex environmental samples, *Environ. Technol. Lett.*, 8, 339, 1987.
26. **Rona, E., Hood, D. W., Muse, L., and Buglio, B.,** Activation analysis of manganese and zinc in seawater, *Limnol. Oceanogr.*, 7, 201, 1962.
27. **Hood, D. W.,** The Chemistry and Analysis of Trace Metals in Seawater, U.S. Atomic Energy Commission Rep., TID 23295, Washington, D. C., 1966.
28. **Guy, R. D. and Chakrabarti, C. L.,** Studies of metal-organic interactions in model systems pertaining to natural waters, *Can. J. Chem.*, 54, 2600, 1976.
29. **Truitt, R. E. and Weber, J. H.,** Determination of complexing capacity of fulvic acid for copper (II) and cadmium (II) by dialysis titration, *Anal. Chem.*, 53, 337, 1981.
30. **Truitt, R. E. and Weber, J. H.,** Copper (II)- and cadmium (II)-binding abilities of some New Hampshire freshwaters determined by dialysis titration, *Environ. Sci. Technol.*, 15, 1981, 1204.
31. **Benes, P. and Steinnes, E.,** In-situ dialysis for the determination of the state of trace elements in natural waters, *Water Res.*, 8, 947, 1974.
32. **Benes, P. and Steinnes, E.,** Migration forms of trace elements in natural fresh waters and the effect of the water storage, *Water Res.*, 9, 741, 1975.
33. **Benes, P.,** Semicontinuous monitoring of truly dissolved forms of trace elements in streams using dialysis in-situ. I. Principle and conditions, *Water Res.*, 14, 511, 1980.
34. **Benes, P. and Majer, V.,** *Trace Chemistry of Aqueous Solutions*, Academia, Prague, 1980.

35. **Morrison, G. M. P.,** Bioavailable metal uptake rate determination in polluted waters by dialysis with receiving resins, *Environ. Technol. Lett.*, 8, 393, 1987.
36. **Hart, B. T. and Davies, S. H. R.,** A new dialysis-ion exchange technique for determining the forms of trace metals in water, *Aust. J. Mar. Freshwater Res.*, 28, 105, 1977.
37. **Morrison, G. M. P.,** Dialysis with receiving resins for the determination of bioavailable metal in natural waters, in Proc. of 9th Australian Symp. on Analytical Chemistry, Royal Australian Chemical Institute, Sydney, 1987, 179.
38. **Mayer, L. M.,** Chemical water sampling in lakes and sediments with dialysis bags, *Limnol. Oceanogr.*, 21, 909, 1976.
39. **Hesslein, R. H.,** An in situ sampler for close interval pore water studies, *Limnol. Oceanogr.*, 21, 912, 1976.
40. **Carignan, R.,** Interstitial water sampling by dialysis: methodological notes, *Limnol. Oceanogr.*, 29, 667, 1984.
41. **Carignan, R., Rapin, F., and Tessier, A.,** Sediment porewater sampling for metal analysis: a comparison of techniques, *Limnol. Oceanogr.*, 49, 2493, 1985.
42. **Blaedel, W. J. and Haupert, T. J.,** Exchange equilibrium through ion exchange membranes, *Anal. Chem.*, 38, 1035, 1966.
43. **Cox, J. A. and Twardowski, Z.,** Tubular flow Donnan dialysis, *Anal. Chem.*, 52, 1503, 1980.
44. **Guy, R. D. and Bourque, C.,** Nafion membrane tubing for metal analysis and speciation of natural waters, in *Proc. 3rd Int. Conf. on Heavy Metals in the Environment,* CEP Consultants, Edinburgh, 1981, 277.
45. **Wilson, R. L. and DiNunzio, J. E.,** Enrichment of nickel and cobalt in natural hard waters by Donnan dialysis, *Anal. Chem.*, 53, 692, 1981.
46. **Guy, R. D., Bourque, C. L., and Dicaire, J.,** Ion exchange membranes for metal ion speciation in natural waters, in *Proc. 4th Int. Conf. on Heavy Metals in the Environment,* CEP Consultants, Edinburgh, 1983, 1223.
47. **Cox, J. A., Slonawska, K., Gatchell, D. K., and Hiebert, A. G.,** Metal speciation by Donnan dialysis, *Anal. Chem.*, 56, 650, 1984.
48. **Figura, P. and McDuffie, B.,** Use of Chelex resin for determination of labile trace metal fractions in aqueous ligand media and comparison of the method with anodic stripping voltammetry, *Anal. Chem.*, 51, 120, 1979.
49. **Figura, P. and McDuffie, B.,** Determination of labilities of soluble trace metal species in aqueous environmental samples by anodic stripping voltammetry and Chelex column and batch methods, *Anal. Chem.*, 52, 1433, 1980.
50. **Fischer, L.,** *An Introduction to Gel Chromatography,* North-Holland, Amsterdam, 1971.
51. **Gjessing, E. T.,** Use of Sephedex gel for the estimation of molecular weight of humic substances in natural waters, *Nature (London),* 208, 1091, 1965.
52. **Steinberg, C.,** Species of dissolved metals derived from oligotropic hard water, *Water Res.*, 14, 1239, 1980.
53. **Mantoura, R. F. C. and Riley, J. P.,** The use of gel filtration in the study of metal binding by humic acids and related compounds, *Anal. Chim. Acta.*, 78, 193, 1975.
54. **Hirata, S.,** Stability constants for the complexes of transition metal ion with fulvic and humic acids in sediments measured by gel filtration, *Talanta,* 28, 809, 1981.
55. **Bender, M. E., Matson, W. R., and Jordan, R. A.,** On the significance of metal complexing agents in secondary sewage effluents, *Environ. Sci. Technol.*, 4, 520, 1980.
56. **Rossing, A. C., Sterritt, R. M., and Lester, J. N.,** The influence of process parameters on the removal of heavy metals in activated sludge, *Water Air Soil Pollut.*, 17, 185, 1982.
57. **Sugai, S. F. and Healy, M. L.,** Voltammetric studies of the organic association of copper and lead in two Canadian inlets, *Mar. Chem.*, 6, 291, 1978.
58. **Gerritse, R. G. and Salomons, W.,** The chromatographic behaviour of heavy metal species in Sephedex gel columns, *Environ. Technol. Lett.*, 4, 463, 1955.
59. **Ishiwatari, R., Hamana, H., and Machihara, T. J.,** Isolation and characterization of polymeric organic materials in a polluted river, *Water Res.*, 14, 1257, 1980.
60. **Dressler, M.,** Extraction of trace amounts of organic compounds from water with porous organic polymers, *J. Chromatogr.*, 165, 167, 1979.
61. **Mantoura, R. F. C. and Riley, J. P.,** The analytical concentration of humic substances from natural waters, *Anal. Chim. Acta,* 76, 97, 1975.
62. **Leonard, J. D. and Crewe, N.,** Study on the extraction of organic compounds from seawater with XAD-2 resin, *Mar. Chem.*, 12, 222, 1983.
63. **Mackey, D. J.,** An investigation of the suitability of Amberlite XAD-1 resin for studying trace metal speciation in seawater, *Mar. Chem.*, 11, 169, 1982.
64. **Mackey, D. J.,** The adsorption of simple trace metal cations on Amberlite XAD-1 and XAD-2, *J. Chromatogr.*, 236, 81, 1982.

65. **Sugimura, Y., Suzuki, Y., and Miyake, Y.,** Chemical forms of minor metallic elements in the ocean, *J. Oceanogr. Soc. Jpn.*, 34, 93, 1978.
66. **Sugimura, Y., Suzuki, Y., and Miyake, Y.,** The dissolved organic iron in seawater, *Deep-Sea Res.*, 25, 309, 1978.
67. **Florence, T. M. and Batley, G. E.,** A new scheme for chemical speciation of copper, lead, cadmium and zinc in seawater, in *Proc. 3rd Int. Conf. on Heavy Metals in the Environment,* CEP Consultants, Edinburgh, 1981, 599.
68. **Hunt, G. and Pangori, N.,** Potential contamination from the use of synthetic adsorbents in air sampling procedures, *Anal. Chem.*, 54, 369, 1982.
69. **James, H. A., Steel, C. D., and Wilson, I.,** Impurities arising from the use of XAD-2 resin for the extraction of organic pollutants in drinking water, *J. Chromatogr.*, 208, 89, 1981.
70. **Fu, T. and Pocklington, R.,** Quantitative adsorption of organic matter from seawater in solid matrices, *Mar. Chem.*, 13, 255, 1983.
71. **Mills, G. L. and Quinn, J. G.,** Isolation of dissolved organic matter and copper organic complexes from estuarine water using reverse-phase liquid chromatography, *Mar. Chem.*, 10, 93, 1981.
72. **Mills, G. L., Hanson, A. K., Quinn, J. G., Cammela, W. R., and Chasteen, N. D.,** Chemical studies of copper-organic complexes isolated from estuarine waters using C_{18} reverse-phase liquid chromatography, *Mar. Chem.*, 11, 355, 1982.
73. **Kremling, K., Wenck, A., and Osterroht, C.,** Investigations of dissolved copper-organic substances in Baltic waters, *Mar. Chem.*, 10, 209, 1981.
74. **Mackey, D. J.,** HPLC analysis of metal-organics in seawater — interference effects attributed to stationary-phase free silanols, *Mar. Chem.*, 16, 105, 1985.
75. **Donet, J. R., Statham, P. J., and Bruland, K. W.,** An evaluation of a C_{18} solid phase extraction technique for isolating metal-organic complexes from Central North Pacific ocean waters, *Mar. Chem.*, 18, 85, 1986.
76. **Douglas, G. S., Mills, G. L., and Quinn, J. G.,** Organic copper and chromium complexes in the interstitial waters of Narragansett Bay sediments, *Mar. Chem.*, 19, 161, 1986.
77. **Mills, G. L., McFadden, E., and Quinn, J. G.,** Chromatographic studies of dissolved organic matter and copper-organic complexes isolated from estuarine waters, *Mar. Chem.*, 20, 313, 1987.
78. **Becher, G., Oestvold, G., Paus, P., and Seip, H. M.,** Complexation of copper by aquatic humic matter studies by reverse-phase liquid chromatography and atomic absorption spectroscopy, *Chemosphere,* 12, 1209, 1983.
79. **Derenbach, J. B., Ehrhardt, M., Osterroht, C., and Petrick, G.,** Sampling of dissolved organic material from seawater with reverse-phase techniques, *Mar. Chem.*, 6, 351, 1978.
80. **Low, G. K.-C.,** Stationary phases for HPLC — some present problems and future trends, *Clinical Biochemist Monograph,* Australian Association of Clinical Biochemistry, Sydney, 1986, 12.
81. **Van den Berg, C. M. G.,** Determination of the complexing capacity and conditional stability constants of complexes of copper (II) with natural organic ligands in seawater by cathodic stripping voltammetry of copper-catechol complex ions, *Mar. Chem.*, 15, 1, 1984.
82. **Kerfoot, W. B. and Vaccaro, R. F.,** Adsorptive extraction for analysis of copper in sea water, *Limnol. Oceanogr.*, 18, 689, 1973.
83. **Marchand, M.,** Physicochemical forms of cobalt, manganese, zinc, chromium and iron in a sea water with and without organic material, *J. Cons. Cons. Int. Explor. Mer,* 35, 130, 1974.
84. **Filby, R. H., Shah, K. R., and Funk, W. H.,** Role of neutron activation analysis in the study of heavy metal pollution of a lake-river system, in *Proc. 2nd Int. Conf. Nuclear Methods in Environmental Research,* Vogt, J. R. and Meyer, W., Eds., National Technical Information Service, Springfield, VA, 1974, 10.
85. **Pankow, J. R., Leta, D. P., Lin, J. W., Ohl, S. E., Shum, W. P., and Janauer, G. E.,** Analysis for chromium traces in the aquatic ecosystem. II. A study of Cr(III) and Cr(VI) in the Susquehanna River Basin of New York and Pennsylvania, *Sci. Total Environ.*, 7, 17, 1977.
86. **Allen, H. E., Crossier, M. L., and Brisbin, T. D.,** Metal speciation in aquatic environments, in *Toxicity to Biota of Metal Forms in Natural Waters, Proceedings of a Workshop,* Andrew, R. W., Hodson, P. V., and Konasewich, D. E., Eds., Great Lakes Regional Office of the Interior Joint Committee, Windsor, Ontario, 1975, 33.
87. **Elderfield, H.,** Chromium speciation in seawater, *Earth Planet. Sci. Lett.*, 9, 10, 1970.
88. **Sibley, T. H. and Morgan, J. J.,** Equilibrium speciation of trace metals in freshwater:seawater mixtures, in *Proc. 1st Int. Conf. on Heavy Metals in the Environment,* Vol. 1, Hutchison, T., Ed., University of Toronto, Ontario, 1975, 319.
89. **Zorkin, H. G., Grill, E. J., and Lewis, A. G.,** An ion-exchange procedure for quantifying biologically active copper in seawater, *Anal. Chim. Acta,* 183, 163, 1986.
90. **Riley, J. P. and Taylor, D.,** Chelating resins for the concentration of trace elements from seawater and their analytical use in conjunction with atomic absorption spectrophotometry, *Anal. Chim. Acta.*, 40, 479, 1968.

91. **Florence, T. M. and Batley, G. E.,** Removal of trace metals from seawater by a chelating resin, *Talanta,* 22, 201, 1975.
92. **Florence, T. M. and Batley, G. E.,** Trace metal species in seawater. I. Removal of trace metals from seawater by a chelating resin, *Talanta,* 23, 179, 1978.
93. **Florence, T. M.,** Trace metal species in fresh waters, *Water Res.,* 11, 681, 1977.
94. **Florence, T. M.,** The speciation of trace elements in waters, *Talanta,* 39, 345, 1982.
95. **Buckley, J. A.,** Preparation of Chelex-100 resin for batch treatment of sewage in river water at ambient pH and alkalinity, *Anal. Chem.,* 57, 1489, 1985.
96. **Hart, B. T. and Davies, S. H. R.,** A batch method for the determination of ion-exchangeable trace metals in natural waters, *Aust. J. Mar. Freshwater Res.,* 28, 397, 1977.
97. **Hart, B. T. and Davies, S. H. R.,** Trace metal speciation in three Victorian lakes, *Aust. J. Mar. Freshwater Res.,* 32, 175, 1981.
98. **Harrison, R. M. and Laxen, D. P. H.,** Physico-chemical speciation of lead in drinking water, *Nature (London),* 286, 791, 1980.
99. **Laxen, D. P. H. and Harrison, R. M.,** Physico-chemical speciation of selected metals in the treated effluent of a lead-acid battery manufacturer and in the receiving river, *Water Res.,* 17, 71, 1983.
100. **Wood, A. M., Evans, D. W., and Alberts, J. J.,** Use of an ion exchange technique to measure complexing capacity on the continental shelf of the south-western United States and in the Sargasso Sea, *Mar. Chem.,* 13, 305, 1983.
101. **Mackey, D. J.,** The strong complexing capacity of seawater — an investigation of south-eastern Australian coastal waters, *Mar. Chem.,* 7, 73, 1983.
102. **Sunda, W. G.,** Measurement of manganese, zinc and cadmium complexation in seawater using Chelex ion exchange equilibrium, *Mar. Chem.,* 14, 365, 1984.
103. **Buckley, J. A., Yoshida, G. A., Wells, N. R., and Aquino, R. T.,** Toxicities of total and Chelex-labile cadmium to salmon in solutions of natural water and diluted sewage with potentially different cadmium complexing capacities, *Water Res.,* 19, 1549, 1985.
104. **Ryan, D. R. and Weber, J. H.,** Comparison of chelating agents immobilized on glass with Chelex-100 for removal and preconcentration of trace copper (II), *Talanta,* 32, 859, 1985.
105. **Florence, T. M. and Stauber, J. L.,** Toxicity of copper to the marine diatom *Nitzschia closterium, Aquat. Toxicol.,* 8, 11, 1986.
106. **Zhang, M.-P. and Florence, T. M.,** Unpublished results, 1986.
107. **Zhang, M.-P. and Florence, T. M.,** A novel adsorbent for the determination of the toxic fraction of copper in natural waters, *Anal. Chim. Acta,* 197, 137, 1987.
108. **Stauber, J. L. and Florence, T. M.,** Reversibility of copper-thiol binding in *Nitzschia closterium* and *Chlorella pyrenoidosa, Aquat. Toxicol.,* 8, 223, 1986.
109. **Slowey, J. F., Jeffrey, L. M., and Hood, D. W.,** Evidence for organic complexed copper in seawater, *Nature (London),* 214, 377, 1967.
110. **Stiff, M. J.,** The chemical states of copper in polluted fresh water and a scheme of analysis to differentiate them, *Water Res.,* 5, 585, 1971.
111. **Chiou, C. T., Freed, V. H., Schmedding, D. W., and Kohnert, R. L.,** Partition coefficient and bioaccumulation of selected organic chemicals, *Environ. Sci. Technol.,* 11, 475, 1977.
112. **Florence, T. M.,** Development of physico-chemical speciation procedures to investigate the toxicity of copper, lead, cadmium and zinc towards aquatic biota, *Anal. Chim. Acta,* 141, 73, 1982.
113. **Florence, T. M., Lumsden, E. G., and Fardy, J. J.,** Evaluation of some physico-chemical techniques for the determination of the fraction of dissolved copper toxic to the marine diatom *Nitzschia closterium, Anal. Chim, Acta,* 151, 281, 1983.
114. **Wallace, G. T.,** The association of copper, mercury and lead with surface-active organic matter in coastal seawater, *Mar. Chem.,* 11, 379, 1982.
115. **Armstrong, F. A. J. and Tibbitts, S.,** Photochemical combustion of organic matter in seawater for nitrogen, phosphorus and carbon determination, *J. Mar. Biol. Assoc.,* 48, 143, 1968.
116. **Batley, G. E. and Farrar, Y. J.,** Irradiation techniques for the release of bound heavy metals in natural waters and blood, *Anal. Chim. Acta,* 99, 283, 1978.
117. **Campbell, P. G. C., Bisson, M., Bougie, A., Tessier, A., and Villeneuve, J. P.,** Speciation of aluminium in acidic freshwaters, *Anal. Chem.,* 55, 2296, 1983.
118. **Batley, G. E. and Gardner, D.,** Copper, lead and cadmium speciation in some estuarine and coastal marine waters, *Estuarine Coastal Mar. Sci.,* 7, 59, 1978.
119. **Blutstein, H. and Smith, J. D.,** Distribution of species of Cu, Pb, Zn and Cd in a water profile in the Yarra River estuary, *Water Res.,* 12, 119, 1178.
120. **Batley, G. E.,** Use of Teflon components in photochemical reactors, *Anal. Chem.,* 56, 3261, 1984.
121. **Batley, G. E.,** Differential pulse polarographic determination of selenium, *Anal. Chim. Acta,* 187, 109, 1986.

122. **Brockbank, C. I., Batley, G. E., and Low, G. K.-C.,** Photochemical decomposition of arsenic species in natural waters, *Environ. Technol. Lett.*, 1988, in press.
123. **Batley, G. E. and Florence, T. M.,** A novel scheme for the classification of heavy metal species in natural waters, *Anal. Lett.*, 9, 679, 1978.
124. **Lecomte, J., Mericam, P., and Astruc, M.,** A new scheme for the speciation of Pb and Cd in a polluted river water, in *Proc. 3rd Int. Conf. on Heavy Metals in the Environment,* CEP Consultants, Edinburgh, 1981, 678.
125. **Hasle, J. R. and Abdullah, M. I.,** Analytical fractionation of dissolved copper, lead and cadmium in coastal seawater, *Mar. Chem.*, 10, 487, 1981.
126. **Sunda, W.,** Personal communcations.
127. **Batley, G. E. and Florence, T. M.,** Unpublished results.

Chapter 4

ELECTROCHEMICAL TECHNIQUES FOR TRACE ELEMENT SPECIATION IN WATERS

T. Mark Florence

TABLE OF CONTENTS

I. INTRODUCTION

Metals in natural waters may be complexed by a wide variety of ligands. Copper in seawater, for example, is believed to exist mainly as undefined, highly stable complexes, the principal ligands perhaps being porphyrins, siderophores, or metallothioneins.[1-6] In freshwaters, fulvic, humic, and tannic acids are often the dominant ligands, and in both seawater and freshwater, adsorption of heavy metal ions on colloidal particles of hydrated metal oxides (e.g., Fe_2O_3, SiO_2, Al_2O_3 and MnO_2) provides a major class of dissolved metal species (Chapter 2).[1,7] Speciation analysis aims to determine the different physicochemical forms of an element that together make up its total concentration in a water sample. All these species can coexist and may or may not be in thermodynamic equilibrium with one another.[5] An ionic metal spike added to a filtered water sample may take times ranging from hours to months to equilibrate with the natural pool of metal species.[1,8-10]

As discussed in Chapter 2, there are two main reasons for studying the speciation of elements in waters — to understand either the biological or the geochemical cycling of the elements.[11-14] Biological cycling includes bioaccumulation, bioconcentration, bioavailability, and toxicity; and geochemical cycling involves the transport, adsorption, and precipitation of the element in the water system. No meaningful interpretation of either biological or geochemical cycling can be made without speciation information.[15-17] Each different physicochemical form of and element has a different toxicity (Chapter 2), so analysis of a water sample for total metal concentration alone does not provide sufficient information to predict toxicity. Lipid-soluble metal complexes may be particularly toxic forms of heavy metals because they can diffuse rapidly through a biomembrane and carry both metal and ligand into the cell.[16,18-21] Alkylmercury compounds, copper xanthates (flotation agents), and copper oxinate (fungicide) are examples of lipid-soluble metal complexes.

Variation in the speciation of an element will also affect its degree of adsorption on suspended matter, its rate of transfer to the sediment, and its overall transport in a water system. Speciation analysis will therefore assist in the prediction of the distance over which a river will be affected by effluent discharged from a point source.[20]

Electroanalysis is a powerful technique for the study of trace element speciation and has been applied to (or is potentially applicable to) over 30 elements: Ag, As, Au, Bi, Br, Cd, Cl, Co, Cr, Cu, Eu, Fe, Ga, Hg, I, In, Mn, Mo, Ni, Pb, Pt, Re, S, Sb, Se, Sn, Tl, U, V, W, Yb, and Zn.[1,22-24] Four metals of prime environmental concern, copper, lead, cadmium, and zinc, can be determined simultaneously and with great sensitivity. Moreover, the redox potential of an electrode can be varied accurately, precisely, and continuously over a wide potential range, and the study of the kinetics of metal complex dissociation at an electrode is supported by well-established theory.[25-31] Of all trace element speciation methods available at present, electroanalysis appears to provide the best opportunity for experimentally modeling the bioavailability of elements and their complexes with organic and inorganic ligands.

Electroanalytical speciation measurements require exceptionally clean working conditions because there is a constant risk of contamination when working at such low metal concentrations.[1,3,32] A clean room, or at least a laminar-flow clean air cupboard, is essential for these analyses. Electrochemical techniques have an important advantage in that the sample requires much less handling and is in contact with fewer potential sources of contamination than when other speciation methods, such as solvent extraction, ultrafiltration, or dialysis, are used.

II. RANGE OF APPLICABILITY OF ELECTROCHEMICAL SPECIATION METHODS

Electrochemical techniques can be used to provide speciation information based on labile/inert discrimination, redox state, and half-wave potential measurements. The techniques are applicable to metals, nonmetals, colloidal particles, and organic compounds.

A. Labile/Inert Discrimination

The determination of labile (i.e., reactive) metal involves the measurement of the concentration of metal in the water sample that can be reduced at, and deposited into, a mercury electrode from a stirred solution. Labile metal is usually expressed as a percentage of total dissolved metal, and the difference between total and labile metal is termed "inert" or "unreactive" metal. Some electrochemical parameters that affect the percentage of labile metal are: deposition potential, electrode rotation (or stirring) rate, mercury drop diameter, pulse frequency, pH, temperature, and buffer composition. Under certain conditions, labile metal has been found to correlate well with the toxic fraction of metal.[11,33]

Labile metal comprises free metal ion and metal that can dissociate in the double layer from complexes or colloidal particles and, hence, be deposited in the mercury electrode.[4,5,34] For natural waters, anodic stripping voltammetry (ASV) is the technique usually used, and it has been applied to labile/inert measurements of the following elements; Cu, Pb, Cd, Zn, Mn, Cr, Tl, Sb, and Bi.[1] Heavy metal "pseudocolloids", i.e., colloidal particles of Fe_2O_3, MnO_2, humic acid, etc. with adsorbed heavy metal ions,[35] can be treated as a special type of metal complex and may contribute significantly to some labile metal measurements.

Labile/inert discrimination for some elements may also be made by chemical, rather than electrochemical, exchange. This approach is particularly advantageous for metals such as iron, which are difficult to determine by direct ASV, and where concentrations are too low for polarography. In one procedure,[36] labile iron was determined by treating the sample with bismuth-EDTA, and the bismuth, liberated by chemical exchange (Equation 1), was measured with high sensitivity by ASV.

$$Fe^{3+} + BiY^- \rightarrow FeY^- + Bi^{3+} \qquad (1)$$

Total iron was then determined in the same way, but after heating the acidified water sample to convert all iron to a reactive state.

B. Redox State

Determination of the redox state of an element in solution is an important speciation measurement because it can drastically affect toxicity, adsorptive behavior, and metal trans-

Table 1
TOXICITY AND ELECTROCHEMICAL LABILITY OF SOME SPECIES IN NATURAL WATERS[a]

Species	Toxicity	Electrochemical lability
Arsenic		
(III)	High	High
(V)	Low	Low
Chromium		
(III)	Low	Low
(VI)	High	High
Thallium		
(I)	High	High
(III)	Low	Low
Cu^{2+}	High	High
$CuCl_2$	High	High
$CuCO_3$	High	High
Cu^{2+}-fulvic acid	Low	Low
Cu^{2+}/humic-Fe_2O_3	Medium	Medium
Cu^{2+}-DMP[b]	High	Low

[a] Data derived from References 1, 3, and 10.
[b] DMP = 2,9-dimethyl-1,10-phenanthroline.

port (Table 1). Polarography and/or ASV have been used to distinguish between Fe(III)/(II),[37] Cr(VI)/(III),[38] Tl(III)/(I),[28,39] Sn(IV)/(II),[28,40] Mn(IV)/(II),[41] Sb(V)/(III),[28] As(V)/(III),[42] Se(VI)/(IV),[43] V(V)/(IV),[44] Eu(III)/(II),[44] U(VI)/(IV),[45] and I(V)/(−I)[46]. Whereas chromium (VI) (chromate) is anionic and highly toxic, chromium (III) is nontoxic and may exist as anionic or cationic hydrolyzed or organic species.[47] For some other elements, however, including thallium, arsenic, and antimony, the lower valency state is the more toxic.[48] The manganese (IV)/(II) discrimination[49] measurement is important because fine MnO_2 particles cause problems in water supply treatment plants by clogging filters.[41]

For several elements, redox state speciation is actually a special case of labile/inert discrimination, since one valency state is electrochemically active and the other inactive within the potential range of the electrode. Unreactive valency states of some elements under certain conditions are As(V), Cr(III), Mn(IV), Sb(V), Sn(IV), and Tl(III). Determination of the labile valency state of the element in the presence of these unreactive forms can be made by a simple ASV or polarographic measurement.[28] Total metal can then be determined after chemical treatment of the sample (e.g., chemical reduction), and the concentration of the inert valency state determined by difference. Some metal ions that are electrochemically inert because they are extensively hydrolyzed in most media, e.g., Sb(V) and Sn(IV), become labile when the sample is made strongly acidic.[28] Figure 1 shows the ASV behavior of Sb(III) and Sb(V) as a function of acidity. Antimony (III) can be determined in 0.2 *M* HCL, and total antimony in 6 to 8 *M* HCL, then Sb(V) by difference. Alternatively, and preferably, total antimony can be measured after reduction of Sb(V) to (III) using hydrazine hydrochloride.[28]

Electrochemical methods measure valency state directly, which is preferable to using ion exchange methods to determine ionic charge, as these may often lead to erroneous interpretations. A commonly used method for distinguishing chromium (VI) from chromium (III) is to pass the sample through a column of anion exchange resin. Chromium (VI) as CrO_4^{2-} is adsorbed, whereas chromium (III) is assumed to pass through the column as cationic species.[38,50] It has been suggested, however, that anionic $Cr(OH)_4^-$ is the most common

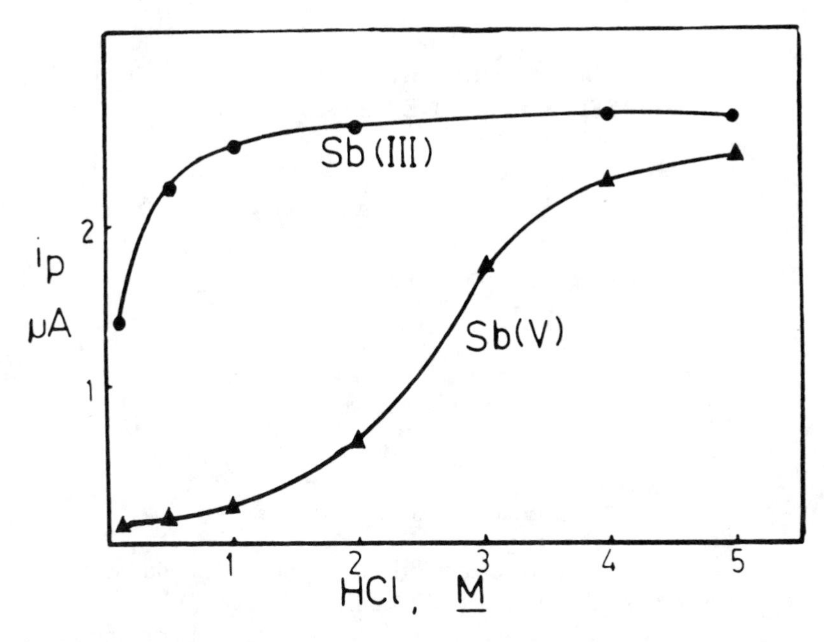

FIGURE 1. Effect of HCl concentration on the ASV peak heights of Sb(III) and Sb(V).

form of chromium in natural waters.[51] In the polarographic method for chromium speciation,[38] chromium (VI) and total chromium are determined sequentially in acetate buffer with half-wave potentials of −0.3 V and −1.8 V vs. sce, respectively.

C. Half-Wave Potential Shifts

Shifts in the polarographic half-wave potential or ASV peak potential of metal ions in the presence of complexing agents can provide information about the thermodynamic stability of complexes in solution.[44] However, quantitative deductions from these shifts, which have a sound theoretical basis for well-defined experimental solutions containing one or, at the most, two ligands, are inapplicable to natural or polluted waters which may have many unknown ligands and several metals. Under these conditions, quantitative interpretation of the shift is impossible, although some qualitative deductions can sometimes be made. For example, the ASV peak potential of copper in seawater is about 200 mV more negative than the peak in nitrate or acetate media. This shift reflects the relatively high stability of copper(I)-chloro complexes compared with those of copper(II).[52] In high-chloride media, copper is stripped from the electrode in a one-electron reaction to form copper(I)-chloro complexes, whereas in nitrate solution, copper(II) is produced in a two-electron stripping step.

D. Limitations of Electrochemical Speciation Techniques

One of the main limitations of electrochemical speciation methods, the inability to measure the concentrations of individual ionic species, is common to most speciation techniques. Ion-selective electrode potentiometry (ISE) is the only method that can measure the activity of an individual ion, but the applicability of ISE to water analysis is severely limited by its poor sensitivity. Other electrochemical techniques, such as polarography and ASV, are dynamic systems which draw current through the solution and disturb ionic equilibria. It is not possible, for example, to use ASV to distinguish between labile cadmium species such as Cd^{2+}, $CdSO_4$, $CdCl^+$, and $CdCO_3$, which may coexist in a river water sample. A single ASV peak is obtained for a mixture of these cadmium species. However, other speciation methods, including ion exchange chromatography, solvent extraction, dialysis, and ultra-

Table 2
ELECTROCHEMICAL LABILITY CLASSIFICATION OF METAL SPECIES[a]

Degree of lability	Examples
Labile	$CdCO_3$, $CuCO_3$, $PbCl_2$, Cd-NTA, Cu-glycine, Zn-cysteine, Cu-citrate
Quasilabile	$PbCO_3$, $ZnCO_3$, Cu-cysteine, Zn-fulvic acid, Cd-tannic acid, Cu^{2+}/humic-Fe_2O_3, Cu-NTA
Inert	Pb-EDTA, $Pb_2(OH)_2CO_3$, Cu-tannic acid,[b], Cu-APDC,[c] Cu-fulvic acid,[b] Zn-tannic acid

[a] Data derived from References 10, 22 to 24, 71, and 80.
[b] Seawater, pH 8.2.
[c] APDC = ammonium pyrrolidinedithiocarbamate.

filtration also disturb the natural ionic equilibria in a water sample during the separation process, and all suffer from the same lack of specificity (Chapter 3).[1]

Direct electrochemical speciation procedures are limited to measuring gross behavioral differences of groups of species (Table 2). This applies to the usual labile/inert discrimination and to the effect of deposition potential on ASV peak height.[29,53,54] Other groups of species can be determined by ASV after chemical treatment of the sample (e.g., UV irradiation, acidification),[1-3,55] after physical separations (ion exchange, ultrafiltration, etc.),[56,57] or after chemical exchange reactions.[36] Although some deductions can be made about the nature of the species that are likely to occupy these behavioral "boxes", exact conclusions cannot be drawn.[58] Since results from all these speciation procedures are operationally defined, it is most important when publishing speciation methods to report all details of the analysis so that results from different laboratories can be compared.

There has been considerable confusion in the literature over the ability of ASV to measure the "existing" or "natural" trace element speciation in a water sample. It is often specified[59-62] that buffer should not be added to a water sample before measurement of speciation by ASV, so as to avoid disturbing the natural ionic equilibria. However, because ASV is a dynamic technique, it cannot possibly measure the "natural" speciation, since the very act of measurement disturbs the equilibrium.[3,63] If the aim of the determination is to estimate the bioavailable fraction of the metal, some pH other than the natural pH of the water may well give the best correlation between ASV-labile metal and bioavailability (Section VII). The purpose of the measurement should always be kept in mind when designing a speciation procedure. For some purposes, however, it may be desirable to avoid changing the pH of the sample.[62] Seawater can be analyzed without the addition of buffer, but some fresh waters have too low an ionic strength and are too poorly buffered to be analyzed directly by ASV.[24,64] It is possible to buffer the water and maintain its original pH by bubbling N_2/CO_2 mixtures of controlled composition through the sample.[65,66] This is rather inconvenient, and bicarbonate buffers are poorly poised and contribute little to the conductivity of the sample.

An important potential interference in ASV, polarography, and other electrochemical techniques is the adsorption of organic matter on the mercury electrode.[3,61,67] An adsorbed layer of organic matter may hinder the diffusion of metal ions and thus diminish or eliminate the diffusion current and cause a nonlinear relationship between stripping current and deposition time.[68] Alternatively, adsorption/desorption processes by organic dipoles on the mercury surface can yield "tensammetric" peaks when high-frequency (AC or pulse) voltammetric peaks are used.[69,70] These tensammetric adsorption waves have no faradaic component, but in ASV are often mistaken for metal stripping peaks, since in natural waters they may appear at potentials similar to those found for cadmium, lead, or copper.[70] They can be readily distinguished from metal peaks because they are absent if a simple DC scan is used, their peak height is seldom proportional to deposition time, peak potential is very sensitive to pH, and they disappear when the sample is UV irradiated.[70] Fortunately, tensammetric peaks are uncommon in water analysis, but analytical chemists need to be aware of their existence.

Interference by adsorbed organic matter may be a more frequent problem in ASV speciation analysis, although it is often difficult to determine if a metal wave is diminished because of physical interference to diffusion, by formation of an inert organic complex, or by a combination of the two processes. Various techniques can be used to address this problem. The presence of an adsorbed film can, in some cases, be detected by multitime domain differential pulse ASV,[71] where pulse currents are measured for a range of pulse widths, and the change in current with pulse width used to detect adsorbed films, which markedly affect the current/time relationships. Organic films on glassy carbon and platinum electrodes can be removed with pulsed laser light, which greatly extends electrode life.[72] Perhaps the most useful technique available at present for elimination interference by adsorbed organic films is to cover the electrode with a permselective coating, such as Nafion,[73] which would allow unimpeded diffusion of the metal ion or complex (Section IV.G).

In many practical cases, however, the measured concentrations of ASV-labile metal can only be operationally defined by the instrumental and solution conditions used, and, in most instances, little information can be deduced about the electrode processes involved.

For standardizing ASV-labile metal measurements, it is important to use an ionic metal-peak height calibration curve, rather than to attempt to quantify the results by standard addition (''spiking'') of ionic metal to the sample. In many natural waters, excess organic matter (e.g., fulvic acid) in the sample will complex the metal spike, and the increase in peak height will be related to total, rather than labile, metal. If the concentration of the spike is high, i.e., at least 20 times that of the complexing agents in the sample, then the peak height difference between the first and second spikes can be used to calculate labile metal. These high spikes, however, may lead to metal contamination of the cell. It is more accurate to use a matched matrix with standard metal concentrations similar to that of the sample.

A special type of interference occurs in ASV as a result of intermetallic formation in the mercury electrode.[1] These intermetallic compounds cause depression of the stripping peaks and shifts in peak potentials. The most common interference is the depression of the zinc wave by an excess of copper. In practical water analysis, however, this is rarely a problem, because metal concentrations are low and zinc is usually in excess of copper.[74]

E. Speciation Schemes Combining Electrochemical and Other Techniques

Speciation information obtained from direct electrochemical analysis (e.g., labile/inert discrimination) can be supplemented by ASV or other measurements after various preliminary treatments of the sample. In this case, electroanalysis is simply used as a highly sensitive method of analysis. Some important preliminary speciation steps are the following.

UV irradiation to destroy organic matter[8,74-76] — If the sample is irradiated at natural pH, only metal associated with organic matter will be liberated, and the increase in labile metal compared with the unirradiated sample represents metal bound in inert organic complexes or to organic colloids.[63,65,77,78] When the sample is acidified ($0.02M HNO_3$) before irradiation, all forms of metal, including inorganic colloids, are converted to labile species, and total metal is obtained.[75,77]

Determination of lipid-soluble complexes — Lipid-soluble metal species are likely to be highly toxic.[11,16,18,79] Extraction of a water sample with *n*-octanol or 20% *n*-butanol in hexane, or passage of the water through a column of Bio-Rad® SM2 resin, will remove the

Table 3
SPECIATION SCHEME FOR COPPER, LEAD, CADMIUM, AND ZINC IN WATERS (FILTRATE ANALYSIS)

Sample (Unacidified): Filter Through a 0.45-μm Membrane Filter; Reject Particulates and Store Filtrate Unacidified at 4°C

Aliquot No.	Operation	Interpretation
1. (20 ml)	Acidify to 0.05 *M* HNO_3, add 0.1% H_2O_2 and UV irradiate for 8 h, then ASV[a]	Total metal
2. (10 ml)	ASV at natural pH for seawater add 0.025 *M* acetate buffer, pH 4.7 for freshwaters	ASV-labile metal
3. (20 ml)	UV irradiate with 0.1% H_2O_2 at natural pH, then ASV[b]	(3)-(2) = organically bound labile metal
4. (20 ml)	Pass through small column of Chelex® 100 resin; ASV on effluent[c]	Very strongly bound metal
5. (20 ml)	Extract with 5 ml of hexane-20% *n*-butanol; ASV on acidified, UV-irradiated aqueous phase[d]	(1)-(5) = lipid-soluble metal

[a] Bring to pH 4.7 with acetate buffer.
[b] Not valid if Fe > 100 μg l^{-1}.
[c] Optional step.
[d] Solvent dissolved in aqueous phase must be removed first.

lipid-soluble fraction of the metal.[77] Analysis of the aqueous phase or column effluent by ASV, and subtraction from total metal, gives lipid-soluble metal.

Chelating resin separation — Metal that cannot be removed from a water sample by a column of Bio-Rad® Chelex®-100 chelating resin represents metal bound in highly stable or inert complexes, or associated with colloidal particles.[76,77,80,81] However, the resin may remove some metal from colloidal particles.[77]

Ultrafiltration and dialysis — These techniques separate species on the basis of molecular size, and both can provide useful information about the size distribution of metal complexes and colloids,[1,3] although contamination can be a problem. In general, the smaller a metal complex, the higher is its biological activity.

Several comprehensive speciation schemes combining ASV and these preliminary treatments have been proposed (Chapter 3).[1-3,8,22,82] The scheme used at present in our laboratory is shown in Table 3.

F. Electrodeposition Prior to Carbon Furnace Atomic Absorption Spectrometry

A novel application of electrochemical techniques to speciation studies is the controlled-potential electrodeposition of trace metals onto graphite furnace tubes, which are then transferred to an atomic absorption spectrometer for electrothermal measurement.[38,83,84] The advantage of this technique is that elements such as chromium, nickel, and cobalt, which are difficult to determine by direct ASV, can be concentrated from solution by electrolysis, using labile/ inert discrimination, and determined with good precision. The furnace atomization simply replaces the ASV stripping step. The flow-through electrolysis cell used by

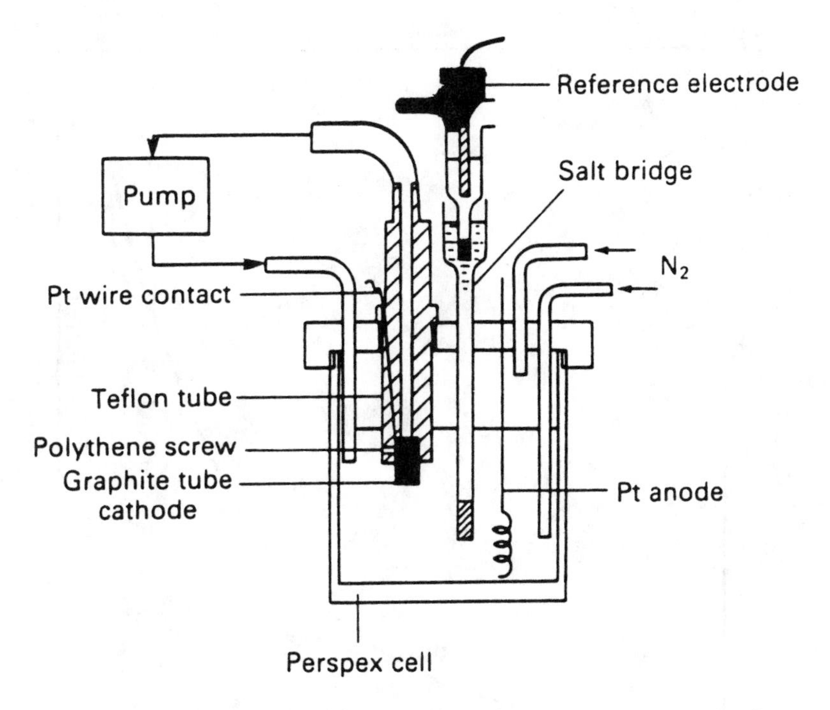

FIGURE 2. Flow-through cell for electrodeposition on an AAS graphite furnace tube. (From Batley, G. E. and Matousek, J. P., *Anal. Chem.*, 49, 2031, 1977. With permission.)

Batley and Matousek[84] is shown in Figure 2; it was successfully applied to the discrimination of chromium (VI) and (III) and the labile/inert forms of nickel and cobalt in natural waters. Another type of flow-through cell using a graphite furnace tube was described for the *in situ* determination of lead and cadmium in seawater.[85]

III. THEORY OF LABILE/INERT DISCRIMINATION

A. The Electrodeposition Step

The dissociation of a 1:1 complex formed between a divalent metal ion, M, and a ligand, L, and the subsequent reduction of M^{2+} at a mercury electrode may be represented by the equilibria,

$$ML \underset{k_f}{\overset{k_d}{\rightleftharpoons}} M^{2+} + L^{2-} \tag{2}$$

$$\beta_1 = [ML]/[M][L] = k_f/k_d \tag{3}$$

$$M^{2+} + 2\,e \rightarrow M^0(Hg) \tag{4}$$

These reactions are shown diagramatically in Figure 3. When the complex, ML, is not itself directly reducible, the electrolysis (faradaic) current is due solely to the reduction of M^{2+} ions dissociated from ML (Equation 2 and 4). This process leads to a kinetically controlled current, and i_k/i_d, the ratio of the kinetic current, i_k, to the diffusion current, i_d, is an index of the lability of the complex. The diffusion current is the current observed for the same concentration of metal ion, but in the absence of ligand. In the absence of kinetic control, $i_k/i_d = 1$. Turner and Whitfield[25] calculated that for ASV at a thin mercury film electrode (TMFE),

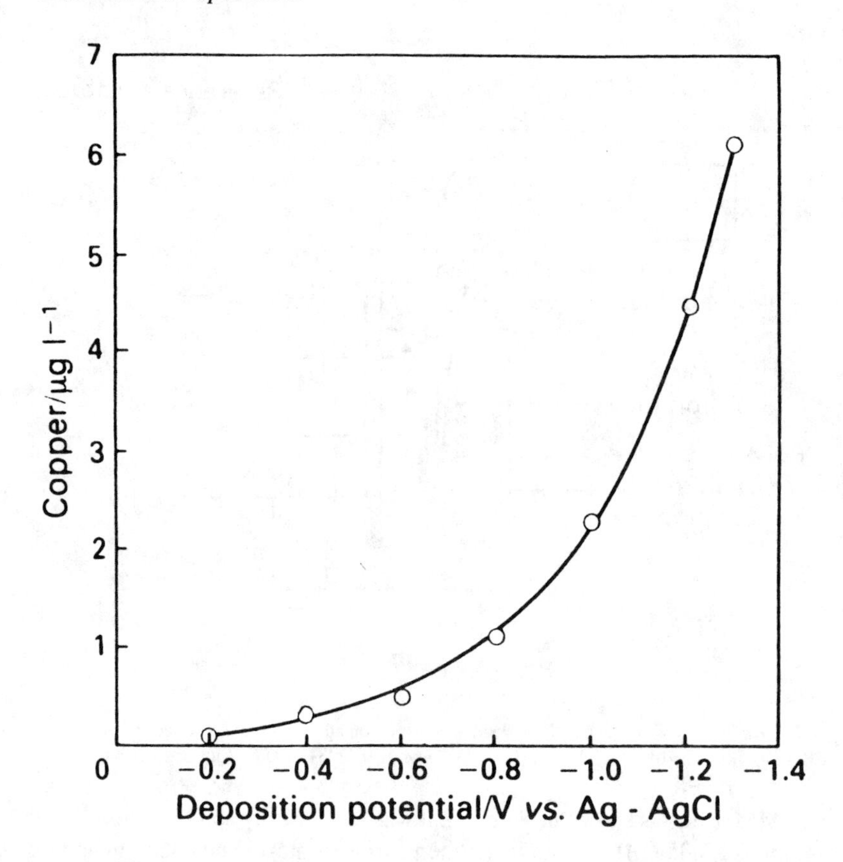

FIGURE 3. Pseudopolarogram of copper in sewage plant effluent water.

$$i_k/i_d = (1 + \sigma\eta^{-1}\tanh\eta)^{-1} \quad (5)$$

and at a hanging mercury drop electrode (HMDE),

$$i_k/i_d = [r_o(r_o + \delta)^{-1} + \sigma r_o(r_o\eta\coth\eta + \delta)^{-1}]^{-1} \quad (6)$$

where $\sigma = k_f[L]/k_d$; $\eta = \delta D^{-1/2}(k_d + k_f[L])^{-1/2}$; δ = diffusion layer thickness (cm); D = diffusion coefficient of the metal ion ($cm^2\ s^{-1}$); and r_o = radius of HMDE (cm). When $i_k/i_d > 0.99$ (i.e., a highly labile complex), it can be shown that,[86]

$$\sigma/\delta[D/k_d(1 + \sigma)] < 10^{-2} \quad (7)$$

Turner and Whitfield[86] suggested the criteria shown in Table 4 for the definition of labile, quasi (or partially)-labile, and inert (nonlabile) complexes in ASV analysis at a rotating disk TMFE. The calculations assume that the ligand, L, is in large excess.

Davison[26] calculated the following criterion for a labile complex using ASV at a rotating disk TMFE, assuming a diffusion coefficient of $1 \times 10^{-5}\ cm^2\ s^{-1}$ for the metal ion.

$$\text{labile } (i_k/i_d > 0.90)\text{: } \beta_1^{3/2}[L]/\delta k_f(1 + \beta_1[L]) < 2.85 \times 10^3\ cm^{-1}\ s \quad (8)$$

For both ASV and polarography, the lability of a complex depends not only on its dissociation kinetics, but also on the effective measurement time which, in the case of the constant electrolysis time of ASV, depends on the time the complex molecule is resident in

Table 4
ELECTROCHEMICAL LABILITY CRITERIA FOR LEAD COMPLEXES

Description	Lability criterion	Lead complexes concerned
Labile	$i_k/i_d > 0.99$	$PbCl^+$, $PbSO_4$
	$\log(\beta_1[L]) < 2$	
Quasilabile	$i_k/i_d < 0.99$	$PbCO_3$, $PbOH^+$, Pb-humic
	$i_k/i_d > (1+\sigma)^{-1}$	acid
Inert	$i_k/i_d = (1+\sigma)^{-1}$	Pb-EDTA

Note: β_1 = stability constant for 1:1 complex; [L] = concentration of ligand; i_k = kinetic current; i_d = diffusion current; $\sigma = \beta_1[L]$.

From Florence, T. M. and Batley, G. E., *Talanta*, 22, 201, 1975. With permission.

the diffusion (or reaction) layer, and this residence time depends in turn on δ, the thickness of the diffusion layer. The larger the value of δ, the longer the residence time of the complex in the diffusion layer, the greater the opportunity for dissociation and deposition of metal in the electrode, and hence the higher the fraction of labile metal.[26] The thickness of the diffusion layer is governed principally by rotation rate for a rotation disk electrode (RDE) or by the rate of solution stirring for a HMDE. The Levich equation can be used to calculate δ (cm) at a RDE.[26]

$$\delta = 1.62\ D^{1/3}\omega^{-1/2}\nu^{1/6} \tag{9}$$

where D is the diffusion coefficient ($cm^2\ s^{-1}$), ω is the electrode rotation rate (radians s^{-1}), and ν is the kinematic viscosity of the electrolyte (Stokes). For rotation speeds in the range 10^2 to 10^4 rpm, values for δ of 5×10^{-3} to 5×10^{-4} cm, respectively, are obtained. Kinetic control in ASV can be studied by measuring the effect of ω on i_k. A constant value of $i_k\omega$ is obtained in the absence of kinetic control, but decreases with increasing ω if kinetic effects are significant. Calculation of δ at a HMDE using a magnetic stirring bar for solution stirring is difficult because of the ill-defined hydrodynamic conditions. However, δ can be determined experimentally for a HMDE by measuring the DC diffusion current, i_d in the stirred solution.[26]

$$i_d = nFADC/\delta \tag{10}$$

where n is the number of electrons involved in the electrode reaction, F is the Faraday, A is the electrode area, and C is the concentration of electroreducible species. A typical value for δ at a HMDE is 2×10^{-3} cm. The diffusion layer thickness at a dropping (or static) mercury electrode can be estimated from[87]

$$\delta = (\pi D t_e)^{1/2} \tag{11}$$

where t_e is the effective measurement time, i.e., the drop time for maximum current DC polarography or, for pulse techniques, the duration of the applied pulse before the current is sampled plus the mean of the sampling interval.[26] For example, if the current is sampled for 20 ms, 40 ms into the life of the pulse, the average measurement time is 50 ms. In the case of AC modulation, t_e is the inverse of the frequency (Hz). The reaction layer thickness,

μ, which may be less than the diffusion layer thickness, is given from reaction layer theory as[29]

$$\mu = (Dk_f^{-1})^{1/2} \tag{12}$$

Van Leeuwen[29] calculated from reaction layer theory, assuming a large excess of ligand, that the following conditions apply for polarographic lability measurements:

$$\text{labile complex:} \quad k_d k_f^{-1/2} t_e^{1/2} \gg 1 \tag{13}$$

$$\text{quasilabile complex:} \quad k_d k_f^{-1/2} t_e^{1/2} \approx 1 \tag{14}$$

$$\text{inert complex:} \quad k_d k_f^{-1/2} t_e^{1/2} \ll 1 \tag{15}$$

For polarographic conditions, it can also be shown that[26]

$$i_k/i_d = \pi k_f t_e / \beta_1^2 [L] \tag{16}$$

It is apparent that ASV lability, determined during the deposition step, depends solely on the kinetic parameters of the metal complex dissociation, the concentration of excess ligand, and the diffusion layer thickness, which in turn is a function of the rate of stirring of the solution or the rotation speed of the electrode. In polarography the diffusion (or reaction) layer thickness is governed by the drop time or, in the case of pulse techniques, by the pulse width and current sampling times. The value of i_k/i_d is not affected by factors such as deposition time, sample volume, or cell volume.[24] Deposition time may only affect i_k/i_d in the special case where and adsorbed substance interferes in a nonlinear manner with the rate of electrodeposition.

An implicit assumption in the preceding discussions is that the metal complex, ML, is not directly reducible. However, where electrons are added directly to the complex without its initial dissociation in the diffusion layer, direct electrochemical reduction of some complexes is known to occur, especially at very negative potentials.[5,53,77,88] The presence of such complexes in a sample can be detected from the effect of ASV deposition potential, E_d, on peak current; the peak current will increase continuously with E_d (Figure 3) instead of increasing from zero to a limiting value over a small range of E_d (Section V.E.). To minimize the chance of directly reducible complexes contributing to the ASV-labile measurement, the deposition potential should be just sufficiently negative to yield the maximum peak current for the free metal ion in that medium, i.e., just on the plateau of the relevant pseudopolarogram (Section V.E). For this reason it is preferable in speciation analysis to determine each element separately, using the minimum deposition potential, rather than measuring copper, lead, cadmium, and zinc simultaneously with a deposition potential of -1.3 V vs. sce.

Reducible metal ions adsorbed on colloidal particles of humic acid, hydrated iron oxide, etc. can be treated as a special type of metal complex in the preceding theoretical discussions, and metal ions dissociated from pseudocolloids at the solution/diffusion layer boundary may contribute to kinetic currents.[1,76,77] Although the involvement of pseudocolloids in metal deposition has been questioned,[62] there seems no reason why ions could not dissociate from these particles at the diffusion layer boundary under the influence of the potential gradient. Metal ions are known to dissociate from colloidal particles as a result of the concentration gradient across a dialysis or ultrafiltration membrane,[89] and metal bonding to the particle is unlikely to be stronger than that involved in some ASV-labile molecular complexes such as Cu-NTA.[11,77]

Olson and Shuman[90] found that Cu^{2+} dissociates from some humic acid binding sites with first-order rate constants as high as 40 s^{-1}. Lumsden and Florence[91] showed that some of the Cu^{2+} adsorbed on the marine diatom *Nitzschia closterium* was available to a mercury electrode during ASV deposition, and Goncalves et al.[92] reported that the surface of bacteria bind copper ions more strongly than hydrous oxides. Pseudocolloids, however, cannot act as directly reducible metal complexes because the diffusion coefficients of colloidal particles are so small (10^{-7} cm^2 s^{-1}) that they would not contribute significantly to the peak current.[5,93]

B. The ASV Stripping Step

The preceding discussion has considered only the effect of deposition parameters on the amount of deposited metal and on kinetic currents in ASV speciation analysis. Ideally, the relative heights of the stripping peaks for labile and total metal in the sample, and hence the calculated percentage of labile metal, should be controlled solely by the preliminary electrodeposition step, i.e., by the amount of metal deposited in the electrode. However, under certain circumstances the kinetics of the stripping process (electrooxidation), especially when pulse techniques are used, may have a significant, even dominant, effect on the stripping peak height. This can occur if a complexing agent present in the sample solution, but not in the standard, affects stripping chemistry or kinetics.[25,30] This situation could arise from a number of causes. If a ligand in the sample solution stabilizes an intermediate valency state of the metal, leading to smaller number of electrons being involved in the electrochemical oxidation, lower stripping peaks will result

$$\text{Standard:} \quad Cu^0 \rightarrow Cu^{2+} + 2\,e \qquad (17)$$

$$\text{Sample:} \quad Cu^0 + 2\,Cl^- \rightarrow CuCl_2^- + e \qquad (18)$$

This has been observed for the ASV determination of copper in the presence of chloride and some other ligands.[52,67,94] Figure 4 shows the double stripping peaks found for copper in the presence of NTA.[95] The more positive peak has the peak potential found for copper in the absence of NTA. This double-peak phenomenon is probably the result of NTA in solution altering the Cu^{2+}/Cu^0 redox potential and leading to some of the copper stripping at a more negative potential. However, as copper ions enter solution in the diffusion layer, the Cu/NTA ratio increases sufficiently to prevent more copper being oxidized.

The presence of complexing agents or surface-active substances may also affect the kinetics of stripping and lead to a change in peak height, especially when pulse techniques are used.[96] Buffle[30] showed how the large surface excess of the oxidized metal ion (compared with the bulk solution), present during the initial stages of stripping, can cause precipitation and other chemical reactions at the electrode surface that might affect the stripping peak current.

Perhaps the most useful method for ensuring that the ASV-labile measurement is controlled only by the deposition step is to use medium exchange where the sample solution, after electrodeposition, is replaced by a new supporting electrolyte in which stripping is carried out.[97,98] The new electrolyte would be chosen to yield reversible, reproducible stripping peaks for the element under study.

A recent study[99] showed that in both synthetic freshwaters and natural river water samples, significant differences were found in some ASV-labile fractions when medium exchange was used (Table 5). For the synthetic mixtures, copper-chloride, copper-humic acid, and zinc-tannic acid showed the greatest change with medium exchange, while large differences were found for zinc, cadmium, and lead in the river waters. It is apparent that medium exchange should be used routinely for ASV-labile determinations to eliminate stripping kinetics affecting the measurement.

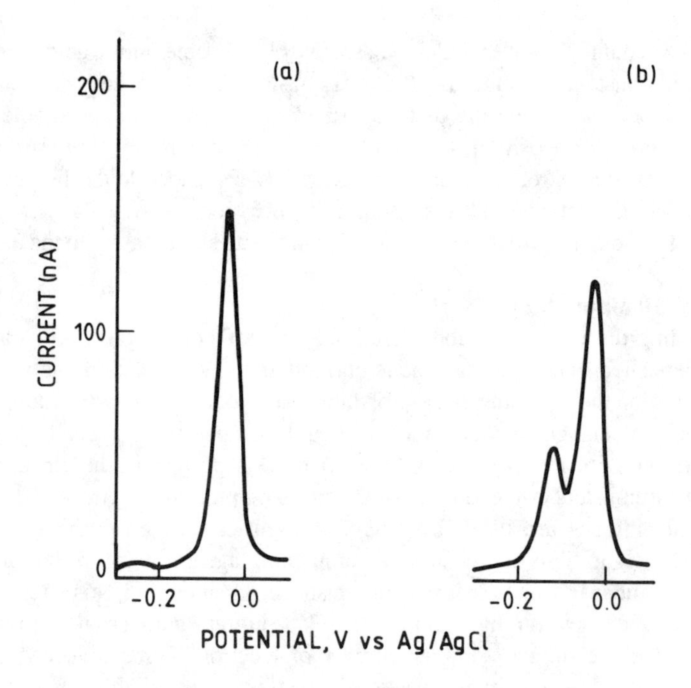

FIGURE 4. Effect of NTA on the stripping voltammetry of copper at the HMDE. (a) 2×10^{-7} *M* Cu; (b) 2×10^{-7} *M* Cu + 1×10^{-6} *M* NTA.

Table 5
EFFECT OF MEDIUM EXCHANGE ON ASV-LABILE METALS IN SOUTH CREEK WATER (PRISTINE)

Metal	Total conc. ($\mu g\ l^{-1}$)	ASV-labile (%)	
		Before	After medium exchange
Zn	1.8	69	100
Cd	0.11	57	100
Pb	0.23	<7	42
Cu	2.6	17	27

C. Comparison of the Kinetics of Dissociation of Metal Complexes at an Electrode and a Biomembrane

For the study of aquatic toxicity by metals, the electrochemical and solution parameters should be chosen so that the ASV-labile fraction of total dissolved metal is similar to the toxic fraction.[11,14] Hydrophilic heavy metal ions are believed to be transported across the hydrophobic space of a biomembrane by the ''shuttle'' process of facilitated diffusion (or ''host-mediated transport''), where a receptor molecular (e.g., a protein) on the outer membrane surface binds a metal ion (see Chapter 2).[100,101] The hydrophobic metal-receptor complex then diffuses to the interior of the membrane and releases the metal ion into the cytosol where it is trapped, perhaps by reaction with a thiol compound.[102] The receptor then diffuses back to the outer surface of the membrane, ready to collect another metal ion.[11,15-18] Alternatively, if the metal complex is lipid soluble, the much more rapid process of direct diffusion can take place. Direct diffusion is basically different from facilitated diffusion, not only because it is faster, but because the ligand is also transported into the cytosol.[11]

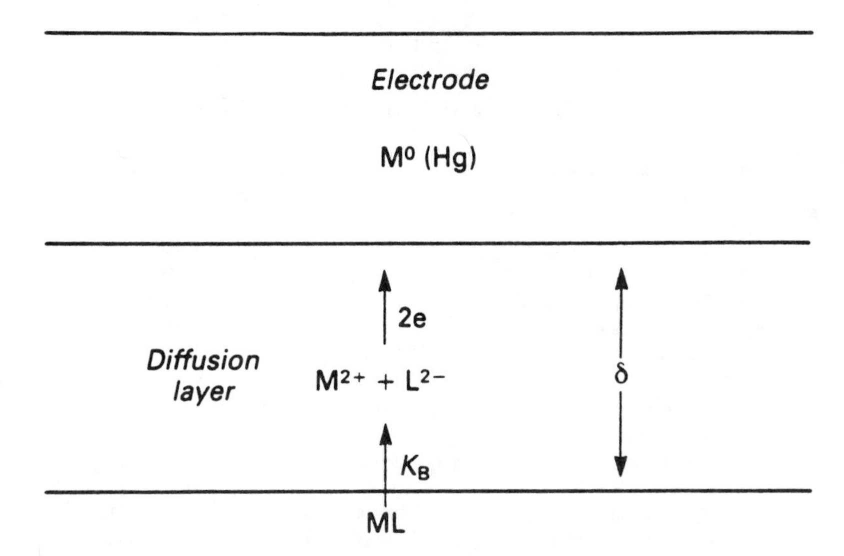

FIGURE 5. Diagrammatic representation of the reduction of a metal complex at a mercury electrode. The degree of dissociation of the metal complex, ML, at the electrode (and hence the lability of the complex) increases with increasing K_B and increasing δ.

The fraction of total metal in solution that can be transported across a membrane surface is equivalent to the bioavailable fraction. This in turn depends on the relative affinity of the metal for solution ligands and the receptor molecule, or on the solution/membrane partition coefficient for a lipid-soluble complex.

The process of metal accumulation in an organism, by dissociation of a metal complex at a membrane surface, facilitated diffusion of the metal through the membrane, and deposition in the cytosol, has obvious similarities to the process of ASV electrodeposition (Figure 5), where the metal complex dissociates at the diffusion layer boundary and the metal ion travels through the diffusion layer to the electrode where the metal is deposited.

IV. ELECTRODES FOR SPECIATION MEASUREMENTS

The electrodes used most often for routine speciation measurements in natural waters are the hanging mercury drop electrode (HMDE), the thin mercury film (on glassy carbon) electrode (TMFE), and the dropping (or static) mercury electrode (DME). Other electrode systems have been used mainly in research or for special applications. Ion-selective electrodes will not be reviewed here because they are usually insufficiently sensitive for speciation analysis in natural waters, although they may have application in polluted waters.

A. Hanging Mercury Drop Electrode (HMDE)

The HMDE is widely used in ASV and speciation analysis. Use of simple DC techniques leads to stripping peaks with a drawn-out shape, due to the slow diffusion of metal from the interior to the surface of the mercury drop. For this reason it is necessary to use high-frequency (pulse or AC) waveforms for ASV at the HMDE.[23] The high frequency techniques respond only to dissolved metal at the surface of the mercury drop, and so the stripping peaks have a sharp, theoretical shape. With a 15-min deposition time, the limit of detection for lead using differential pulse ASV (DPASV) at the HMDE was found to be 5×10^{-11} *M*, based on noise levels in a Princeton Applied Research Model 174 voltammeter.[28] However, reagent blanks usually increase this limit substantially.

The use of a static mercury drop electrode (SMDE, Section IV.A) instead of the older

Table 6
RELATIVE SENSITIVITY OF SOME ELECTROCHEMICAL TECHNIQUES

Electrochemical technique[a]	Limit of detection for lead (*M*)[25,81,136]
DC polarography (DME)	2×10^{-6}
DC polarography (SMDE)	1×10^{-7}
DP polarography (DME)	8×10^{-8}
DP polarography (SMDE)	1×10^{-7}
DP anodic stripping voltammetry (HMDE)	2×10^{-10}
SW anodic stripping voltammetry (HMDE)	1×10^{-10}
DC anodic stripping voltammetry (TMFE)	5×10^{-11}
DP anodic stripping voltammetry (TMFE)	1×10^{-11}
SW anodic stripping voltammetry (TMFE)	5×10^{-12}

[a] DC = direct current; DP = differential pulse; SW = square wave; DME = dropping mercury electrode; SMDE = static mercury drop electrode; HMDE = hanging mercury drop electrode; TMFE = thin mercury film electrode.

micrometer screw-type hanging mercury electrode system, greatly improves the reproducibility, reliability, and simplicity of ASV at a HMDE, because the mercury drop is formed automatically and its size is very reproducible.[87]

B. Thin Mercury Film Electrode (TMFE)

While the use of a HMDE or a SMDE may offer somewhat better reproducibility than a TMFE, especially for zinc,[28] the rotated TMFE is much more sensitive (Table 6). This higher sensitivity is essential for high-purity samples such as open-ocean seawater, which cannot be analyzed at the HMDE.[103,104] While pulse techniques are essential for a HMDE, ASV at a TMFE can conveniently be carried out using a simple DC scan because the mercury film is so thin that dissolved metal is stripped from the film very quickly. Use of differential pulse modulation at a TMFE decreases the limit of detection by a factor of about five over a DC scan.[105] The glassy carbon TMFE can be rotated (1000 to 3000 rpm) or the solution stirred, although rotation gives more precise results.[5]

The substrate used for a TMFE is nearly always glassy carbon, polished to a mirror finish with diamond or alumina dust.[36,106] Glassy carbon is a commercially available, synthetic substance, almost as hard as diamond, but with good electrical conductivity and a wide potential range.[105,107] Like all forms of carbon, however, glassy carbon is rapidly attacked by free halogens,[108] so the electrode should never be polarized in the positive region when the solution contains halide ions. In chloride media, a film of mercurous chloride (calomel) forms on a TMFE if the electrode potential, E, is more positive than

$$E = +0.026 - 0.0296 \log_{10}[Cl^-]^2 \text{ V vs. sce} \quad (19)$$

This film of calomel seriously degrades the performance of the electrode and is difficult to remove (ethanol is the best solvent).[105] Because even low chloride concentrations lead to calomel formation, in general a TMFE should not be polarized at potentials more positive than 0 V vs. sce.[105] When first prepared, a glassy carbon electrode should be polished metallographically (diamond dust) and thereafter should only require polishing with wet and dry filter paper (e.g., Whatman® no. 541) after each analysis. If the electrode becomes contaminated with organic matter or metal hydroxides, wiping with filter paper soaked in ethanol or 2 *M* HNO_3, respectively, will usually restore the surface. The mercury film should be removed by wiping with filter paper, and not by anodic polarization,[109] as this will degrade the surface if chloride is present.[105]

The mercury film may be electrodeposited on the glassy carbon substrate by two methods — preformed or *in situ* deposition. The preformed method consists of electrodepositing a film of mercury from a stirred mercuric nitrate solution (1×10^{-4} *M*, pH 3 to 5, −0.6 V vs. sce for 5 to 10 min). The plated electrode is then washed briefly and immediately used for analysis of the deaerated sample. A fresh film must be deposited for each sample. The *in situ* technique simply involves adding an aliquot of 1×10^{-2} *M* $Hg(NO_3)_2$ (kept in a dark bottle at pH 3 to prevent autoreduction to Hg[I]) to each sample to give a final concentration of 2 to 4×10^{-5} *M* Hg^{2+}. During the deposition step of ASV, trace metals and Hg(II) are reduced simultaneously and codeposited, forming a very thin film of a dilute amalgam on the electrode. Measurements are usually made on the second or third deposition/stripping cycle, as the first deposition is needed to condition the electrode.[106] The mercury film thickness, ℓ (cm), may be calculated from[28]

$$\ell = 2.43 \times 10^{-11}\ \mathrm{it/r^2} \tag{20}$$

where i is the limiting mercuric ion deposition current (μA), t is the deposition time (s), and r is the radius of the electrode surface (cm). Typical mercury film thicknesses used in the *in situ* technique are 5 to 10×10^{-6} cm. The *in situ* deposition method is much simpler than preforming a new film for each sample, and it avoids the danger of oxidation of the preformed film before it can be transferred to the deaerated sample. Oxidized films give erratic results, especially for copper.

It has often been claimed[59-61] that the *in situ* mercury film cannot be used for speciation studies because the addition of mercuric ions to the sample will change the "natural" speciation and cause an increase in labile metal as a result of Hg^{2+} exchanging with a metal complex, ML, liberating free metal ion, M^{2+} (Section II.D).

$$\mathrm{ML} + \mathrm{Hg^{2+}} \rightarrow \mathrm{HgL} + \mathrm{M^{2+}} \tag{21}$$

Certainly, mercury(II) forms very stable complexes with many ligands, and the exchange reaction (Equation 21) may readily occur with labile metal complexes in natural waters. However, it is unlikely that this exchange reaction would significantly affect many ASV speciation results.[63] If the complex, ML, is sufficiently labile to undergo significant chemical exchange with Hg^{2+} during the period of the analysis (10 to 20 min), then it may also dissociate at the electrode surface and yield labile metal. If this occurred, the addition of Hg^{2+} would have no effect on the measured concentration of ASV-labile metal. Recent research[110] using both natural waters and synthetic water containing various ligands showed that 2 to 4×10^{-5} *M* Hg^{2+} had no effect on the determination of ASV-labile Zn, Cd, Pb, and Cu. Low concentrations (0.02 to 0.05 *M*) of acetate also had a neglible effect on the measurement.

A special application of a glassy carbon TMFE is its use in a microcell using a 13-mm-diameter membrane disk to adsorb the sample (15 μl).[111] The membrane disk with absorbed sample (containing Hg^{2+}) is dropped into the cavity of a Perspex block. The base of the cavity has flush-fitting silver disks acting as auxiliary and reference electrodes. The glassy carbon working electrode is mounted in a Teflon® rod which makes a sliding fit in the Perspex block. When the cell is screwed together, the membrane disk is compressed between the glassy carbon electrode and the two silver electrodes. An O-ring seals the cell. Oxygen is removed by applying a potential of −1.4 V vs. Ag/AgCl for 20 s. Conventional ASV-labile measurements can then be made. Disks of an ultrafiltration membrane or Chelex®-100 paper can be placed between the glassy carbon electrode and the sample disk to provide

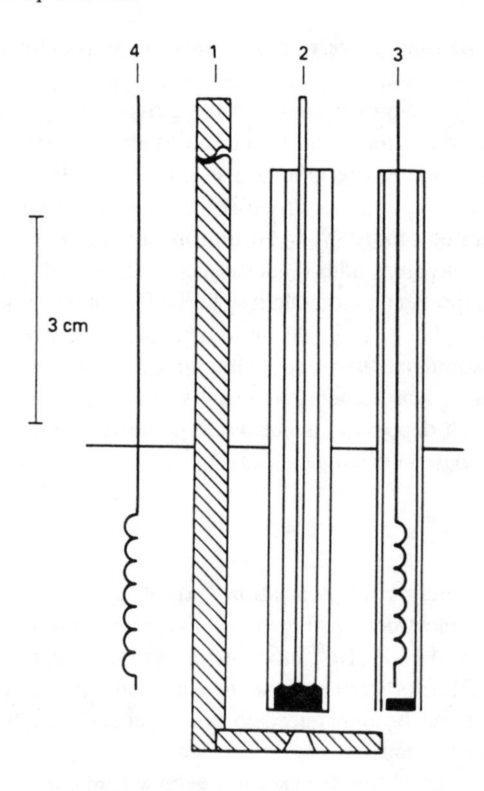

FIGURE 6. Jet stream electrode. (From Magjer, T. and Branica, M., *Croat. Chem. Acta*, 49, L1, 1977. With permission.)

additional speciation measurements.[103] A similar filter paper ASV cell using a mercury pool electrode has also been described.[112]

C. Jet Stream Mercury Film Electrode

A new method for transporting sample to the surface of a glassy carbon TMFE was described by Magjer and Branica.[113] In this technique, instead of rotating the electrode or stirring the solution, a flat disk having a conically shaped hole is positioned below the glassy carbon electrode (Figure 6). The disk is then vibrated at high frequency in a vertical plane, forcing solution onto the electrode surface in a jet stream. The sensitivity of the electrode is critically dependent on the geometries of the vibrating disk and the conical hole, but if these parameters are optimized, higher sensitivity than rotation or stirring can be achieved.[114]

D. Flow-Through Cells

A variety of flow-through cells designed for on-line stripping analysis have been described.[115-120] These cells often have dual mercury-plated glassy carbon or reticulated vitreous carbon[117] electrodes with independent potential control for removal of dissolved oxygen or interfering elements at the upstream electrode.[119] Labile/inert speciation measurements are possible with these electrodes, although they have seldom been used for this purpose. Flow-through cells used for high performance liquid chromatography could also be used for speciation measurements in a closed-loop system.[84] The wall-jet electrode, in which a jet of the sample impinges on the working electrode,[121] should provide excellent sensitivity and, because of its rapid hydrodynamic characteristics, could yield data on the dissociation kinetics of metal complexes.

E. Streaming Mercury Electrode (SME)

The SME, first used by Heyrovsky for oscillographic polarography,[44] involves forcing a thin stream of mercury through a short path of the test solution, using a mercury reservoir to provide the necessary pressure. The electrode thus consists of a short, rapidly changing cylinder of mercury. A modified SME which uses less mercury and gives more reproducible results has been described.[122]

A unique characteristic of the SME is that the electrode is being constantly and very rapidly renewed, so that only fast electrochemical reactions are registered and, most importantly, substances which adsorb on mercury have little or no effect.[122] Although the use of the DME with high frequency techniques can also discriminate against slow electrode reactions, organic matter (e.g., humic acid) adsorbed on the electrode can seriously affect results. However, with the SME, the electrode is being renewed so rapidly that adsorption has little chance to occur and so has a negligible effect on electrode kinetics.[122] The SME has not yet been applied to speciation measurements, but it may prove especially useful, in conjunction with differential pulse or square wave modulation, for measuring free metal ion in the presence of metal complexes and surface active substances.

F. Microelectrodes

Electrodes consisting of minute carbon fibers or noble metal wires (5 to 10 μm diameter, 0.1 to 0.3 cm length), either bare or mercury coated, are finding increased use in electroanalytical chemistry. The electrodes exhibit low background current and, because of the extremely low cell current, the iR drop in the solution is negligible.[123-133] These low cell currents provide an analysis which is essentially nondestructive, so *in vivo* analysis, e.g., in the brain, can be made without damage to the animal.[134] Analysis can be carried out in the absence of supporting electrolyte and in aprotic organic solvents.[124] A two-electrode system may be used, thus avoiding the need for a potentiostat which is expensive and is a major source of electronic noise.[124]

Microelectrodes constructed of carbon fiber, platinum, gold, or mercury have considerable potential for speciation analysis *in vivo*.[134-136] The more general application of microelectrodes in small, microprocessor-based, battery-operated, portable electroanalytical instruments for clinical analysis is limited only by the imagination of researchers.[137]

G. Chemically Modified Electrodes

Electrodes which have been coated with a chemical which alters their characteristics are now quite widely used in electrochemistry, and some systems have been applied to electroanalysis.[138] The substrate may be platinum,[139] carbon paste,[140] or glassy carbon.[141] Polymers such as poly(4-vinylpyridine)[142] and vinylferrocene can be electrodeposited on a platinum electrode, or groups such as $-Si(CH_2)_3NHCOCOOH$ directly bonded to oxide groups on glassy carbon.[141]

Various polymers may be used to cover a glassy carbon or platinum electrode to inhibit fouling of the electrode surface by organic compounds in the sample solution.[143,144] Stewart and Smart[145] showed that a glassy carbon TMFE covered with a dialysis membrane gave excellent results for the ASV determination of cadmium. Wang and co-workers[146-148] used cellulose acetate films, base hydrolyzed to control the permeability, on a mercury film-glassy carbon substrate for ASV and CSV of trace metals, and showed that the cellulose-covered electrode tolerated much higher concentrations of surface-active materials, including gelatine, agar, and Triton® X-100.

Nafion (E. I. du Pont de Nemours) is a noncross-linked perfluorosulfonate cation-exchange resin which, when coated on a glassy carbon electrode, provides excellent discrimination against large organic molecules.[149-152] Nafion, being a cation exchanger, is highly permeable to metal cations, but prevents fouling of the electrode surface by organic matter such as humic acid or proteins.[149] This characteristic of Nafion-coated glassy carbon electrodes allowed the direct ASV determination of cadmium, lead, and copper in whole blood, urine, and sweat.[153] Small, positively charged hydrophobic organic molecules, e.g., nitrophenols, can also be determined selectively using a Nafion-coated glassy carbon electrode,[150] and Wang and Tuzhi[154] showed that a bilayer coating, consisting of cellulose acetate atop a Nafion film, can provide even greater discrimination in favor of small molecules. These workers were able to determine dopamine in the presence of norepinephrine or epinephrine which, being larger molecules, are excluded by the cellulose acetate film. Use of a Nafion, or similar coating, on an electrode, endows the electrode with built-in medium exchange (Section III.B), since the deposited metal is stripped not into the solution, but into the Nafion layer, which excludes complexing ligands present in the sample solution.

Other electrode coatings can be prepared by gamma-irradiation of monomers to form immobilized polymers on the electrode surface.[144] The radiation dose can be varied to control the degree of cross-linking and, hence, the permeability of the films. Poly(acrylonitrile), poly(dimethyldiallylammonium chloride), poly(ethyleneimine) and poly(4-vinylpyridine) have been used successfully as electrode coatings for trace analysis.[144]

V. ELECTROCHEMICAL TECHNIQUES FOR SPECIATION

A. Polarography

In natural waters, even using differential pulse or square wave modulation, polarography is not sufficiently sensitive for speciation measurements of most elements. Many of the elements of interest (e.g., the toxic elements) are present in the range 10^{-10} to 10^{-8} *M*, whereas polarography is limited to concentrations about 10^{-7} *M*. For iodine speciation in seawater, however, polarography is an ideal technique for determining iodate/iodine ratios.[46] Iodate is usually present in seawater at concentrations of about 3×10^{-7} *M*, and because its reduction involves six electrons, a large and sharp polarographic peak is produced. Polarography may also find application for speciation studies of polluted waters which have much higher metal concentrations. The technique is especially useful for valency state discrimination (Section II.B).

The development of the static mercury drop electrode (SMDE) has greatly simplified and improved polarographic analysis (Section IV.A).[155] Whereas the conventional (Heyrovsky) dropping mercury electrode produces a gravity-fed mercury drop of continuously changing area, the SMDE has a constant area when the current-voltage curve is recorded, thus essentially eliminating charging current due to drop growth.[87] This advantage of the SMDE is achieved by using a wide-bore capillary, through which mercury flow is controlled by a valve which can be opened for variable times. This allows drops of different size to form very quickly. The voltage scan is applied after the valve has been closed and the drop is stationary. After completion of the scan, the drop is mechanically detached. Because the charging current is so small with a SMDE, the advantages of pulse techniques over a simple DC scan are only marginal (Table 6).[87] The Metrohm® SMDE has outstanding performance, and uses inexpensive, wide-bore capillary tubing for the electrode. Bond et al.[156] designed an efficient, high capacity flow-through cell for use with the EG & G® PAR Model 303 SMDE.

B. Anodic Stripping Voltammetry (ASV)

ASV is the most widely applicable electrochemical technique for trace element speciation in waters.[23] Because of the "built-in" concentration step in ASV, extremely high sensitivity can be obtained. At present, the most sensitive commercially available ASV technique is square wave stripping at a glassy carbon TMFE (Table 6).[157] In an unmodified (i.e., no pretreatment) sample such as open-ocean seawater, metal concentrations below 10^{-11} *M* can be determined, although for many analyses, the limit of detection is set by the blank, and not by the intrinsic sensitivity of the technique.[28] Differential pulse voltammetry is a factor of two or three less sensitive than the square wave method,[157] and a HMDE is five to ten times less sensitive than a TMFE.

Dissolved oxygen is a serious interference in ASV, and care must be taken to remove it completely. Ideally, the ASV cell should be under a positive pressure of oxygen-free nitrogen, but if this is not possible, the cell should be sealed as well as possible, and a rapid flow of inert gas maintained at all times. If the gas flow is too vigorous, however, solution spray in the cell may cause memory effects. Mechanically detached DMEs pose a special problem, because the cell must have a slot to allow for movement of the electrode. It is better to use a high quality grade of oxygen-free nitrogen than to complicate the system (with the possibility of air leaks through the tubing connectors) by installing an oxygen scrubbing system. Oxygen contamination is much more likely to originate from air ingress into the cell or through tubing than from impurity in the sparging gas.[158]

Dissolved oxygen can cause an apparent increase in the copper and lead stripping peaks[85,112,158] and, in unbuffered solutions, a decrease in the cadmium peak as a result of the consumption of hydrogen ion at the electrode surface.

$$O_2 + 4\,H^+ + 4\,e \rightarrow 2\,H_2O \tag{22}$$

In many supporting electrolytes, oxygen contamination is manifested by a broadening of the copper stripping peaks.[158]

C. Cathodic Stripping Voltammetry (CSV)

CSV involves the cathodic stripping of an insoluble film, usually the mercury salt of the analyte (H_2L) deposited on the working electrode.

$$\begin{aligned} &\text{deposition:} \quad Hg + L^{2-} \rightarrow HgL + 2\,e \\ &\text{stripping:} \quad HgL + 2\,e \rightarrow Hg + L^{2-} \end{aligned} \tag{23}$$

CSV has not yet found a great deal of application in trace element speciation. Arsenic(III) and selenium(IV) can be determined in the presence of their higher valency states,[1,3] and sulfide can be measured in a large excess of other inorganic or organic sulfur compounds,[93,159].

The use of surface-active metal complexes in CSV has produced sensitive methods for some trace elements.[160,161] Sometimes termed adsorptive stripping voltammetry, this technique involves the adsorption of an organic ligand on a mercury electrode, followed by reaction of a metal ion with the adsorbed ligand to form a metal complex, then cathodic stripping of the film to produce a metal amalgam.[162] The determination of copper by adsorptive stripping of its catechol complex can be described by

$$C_6H_4(OH)_2 + Hg(\text{electrode}) \rightarrow Hg....(OH)_2\text{–}C_6H_4... \tag{24}$$

$$Hg...(OH_2)\text{–}C_6H_4 + Cu^{2+} \rightarrow Hg....C_6H_4(O)_2Cu + 2\,H^+... \tag{25}$$

$$Hg...C_6H_4(O_2)Cu + 2\,e^- + 2\,H^+ \rightarrow Hg(Cu^0)\text{–}C_6H_4(OH)_2... \tag{26}$$

An alternative procedure is to measure the reduction of the ligand, rather than the metal, in the adsorbed complex. This procedure was originally used for the determination of total and reactive aluminium in waters using linear scan voltammetry and the di-*o*-hydroxyazo dye Solochrome Violet RS.[163] Very sharp, peak-shaped voltammograms were obtained with a limit of detection of 0.2 μg Al per liter as a result of adsorption of the aluminium-dye complex on the mercury drop.[164] Similar methods were described for the lanthanides using

Solochrome Violet RS,[165] and for aluminum[166] and zirconium[167] using 1,2-dihydroxyanthraquinone sulfonate.

Adsorptive stripping procedures were developed for nickel and cobalt by Pihlar et al.[168] using dimethylglyoxime, which allowed determination of these metals in waters at concentrations below 0.1 $\mu g\ l^{-1}$. Kalvoda and co-workers further expanded the technique.[160,161] Van den Berg then followed with highly sensitive methods for nickel and cobalt (dimethylglyoxime), copper, iron, vanadium and uranium (catechol), uranium and molybdenum (oxine), zinc (APDC), and aluminum (anthraquinone sulfonate).[167-177] Wang and Mahmoud also used adsorptive stripping with Solochrome Violet RS for the determination of aluminum,[178] the lanthanides,[179] and titanium,[180] and Eriochrome® Black T for manganese.[181]

A problem in the application of adsorptive stripping procedures to the determination of trace elements in waters is that natural organic matter can adsorb on the electrode and compete with the complexing ligand for adsorption sites. Wang et al.[147] showed that this difficulty can be largely overcome by using a protective coating of cellulose acetate on a glassy carbon electrode. The cellulose acetate coating allows the small complexing ligand (e.g., catechol) to diffuse through, but excludes larger molecules, such as humic acid.

Metal concentrations as low as 10^{-10} *M* can be determined by adsorptive stripping voltammetry using a short deposition time,[182] and labile and inert metal species in a water sample can be distinguished on the basis of their reactivity with the organic ligand. This technique has considerable promise for speciation analysis and further applications will certainly appear.

D. Potentiometric Stripping Analysis (PSA)

PSA, largely developed by Jagner,[183] uses the same initial step as ASV, i.e., metal is deposited into a TMFE at a controlled potential. However, instead of applying a voltage ramp to oxidize and strip the metal, a chemical oxidant (O) in solution is allowed to diffuse to the electrode to oxidize the deposited metal, and the potential of the working electrode is followed as a function of time.[184]

$$\text{deposition:} \quad M^{2+} + 2\,e \rightarrow M^{0}(Hg) \tag{27}$$

$$\text{oxidation:} \quad M^{0}(Hg) + O \rightarrow M^{2+} + R^{2-} \tag{28}$$

Well-separated potential-time steps are obtained as the metals are successively oxidized by oxidants such as dissolved oxygen or mercuric ion. PSA is much less affected by adsorbed organics than is ASV,[183,185] and redox compounds do not interfere with the analysis.

Arsenic(III) has been determined in the presence of arsenic(V) by PSA,[186] but there has been little other interest in applying the technique to speciation analysis.

E. Pseudopolarography

A pseudopolarogram is a plot of ASV stripping peak current vs. deposition potential. The half-wave potential (E_c) of a pseudopolarogram of a metal is related to (but not identical to) the polarographic half-wave potential ($E_{1/2}$). The value of $E_{1/2} - E_c$ becomes increasingly positive as the rate constant of the electrochemical reaction increases.[187] Pseudopolarograms may have a classical polarographic shape, or the peak height may increase continuously with deposition potential (Figure 4). This latter behavior implies that metal complexes are present that are directly reduced, i.e., they diffuse intact to the electrode surface without first dissociating in the diffusion layer to metal ion and ligand.[77]

Brown and Kowalski[187] demonstrated the application of pseudopolarography to a study of the speciation of arsenic, cadmium, and lead in various natural waters. Valenta[188] used pseudopolarography to identify lead carbonato complexes in seawater, while Bubic and Branica[189] used the same technique to study the ionic state of cadmium in seawater.

F. Modulation Waveforms

Modulating the DC voltage ramp with various waveforms provides increased sensitivity in ASV, especially when a mercury drop electrode is used.[181] At present, the most commonly used modulation waveforms in stripping analysis are differential pulse and square wave. However, AC and staircase waveforms have also been used.[190,191] The use of microcomputers[23,192,193] in electrochemical instrumentation allows a wide range of waveforms to be applied to the cell to optimize the analysis, in terms of sensitivity and selectivity, for a particular sample type. Square wave and staircase voltammetry have the advantage over differential pulse that much faster scan rates can be used,[191] up to 2 $V s^{-1}$ with a square wave frequency of 200 Hz, so that a complete voltammogram can be obtained in less than 2 s, and on a single drop in polarography.[157,194] Differential pulse voltammetry cannot utilize scan rates in excess of 5 $mV s^{-1}$, so that scanning from the zinc to the copper ASV peaks takes at least 4 min. A disadvantage of pulse and square wave techniques is that they are more affected than linear scan voltammetry by substances which adsorb on the mercury electrode,[28] although normal pulse is much less affected than differential pulse. Adsorbed layers interfere seriously in differential pulse ASV because of the multiple redox reactions that occur at the electrode during deposition.[195] It must be appreciated that different modulation waveforms will give different results in labile/inert ASV determinations.

VI. SOME SPECIATION RESULTS USING ELECTROCHEMICAL TECHNIQUES

A. Collection and Preservation of Water Samples for Speciation Measurements

Detailed instructions have been given in Chapter 1 for the contamination-free collection of water samples for trace element speciation analysis.[1,3,32,196] In general, samples should be collected in linear polyethylene bottles which are initially acid cleaned, then reserved for collecting the same type of water. Special procedures are required for some elements such as mercury and iodine.[1] The collected water sample cannot, of course, be preserved by adding acid, since this alters the element speciation. Freezing of water samples is also prohibited for trace heavy metal speciation because concentration of the solutes during the freezing process may cause hydrolysis of metal ions and other reactions which are irreversible, or only slowly reversible, on thawing.[3] The safest preservation procedure is to filter the sample immediately after collection and store the filtered sample at 4°C. The concentrations of copper, lead, cadmium, and zinc in both freshwater and seawater samples remained unchanged for several months under these conditions.[3,76,197] Reports of serious adsorption of these metals onto polyethylene containers can be traced to the use of ionic spikes (either stable or radioactive) in the water samples to measure such losses.[3] Whereas ionic metal rapidly partitions to the walls of the plastic container, the naturally present metal in pristine water samples, very little of which is in the ionic form, has a low affinity for both polyethylene and glass.[3,76] In polluted waters, however, ionic metal may persist close to the source of pollution or when the complexing capacity of the water is exceeded. In such cases, losses may occur on storage. On the other hand, storage of ultrapure waters in polyethylene containers may lead to zinc contamination from the plastic.[3,198]

Filtration and any other manipulation of a water sample should be carried out in a clean room or a clean air cupboard (Chapter 1). Electroanalysis should also be done in a clean room, glass electrolysis cells should be siliconized, and the cell and electrodes should be rinsed copiously with high-purity water.[199]

B. Selected Electrochemical Speciation Results for Some Trace Elements in Waters

1. Copper

Computer chemical models for the speciation of inorganic copper in seawater predict that the carbonato and hydroxy complexes are the dominant species.[1,3,200] The computed distri-

Table 7
CONCENTRATION AND ASV-LABILE FRACTIONS OF DISSOLVED METALS IN SURFACE SEAWATER

Metal	Concentration (ng l^{-1}) and labile fraction (%) in parentheses[1,3]	
	Open ocean	Near shore
Cu	120	350 (45)
Pb	14	250 (25)
Cd	15	75 (85)
Zn	10	1500 (50)
Ni	150	500 (70)
Fe	750	3500 (<20)
Mn	500	1500 (20)

bution of these complexes varies widely with the models used,[3,201] but the latest[199] calculations indicate that in seawater at pH 8.2, 25°C, and a total alkalinity of 2.3 meq kg^{-1}, inorganic copper exists as $CuCO_3^0$ (82%), $CuOH^+$ + $Cu(OH)_2^0$ (6.5%), $Cu(OH)(CO_3)^-$ (6.3%), $CuHCO_3^+$ (1.0%), and Cu^{2+} (2.9%). These species are all believed to be ASV labile.[27,201] In a typical freshwater, more than 90% of inorganic copper should be present as $CuCO_3$, although some is likely to be associated with colloidal particles of hydrated iron oxide.[1,3,52,202,203]

Coastal surface seawater usually has 40 to 60% of total copper (total copper in surface Pacific water off Sydney is 0.3 to 0.8 μg l^{-1})[197] present as inert organic complexes.[3,12] These complexes are so stable that they pass essentially unchanged through columns of iminodiacetate (Chelex®-100) or thiol resins.[77] It has been suggested that the copper-binding ligands are siderophores, metallothioneins, or porphyrins.[16,77] In unpolluted seawater, ASV-labile copper is usually less than 50% of total dissolved copper, even at a pH as low as 4.7 (Table 7).[3,77,197] Most of the inert copper is organically bound, but a significant fraction is inorganic, probably adsorbed on colloidal particles of hydrated iron oxide, which are perhaps coated with humic acid.[8,94,204,205] Most freshwater streams also have little ASV-labile copper, and the percentage of organically bound copper is usually high.[76,206-208]

Industrially polluted waters sometimes exhibit copper pseudopolarograms (Section V.E) which do not have a plateau, but which give continuously increasing peak currents with increase in deposition potential (Figure 3). Such behavior indicates the presence of directly reducible copper complexes.

The determination of the activity of free cupric ion using the copper ion-selective electrode is unreliable in chloride media.[209]

2. *Lead*

Computer modeling of freshwaters suggests that carbonato species, e.g., $PbCO_3$ and $Pb_2(OH)_2CO_3$ are the main (>90%) inorganic lead species,[1,3] while in seawater, speciation is divided between carbonato complexes (83%) and chloro species (11%).[210] Calculations by Turner and Whitfield[86] and by Valenta[188] suggest that in seawater the carbonato and hydroxy complexes are only partially ASV labile.

Unlike copper, lead has a stronger affinity for some inorganic adsorbents, especially iron oxide, than for organic ligands, and it is likely that in most natural waters with pH above 7, a significant fraction of the lead is associated with hydrated Fe_2O_3.[1,3] Batley and Gardner[197] found that in seawater, 40 to 80% of dissolved lead was present in the inorganic colloid fraction, while in some low-pH (pH 6.0) freshwaters, most lead appeared as an electroinactive inorganic molecular species, possibly $Pb_2(OH)_2CO_3$.[76] Most natural waters have little ASV-labile lead (Table 7).[34,211]

Alkyllead species in natural waters may be determined by ASV[212] after selective organic-phase extraction.[213]

3. *Cadmium*

In seawater, cadmium is computed to exist as the $CdCl^+$ and $CdCl_2^0$ complexes (92%), while in river water, the dominant inorganic forms are Cd^{2+} and $CdCO_3$, depending on pH.[1,3,210] A high proportion (>70%) of cadmium is ASV labile in both seawater[197] and freshwaters (Table 7).[76] Because cadmium ions are adsorbed on colloidal particles at only relatively high pH values,[1,3] very little cadmium is present as pseudocolloids. In anoxic waters, cadmium may exist as nonlabile $CdHS^+$.[3,197,210]

Cadmium contamination during analysis can occur via rubber O-rings or seals and color-code markings on pipettes.[1,3]

4. *Zinc*

The main zinc species computed to be present in seawater are Zn^{2+} (27%), chloro complexes (47%), and $ZnCO_3$ (17%), while in freshwaters the dominant inorganic forms are Zn^{2+} (50%) and $ZnCO_3$ (38%).[1,3,210] The carbonato complexes of zinc, and especially the basic carbonates, may have low ASV lability.[76,214,215] Only about 50% of total zinc in seawater and river water is ASV labile (Table 7) or extractable by ammonium pyrrolidinedithiocarbamate, even though added ionic zinc spikes are completely extractable.[8,76,215] Open ocean water contains as little as 10 ng Zn per liter at the surface,[216] although coastal seawater usually has 0.5 to 2 μg Zn per liter as a result of river inputs and sewage outfalls.[3,203,215,217]

Zinc determinations at the sub-microgram-per-liter level are extremely difficult because of contamination problems which may originate from a variety of sources including paint, skin, clothing, plastics, rubber filter membranes, reagent chemicals, and vapor from copying machines.[1,215] The HMDE generally produces more precise results for zinc than does a TMFE, because small changes in hydrogen overpotential on a mercury-coated glassy carbon electrode affect the efficiency of zinc electrodeposition.

High copper concentrations depress the zinc ASV stripping peak as a result of the formation of intermetallic compounds in the mercury.[218] This interference, however, is rare in natural water analysis.

5. *Manganese*

The natural water chemistry of manganese is dominated by nonequilibrium behavior.[3] Oxidation of Mn(II) to Mn(IV), i.e., to MnO_2, is thermodynamically favored in seawater and high-pH freshwaters, but the oxidation is extremely slow unless catalytic bacteria are present.[219] Colloidal MnO_2 is troublesome in water treatment plants because it blocks filters and causes discoloration. Both polarography[220] and ASV[49] have been used to determine labile manganese (Mn^{2+} and Mn(II) complexes) in the presence of electroinactive MnO_2. Manganese(III), formed from the oxidation of Mn(II) by algae-produced superoxide radical ($O_2^{\dot{-}}$), may also be present.[219]

Knox and Turner[41] found that in samples from the Tamar Estuary (SW England), polarographically detectable manganese varied, over a 6-month period, from <10% up to 100% of total manganese (31 to 252 μg l^{-1}) (Table 7).

VII. CORRELATION BETWEEN ASV-LABILE MEASUREMENTS AND TOXICITY

As discussed in Chapter 2, variations in the speciation of trace elements can dramatically change their toxicity. Most studies of the toxicity of heavy metals to fish and other aquatic organisms have shown that the free (hydrated) metal ion is the most toxic form,and that toxicity is related to the activity of free metal ion rather than to total metal concentration.[3,11,14,16,17,221-226] Toxicity usually decreases with increasing water hardness or salinity, presumably because of increased metal complexing by inorganic ligands.[15,227] Nature has provided aquatic animals with effective defenses against ingested heavy metals, which are eliminated via the gut[101,228] or detoxified in the liver, kidneys, and spleen by a group of high-sulfur proteins, the metallothioneins, which are synthesized in these organs in response to a heavy metal challenge.[16] These defenses allow the animal to cope with quite high levels of heavy metals in the food chain and sediment; toxicity occurs only with "spillover", i.e., when the metal intake exceeds the ability of the body to synthesize metallothionein. Evolution has not, however, equipped animals to tolerate free metal ion in the water which contacts their gills or other exposed biomembranes.[16] Unpolluted seawater or river water contains very little free metal ion, most of the dissolved metal being present as nontoxic complexes (e.g., with fulvic acid) or adsorbed on colloidal particles (e.g., humate-coated Fe_2O_3 or fibrils[229]). Natural waters use these detoxification mechanisms to convert free metal ions to nontoxic forms, but considerable damage can be caused close to the source of pollution if the complexing capacity (Section VIII) of the water is exceeded.

Copper(II) ions bind initially to marine phytoplankton with a stability constant, log β_1, in the range 10 to 12, complexing apparently occurring via protein amino and carboxylic acid groups.[230] Copper is then transported across the membrane by a carrier protein (facilitated diffusion),[231,232] where it reacts with a thiol (possibly glutathione) in the cytosol or on the interior surface of the membrane, and is reduced to copper(I).[230] Reaction with cytosolic thiols and thiol-containing enzymes may be a common toxic effect of heavy metals, although deactivation of enzymes such as catalase by metal substitution may also be involved.[201,230]

Although there is considerable evidence that free metal ion is the most toxic metal form, the situation is not completely clear.[15,17] Some studies suggest that other species, such as the copper-hydroxy complex[232] and the copper complexes of citrate and ethylenediamine,[233] are also toxic. In addition, lipid-soluble copper[11,16,79] and mercury[16] complexes are extremely toxic, and a step to measure lipid-soluble metal complexes should be included in all trace element speciation schemes for polluted waters (Section II.E).

Attempts to use ASV-labile measurements to determine the toxic fraction of a metal have met with varied success.[3,11,33,223,234,235] Young et al.,[33] using larval shrimp as a test species, found good correlation between ASV-labile copper and toxicity, whereas Srna et al.[234] reported that ASV gave values which were only half those measured by bioassay. Florence et al.[11] found that ASV-labile copper, determined in seawater using a low deposition potential, correlated well with copper toxicity towards the marine diatom *Nitzschia closterium* when natural complexing agents, including fulvic, humic, tannic, and alginic acids, and hydrated iron oxide, were present in the growth medium. However, when synthetic ligands such as nitrilotriacetic acid (NTA), oxine, or ethylxanthate were present, there was no sensible correlation (Table 8). The fraction of total dissolved copper removed by a column of Chelex®-100 resin grossly overestimated the toxic fraction (Chapter 3),[11] although an adsorbent column consisting of aluminum hydroxide deposited in a cation-exchange resin column removes a fraction of copper from both seawater and freshwater that is very similar to the toxic fraction determined by algal assay.[236] Hering et al.[237] found good correlation between bacterial bioassay and fixed-potential amperometry.

ASV-labile metal may therefore be a simple and reasonable method for measuring the toxic fractions of metals in natural waters, but could be inapplicable if synthetic ligands are present.

Table 8
CORRELATION BETWEEN ASV-LABILE AND TOXIC FRACTIONS OF COPPER IN SEAWATER USING THE MARINE DIATOM *NITZSCHIA CLOSTERIUM*

Ligand and concentration[a]	Copper ($M \times 10^7$)	ASV-labile fraction[b] (%)		Toxic fraction[c] (%)
		−0.6 V	−1.3 V	
Fulvic acid, 1×10^{-5} *M*	3.2	1.5	2.9	7.5
Tannic acid, 6×10^{-7} *M*	3.2	5.5	10.5	12.5
Iron-humic colloid[d]	3.2	70	74	60
NTA,[e] 2×10^{-5} *M*	3.2	100	100	20
LAS,[f] 0.5 mg l^{-1}	3.2	65	100	25
Oxine, 5×10^{-8} *M*	0.32	64	100	>100
DMP,[g] 5×10^{-8} *M*	0.32	2.5	—	>100
Ethylxanthate, 2×10^{-6} *M*	3.2	10.5	48	>100

[a] Data from Reference 10.
[b] pH 8.2, with deposition potential of −0.6 V or −1.3 V vs. Ag/AgCl.
[c] Fraction of added copper appearing toxic compared with ligand-free solution.
[d] 1.0 mg Fe per liter + 5.3 mg humic acid per liter.
[e] NTA = nitrilotriacetic acid.
[f] LAS = linear alkylbenzene sulfonate.
[g] DMP = 2,9-dimethyl-1,10-phenanthroline.

VIII. COMPLEXING CAPACITY

Natural waters contain a variety of metal complexing agents, including fulvic, humic, and tannic acids, lignin, and colloidal particles of Fe_2O_3, Al_2O_3, and MnO_2.[1,238] Polluted waters may contain additional natural and synthetic compounds. These ligands are usually well in excess of the metals present, and determination of this excess ''metal complexing capacity'' is an important water quality parameter because it is a measure of the concentration of heavy metal that can be discharged to a waterway before free metal ion appears.[1,3,239-251]

Complexing capacity is determined by titrating the water sample with a heavy metal ion; copper(II) is usually chosen as the titrant because it is a common heavy metal ion, highly toxic to aquatic organisms.[68] Complexing capacity is then defined as the concentration of cupric ion (moles per liter) that must be added to a water sample before free Cu^{2+} appears. It reflects the concentration of organic and inorganic substances in the water sample, both molecular and colloidal, that bind (and detoxify) copper ions. Near-shore surface seawater has a copper complexing capacity of about 2×10^{-8} *M*, while river waters range from 1 to 50×10^{-8} *M* (Table 9).

Methods used to measure complexing capacity include bioassays, ion exchange on resins or MnO_2, ion selective electrode potentiometry, copper salt solubilization, chemical exchange, amperometry, and voltammetry.[1,3,237,239,241,252-254] Of these methods, voltammetry, using an ASV titration, has been most widely applied. ASV titration consists of adding aliquots of a standard copper solution to the sample and measuring the copper ASV peak until the slope of the peak current-copper concentration curve increases to that found for ionic copper (Figure 7). Assuming a 1:1 copper:ligand complex, the complexing capacity (C) and the apparent stability constant (*K) can be found from a plot of the relationship[240]

$$[Cu]/(Cu_T - [Cu]) = [Cu]/C + 1/{*K}\ C \qquad (29)$$

Table 9
COMPLEXING CAPACITY OF SOME NATURAL WATERS

Source of water	Complexing capacity[a] ($M \times 10^6$ Cu^{2+})	pH[b]	log *K[c]
Lake Ontario	0.34	7.4	8.6
Chapel Hill Lake	31	6.0	5.0
Swiss Lakes	2.7	8.8	10.9
Newport River	0.87	7.0	9.7
Neuse River	0.21	6.8	9.5
Magela Creek, Australia	0.10—0.46	6.0	7.6
Pacific Ocean, coastal	0.02—0.2	4.8	—

[a] ASV titration with Cu^{2+}.[1,191]
[b] pH of titration.
[c] Conditional stability constant of copper complex.

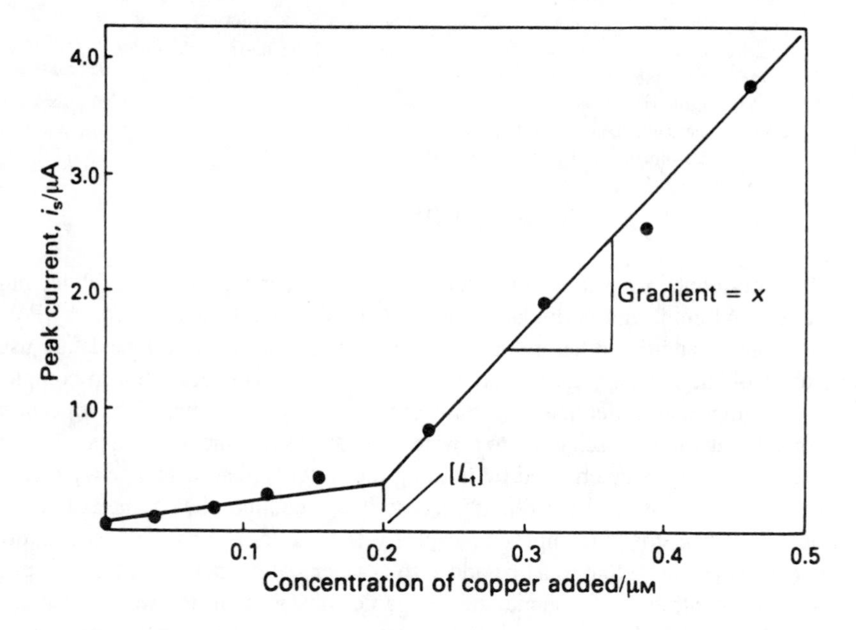

FIGURE 7. Complexing capacity titration of a natural freshwater.

where [Cu] is the concentration of free cupric ion, and Cu_T is total copper concentration. Some typical values for C and *K are shown in Table 9.

There are several problems associated with the ASV titration method for determining complexing capacity: (1) some copper complexes, such as Cu-NTA, although thermodynamically stable, are kinetically labile and dissociate extensively in the diffusion layer, the complex appearing as free metal ion; these kinetic currents can be corrected for to some extent,[202,255,256] but the procedure required is quite complex;[17] (2) organic matter adsorbed on the electrode may cause a depression of the metal ASV peak by hindering electrodeposition, even though no actual complex formation takes place; and (3) formation of the copper complex may be slow, and several hours may need to be allowed between the additions of copper titrant; it is often better to add increasing aliquots of standard copper solution to a series of flasks containing a fixed volume of sample, and allow to stand overnight before ASV measurement.

The problem of electrode fouling by organics could be minimized by the use of linear scan voltammetry at a rapidly dropping mercury electrode or a streaming mercury electrode, providing the samples had a sufficiently high complexing capacity. Interference from adsorbed organic matter increases in the order: differential pulse polarography (DME) $<$ normal pulse ASV (HMDE) $<$ linear scan ASV (TMFE) $<$ differential pulse ASV (HMDE).[195,257] In ASV, the electrode is exposed to the organic matter for the period of the deposition time, whereas in polarography, exposure is only for the drop time. In linear scan voltammetry, the metal ion has to cross the adsorbed organic layer only once during deposition, whereas in differential pulse techniques, where multiple redox reactions are involved, many crossings of the adsorbed layer must occur (Section V.F). Electrode coated with Nafion, cellulose acetate, or other protective films are also useful for ASV complexing capacity titrations. Sharper titration breaks are obtained because the coatings minimize interference by non-complexing, surface-active organics.[258]

Van den Berg[250] has described a ligand-exchange CSV method for measuring complexing capacity, based on competition between the natural ligands and catechol for copper ions, followed by cathodic stripping of the adsorbed copper-catechol complex (Section V.C). An estimate of the copper-ligand conditional stability constant can also be obtained. This procedure has the advantages that the problem of dissociation of the copper complex in the double layer is eliminated, and interference by adsorption of organics may be less severe if the copper-catechol complex is preferentially adsorbed. However, because of the high stability constant of the copper-catechol complex,[195] the method would measure only those ligands which form relatively stable complexes with copper. Waite and Morel[249] described a novel amperometric titration method for complexing capacity using copper(II) as titrant and high chloride media to stabilize copper(I).

Ion-selective electrodes measure the activity of free, hydrated metal ion, and no other species. If a copper ion-selective electrode is calibrated with a standard $CuSO_4$ solution in noncomplexing media (e.g., nitrate or perchlorate), then even simple complexes such as $CuCl^+$, $Cu(OH)_2$, and $CuCO_3$ will be included in the complexing capacity measurement, since they are not sensed by the electrode. This explains why literature results for complexing capacity determined by copper(II) titrations using an ion-selective electrode for end point detection are often much higher than ASV values for similar waters.[1,240]

Before any method for measuring complexing capacity is chosen over others, it should be shown that it gives a reasonable correlation with bioassay techniques, otherwise it will have little value for ecotoxicological studies. The ideal method for measuring complexing capacity would be one where the affinity of the analytical probe for the metal ion titrant would be the same as that of a biomembrane, e.g., the gill of a fish, for the metal ion. If this ideal situation could be achieved, the equilibrium constant for the reaction,

$$ML + P \rightleftharpoons MP + L \tag{30}$$

where P is the analytical probe, and ML is the complex formed between the metal titrant and the natural ligand, would be the same as the constant for the reaction,

$$ML + B \rightleftharpoons MB + L \tag{31}$$

where B is a biomembrane.

IX. CONCLUSIONS AND RECOMMENDATIONS FOR FUTURE RESEARCH

Speciation analysis is essential to an understanding of the biological and geochemical cycling of trace elements; simple total element analysis provides little information about these processes. Dividing trace elements in waters into different behavioral classes (speciation "boxes") is a difficult task when the total concentration is at, or below, the microgram-per-liter level. Electroanalysis, especially ASV, is perhaps the most powerful technique available for this exacting branch of analytical chemistry. It must be appreciated, however, that ASV and polarography are dynamic techniques and cannot possibly measure the "natural" speciation of a trace element in a water sample, because the measurement itself disturbs the equilibrium. All electrochemical speciation results are, therefore, operationally defined. This characteristic of electroanalysis may actually be an advantage, since the interaction of a trace metal with a biomembrane is also a dynamic process, and it should be possible to choose the solution and electrochemical parameters so that the kinetics of electrodeposition are similar to the rate of uptake of a trace metal by a biological system. An important point here is that the effective measurement time of different electrochemical techniques (ASV, polarography, linear scan, pulse) varies considerably, and hence the kinetic contribution of metal complexes to the analytical signal will also vary with the method used. Full analytical details (including calculation of the diffusion layer thickness) must therefore be reported in all published research on trace element speciation.[24]

Two broad areas of the methodology of electrochemical speciation analysis of waters would benefit from further research. These are

1. The relationship between trace element speciation and aquatic toxicity. Much of the interest in speciation stems from the knowledge that the toxicity of different physicochemical forms of an element varies enormously, and the speciation analysis could possibly be used to estimate the potential toxicity of a water system. There is little point in developing, from a purely chemical viewpoint, "new" speciation schemes without consideration of their application. If the ultimate aim is directed towards ecotoxicology, then development of the speciation method should be carried out in parallel with bioassays in an attempt to achieve the best correlation. On the other hand, if the research aim is to study geochemical cycling, then the speciation procedure should be tailored to mimic, as closely as possible, the relevant adsorption and precipitation processes. These remarks also apply to the development of new methods for measuring complexing capacity; bioassays must be carried out hand in hand with the chemistry to ensure the relevance of the data.
2. Electrochemical speciation measurements may often be affected by extraneous substances, especially surface-active compounds, in solution. In ASV trace metal speciation, the interpretation of the results is greatly simplified if it can be assumed that the deposition step alone controls the results, i.e., that the kinetics of metal deposition controls the magnitude of the stripping peak, and that stripping kinetics are unaffected by ligands that are present in the sample, but not the standard solution. Perhaps the best way to ensure that this situation exists is to use a medium exchange technique. This involves depositing metals from the sample solution, but replacing it with a simple electrolyte (e.g., acetate buffer) before the stripping step. Research is required to design better cells for medium exchange and to determine the usefulness and application of the procedure.[99]

Adsorption of surface-active substances (e.g., humic matter) from the sample solution onto the mercury electrode is one of the most serious complications in electrochemical speciation measurements. Both deposition and stripping currents may be decreased in a non-predictable manner, and there is often a nonlinear relationship between peak current and deposition time. Because the build-up of an adsorption layer on an electrode is a relatively slow process, adsorption has less influence when short ASV deposition times (and short drop times in polarography) are used. To overcome the problem of adsorption, the streaming mercury electrode (SME) should be investigated for speciation analysis when total metal concentrations are sufficiently high to allow its use.[24] Because of the extremely rapid renewal of the electrode in a SME, adsorptive processes and metal complexes with slow dissociation kinetics usually have little effect on the diffusion current. The SME may be especially useful for complexing capacity titrations.

However, the most promising technique for the elimination of interference by adsorption is to cover the thin mercury film electrode with a permselective coating of Nafion, cellulose acetate, or similar materials. This type of covered electrode will have extensive applications in ASV analysis of waters and biological fluids and complexing capacity measurements, and will be particularly useful in flow-through cells for continuous monitoring of waters where electrode fouling is a vexing problem.[148] The combination of permselective-coated electrodes and small, portable, microprocessor-based, battery-operated electroanalytical instruments will greatly expand the applications of electroanalysis to organic compounds as well as trace inorganics.

REFERENCES

1. **Florence, T. M. and Batley, G. E.,** Chemical speciation in natural waters, *CRC Crit. Rev. Anal. Chem.*, 9, 219, 1980.
2. **De Mora, S. J. and Harrison, R. M.,** Physicochemical speciation of inorganic compounds in environmental media, *Hazard Assess. Chem. Curr. Dev.*, 3, 1, 1984.
3. **Florence, T. M.,** The speciation of trace elements in waters, *Talanta*, 29, 345, 1982.
4. **Florence, T. M.,** Electrochemical approaches to metal speciation, *Anal. Proc.*, 20, 552, 1983.
5. **Florence, T. M.,** Recent advances in stripping analysis, *J. Electroanal. Chem.*, 168, 207, 1984.
6. **Burton, J. D.,** Physicochemical limitations in experimental investigations, *Philos. Trans. R. Soc. London Ser. B*, 286, 443, 1979.
7. **Tipping, E.,** Some aspects of the interactions between particulate oxides and aquatic humic substances, *Mar. Chem.*, 18, 161, 1986.
8. **Batley, G. E. and Florence, T. M.,** Determination of the chemical forms of dissolved cadmium, lead and copper in seawater, *Mar. Chem.*, 4, 347, 1976.
9. **Catanzaro, E. J.,** Some relationships between exchangeable copper and lead and particulate matter in a sample of Hudson River water, *Environ. Sci. Technol.*, 10, 386, 1976.
10. **Allen, H. E., Boolayangoor, C., and Noll, K. E.,** Changes in physicochemical forms of lead and calcium added to freshwater, *Environ. Int.*, 7, 337, 1982.
11. **Florence, T. M., Lumsden, B. G., and Fardy, J. J.,** Evaluation of some physicochemical techniques for the determination of the fraction of dissolved copper toxic to the marine diatom *Nitzschia closterium*, *Anal. Chim. Acta*, 151, 281, 1983.
12. **Mackey, D. J.,** Metal organic complexes in seawater — an investigation of naturally occurring complexes of Cu, Zn, Fe, Mg, Ni, Cr, Mn, and Cd using high-performance liquid chromatography with atomic fluorescence detection, *Mar. Chem.*, 13, 169, 1983.
13. **Morel, F. and Morgan, J.,** A numerical method for computing equilibria in aqueous chemical systems, *Environ. Sci. Technol.*, 6, 58, 1972.
14. **Whitfield, M. and Turner, D. R.,** Critical assessment of the relationship between biological thermodynamic and electrochemical availability, in *Chemical Modelling in Aqueous Systems*, Jenne, E. A., Ed., ACS Symp. Ser. 93, American Chemical Society, Washington, D.C., 1979, 657.
15. **Borgmann, U.,** Metal speciation and toxicity of free metal ions to aquatic biota, in *Aquatic Toxicology*, Nriagu, J. O., Ed., John Wiley & Sons, New York, 1983, 47.
16. **Florence, T. M.,** Trace element speciation and aquatic toxicology, *Trends Anal. Chem.*, 2, 162, 1983.

17. **Turner, D. R.,** Relationships between biological availability and chemical measurements, *Met. Ions Biol. Syst.*, 18, 137, 1984.
18. **Florence, T. M., Lumsden, G. G., and Fardy, J. J.,** Algae as indicators of copper speciation, in *Complexation of Trace Metals in Natural Waters,* Kramer, C. J. and Duinker, J. C., Eds., Martinus Nijhoff, The Hague, 1984, 411.
19. **Florence, T. M., Stauber, J. L., and Mann, K.,** The reaction of copper-2,9-dimethyl-1,10-phenanthroline with hydrogen peroxide, *J. Inorg. Biochem.*, 24, 243, 1985.
20. **Chapman, B. M., Jones, D. R., and Jung, R. F.,** Processes controlling metal ion attenuation in acid mine drainage streams, *Geochim. Cosmochim. Acta,* 47, 1957, 1983.
21. **Stauber, J. L. and Florence, T. M.,** Mechanism of toxicity of ionic copper and copper complexes to algae, *Mar. Biol.*, 94, 511, 1987.
22. **Batley, G. E.,** Electroanalytical techniques for the determination of heavy metals in seawater, *Mar. Chem.*, 12, 107, 1983.
23. **Bond, A. M.,** *Modern Polarographic Techniques in Analytical Chemistry,* Marcel Dekker, New York, 1980.
24. **Florence, T. M.,** Electrochemical approaches to trace element speciation in waters, a review, *Analyst,* 111, 489, 1986.
25. **Turner, D. R. and Whitfield, M.,** The reversible electrodeposition of trace metal ions from multi-ligand systems. I. Theory, *J. Electroanal. Chem.*, 103, 73, 1979.
26. **Davison, W.,** Defining the electroanalytically measured species in a natural water sample, *J. Electroanal. Chem.*, 87, 395, 1978.
27. **Davison, W. and Whitfield, M.,** Modulated polarographic and voltammetric techniques in the study of natural water chemistry, *J. Electroanal. Chem.*, 75, 763, 1977.
28. **Batley, G. E. and Florence, T. M.,** An evaluation and comparison of some techniques of anodic stripping voltammetry, *J. Electroanal. Chem.*, 55, 23, 1974.
29. **Van Leeuwen, H. P.,** Kinetic classification of metal complexes in electroanalytical speciation, *J. Electroanal. Chem.*, 99, 93, 1979.
30. **Buffle, J.,** Calculation of the surface concentration of the oxidized metal during the stripping step in the anodic stripping techniques and its influence on speciation measurements in natural waters, *J. Electroanal. Chem.*, 125, 273, 1981.
31. **Zirino, A. and Kouvanes, S. P.,** Anodic stripping peak currents: electrolysis potential relationships for reversible systems, *Anal. Chem.*, 49, 56, 1977.
32. **Batley, G. E. and Gardner, D.,** Sampling and storage of natural waters, *Water Res.*, 11, 745, 1977.
33. **Young, J. S., Gurtisen, J. M., Apts, C. W., and Crecelius, E. A.,** The relationship between the copper complexing capacity of seawater and copper toxicity in shrimp zoeae, *Mar. Environ. Res.*, 2, 265, 1979.
34. **Chau, Y. K. and Lum Shue-Chan, K.,** Determination of labile and strongly bound metals in lake water, *Water Res.*, 8, 383, 1974.
35. **Benes, P. and Majer, V.,** *Trace Chemistry of Aqueous Solutions,* Elsevier, Amsterdam, 1980.
36. **Florence, T. M.,** Determination of iron by anodic stripping voltammetry, *J. Electroanal. Chem.*, 26, 293, 1970.
37. **Leon, L. E. and Sawyer, D. T.,** Simultaneous determination of iron(II) and iron(III) at micromolar concentrations by differential pulse polarography, *Anal. Chem.*, 53, 706, 1981.
38. **Batley, G. E. and Matousek, J. P.,** Determination of chromium speciation in natural waters by electrodeposition on graphite tubes for electrothermal atomization, *Anal. Chem.*, 52, 1570, 1980.
39. **Batley, G. E. and Florence, T. M.,** Determination of thallium in natural waters by anodic stripping voltammetry, *J. Electroanal. Chem.*, 61, 205, 1975.
40. **Phillips, S. L. and Shain, I.,** Application of stripping analysis to the trace determination of tin, *Anal. Chem.*, 37, 262, 1962.
41. **Knox, S. and Turner, D. R.,** Polarographic measurement of manganese(II) in estuarine waters, *Estuarine Coastal Mar. Sci.*, 10, 317, 1980.
42. **Henry, F. T., Kirch, T. O., and Thorpe, T. M.,** Determination of trace level arsenic(III), arsenic(V), and total inorganic arsenic by differential pulse polarography, *Anal. Chem.*, 51, 215, 1979.
43. **Hamilton, T. W., Ellis, J., and Florence, T. M.,** Determination of selenium and tellurium in electrolytic copper by anodic stripping voltammetry at a gold film electrode, *Anal. Chim. Acta,* 110, 87, 1979.
44. **Heyrovsky, J. and Kuta, J.,** *Principles of Polarography,* Academic Press, New York, 1966.
45. **Kubota, H.,** Determination of the stoichiometry of uranium dioxide. Polarographic determination of uranium(VI) in uranium dioxide, *Anal. Chem.*, 32, 610, 1960.
46. **Butler, E. C. V. and Smith, J. D.,** Iodine speciation in sea waters: the analytical use of ultra-violet photo-oxidation and differential pulse polarography, *Deep-Sea Res.*, 27, 489, 1980.
47. National Academy of Sciences, Chromium, Committee on Biological Effects of Atmospheric Pollutants, Division of Medical Sciences, National Research Council, U.S. National Academy of Sciences, Washington, D.C., 1974.

48. **Mertz, W. and Cornatzer, W. E.,** *Newer Trace Elements in Nutrition,* Marcel Dekker, New York, 1971.
49. **O'Halloran, R. J.,** Anodic stripping voltammetry of manganese in seawater at a mercury film electrode, *Anal. Chim. Acta,* 140, 51, 1982.
50. **Orvini, E., Zerlina, T., Gallorini, M., and Speziali, M.,** Determination of chromium(VI) and chromium(III) traces in natural waters by neutron activation analysis, *Radiochem. Radioanal. Lett.,* 43, 173, 1980.
51. **Sibley, T. H. and Morgan, J. J.,** Equilibrium speciation of trace metals in freshwater: seawater mixtures, in *Proc. 1st Int. Conf. on Heavy Metals in the Environment,* Vol. 1, Hutchison, T., Ed., University of Toronto, Toronto, 1975, 319.
52. **Nelson, A. and Mantaoura, R. F. C.,** Voltammetry of copper species in estuarine waters. I. Electrochemistry of copper species in chloride media, *J. Electroanal. Chem.,* 164, 237, 1984.
53. **Figura, P. and McDuffie, B.,** Use of Chelex resin for determination of labile trace metal fractions in aqueous ligand media and comparison of the method with anodic stripping voltammetry, *Anal. Chem.,* 51, 120, 1979.
54. **Wang, J.,** *Stripping Analysis, Principles, Instrumentation, and Applications,* VCH Publishers, Deerfield Beach, FL, 1985.
55. **Batley, G. E.,** Current status of metal speciation studies in natural waters, in *Trace Element Speciation in Surface Waters and its Ecological Implications,* Leppard, G. G., Ed., Plenum Press, New York, 1983, 17.
56. **De Mora, S. J. and Harrison, R. M.,** The use of physical separation techniques in trace metal speciation studies, *Water Res.,* 17, 723, 1983.
57. **Steinnes, E.,** Physical separation techniques in trace element speciation studies, in *Trace Element Speciation in Surface Waters and its Ecological Implications,* Leppard, G. G., Ed., Plenum Press, New York, 1983, 37.
58. **Brugmann, C.,** Electrochemical speciation of trace metals in sea water, *Sci. Total Environ.,* 37, 41, 1984.
59. **Skogerboe, R. K., Wilson, S. A., and Osteryoung, J. G.,** Scheme for classification of heavy metal species in natural waters. Comments, *Anal. Chem.,* 52, 1960, 1980.
60. **Laxen, D. P. and Harrison, R. M.,** Cleaning methods for polythene containers prior to the determination of trace metals in freshwater samples, *Anal. Chem.,* 53, 345, 1981.
61. **Kramer, C. J. M., Yu, G. H., and Duinker, J. C.,** Possibilities for misinterpretation in ASV-speciation studies of natural waters, *Fresenius Z. Anal. Chem.,* 317, 383, 1984.
62. **Goncalves, M. L. S., Sigg, L., and Stumm, W.,** Voltammetric methods for distinguishing between dissolved and particulate metal ion concentrations in the presence of hydrous oxides. A case study on lead(II), *Environ. Sci. Technol.,* 19, 171, 1985.
63. **Florence, T. M. and Batley, G. E.,** Heavy metal species in natural waters — reply to R. Skogerboe et al., *Anal. Chem.,* 52, 1962, 1980.
64. **Copeland, T. R., Christie, J. H., Skogerboe, R. K., and Osteryoung, R. A.,** Effect of supporting electrolyte concentration in pulsed stripping voltammetry at the thin film mercury electrode, *Anal. Chem.,* 45, 995, 1973.
65. **Laxen, D. P. H. and Harrison, R. M.,** A scheme for the physicochemical speciation of trace metals in freshwater samples, *Sci. Total Environ.,* 19, 59, 1981.
66. **Nurnberg, H. W. and Raspor, B.,** Applications of voltammetry in studies of the speciation of heavy metals by organic chelators in sea water, *Environ. Technol. Lett.,* 2, 457, 1981.
67. **Nelson, A.,** Voltammetry of copper species in estuarine waters, *Anal. Chim. Acta,* 169, 273, 1985.
68. **Bhat, G. A., Saar, R. A., Smart, R. B., and Weber, J. H.,** Titration of soil-derived fulvic acid by copper(II) and measurement of free copper(II) by anodic stripping voltammetry and copper(II) selective electrodes, *Anal. Chem.,* 53, 2275, 1981.
69. **Jacobsen, E. and Lindseth, H.,** Effects of surfactants in differential pulse polarography, *Anal. Chim. Acta,* 86, 123, 1976.
70. **Batley, G. E. and Florence, T. M.,** The effect of dissolved organics on the stripping voltammetry of seawater, *J. Electroanal. Chem.,* 72, 121, 1976.
71. **Bond, A. M., Greenhill, H. B., Heritage, I. D., and Reust, J. B.,** Development of a microprocessor-based electrochemical instruments in interfaced to a microcomputer system for differential pulse stripping voltammetry in different time domains, *Anal. Chim. Acta,* 165, 209, 1984.
72. **Poon, M. and McCreery, R. L.,** Repetitive in situ renewal and activation of carbon and platinum electrodes: applications to pulse voltammetry, *Anal. Chem.,* 59, 1615, 1987.
73. **Hoyer, B., Florence, T. M., and Batley, G. E.,** Application of Nafion-coated glassy carbon electrodes in anodic stripping voltammetry, Proc. 9th Australian Symp. on Analytical Chemistry, Royal Australian Chemistry Institute, Sydney, 1987, 565.
74. **Florence, T. M.,** Determination of bismuth in marine samples by anodic stripping voltammetry, *J. Electroanal. Chem.,* 49, 253, 1974.

75. **Batley, G. E. and Farrar, Y. J.,** Irradiation techniques for the release of bound heavy metals in natural waters and blood, *Anal. Chim. Acta,* 99, 283, 1978.
76. **Florence, T. M.,** Trace metal species in fresh waters, *Water Res.,* 11, 681, 1977.
77. **Florence, T. M.,** Developments of physicochemical speciation procedures to investigate the toxicity of copper, lead, cadmium and zinc towards aquatic biota, *Anal. Chim. Acta,* 141, 73, 1982.
78. **Blutstein, H. and Smith, J. D.,** Distribution of species of Cu, Pb, Zn and Cd in a water profile of the Yarra River estuary, *Water Res.,* 12, 119, 1978.
79. **Ahsanullah, M. and Florence, T. M.,** Toxicity of copper to the marine amphipod *Allorchestes compressa* in the presence of water and lipid-soluble ligands, *Mar. Biol.,* 84, 41, 1984.
80. **Florence, T. M. and Batley, G. E.,** Removal of trace metals from seawater by a chelating resin, *Talanta,* 22, 201, 1975.
81. **Florence, T. M. and Batley, G. E.,** Trace metal speciation in seawater: removal of trace metals from seawater by a chelating resin, *Talanta,* 23, 179, 1976.
82. **Figura, P. and McDuffie, B.,** Determination of labilities of soluble trace metal species in aqueous environmental samples by anodic stripping voltammetry and chelex column and batch methods, *Anal. Chem.,* 52, 1433, 1980.
83. **Thomassen, Y., Larsen, B. V., Langmyhr, F. J., and Lund, W.,** The application of electrode position techniques to flameless atomic absorption spectrometry. IV. Separation and preconcentration on graphite, *Anal. Chim. Acta,* 83, 103, 1976.
84. **Batley, G. E. and Matousek, J. P.,** Determination of heavy metals in seawater by atomic absorption spectrometry after electrodeposition on pyrolytic graphite coated tubes, *Anal. Chem.,* 49, 2031, 1977.
85. **Batley, G. E.,** *In situ* electrodeposition for the determination of lead and cadmium in sea water, *Anal. Chim. Acta,* 124, 121, 1981.
86. **Turner, D. R. and Whitfield, M.,** The reversible electrodeposition of trace metal ions from multi-ligand systems. II. Calculations on the electrochemical availability of lead at trace levels in seawater, *J. Electroanal. Chem.,* 103, 61, 1979.
87. **Bond, A. M. and Jones, R. D.,** The analytical performance of direct current, normal pulse and differential pulse polarography with static mercury drop electrodes, *Anal. Chim. Acta,* 121, 1, 1980.
88. **Tuschall, J. R. and Brezonik, P. L.,** Evaluation of the copper anodic stripping voltammetry complexometric titration for complexing capacities and conditional stability constants, *Anal. Chem.,* 53, 1986, 1981.
89. **Guy, R. D. and Chakrabarti, C. L.,** Analytical techniques for speciation of trace metals, in *Proc. 1st Int. Conf. on Heavy Metals in the Environment,* Vol. 1, Hutchison, T., Ed., University of Toronto, Toronto, 1975, 319.
90. **Olson, D. L. and Shuman, M. S.,** Copper dissociation from estuarine humic materials, *Geochim. Cosmochim. Acta,* 49, 1371, 1985.
91. **Lumsden, B. R. and Florence, T. M.,** A new algal assay procedure for the determination of the toxicity of copper species in seawater, *Environ. Technol. Lett.,* 4, 271, 1983.
92. **Goncalves, M. D. L. S., Sigg, L., Reutlinger, M., and Stumm, W.,** Metal ion binding by biological surfaces: voltammetric assessment in the presence of bacteria, *Sci. Total Environ.,* 60, 105, 1987.
93. **Florence, T. M.,** Degradation of protein disulphide bonds in dilute alkali, *Biochem. J.,* 189, 507, 1980.
94. **Florence, T. M. and Batley, G. E.,** Determination of copper in seawater by anodic stripping voltammetry, *J. Electroanal. Chem.,* 75, 791, 1977.
95. **Morrison, G. M. P.,** Private communication, 1987.
96. **Reignier, M. and Buess-Herman, C.,** On the coupling of anodic stripping with electrocatalysis: application to the determination of copper traces, *Fresenius Z. Anal. Chem.,* 317, 257, 1984.
97. **Ariel, M., Eisner, U., and Gottesfeld, S.,** Trace analysis by anodic stripping voltammetry. II. The method of medium exchange, *J. Electroanal. Chem.,* 7, 307, 1964.
98. **Desimoni, E., Palmisano, F., and Sabbatini, L.,** Simultaneous determination of tin and lead at the parts-per-billion level by coupling differential pulse anodic stripping voltammetry with a matrix exchange method, *Anal. Chem.,* 52, 1889, 1980.
99. **Florence, T. M. and Mann, K. J.,** The use of medium exchange in anodic stripping voltammetry, *Anal. Chim. Acta,* 200, 305, 1987.
100. **Langston, W. J. and Bryan, G. W.,** The relationships between metal speciation in the environment and bioaccumulation in aquatic organisms, in *Complexation of Trace Metals in Natural Waters,* Kramer, C. J. and Duinker, J. C., Eds., Martinus Nijhoff, The Hague, 1983, 375.
101. **Boudou, A., Georgescauld, D., and Desmazes, J. P.,** Ecotoxicological role of the membrane barriers in transport and bioaccumulation of mercury compounds, in *Aquatic Toxicology,* Nriagu, J. O., Ed., John Wiley & Sons, New York, 1983, 117.
102. **Williams, R. J. P.,** Natural selection of the chemical elements, *Proc. R. Soc. London Ser. B,* 213, 361, 1981.

103. **Mart, L.,** Prevention of contamination and other accuracy risks in voltammetric trace metal analysis of natural waters. I. Preparatory steps, filtration and storage of water samples, *Fresenius Z. Anal. Chem.*, 296, 350, 1979.
104. **Mart, L., Nurnberg, H. W., and Valenta, P.,** Prevention of contamination and other accuracy risks in voltammetric trace metal analysis of natural waters. III. Voltammetric ultratrace analysis with a multicell system designed for clean bench working, *Fresenius Z. Anal. Chem.*, 300, 350, 1980.
105. **Florence, T. M.,** Comparison of linear scan and differential pulse anodic stripping voltammetry at a thin mercury film glassy carbon electrode, *Anal. Chim. Acta*, 119, 217, 1980.
106. **Florence, T. M.,** Anodic stripping voltammetry with a glassy carbon electrode mercury-plated in situ, *J. Electroanal. Chem.*, 27, 273, 1970.
107. **Van der Linden, W. E. and Dieker, J. W.,** Glassy carbon as electrode material in electroanalytical chemistry, *Anal. Chim. Acta*, 119, 1, 1980.
108. **Cowland, F. C. and Lewis, J. C.,** Vitreous carbon — a new form of carbon, *J. Mater. Sci.*, 2, 507, 1967.
109. **Clem, R. G.,** Cause of loss of hydrogen-overvoltage on graphite electrodes used for anodic stripping voltammetry, *Anal. Chem.*, 47, 1778, 1975.
110. **Mann, K. J. and Florence, T. M.,** Trace element speciation by anodic stripping voltammetry: the effects of added mercuric and acetate ions, *Sci. Total Environ.*, 60, 67, 1987.
111. **Stauber, J. L. and Florence, T. M.,** The determination of trace metals in sweat by anodic stripping voltammetry, *Sci. Total Environ.*, 60, 263 1987.
112. **Yoshimura, T. and Okazaki, S.,** Studies on porous layer electrochemistry using a filter paper. II. Fundamental and analytical applications, *Fresenius Z. Anal. Chem.*, 316, 777, 1983.
113. **Magjer, T. and Branica, M.,** A new electrode system with efficient mixing of electrolyte, *Croat. Chem. Acta*, 49, L1, 1977.
114. **Kramer, C. J. M., Guo-Hui, Y., and Duinker, J. C.,** Optimization and comparison of four mercury working electrodes in speciation studies by differential pulse anodic stripping voltammetry, *Anal. Chim. Acta*, 164, 163, 1984.
115. **Heineman, W. R.,** Analytical electrochemistry: methodology and applications of dynamic techniques, *Anal. Chem.*, 52, 139R, 1980.
116. **Seitz, W. R., Jones, R., Klatt, L. N., and Mason, W. P.,** Anodic stripping voltammetry at a tubular mercury-covered graphite electrode, *Anal. Chem.*, 45, 870, 1973.
117. **Blaedel, W. J. and Wang, J.,** Anodic stripping voltammetry at a reticulated mercury vitreous carbon electrode, *Anal. Chem.*, 51, 1724, 1979.
118. **Lieberman, S. H. and Zirino, A.,** Anodic stripping voltammetry of zinc in seawater with a tubular mercury-graphite electrode, *Anal. Chem.*, 46, 20, 1974.
119. **Wang, J. and Dewald, H. D.,** Dual coulometric-voltammetric cells for on-line stripping analysis, *Anal. Chem.*, 55, 933, 1983.
120. **Bixler, J. W., Bond, A. M., Lay, P. A., Thormann, W., Van den Busch, P., Fleischmann, M., and Pons, B. S.,** Instrumental configurations for the determination of sub-micromolar concentrations of electroactive species with carbon, gold, and platinum microdisk electrodes in static and flow-through cells, *Anal. Chim. Acta*, 187, 67, 1986.
121. **Gunasingham, H. and Fleet, B.,** Wall-jet electrode in continuous monitoring voltammetry, *Anal. Chem.*, 55, 1409, 1983.
122. **Florence, T. M. and Farrar, Y. J.,** Polarography of azobenzene and its *p*-sulphonic acids, *Aust. J. Chem.*, 17, 1085, 1967.
123. **Cushman, M. R., Bennett, B. G., and Anderson, C. W.,** Electrochemistry of carbon fibers. I. Characteristics of the mercury film carbon filter electrode in differential pulse anodic stripping voltammetry, *Anal. Chim. Acta*, 130, 323, 1981.
124. **Bond, A. M., Fleischmann, M., and Robinson, J.,** Electrochemistry in organic solvents without supporting electrolyte using platinum microelectrodes, *J. Electroanal. Chem.*, 168, 299, 1984.
125. **MacFarlane, D. R. and Wong, D. K. Y.,** Thin-ring ultra-microelectrodes, *J. Electroanal. Chem.*, 185, 197, 1985.
126. **Shulze, G. and Frenzel, W.,** Potentiometric stripping analysis and anodic stripping voltammetry with carbon fiber electrodes, *Anal. Chim. Acta*, 159, 95, 1984.
127. **Baranski, A. S.,** Rapid anodic stripping analysis with ultra microelectrodes, *Anal. Chem.*, 59, 662, 1987.
128. **Howell, J. O. and Wightman, R. M.,** Ultrafast voltammetry and voltammetry in highly resistive solutions with microvoltammetric electrodes, *Anal. Chem.*, 56, 524, 1984.
129. **Jennings, V. J. and Morgan, J. E.,** Use of a single carbon fibre electrode for chronopotentiometric stripping analysis, *Analyst*, 110, 121, 1985.
130. **Golas, J. and Osteryoung, J.,** Electrodeposition and anodic stripping of silver on single carbon fibers, *Anal. Chim. Acta*, 192, 225, 1987.

131. **Edmonds, T. E.,** Electroanalytical applications of carbon fibre electrodes, *Anal. Chim. Acta,* 175, 1, 1985.
132. **Sottery, J. P. and Anderson, C. W.,** Short-pulse rapid-scan stripping voltammetry of a thin mercury film carbon fiber electrode, *Anal. Chem.,* 59, 140, 1987.
133. **Baranski, A. S. and Quon, H.,** Potentiometric stripping determination of heavy metals with carbon fiber and gold microelectrodes, *Anal. Chem.,* 58, 407, 1986.
134. **Wightman, R. M.,** Microvoltammetric electrodes, *Anal. Chem.,* 53, 1125A, 1981.
135. **Ponchon, J. L., Cespuglio, R., Gonon, F., Jouvet, M., and Pujol, J.-F.,** Normal pulse polarography with carbon fiber electrodes for in vitro and in vivo determination of catecholamines, *Anal. Chem.,* 51, 1483, 1979.
136. **Wehmeyer, K. R. and Wightman, R. M.,** Cylic voltammetry and anodic stripping voltammetry with mercury ultra microelectrodes, *Anal. Chem.,* 57, 1989, 1985.
137. **Ewing, A. G., Withnell, R., and Wightman, R. M.,** Instrument design for pulse voltammetry with microvoltammetric electrodes, *Rev. Sci. Instrum.,* 52, 454, 1981.
138. **Guadalupe, A. R. and Abruna, H. D.,** Electroanalysis with chemically modified electrodes, *Anal. Chem.,* 57, 142, 1985.
139. **Takeuchi, E. S. and Osteryoung, J.,** Preparation and investigation of polymer-modified electrodes by square wave voltammetry, *Anal. Chem.,* 57, 1768, 1985.
140. **Cheek, G. T. and Nelson, R. F.,** Applications of chemically modified electrodes to analysis of metal ions, *Anal. Lett.,* 11, 393, 1978.
141. **Miwa, T., Jin, L. T., and Mizuike, A.,** Differential-pulse anodic stripping voltammetry of copper with a chemically-modified glassy carbon electrode, *Anal. Chim. Acta,* 160, 135, 1984.
142. **Wang, J., Golden, T., and Tuzhi, P.,** Poly(4-vinylpyridine)-coated glassy carbon flow detectors, *Anal. Chem.,* 59, 740, 1987.
143. **Murray, R. W., Ewing, A. G., and Durst, R. A.,** Chemically modified electrodes. Molecular design for electro analysis, *Anal. Chem.,* 59, 379A, 1987.
144. **Castro, E. S., Huber, E. W., Villarroel, D., Galiatsatos, C., Mark, J. E., Heineman, W. R., and Murray, P. T.,** Electrodes with polymer network films formed by γ-irradiation cross-linking, *Anal. Chem.,* 59, 134, 1987.
145. **Stewart, E. E. and Smart, R. B.,** Differential pulse anodic stripping voltammetry of cadmium(II) with a rotating membrane-covered mercury film electrode, *Anal. Chem.,* 56, 1131, 1984.
146. **Wang, J. and Hutchins, L. D.,** Thin-layer electrochemical detector with a glassy carbon electrode coated with a base-hydrolyzed cellutosic film, *Anal. Chem.,* 57, 1536, 1985.
147. **Wang, J., Banakdar, M., and Pack, M. M.,** Glassy carbon electrodes coated with cellulose acetate for adsorptive stripping voltammetry, *Anal. Chim. Acta,* 192, 215, 1987.
148. **Wang, J. and Hutchins-Kumar, L. D.,** Cellulose acetate coated mercury film electrodes for anodic stripping voltammetry, *Anal. Chem.,* 58, 402, 1986.
149. **Hoyer, B., Florence, T. M., and Batley, G. E.,** Application of polymer-coated glassy carbon electrodes in anodic stripping voltammetry, *Anal. Chem.,* 59, 1608, 1987.
150. **Matsue, T., Akiba, U., and Osa, T.,** Regioselective electrode system with a poly(perfluorosulfonic acid)-coated electrode based on cyclodextin complexation, *Anal. Chem.,* 58, 2096, 1986.
151. **Martin, C. R., Rhoades, T. A., and Ferguson, J. A.,** Dissolution of perfluorinated ion containing polymers, *Anal. Chem.,* 54, 1639, 1982.
152. **Moore, R. B. and Martin, C. R.,** Procedure for preparing solution-cast perfluorosulfonate ionomer films and membranes, *Anal. Chem.,* 58, 2569, 1986.
153. **Hoyer, B. and Florence, T. M.,** Application of polymer-coated electrodes to the direct determination of trace metals in body fluids by anodic stripping voltammetry, *Anal. Chem.,* 59, 2839, 1987.
154. **Wang, J. and Tuzhi, P.,** Selectivity and sensitivity improvements at perfluorinated ionomer/cellulose acetate bilayer electrodes, *Anal. Chem.,* 58 3257, 1986.
155. **Peterson, W. M.,** Static mercury drop electrode, *Am. Lab.,* 11, 69, 1979.
156. **Bond, A. M., Hudson, H. A., and Van den Bosch, P. A.,** High flow rate cells for continuous monitoring of low concentrations of electroactive species by polarography and stripping voltammetry at the static mercury drop electrode, *Anal. Chim. Acta,* 127, 121, 1981.
157. **Borman, S. A.,** New electroanalytical pulse techniques, *Anal. Chem.,* 54, 698A, 1982.
158. **Batley, G. E.,** Interferences in the determination of copper in natural waters by anodic stripping voltammetry, *Anal. Chim. Acta,* 189, 371, 1986.
159. **Shimizu, K. and Osteryoung, R. A.,** Determination of sulfide by cathodic stripping voltammetry of silver sulfide films at a rotating silver disk electrode, *Anal. Chem.,* 53, 584, 1981.
160. **Kalvoda, R.,** Adsorptive accumulation in stripping voltammetry, *Anal. Chim. Acta,* 138, 11, 1982.
161. **Lam, N. K., Kalvoda, R., and Kopanica, M.,** Determination of uranium by adsorptive stripping voltammetry, *Anal. Chim. Acta,* 154, 79, 1983.
162. **Van den Berg, C. M. G.,** The determination of trace metals in sea-water using cathodic stripping voltammetry, *Sci. Total Environ.,* 49, 89, 1986.

163. **Florence, T. M.,** Detection of aluminium in thorium compounds by linear-sweep oscillographic polarography, *Anal. Chem.*, 34, 496, 1962.
164. **Florence, T. M. and Belew, W. L.,** Interpretation of the polarographic behaviour of metal-solochrome violet RS complexes, *J. Electroanal. Chem.*, 21, 157, 1969.
165. **Florence, T. M. and Aylward, G. H.,** Electrochemical studies on eriochrome violet B and its lanthanide complexes. I. Behaviour of eriochrome violet-B and its lanthanide complexes of the dropping mercury electrode: DC, AC, and cathode ray polarography, *Aust. J. Chem.*, 15, 65, 1962.
166. **Van den Berg, C. M. G., Murphy, K., and Riley, J. P.,** The determination of aluminium in seawater and freshwater by cathodic stripping voltammetry, *Anal. Chim. Acta*, 188, 177, 1986.
167. **Zittel, H. E. and Florence, T. M.,** Voltammetric method for determination of zirconium, *Anal. Chem.*, 39, 355, 1967.
168. **Pihlar, B., Valenta, P., and Nurnberg, H. W.,** New high-performance analytical procedure for the voltammetric determination of nickel in routine analysis of waters, biological materials, and food, *Fresenius Z. Anal. Chem.*, 307, 337, 1981.
169. **Van den Berg, C. M. G.,** Direct determination of sub-nanomolar levels of zinc in seawater by cathodic stripping voltammetry, *Talanta*, 31, 1069, 1984.
170. **Van den Berg, C. M. G.,** Determination of copper in seawater by cathodic stripping voltammetry of complexes with catechol., *Anal. Chim. Acta*, 164, 195, 1984.
171. **Van den Berg, C. M. G.,** Determining trace concentrations of copper in water by cathodic film stripping voltammetry with adsorptive collection, *Anal. Lett.*, 17, 2141, 1984.
172. **Van den Berg, C. M. G. and Huang, Z. Q.,** Determination of iron in seawater using cathodic stripping voltammetry preceded by adsorptive collection with the hanging mercury drop electrode, *J. Electroanal. Chem.*, 177, 269, 1987.
173. **Van den Berg, C. M. G. and Huang, Z. Q.,** Determination of uranium(VI) in sea water by cathodic stripping voltammetry of complexes with catechol., *Anal. Chim. Acta*, 164, 209, 1984.
174. **Van den Berg, C. M. G. and Huang, Z. Q.,** Direct electrochemical determination of dissolved vanadium in seawater by cathodic stripping voltammetry with the hanging mercury drop electrode, *Anal. Chem.*, 56, 2383, 1984.
175. **Van den Berg, C. M. G.,** Direct determination of molybdenum in sea water by adsorption voltammetry, *Anal. Chem.*, 57, 1532, 1985.
176. **Van den Berg, C. M. G.,** Chemical speciation in aqueous solution, *Anal. Proc.*, 21, 359, 1987.
177. **Van den Berg, C. M. G. and Nimmo, M.,** Direct determination of uranium in water by cathodic stripping voltammetry, *Anal. Chem.*, 59, 924, 1987.
178. **Wang, J., Farios, P. A., and Mahmoud, J. S.,** Stripping voltammetry of aluminium based on adsorptive accumulation of its solochrome violet RS complex at the static mercury drop electrode, *Anal. Chim. Acta*, 172, 57, 1985.
179. **Wang, J., Farias, P. A., and Mahmoud, J. S.,** Trace determination of lanthanum, cerium, and praseodymium based on adsorptive stripping voltammetry, *Anal. Chim. Acta*, 171, 215, 1985.
180. **Wang, J. and Mahmoud, J. S.,** Stripping voltammetry with adsorptive accumulation for trace measurements of titanium, *J. Electroanal. Chem.*, 208, 383, 1986.
181. **Wang, J. and Mahmoud, J. S.,** Stripping voltammetry of manganese based on chelate adsorption of the hanging mercury drop electrode, *Anal. Chim. Acta*, 182, 147, 1986.
182. **Gilbert, M. J. and Powell, H. K.,** Cobalt analysis by adsorption voltammetry, in Proc. of 9th Australian Symp. on Analytical Chemistry, Royal Australian Chemistry Institute, Sydney, 1987, 573.
183. **Jagner, D.,** Potentiometric stripping analysis. A review, *Analyst*, 107, 593, 1982.
184. **Labar, C. and Lamberts, L.,** Potentiometric stripping analysis at a stationary electrode, *Anal. Chim. Acta*, 132, 23, 1981.
185. **Jagner, D., Josefson, M., and Westerlund, S.,** Simultaneous determination of cadmium and lead in urine by means of computerized potentiometric stripping analysis, *Anal. Chim. Acta*, 128, 155, 1981.
186. **Jagner, D., Josefson, M., and Westerlund, S.,** Determination of arsenic(III) by computerized potentiometric stripping analysis, *Anal. Chem.*, 53, 2144, 1981.
187. **Brown, S. D. and Kowalski, B. R.,** Pseudopolarographic determination of metal complex stability constants in dilute solution by rapid scan anodic stripping voltammetry, *Anal. Chem.*, 51, 2133, 1979.
188. **Valenta, P.,** Voltammetric studies on trace metal speciation in natural waters. I, in *Trace Element Speciation in Surface Waters and its Ecological Implications*, Leppard, G. G., Ed., Plenum Press, New York, 1983, 49.
189. **Bubic, S. and Branica, M.,** Voltammetric characterization of the ionic state of cadmium present in seawater, *Thalassia Jugosl.*, 9, 47, 1973.
190. **Barrett, P., Davidowski, L. J., and Copeland, T. R.,** Staircase voltammetry and pulse polarography with a microcomputer-controlled polarograph, *Anal. Chim. Acta*, 122, 67, 1980.
191. **Turner, D. R., Robinson, S. G., and Whitfield, M.,** Automated electrochemical stripping of copper, lead, and cadmium in seawater, *Anal. Chem.*, 56, 2387, 1984.

192. **Kryger, L.**, Microcomputers in electrochemical trace elemental analysis, *Anal. Chim. Acta*, 133, 591, 1981.
193. **Brown, S. D. and Kowalski, B. R.**, Minicomputer-controlled, background-subtracted anodic stripping voltammetry; evaluation of parameters and performance, *Anal. Chim. Acta*, 107, 13, 1979.
194. **Osteryoung, J. G. and Osteryoung, R.**, Square wave voltammetry, *Anal. Chem.*, 57, 101A, 1985.
195. **Varney, M. S., Turner, D. R., Whitfield, M., and Mantoura, R. F. C.**, The use of electrochemical techniques to monitor complexation capacity titrations in natural waters, in *Complexation of Trace Metals in Natural Waters*, Kramer, C. J. M. and Duinker, J. C., Eds., Martinus Nijhoff, The Hague, 1983, 33.
196. **Cescon, P., Scarponi, G., and Moret, I.**, Electrochemical determination of the contamination of seawater samples during storage and filtration, *Sci. Total Environ.*, 37, 95, 1984.
197. **Batley, G. E. and Gardner, D.**, Copper, lead and cadmium speciation in some estuarine and coastal marine waters, *Estuarine Coastal Mar. Sci.*, 7, 1978, 59.
198. **Landy, M. P.**, An evaluation of differential pulse anodic stripping voltammetry at a rotating glassy carbon electrode for the determination of cadmium, copper, lead and zinc in antarctic snow samples, *Anal. Chim. Acta*, 121, 39, 1980.
199. **Bubic, S. and Sipos, L.**, Comparison of different electroanalytical techniques for the determination of heavy metals in sea water, *Thalassia Jugosl.*, 9, 55, 1973.
200. **Symes, J. L. and Kester, D. R.**, Copper(II) interaction with carbonate species based on malachite solubility in perchlorate medium at the ionic strength of seawater, *Mar. Chem.*, 16, 189, 1985.
201. **Allen, H. E. and Brisbin, T. D.**, Protection of bioavailability of copper in natural waters, *Thalassia Jugosl.*, 16, 331, 1980.
202. **Leckie, J. O. and Davis, J. A.**, Aqueous environmental chemistry of copper, in *Copper in the Environment*, Part 1, Nriagu, J. O., Ed., John Wiley & Sons, New York, 1979, 89.
203. **Van den Berg, C. M. G., Buckley, P. J. M., Huang, Z. Q., and Nimmo, M.**, An electrochemical study in the speciation of copper, zinc and iron in two estuaries in England, *Estuarine Coastal Shelf Sci.*, 22, 479, 1986.
204. **Balistrieri, L., Brewer, P. G., and Murray, J. W.**, Scavenging residence times of trace metals and surface chemistry of sinking particles in the deep ocean, *Deep-Sea Res.*, 28, 101, 1981.
205. **Piotrowicz, S. R., Springer-Young, M., Puig, J. A., and Spencer, M. J.**, Anodic stripping voltammetry for evaluation of organic-metal interactions in seawater, *Anal. Chem.*, 54, 1367, 1982.
206. **Baudo, R.**, Chemical speciation of trace elements in the aquatic environments: a literature review, *Mem. Ist. Ital. Idrobiol. Dott. Marco de Marchi*, 38, 463, 1981.
207. **Buffle, J.**, A critical comparison of studies of complex formation between copper(II) and fulvic substances of natural waters, *Anal. Chim. Acta*, 118, 29, 1980.
208. **Negishi, M. and Matsunaga, K.**, Organically-bound copper in lake and river waters in Japan, *Water Res.*, 17, 91, 1983.
209. **Westall, J. C., Morel, F. M. M., and Hume, D. N.**, Chloride interference in cupric ion selective electrode measurements, *Anal. Chem.*, 51, 1792, 1979.
210. **Nordstrom, D. K., Jenne, E. A., and Ball, J. W.**, Redox equilibria of iron in acid mine waters, in *Chemical Modelling in Aqueous Systems*, Jenne, E. A., Ed., ACS Symp. Ser. 93, American Chemical Society, Washington, D.C., 1979, 51.
211. **Benes, P., Koc, J., and Stulik, K.**, The use of anodic stripping voltammetry for determination of the concentration and forms of existence of lead in natural waters, *Water Res.*, 13, 967, 1979.
212. **Bond, A. M., Bradbury, J. R., Hanna, P. J., Howell, G. N., Hudson, H. A., and Strother, S.**, Examination of interferences in the stripping voltammetric determination of trimethyllead in seawater by polarography and mercury-199 and lead-207 nuclear magnetic resonance spectrometry, *Anal. Chem.*, 56, 2392, 1987.
213. **Colombini, M. P., Fuoco, R., and Papoff, P.**, Electrochemical speciation and determination of organometallic species in natural waters, *Sci. Total Environ.*, 37, 61, 1987.
214. **Bernhard, M., Goldberg, E. D., and Piro, A.**, Zinc in sea water, Overview, *Phys. Chem. Sci. Res. Rep.*, 1, 43, 1975.
215. **Florence, T. M.**, Speciation of zinc in natural waters, in *Zinc in the Environment. Part I: Ecological Cycling*, Nriagu, J. O., Ed., John Wiley & Sons, New York, 1983, 199.
216. **Bruland, K. W., Knauer, G. A., and Martin, J. H.**, Zinc in north-east Pacific water, *Nature (London)*, 271, 741, 1978.
217. **Martin, J. H., Knauer, G. A., and Flegal, A. R.**, Distribution of zinc in natural waters, in *Aquatic Toxicology*, Nriagu, J. O., Ed., John Wiley & Sons, New York, 1983, 193.
218. **Shuman, M. W. and Woodward, G. P.**, Intermetallic compound formation between copper and zinc in mercury and its effect on anodic stripping voltammetry, *Anal. Chem.*, 48, 1979, 1976.
219. **Stauber, J. L. and Florence, T. M.**, Interactions of copper and manganese: a mechanism by which manganese alleviates copper toxicity to the marine diatom, *Nitzschia closterium*, *Aquat. Toxicol.*, 7, 241, 1985.

220. **Colombini, M. P. and Fuaco, R.,** Determination of manganese at ng/mL levels in natural waters by differential pulse polarography, *Talanta,* 30, 901, 1983.
221. **Andrew, R. W., Biesinger, K. E., and Glass, G. E.,** Effects of inorganic complexing on the toxicity of copper to *Daphnia Magna, Water Res.,* 11, 309, 1977.
222. **Magnuson, V. R., Harriss, D. K., Sun, M. S., Taylor, D. K., and Glass, G. E.,** Relationships of activities of metal-ligand species to aquatic toxicity, in *Chemical Modellding in Aqueous Systems,* Jenne, E. A., Ed., ACS Symp. Ser. 93, American Chemical Society, Washington, D.C., 1979, 635.
223. **Sunda, W. G., Klaveness, D., and Palumbo, A. V.,** Bioassays of cupric ion activity and copper complexation, in *Complexation of Trace Metals in Natural Waters,* Kramer, C. J. and Duinker, J. C., Eds., Martinus Nijhoff, The Hague, 1983, 393.
224. **Babich, H. and Stotzky, G.,** Influence of chemical speciation on the toxicity of heavy metals to the microbiota, in *Aquatic Toxicology,* Nriagu, J. O., Ed., John Wiley & Sons, New York, 1983, 1.
225. **Peterson, R.,** Influence of copper and zinc on the growth of a freshwater alga, *Scenedesmus quadricauda:* the significance of chemical speciation, *Environ. Sci. Technol.,* 16, 443, 1982.
226. **Hodson, P. V., Borgmann, U., and Shear, H.,** Toxicity of copper to aquatic biota, in *Aquatic Toxicology,* Nriagu, J. O., Ed., John Wiley & Sons, New York, 1983, 307.
227. **Pagenkopf, G. K.,** Gill surface interaction model for trace metal toxicity to fishes: role of complexation, pH, and water hardness, *Environ. Sci. Technol.,* 17, 342, 1983.
228. **Cross, F. A. and Sunda, W. G.,** Relationship between bioavailability of trace metals and geochemical processes in estuaries, in *Estuarine Interactions,* Wiley, M. L., Ed., Academic Press, New York, 1977, 429.
229. **Leppard, G. G., Massalski, A., and Lean, D. R. S.,** Electron-opaque microscopic fibils in lakes: their demonstration, biological derivation and potential significance in the redistribution of cations, *Protoplasma,* 92, 289, 1977.
230. **Florence, T. M. and Stauber, J. L.,** Toxicity of copper complexes to the marine diatom, Nitzschia closterium, *Aquat. Toxicol.,* 8, 11, 1986.
231. **Jennette, K. W.,** The role of metals in carcinogenesis: biochemistry and metabolism, *Environ. Health Perspect.,* 40, 233, 1981.
232. **Albergoni, V. and Piccinni, E.,** Biological response to trace metals and their biochemical effects, in *Trace Element Speciation in Surface Waters and its Ecological Implications,* Leppard, G. G., Ed., Plenum Press, New York, 1983, 159.
233. **Guy, R. D. and Kean, A. R.,** Algae as a chemical speciation monitor — I.A. comparison of algal growth and computer calculated speciation, *Water Res.,* 14, 891, 1980.
234. **Srna, R. F., Garrett, K. S., Miller, S. M., and Thum, A. B.,** Copper complexation capacity of marine water samples from Southern California, *Environ. Sci. Technol.,* 1782, 1980.
235. **Gachter, R., Lum Shue-Chan, K., and Chau, Y. K.,** Complexing capacity of the nutrient medium and its relation to inhibition of algal photosynthesis by copper, *Schweiz. Z. Hydrol.,* 35, 252, 1973.
236. **Zhang, M. and Florence, T. M.,** A novel adsorbent for the determination of the toxic fraction of copper in natural waters, *Anal. Chim. Acta,* 197, 137, 1987.
237. **Hering, J. G., Sunda, W. G., Ferguson, R. L., and Morel, F. M. M.,** A field comparison of two methods for the determination of copper complexation: bacterial bioassay and fixed potential amperometry, in press.
238. **Langford, C. H., Gamble, D. S., Underdown, A. W., and Lee, S.,** Interaction of metal ions with a well characterized fulvic acid, in *Aquatic and Terrestrial Humic Materials,* Christman, R. F. and Gjessing, E. T., Eds., Ann Arbor Science, Ann Arbor, MI, 1983, 219.
239. **Plavsic, M., Krznaric, D., and Branica, M.,** Determination of the apparent copper complexing capacity of sea water by anodic stripping voltammetry, *Mar. Chem.,* 11, 17, 1982.
240. **Hart, B. T.,** Trace metal complexing capacity of natural waters: A review, *Environ. Technol. Lett.,* 2, 95, 1981.
241. **Neubecker, T. A. and Allen, H. E.,** The measurement of complexation capacity and conditional stability constants for ligands in natural waters, *Water Res.,* 17, 1, 1983.
242. **Tuschall, J. R. and Brezonik, P. L.,** Evaluation of the copper anodic stripping voltammetry complexometric titration for complexing capacities and conditional stability constants, *Anal. Chem.,* 53, 1986, 1981.
243. **Shuman, M. S. and Woodward, G. P.,** Chemical constants of metal complexes from a complexometric titration followed with anodic stripping voltammetry, *Anal. Chem.,* 45, 2032, 1973.
244. **Shuman, M. S. and Cromer, J. L.,** Copper association with aquatic fulvic and humic acids. Estimation of conditional formation constants with a titrimetric anodic stripping voltammetry procedure, *Environ. Sci. Technol.,* 13, 543, 1979.
245. **Shuman, M. S. and Michael, L. C.,** Application of the rotated risk electrode to measurement of copper complex dissociation rate constants in marine coastal samples, *Environ. Sci. Technol.,* 12, 1069, 1978.
246. **Shuman, M. S. and Woodward, G. P.,** Stability constants of copper-organic chelates in aquatic samples, *Environ. Sci. Technol.,* 11, 809, 1977.

247. **Bhat, G. A. and Weber, J. H.,** Exchange of comments on evaluation of the copper anodic stripping voltammetry complexometric titration for complexing capacities and conditional stability constants, *Anal. Chem.*, 54, 2116, 1982.
248. **Hirose, K. and Sugimura, Y.,** Role of metal-organic complexes in the marine environment. A comparison of the copper and ligand titration methods, *Mar. Chem.*, 16, 239, 1985.
249. **Waite, T. D. and Morel, F. M. M.,** Characterization of complexing agents in natural waters by copper(II)/copper(I) amperometry, *Anal. Chem.*, 55, 1268, 1983.
250. **Van den Berg, C. M. G.,** Determination of the complexing capacity and conditional stability constants of complexes of copper(II) with natural organic ligands in seawaters by cathodic stripping voltammetry of copper-catechol complex ions, *Mar. Chem.*, 15, 1, 1984.
251. **Wood, A. M., Evans, D. W., and Alberts, J. J.,** Use of an ion exchange technique to measure copper complexing capacity on the continental shelf of the southeastern United States and in the Sargasso Sea, *Mar. Chem.*, 13, 305, 1983.
252. **Davey, E. W., Morgan, M. J., and Erikson, S. J.,** Biological measurement of the copper complexation capacity of sea water, *Limmol. Oceanogr.*, 18, 993, 1973.
253. **Jardim, W. F. and Allen, H. E.,** Measurement of copper complexation by naturally occurring ligands, in *Complexation of Trace Metals in Natural Waters,* Kramer, C. J. M. and Duinker, J. C., Eds., Martinus Nijhoff, The Hague, 1983, 1.
254. **Van den Berg, C. M. G.,** Determination of complexing capacities and conditional stability constants using ion exchange and ligand competition techniques, in *Complexation in Trace Metals in Natural Waters,* Kramer, C. J. M. and Duinker, J. C., Eds., Martinus Nijhoff, The Hague, 1983, 17.
255. **Van Leeuwen, H. P.,** Voltammetric titrations involving metal complexes: effect of kinetics and diffusion coefficients, *Sci. Total Environ.*, 60, 45, 1987.
256. **Buffle, J., Vuilleumier, J. J., Tereier, M. L., and Parthasarathy, N.,** Voltammetric study of humic and fulvic substances. V. Interpretation of metal ion complexation measured by anodic stripping voltammetric methods, *Sci. Total Environ.*, 60, 75, 1987.
257. **Turner, D. R., Barney, M. S., Whitfield, M., Mantoura, R. F. C., and Riley, J. P.,** Electrochemical studies of copper and lead complexation by fulvic acid. II. A critical comparison of potentiometric and polarographic measurements, *Sci. Total Environ.*, 60, 17, 1987.
258. **Morrison, G. M. P. and Florence, T. M.,** *Electroanalysis,* 1988, in press.

Chapter 5

MATHEMATICAL MODELING OF TRACE ELEMENT SPECIATION

T. David Waite

TABLE OF CONTENTS

I. INTRODUCTION

It is clear from the preceding chapters that interest in the speciation of trace elements in aqueous solutions has increased markedly over the last decade. In particular, an awareness of the need to understand natural water processes in greater detail has instigated considerable effort into the study of the distribution of trace elements between different physicochemical forms. Examples of processes where speciation is particularly important include the interaction of trace metals with aquatic organisms (where the free metal ion activity appears to be a major indicator of metal toxicity or availability),[1] the movement of metals in saturated and unsaturated zones (where the interaction of metals with sediments is strongly dependent on properties of the metal species such as size, charge, and associated ligand functionality),[2] and the deposition or dissolution of trace element-bearing minerals under a particular set of solution conditions (such as temperature, pressure, and redox potential).[3]

While advances in analytical techniques are overcoming contamination and sensitivity problems and are enabling considerable species selectivity, it is still often difficult or, in some cases, impossible to determine the concentrations of trace element species of interest or to elucidate the major physiochemical forms present. Calculation of trace element speciation using mathematical models based on fundamental thermodynamic (and, in some cases, kinetic) concepts may enable verification of analytical data or may provide insight into the likely concentrations of species that cannot be quantified either because of sensitivity constraints or experimental difficulties (e.g., high temperature or pressure). Alternatively, the mathematical modeling of reliable speciation data may provide valuable thermodynamic (and possibly kinetic) parameters such as species formation or rate constants for the system of interest.

While application of mathematical models to the elucidation of trace element speciation appears particularly appealing in systems where the required thermodynamic (and kinetic) parameters are known with some certainty, application to complex systems, such as natural waters, clearly presents some difficulties.[4,5] In particular, the presence of ill-defined soluble and particulate organic compounds (humic substances) and inorganic colloidal materials of uncertain mineralogy and surface properties that may markedly influence trace element speciation, often in a time-dependent way, must be accounted for.[6]

II. MATHEMATICAL APPROACHES TO EQUILIBRIUM MODELING

While it is recognized that kinetic factors may play an important role in trace element speciation, all mathematical modeling approaches commence with the problem of finding the equilibrium composition of the system of interest. Natural waters, for example, contain species distributed among an aqueous phase, a gas phase, and several solid phases. Exchange of matter with the surroundings being neglected and ideality being assumed, the thermodynamic treatment of such a closed system is relatively straightforward,[7] and the equilibrium composition is given by the minimum for the Gibbs free energy function or, equivalently, by the mass action laws, both subject to the constraints of the mole balance conditions.[8] Despite the thermodynamic simplicity of this problem, a variety of modeling approaches have been adopted and mathematical algorithms applied.[9-11]

A closed system can be defined by a set of equations expressing the conservation of the chemical components of the system.[11] There is one equation for each component, i.e.,

$$\sum_{i=1}^{N} a_{ki} n_i = b_k; \qquad k = 1,\ldots,M \tag{1}$$

where n_i is the number of moles of species i, N is the total number of species in the system, M is the number of independent components, a_{ki} is the number of moles of component k in one mole of species i, and b_k is the total number of moles of each component k. The general solution to Equation 1, a set of M linear equations in N unknowns, is[11]

$$n_i = n_i^0 + \sum_{j=1}^{R} \nu_{ij} \xi_j; \qquad i = 1,\ldots,N \tag{2}$$

where n_i^0 is any particular solution (e.g., an initial composition), and υ_{ij} is the rate of change of the mole number of the ith species n_i with respect to the reaction parameter ξ_j; i.e.,

$$\left(\frac{\partial n_i}{\partial \xi_i}\right)_{\xi_{k \neq j}} = \nu_{ij}; \qquad i = 1,\ldots,N; \quad j = 1,\ldots,R \tag{3}$$

where the notation $\xi_{k \neq j}$ means all ξ other than the ith.

The Gibbs free energy, G, of the system is

$$G = \sum_{i=1}^{N} n_i \mu_i \tag{4}$$

where the partial molar free energy of species i, μ_i, is defined in terms of its mole fraction X_i and a standard free energy value,

$$\mu_i = \mu_i^0 + RT\ln X_i \tag{5}$$

As noted above, the equilibrium composition is given by the minimum of the Gibbs free energy function, a necessary condition for which is

$$\left(\frac{\partial G}{\partial \xi}\right)_{\xi_{k \neq j}} = 0; \qquad j = 1,\ldots,R \tag{6}$$

Now

$$\left(\frac{\partial G}{\partial \xi_j}\right)_{\xi_{k \neq j}} = \sum_{i=1}^{N} \left(\frac{\partial G}{\partial n_i}\right)_{n_{k \neq i}} \left(\frac{\partial n_i}{\partial \xi_j}\right)_{\xi_{k \neq j}} \qquad j = 1,\ldots,R \tag{7}$$

and

$$\left(\frac{\partial G}{\partial n_i}\right)_{n_{k \neq i}} = \mu_i \tag{8}$$

thus,

$$\sum_{i=1}^{N} \nu_{ij}\mu_i = 0; \qquad j = 1,\ldots,R \tag{9}$$

From Equation 5:

$$\sum_{i=1}^{N} \nu_{ij}\mu_i^0 + RT\sum_{i=1}^{N} \nu_{ij}\ln X_i = 0$$

$$\Delta G^0 = -RT\sum_{i=1}^{N} \nu_{ij}\ln X_i$$

$$= -RT\ln \prod_{i=1}^{N} X_i^{\nu_{ij}}$$

$$= -RT\ln K_j \tag{10}$$

where ΔG^0 is the standard free energy change for the jth equation, and

$$K_j = \prod_{i=1}^{N} X_i^{\nu_{ij}} \tag{11}$$

is the equilibrium constant for stoichiometric equation j, written in terms of mole fractions.

Mathematical equilibrium models are typically categorized according to the approach adopted in implementing the above equations. Thus, models in which the Gibbs free energy of the system (Equation 4) is minimized subject to the mole balance constraints expressed in Equation 1 have been termed "free energy minimization" models, while "equilibrium constant" models assume that the minimum Gibbs free energy has been attained and use the mass law expressions of Equation 11 in arriving at solutions that satisfy the mole balance constraints (Equation 1). While this categorization has been widely used,[9,10,12,13] Smith and Missen[11] suggest that it obscures the basic similarities in algorithms used in the two approaches, and they classified mathematical equilibrium models more in terms of the way the mole balance constraints are used in calculations. Thus, the free energy minimization problem may be converted from an optimization problem constrained by the mole balance conditions to an unconstrained optimization problem by incorporation of the mole balance conditions in the expression to be minimized through application of the mass law expressions. The constrained optimization problem is classified as a "nonstoichiometric" formulation and the unconstrained optimization problem as a "stoichiometric" formulation. The equilibrium constant approach also makes use of the mass law expressions and is considered a "stoichiometric" formulation. The equilibrium constant approach was pioneered by workers such as Brinkley,[14] while White et al.[15] initiated the constrained free energy minimization method. The latter approach has been developed extensively in a series of reports from the RAND Corporation.[16,17]

The constrained optimization problem is typically solved by removal of constraints using Lagrange multipliers.[18] Such a process results in a set of nonlinear equations. Similarly, through application of Taylor series expansion, the unconstrained optimization problem may be converted to a set of nonlinear equations. The equilibrium constant approach results directly in a set of nonlinear equations. In the majority of recently formulated mathematical models, these sets are solved using the well-established Newton-Raphson method.[19]

An alternative method of solution of these equations is that of successive approximations. This method has been used widely in early models adopting the equilibrium constant approach and has been implemented either by "brute force" or "continued fractions."[10,20] The brute force method is the classical approach where mass action expressions are substituted directly into the mass balance conditions and solved for total concentrations which are then compared to the analytical values. In the continued fraction method, the nonlinear equations are rearranged to solve for free ion concentrations which are initially assumed to be equal to the total concentrations.[12,20,21] While the brute force method of solution is relatively slow, the continued fraction method may converge more quickly than the Newton-Raphson method.[20] Further detail on the mathematical approaches used in specific models is given later in this chapter.

III. METHODS OF NONIDEALITY AND TEMPERATURE CORRECTION

In an ideal system, the molar free energy μ_i of any species i depends exclusively on the mole fraction X_i of that species (Equation 5). In a natural water, the assumption that the composition of the system has no effect on μ is no longer valid, particularly at high solute concentrations, and account must be taken of species interactions in any attempt to obtain a realistic model. In addition, the vast majority of thermodynamic data (such as free energy of formation and equilibrium constants) that are essential as input to any mathematical model have been obtained at 1 atm and 25°C, yet species distribution modeling in systems with temperature and pressure other than these standard conditions may be required. The effect of pressure variations on the thermodynamic parameters of solution species are typically relatively small and only need be considered in extreme situations (e.g., deep ocean or hydrothermal fluid modeling). Parameter changes induced by temperature variations may be major even at relatively low temperatures, and suitable corrections must be made if realistic equilibrium modeling is to be undertaken at temperatures other than 25°C.

A. Nonideality Correction

Nonideality corrections are needed in real systems to account for interactions not described by chemical reactions. In dilute solutions such interactions are mostly due to long-range electrostatic forces among ions, while in concentrated solutions specific ion interactions may become important.[22] Such effects are typically accounted for by including the activity coefficient, γ_i, in the molar free energy expression for species i, i.e.,

$$\mu_i = \mu_i^0 + RT\ln X_i + RT\ln\gamma_i \tag{12}$$

In dilute solutions (ionic strength, $I < 0.1$ *M*), long-range electrostatic interactions are adequately described by the extended Debye-Huckel expression for the activity coefficient:

$$\ln\gamma_i = -AZ_i^2 \frac{I^{1/2}}{1 + BaI^{1/2}} \tag{13}$$

where Z_i is the charge number of the ion, A and B are constants determined by the absolute temperature and dielectric constant of the system, and a is an adjustable size parameter introduced in the Debye-Huckel derivation as the distance of closest approach between the center of adjacent ions and corresponding roughly to the radius of the hydrated ion (typically in the range of 3 to 9 Å).

While the Debye-Huckel expression has a solid foundation in classical thermodynamics and electrostatics, it does not account for all interactions among solutes. As can be seen in Figure 1, the Debye-Huckel formula does not accurately predict the activity coefficients of

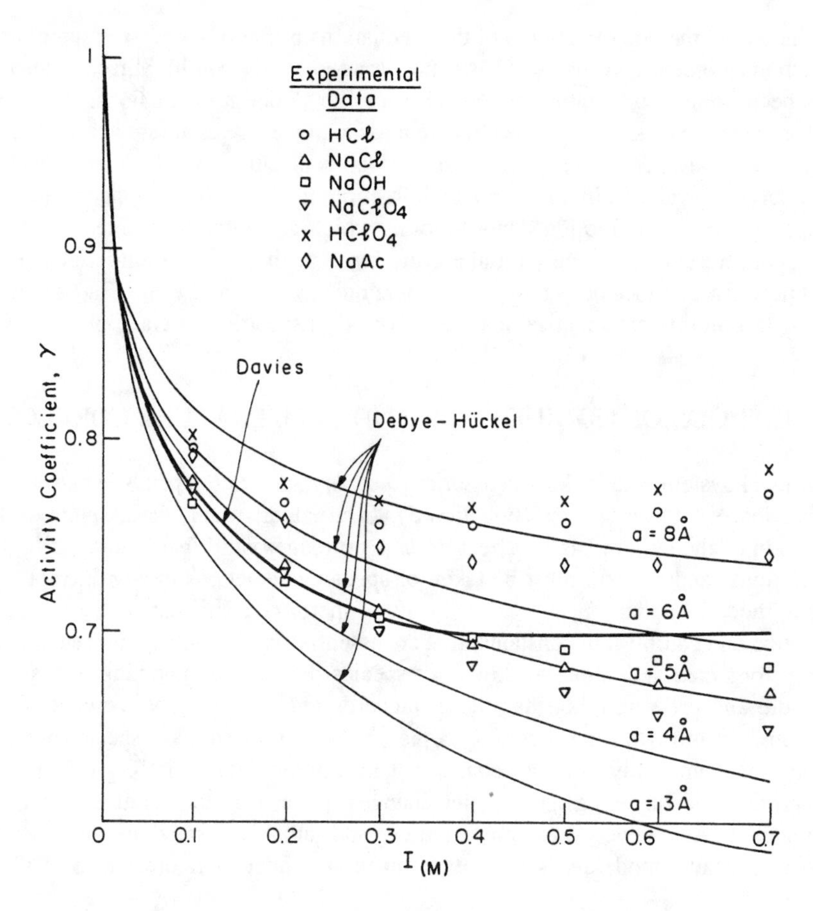

FIGURE 1. Activity coefficients of 1:1 electrolytes as a function of ionic strength. The lines correspond to the predictions of the extended Debye-Huckel (Equation 13) and the Davies formulas (Equation 15) for singly charged ions. (From Morel, F. M. M., *Introduction to Aquatic Chemistry,* John Wiley & Sons, New York, 1983. With permission.)

simple electrolytes at high ionic strengths; even at an ionic strength of 0.1 *M*, the activity coefficient of such an important salt as NaCl is measurably underestimated by the formula. In order to extend the applicability of the Debye-Huckel formula to higher ionic strength systems, a variety of empirical and semiempirical expressions have been proposed. Two such formulas are given below:

$$\ln\gamma_i = -AZ_i^2 \frac{I^{1/2}}{1 + BaI^{1/2}} + \dot{B}I \tag{14}$$

$$\ln\gamma_i = -AZ_i^2 \left(\frac{I^{1/2}}{1 + I^{1/2}} - bI\right) \tag{15}$$

where $\dot{B}$ and b are empirical parameters (b is typically in the range 0.2 to 0.3). The Davies equation[22] (Equation 15) provides reasonable estimates of activity coefficients up to an ionic strength of about 0.7 *M* (see Figure 1).

Accurate modeling of seawater and waters of even higher ionic strength (as is typical of many groundwaters) requires thermodynamic analyses that accounts for specific as well as nonspecific interactions. This is most commonly accomplished through extensions to the

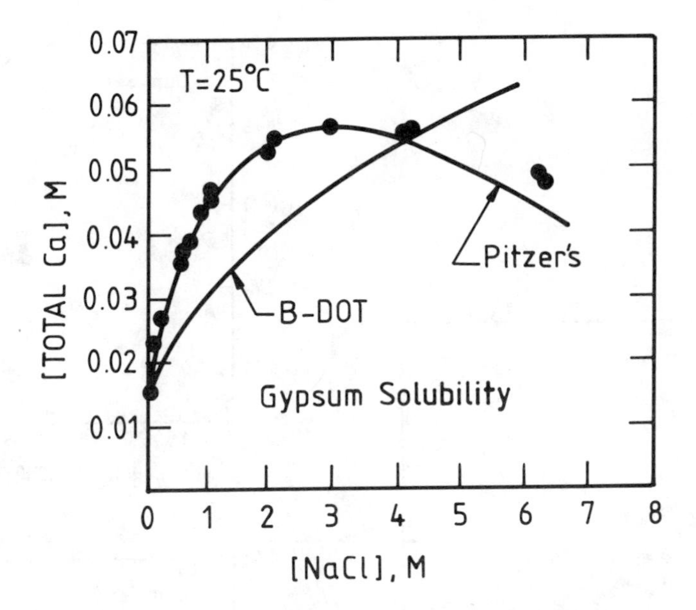

FIGURE 2. Comparison of experimental data for gypsum solubility in brine with data calculated using the Pitzer equation and the "B-dot" form of the modified, extended Debye-Huckel equation (Equation 14). (From Wolery, T. J., Isherwood, D. J., Jackson, K. J., Delany, J. M., and Purgdomenech, I., EQ3/6: Status and Applications, Lawrence Livermore National Laboratory, Livermore, CA, 1984.)

specific ion interaction model of Bronsted[23] and Guggenheim[24] developed by Mayer,[25] Scatchard,[26] and Pitzer.[27-29] The general formula for these extensions consists of a virial expansion of the form

$$\ln\gamma_i = \ln\gamma_{DH} + \sum_j B_{ij}X_j + \sum_j \sum_k C_{ijk}X_jX_k + \dots \quad (16)$$

where the first term is the Debye-Huckel activity coefficient, and X_j is the molar concentration of species j. The second virial coefficients B_{ij} account for specific interactions among pairs of ions, the third virial coefficients C_{ijk} account for specific interactions among three ions and so on. Like γ_{DH}, the higher order virial coefficients are functions of the ionic strength, and successful formulations for these coefficients are at the core of thermodynamic descriptions of concentrated electrolytes. Whitfield and others[30-32] have applied the Pitzer equation[29] successfully to seawater and good agreement with experimental data have been obtained in systems with ionic strengths up to about 6 *M* by Harvie and Weare.[33] The ability of the Pitzer equation is demonstrated in Figure 2, in which experimental and calculated data for gypsum solubility in brine are compared. The inadequacy of a modified Debye-Huckel expression in modeling the experimental data is also apparent. Activity coefficients of several major ionic species in various Australian groundwaters calculated using a Pitzer equation with virial coefficients estimated using the data of Harvie and Weare[33] are shown in Figure 3 and exhibit the increase in activity coefficient observed for salts such as NaCl at high ionic strengths.[34]

Neutral species may also behave in a nonideal manner, exhibiting activity coefficients that deviate from unity. Particular attention has been given to the activity coefficients of dissolved gases such as CO_2, and a simple expression of the form

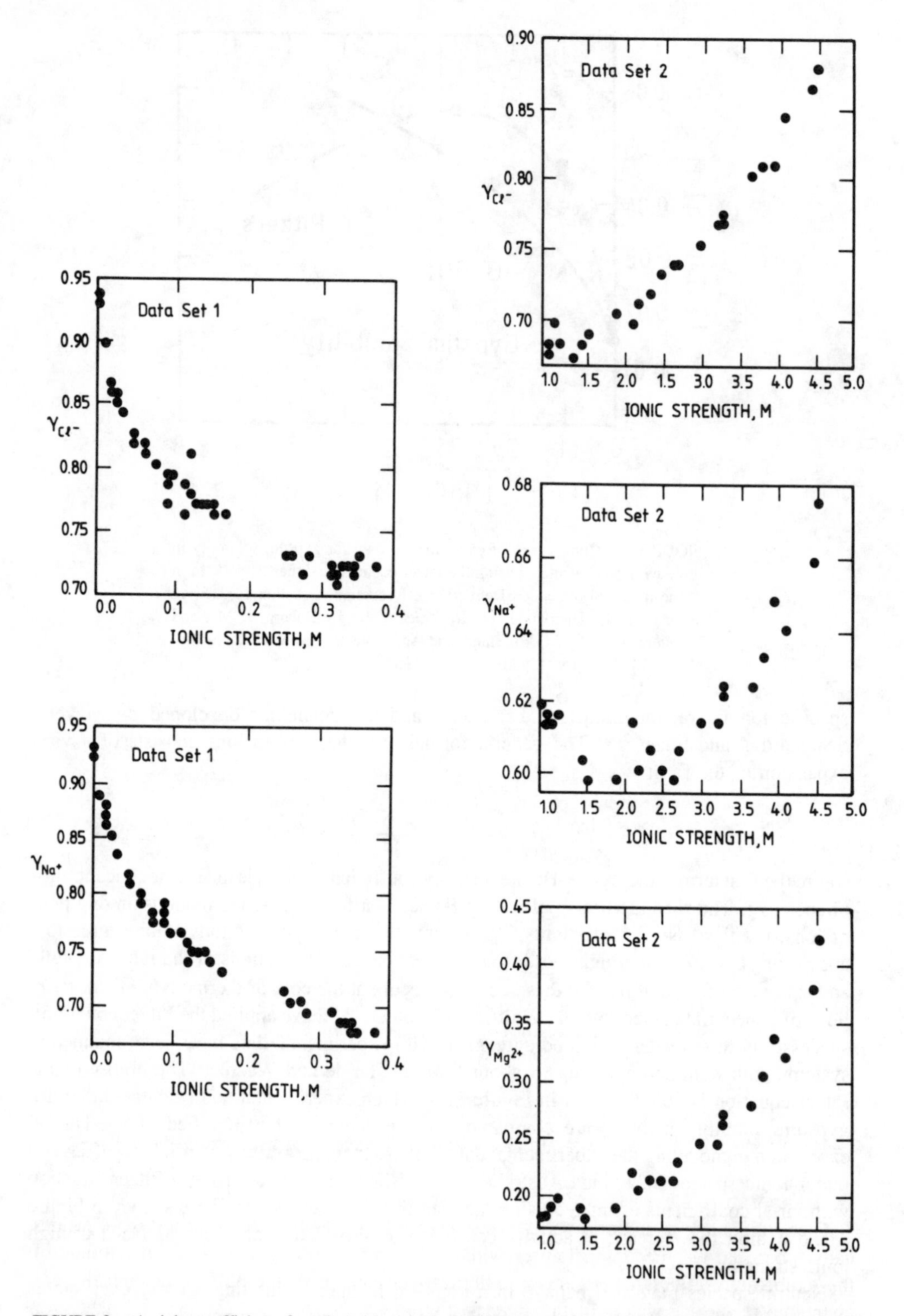

FIGURE 3. Activity coefficients for selected ions in Australian groundwaters calculated using a Pitzer equation with virial coefficients estimated from the data of Harvie and Weare.[33] (From Giblin, A. M., Unpublished data, 1987.)

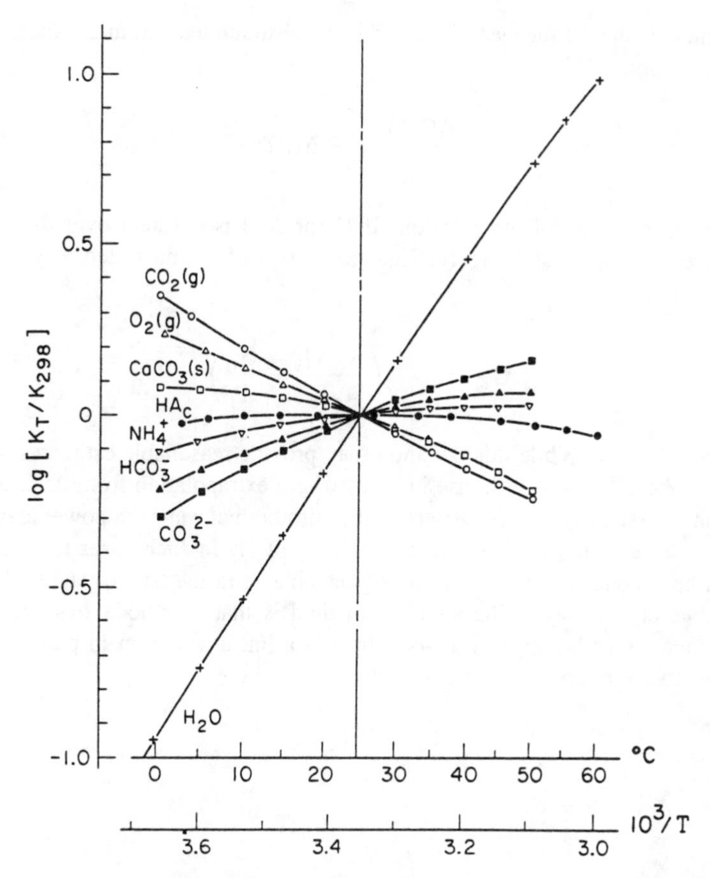

FIGURE 4. Temperature dependency of various equilibrium constants. The reactions and associated equilibrium constants at 25°C are as follows:

CO_2(g): $CO_2(g) = CO_2(aq)$; $\log K_{298} = -1.464$
O_2(g): $O_2(g) = O_2(aq)$; $\log K_{298} = -3.066$
$CaCO_3$(s): $CaCO_3(s) = Ca^{2+} + CO_3^{2-}$; $\log K_{298} = -8.475$
HAc: $HAc = H^+ + Ac^-$; $\log K_{298} = -4.756$
NH_4^+: $NH_4^+ = H^+ + NH_3(aq)$; $\log K_{298} = -9.249$
HCO_3^-: $CO_2 + H_2O = H^+ + HCO_3^-$; $\log K_{298} = -6.351$
CO_3^{2-}: $HCO_3^- = H^+ + CO_3^{2-}$; $\log K_{298} = -10.330$
H_2O: $H_2O = H^+ + OH^-$: $\log K_{298} = -13.9965$

(From Morel, F. M. M., *Introduction to Aquatic Chemistry,* John Wiley & Sons, New York, 1983. With permission.)

$$\ln\gamma_i = C_iI \tag{17}$$

where C_i is referred to as the "salting coefficient", have been used extensively.[35]

B. Temperature Correction

Thermodynamic parameters are typically highly temperature dependent, and 25°C values cannot be used at other temperatures without appropriate corrections.[36] The magnitude of the temperature effects on various equilibrium constants over the 35°C span typically encountered in surface waters is illustrated in Figure 4.

Expressions for the temperature dependence of thermodynamic parameters follow directly from the Gibbs-Helmholtz equation;[37] i.e., at constant pressure

$$\frac{\partial(G/T)}{\partial T} = -H/T^2 \tag{18}$$

where H is the enthalpy of the system, and T is the absolute temperature. Alternatively, for a chemical reaction:

$$\frac{\partial(\Delta G/T)}{\partial T} = -\Delta H/T^2 \tag{19}$$

where ΔH is the enthalpy of the reaction. If H (or ΔH) is constant over the temperature range of interest, then Equation 18 (or Equation 19) may be integrated to yield the van't Hoff isochore:[37]

$$G_{T_2} = -T_2\left(\frac{H_{T_1}\Delta T}{T_1 T_2} + \frac{G_{T_1}}{T_1}\right) \tag{20}$$

where $\Delta T = T_2 - T_1$. While this equation may provide reasonable estimates over narrow temperature ranges,[36] it is of little use in cases where extrapolation from 25°C to relatively high temperatures (such as those experienced in hydrothermal waters or power plant effluents) is required because of the temperature dependency of H. In such cases the effect of temperature can be accounted for by considering the change in the heat capacity of the system in one of a variety of ways. The simplest method is that of Khodakovskiy et al.[38] who pointed out that, to a first approximation, there is a linear variation in partial molar ionic at specific heat with temperature at constant pressure; i.e.,

$$C_p \cong b^1 T \tag{21}$$

$$H_T = \int_{T_1}^{T_2} C_p dT \tag{22}$$

and

$$G_T = -T\int_{T_1}^{T_2} \frac{HT}{T^2}\, dT \tag{23}$$

Thus, having evaluated b^1, one can substitute Equation 21 directly into Equation 22 and 23, and integrate to determine G_T for a species.

A variety of more elaborate methods of temperature correction of thermodynamic parameters have been reported,[39-41] the most useful of which appears to be that of Criss and Cobble[40,42] who showed that the free energy of a reaction may be expressed as

$$\Delta G_{T_2} = \Delta G_{T_1} + \int_{T_1}^{T_2} \Delta C_p dT - T_2\int_{T_1}^{T_2} (\Delta C_p/T)dT - \Delta T \cdot \Delta S_{T_1} \tag{24}$$

where ΔS_{T_1} is the entropy change of the reaction at temperature T_1. Equation 24 approximates to

$$\Delta G_{T_2} = \Delta G_{T_1} - \Delta T \cdot \Delta S_{T_1} + \Delta C_{p}\Big|_{T_1}^{T_2} \{\Delta T - T_2 \ln(T_2/T_1)\} \tag{25}$$

Having obtained ΔT_2 for a reaction, G_{T2} values for individual ionic species can be obtained by back-calculation.

The importance of temperature correction to species distribution and the large differences in thermodynamic parameters predicted by the various methods is apparent from Figure 5, in which calculated boundaries between the major aluminum species Al^{3+}, Al_2O_3(s) and

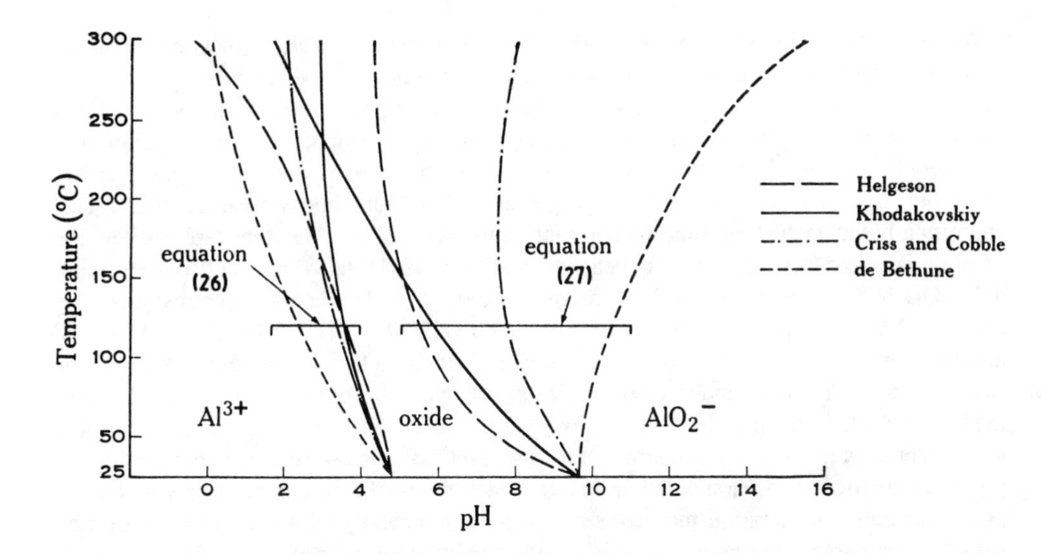

FIGURE 5. Temperature dependency of aluminum speciation as estimated by various methods. (Adapted from Lowson, R. T., *Aust. J. Chem.*, 27, 105, 1974. With permission.)

AlO_2^- are plotted as a function of temperature.[43] The predominance areas shown in Figure 5 are separated by boundaries defined by the following equations:

$$2Al^{3+} + 3H_2O \rightleftharpoons Al_2O_3 + 6H^+ \tag{26}$$

$$Al_2O_3 + H_2O \rightleftharpoons 2AlO_2^- + 2H^+ \tag{27}$$

IV. OVERVIEW OF AVAILABLE MODELS

While the foundations for general-purpose algorithms for the computation of chemical equilibrium were laid by Brinkley[14] in 1947, the first model of chemical speciation in a natural water was the ion-association model of major seawater ions of Garrels and Thompson[45] in 1962. Most models since this time have adopted a similar stoichiometric formulation based on use of equilibrium constants and ion-association concepts. Only in the last few years, with the compilation of extensive free energy data sets and the development of accurate specific ion interaction models (particularly that of Pitzer[27-29]) has the nonstoichiometric constrained optimization approach become popular. A number of reviews of the various codes developed over the last 20 years are available.[12,13,44-46]

The method of successive approximations[10,20] was used in the hand calculations of Garrels and Thompson[47] and was the method of solution used in the early chemical equilibrium computer codes LETAGROP and HALTAFALL by Sillen and co-workers.[48-50] A similar method was used by Perrin[51,52] in the program COMICS and in the modified versions of this program, COMPLEX[53] and SIAS.[54,55] A successive approximation method in conjunction with an "equilibrium constant" formulation was also used in WATCHEM — a program developed to aid in the investigation of iron corrosion processes of wells.[56] This model spawned the WATEQ series of models[57-61] as well as WATSPEC[20] and SOLMNEQ.[62] WATEQ is unique in that the model incorporates a pressure dependence of equilibrium constants, and a number of these models are equipped to handle elevated temperatures. WATEQ2 uses the van't Hoff equation or, where available, analytical expressions for the dependence of log K on temperature, while SOLMNEQ contains a table of equilibrium constants calculated at temperature increments between 0 and 350°C and interpolates between these increments for the actual value used in the modeling.[44]

While the above models have been of considerable use in a variety of areas, they are relatively limited in capability in that while the distribution of species in solution can be calculated and the likelihood of precipitation of solid phases estimated, transfer of mass between two or more phases (e.g., precipitation, dissolution) cannot generally be performed. A variety of ''second-generation'' programs are available which possess this mass transfer capability. One such program is REDEQL, developed by Morel and Morgan in 1972.[63] This program is based on the ''equilibrium constant'' approach and uses Newton-Raphson iteration to minimize the difference between measured and calculated total component concentration. REDEQL has served as the parent for a number of widely used programs including REDEQL2,[64] MINEQL,[65] REDEQL.UMD,[66] and GEOCHEM.[67] The code MINTEQ[68] should also be mentioned here since it uses an approach identical to that of MINEQL, but with the starting estimates for the independent variables enhanced by predictor logic and the thermodynamic data base upgraded from the WATEQ3 code.[61] These programs define the entire species set for a given system in terms of a ''basis set'' of components, the choice of which is critical to efficient solution of the problem. In fact, unless the dominant species are chosen as components, the iteration may converge at an impractically slow rate or even diverge. MINEQL and associated codes are particularly attractive because of their ability to change their basis set to influence the iteration to converge.[65] This ability to transform the basis set as progress is made toward equilibrium is particularly important when certain components (such as trace elements) are present at low concentration. The program MICROQL[69] has also been developed along similar lines to MINEQL, but possesses the added flexibility of choice of basis set component species at will. Because of this last attribute, the thermodynamic data base cannot be fixed, but must be input each time a speciation calculation is performed.[70]

A variety of other speciation programs of form similar to MINEQL, but with extended capability (e.g., temperature correction of thermodynamic data) or increased convergence efficiency have been developed. For example, CHEMEQUIL-2, developed by Tripathi,[70] interfaces the standard Newton-Raphson method with a modified successive approximations method and achieves savings in execution time of one to two orders of magnitude. In addition, considerably lower memory requirements are obtained by using a data structure which enables storage of only non-zero values in the stoichiometric coefficients matrix.

A number of so-called ''reaction path'' codes have been developed which consider states of partial equilibrium attained during the reaction path towards complete equilibrium. The first reaction path model, PATHI, was developed by Helgeson and co-workers[71-73] and has been applied to the study of weathering reactions, diagenesis, metamorphism and hydrothermal metasomatism,[73] the formation of ore deposits,[74] and scaling of conduits caused by precipitation from geothermal brines.[75] Improvements in the efficiency of this type of program have been made by Herrick,[76] which led to the development of FASTPATH, by Wolery,[77] who developed EQ3/6, and by Reed,[78] who developed SOLVEQ. The EQ3/6 model has been adopted for further development and application to problems in geologic disposal of high-level nuclear waste by both the Nevada Nuclear Waste Storage Investigations (NNWSI) and the U.S. Office of Nuclear Waste Isolation (ONWI), and as a result, early versions of this code have been upgraded and extended considerably.[79] The simplified reaction path code PHREEQE,[80] which uses a continued fractions successive approximation method to adjust the concentration of component species and a Newton-Raphson method to solve the model balance equations, has also been used widely, particularly in the interpretation of groundwater chemistry.[81]

A summary of the evolution of some of the more popular natural water speciation models is shown in Figure 6. Most of these models have been formulated using an equilibrium constant approach, but, as mentioned earlier, models based on a nonstoichiometric optimization approach are gaining in popularity. Early programs of this form include those from

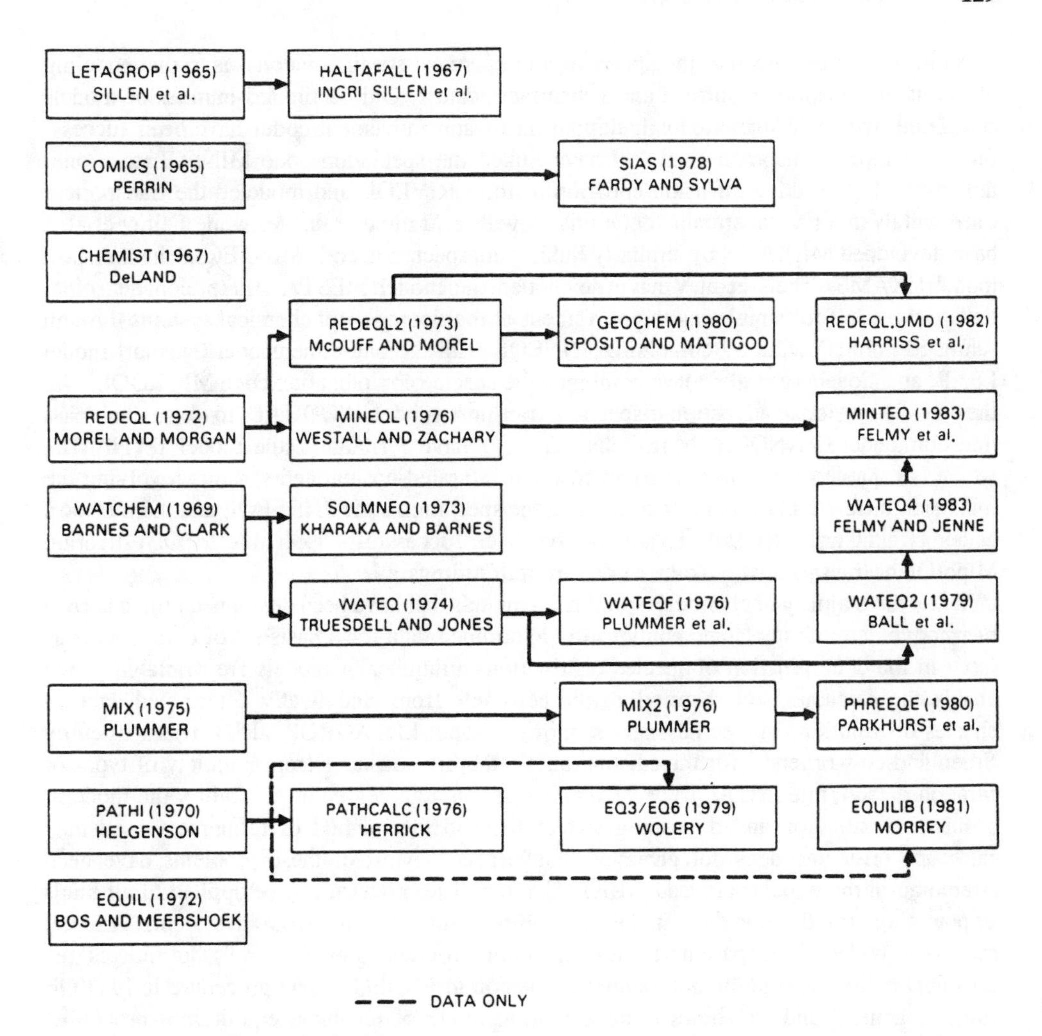

FIGURE 6. Evolution of some of the major natural water chemical speciation codes. (From Kincaid, C. T., Morrey, J. R., and Rogers, J. E., EPRI EA-3417, Vol. 1, Electric Power Research Institute, Palo Alto, CA., November, 1984.)

the RAND Corporation[16,17,82] and Karpov and Kaz'min,[83,84] while Eriksson,[85] Gautam and Sider,[86] Harvie and co-workers,[87,88] and Turnbull and Wadsley[89] have applied this approach to the solution of complex equilibria in aqueous systems in more recent years. The THERMODATA code of Turnbull and Wadsley[89] is particularly flexible and well-documented and is finding increasing application in interpretation of natural water chemistry.

While most of the models referred to above describe only complexation/dissociation reactions in the dissolved phase and, in some cases, precipitation/dissolution of solids (i.e., mass transfer), a number have the added capability of describing adsorption processes, which is likely to be critical in any attempt to describe speciation in the presence of strongly adsorbing solids such as iron, aluminum, or manganese oxyhydroxides. Most of the speciation codes that incorporate adsorption subroutines do so by adopting a site-binding model for the solid-solution interphase.[90] The REDEQL series of codes (REDEQL2,[64] SURFEQL,[91] MICROQL[69]) cater for adsorption processes in this way as do GEOCHEM[67] and ADSORB.[92,93] GEOCHEM also possesses an ion-exchange routine for major ion adsorption, and ADSORB has the capability of considering ternary surface complexes. Adsorption modeling and incorporation of such modeling into general speciation codes is discussed in more detail in a later section.

An important extension of the application of chemical speciation codes is to the modeling of pollutant transport in surface and subsurface waters. Only a limited number of models combining species distribution calculations and water movement codes have been successfully developed. Chapman et al.[94,95] have linked the speciation code MINEQL to a one-dimensional convective-diffusion equation to form RIVEQL and modeled the transport of trace metals in a natural stream contaminated with acid mine drainage, while Felmy et al.[96] have developed MEXAMS by similarly linking the speciation code MINTEQ to the transport model EXAMS. Theis et al.[97] have developed the model FIESTA (finite element solute transport model) for simulating the movement of multicomponent chemical systems through saturated porous media by combining MINEQL with the one-dimensional transport model FEAP, and Cedeberg et al.[98] have combined the surface complexation code MICROQL with the two-dimensional advection-dispersion transport model ISOQUAD to form the mass transport code TRANQL.[99] Narasimhan et al.[100] have formulated the model DYNAMIX which is capable of handling multidimensional saturated-unsaturated systems involving the mixing of different pH waters by combining the speciation code PHREEQE and the transport codes TRUST and TRUMP. This model has been successfully applied to the investigation of pollutant transport away from a uranium mill tailings pile.

While the majority of chemical equilibrium models that have been developed utilize known thermodynamic constants and analytically determined data for a basis set of components as input in the determination of species distribution, a number of models are available which enable the calculation of thermodynamic constants from analytically determined data on species distribution. The earliest program of this type, LETAGROP VRID, was written by Sillen and co-workers[101] for the calculation of stability constants from a variety of types of titration data. While LETAGROP VRID possesses a high degree of flexibility, the mode of problem formulation and data input is tedious, and the method of refinement used may introduce error and does not guarantee convergence. Most of these problems have been overcome in the popular program MINIQUAD.[102] The program can be applied to all kinds of potentiometric titration data, including multireactant and multielectrode systems. Another relatively widely used parameter determination program, FITEQL,[103,104] formulates the chemical equilibrium problem in a similar fashion to MINEQL. The procedure in FITEQL is quite general and can be applied to a wide range of chemical equilibrium problems, including adsorption on charged surfaces. Adjustable parameters may be stability constants of soluble species, stability constants of adsorbed species, total concentrations of components, solubility products, or Henry's Law constant for gases.

V. DETAILED DISCUSSION OF SELECTED MODELS

A. THERMODATA/CHEMIX

1. Program Synopsis

The CSIRO-SGTE (Commonwealth Scientific and Industrial Research Organisation-Scientific Group Thermodata Europe) THERMODATA system consists of a data bank of basic thermodynamic data and a set of compatible programs to carry out thermodynamic calculations with the data.[105] The major THERMODATA programs and their applications are listed in Table 1. The program CHEMIX, for the calculation of species distribution in a complex system, will be of major interest in trace element speciation studies. The program was initially applied to metallurgical problems, but with recent extensive upgrading of the data base, it is now being used widely in the analysis of natural water systems.

THERMODATA/CHEMIX allows a set of possible chemical species and phases present in a chemical system to be easily defined using standard chemical nomenclature. The initial state of the system may also be defined in terms of selected elements or species and their amounts. The basic thermodynamic data for the set of species are retrieved from the data

Table 1
PROGRAMS AND THEIR APPLICATIONS CONTAINED WITHIN THE THERMODATA SYSTEM

Program	Function
FILER	Fitting and storage of basic data
REACT	Properties of species and reactions
SYSTEM	Predominance area diagrams
CHEMIX	Equilibrium calculations on systems
EXERGY	Analysis of efficiency of processes
VAPOUR	Vapor pressure and critical data
ESTIMA	Estimation of thermodynamic data

Table 2
METHODS OF CALCULATING ACTIVITY COEFFICIENTS IN THE CHEMIX PROGRAM OF THERMODATA

Description	Phase type
Polynomial	Any
Linear interpolation	Any
Berthelot equation of state	Gaseous
Virial equation of state	Gaseous
Margules, binary	Condensed
Margules, three-suffix	Condensed
Redlich-Kister, binary	Condensed
Redlich-Kister, general	Condensed
Lupis-Elliott, dilute	Condensed
Extended Debye-Huckel	Aqueous
Pitzer-Whitfield	Aqueous

bank or entered directly. For nonideal solution phases, the activity coefficients of the component species must then be defined as functions of temperature and composition. A number of well-known methods, including the extended Debye-Huckel and Pitzer methods, are provided for this purpose (see Table 2). The program calculates by the Lagrange multiplier method the particular species and their quantities which give the minimum total Gibbs free energy for the system at given temperature and pressure. Following this, enthalpy balances between initial and equilibrium states, the distributions of the elements amoung the equilibrium products, and the thermodynamic properties of the total system may be calculated. In order to be able to rapidly explore the effects of system variables, there are further options for various repetitive or step-wise calculations. There are also options to fix, maximize, or minimize chosen properties of the system by systemically varying others.[105]

Most of the thermodynamic data required in THERMODATA are stored in a computer data bank which is subdivided into separate files on the basis of the type and source of the data (see Table 3). The data bank CPDNPLDAT was leased from the U.K. National Physical Laboratory in 1975 and was used in the initial implementation of THERMODATA, but has now been superceded by the data file CPDSGTDAT, operated under agreement with SGTE. The data file CPDMRLDAT was assembled by CSIRO from a number of published critical compilations[106-109] and the file CPDJANDAT was assembled from the well-known JANAF tabulations.[110] The relatively recent addition of U.S. National Bureau of Standards data on

Table 3
TYPE AND SOURCE OF THERMODYNAMIC DATA CONTAINED WITHIN THERMODATA

File	Data sets	Date	Data type	Substance type
CPDNPLDAT	1713	1980	Hf, S, Cp vs. T	Inorganic species
CPDSGTDAT	1438	1983	Hf, S, Cp vs. T	Inorganic species
CPDJANDAT	1462	1978	Hf, S, Cp vs. T	Inorganic species
CPDMRLDAT	2034	1982	Hf, S, Cp vs. T	General species
CRDNBSDAT	1431	1982	Hf, S, Cp 298K	Aqueous species
CRINPLDAT	600	1975	Tc, Pc, Vc	Liquid species
VAPMRLDAT	500	1976	p vs. T	Solid, liquid species
ANTMRLDAT	300	1976	p vs. T	Solid, liquid species

aqueous species (file CPDNBSDAT) has further extended the applicability of THERMODATA to natural water speciation calculations.[105]

While thermodynamic data over a range of temperatures are available for many species, this is not always the case. The THERMODATA program ESTIMA provides for this eventuality by allowing the estimation of basic thermodynamic data over a range of temperatures using one of a variety of provided methods. Estimation methods included in THERMODATA of major interest in natural water speciation calculations are those of Criss and Cobble,[40] Khodakovskiy,[38] and Barner and Scheuerman,[42] described in Section III.

The first version of the THERMODATA system was written in 1975 in Extended Fortran IV for implementation on a CYBER76 computer. Continuous extension of both the data bank and application programs has occurred since then, culminating in the present implementation, Version V, written in Fortran 77 for VAX and IBM PC computers.

2. *Mathematical Formulation of THERMODATA/CHEMIX*

As mentioned above, THERMODATA uses a Gibbs free energy minimization approach with application of Lagrange multipliers. Using the terminology developed in Section II, the problem may be expressed as

$$\min G = \sum_{i=1}^{N} n_i\mu_i \tag{4}$$

subject to

$$\sum_{i=1}^{N} a_{ki}n_i = b_k \qquad k = 1,\ldots,M \tag{1}$$

The Lagrangian, L, for this constrained optimization problem may be written[18]

$$L = \sum_{i=1}^{N} n_i\mu_i + \sum_{k=1}^{M} \lambda_k \left(b_k - \sum_{i=1}^{N} a_{ki}n_i\right) \tag{28}$$

where $\lambda_1, \ldots, \lambda_M$ are unknown Lagrange multipliers. Then the necessary conditions provide the following set of N + M equations in N + M unknowns ($n_1, n_2, \ldots, n_N, \lambda_1, \lambda_2, \ldots, \lambda_M$):

$$\left(\frac{\partial L}{\partial n_i}\right)_{n_{j\neq i}} = \mu_i - \sum_{k=1}^{M} a_{ki}\lambda_k = 0, \qquad (n_i > 0) \tag{29}$$

and

$$\left(\frac{\partial L}{\partial \lambda_k}\right)_{\lambda_{j\neq k}} = b_k - \sum_{i=1}^{N} a_{ki}n_i = 0 \quad (30)$$

The solution of these equations involves the introduction of an appropriate expression for μ_i (such as that in Equation 12 for dissolved species). Equation 29 is nonlinear, but may be linearized by expansion in a Taylor series around an estimated equilibrium composition up to and including the term of the first order. This is equivalent to making a quadratic approximation to the free energy surface. The estimated species concentrations must be nonnegative, but need not satisfy the mass-balance constraints. The resulting set of linear equations may be solved iteratively.[85]

3. Application of THERMODATA/CHEMIX

Two examples of the application of THERMODATA/CHEMIX to the analysis of natural water speciation problems are presented below. In Example 1, the solubility of gypsum in strong aqueous solution is presented mainly to provide insight into the detailed operation and capabilities of CHEMIX. A complete, annotated input data set and associated output is presented. In Example 2, the results of equilibrium speciation calculations for zinc in groundwaters exhibiting major compositional differences are presented.

a. Example 1. Solubility of Gypsum in Strong Aqueous Solution

In this example we are interested in the final composition of a strong aqueous solution in equilibrium with gypsum. The input data used to define the problem and associated output are given in annotated form in Figure 7. This program is operated through a series of directive commands (a subset of which are defined in Figure 7).

The input and output files are relatively self-explanatory. The directive SYSTEM allows the definition of all the possible phases and component species which may be present in the chemical system at equilibrium. The phase name record defines the name of a phase which may be a gaseous, liquid, solid solution, or a single phase. There are three reserved phase names — GAS, AQUEOUS, and SINGLE, but others may be added. The component record defines the formula of a component of a phase.

The GAMMA directive allows the setting of parameters which relate activity coefficients to concentration, temperature, and pressure for the component species of the phases of a system. In this example the Pitzer model is used to relate activity coefficient to ionic strength and charge. This method allows for the solvent (water) and up to ten neutral species, ten cations, and ten anions. Data are entered as follows:

ij	C1 C2 C3 C4	(binary interactions cation-anion)
ij	C5	(binary interactions cation-cation anion-anion)
ij	C6	(binary interactions neutral-neutral neutral-cation neutral-anion)
ijk	C7	(ternary interaction cation-cation-anion anion-anion-cation)

where i, j, k are the species numbers within the aqueous phase starting from the first solute species. The coefficients are all based on the molality unit.

b. Example 2. Zinc Speciation in Groundwaters

Proper accounting of ionic strength effects is critical in undertaking speciation calculations in many groundwaters and hydrothermal brines where ionic strengths in excess of 1 *M* are common. This is clearly demonstrated in Figure 3 where activity coefficients for various major anions and cations in Australian groundwaters are plotted as a function of ionic strength.[34] Similar calculations of activity coefficients for trace elements must be performed in any assessment of the likely dominant forms present.

```
INPUT

TITLE
Solubility of gypsum in strong aqueous solution

SYSTEM      Enter phases and species in system.
GAS
N2 (G)
O2 (G)
H2 (G)
H2O (G)

AQUEOUS
H2O
NA E-1 (AQ)
H E-1 (AQ)
CA E-2 (AQ)
O H E (AQ)
S O4E2 (AQ)

SINGLE
CA S O6H4
NA2 S O4


FILE        Search specified data files for thermodynamic data.
CPDMRLDAT
CPDNPLDAT

GIBBS       Enter Gibbs energy parameters.
CA S O4 [H2O]2
-429867.

INSPECIES   Enter phases and species for input.
GAS
O2 (G)
N2 (G)

AQUEOUS
H2O

SINGLE
CA S O6H4
NA2 S O4


TEMPERATURE  Enter list of temperatures.
298.15

PRESSURE     Enter list of pressures.
1.
```

```
SINGLE
CA S O6H4
NA2 S O4


PHASES  4   SPECIES  12   ELEMENTS  7 N  O  H  E  Na Ca S

DATAFILE CPDMRLDAT SEARCHED
  2 SPECIES DATASETS MISSING

DATAFILE CPDNPLDAT SEARCHED

GIBBS ENERGY VS TEMPERATURE RELATIONS     UNITS CAL

DGf = A + B*T + C*T*T + D*T*T*T + E/T + F*T*ln T

CA S O6H4

A = -0.42987E+06  B =  0.00000E+00  C =  0.00000E+00  TMIN =  298.15 K
D =  0.00000E+00  E =  0.00000E+00  F =  0.00000E+00  TMAX = 6000.00 K

INITIAL INPUT OF SPECIES

GAS                         MOLE        GRAM     MOLE FRN.   MASS FRN.

N2 (G)                  0.10000E-02 0.28013E-01 0.80000     0.77787
O2 (G)                  0.25000E-03 0.79997E-02 0.20000     0.22213
PHASE  TOTALS           0.12500E-02 0.36013E-01

AQUEOUS                     MOLE        GRAM     MOLE FRN.   MASS FRN.

H2O                      55.520      1000.2      1.0000      1.0000
PHASE  TOTALS            55.520      1000.2

SINGLE                      MOLE        GRAM     MOLE FRN.   MASS FRN.

CA S O6H4                1.0000      172.17
NA2 S O4                0.10000E-05 0.14204E-03

SYSTEM TOTALS            56.521      1172.4

ACTIVITY COEFFICIENT PARAMETERS  FOR PHASE AQUEOUS

MODEL NUMBER 18 NAME PITZER

NA E-1 (AQ)        O H E (AQ)          0.086400  0.253000  0.000000  0.002200
NA E-1 (AQ)        S O4E2 (AQ)         0.019580  1.113000  0.000000  0.002020
H E-1 (AQ)         O H E (AQ)          0.000000  0.000000  0.000000  0.000000
H E-1 (AQ)         S O4E2 (AQ)         0.002700  0.000000  0.000000  0.014700
CA E-2 (AQ)        O H E (AQ)          0.171750  1.200000  0.000000  0.000000
CA E-2 (AQ)        S O4E2 (AQ)         0.200000  2.650000-55.700000  0.000000
```

```
INMOLE        Enter moles of input species.
.00025,.001,55.52,1.,.000001

GAMMA      Enter activity coefficient parameters.
AQUEOUS
18
1 4 .0864      .2530      0.         .00220
1 5 .01958     1.113      0.         .00202
2 4 0.         0.         0.         0.
2 5 .0027      0.         0.         .0147
3 4 .17175     1.2        0.         0.
3 5 .2         2.65       -55.7      0.

EQUILIBRATE   Calculate equilibrium of system.

OUTPUT

Solubility of gypsum in strong aqueous solution

POSSIBLE SPECIES AND PHASES IN SYSTEM

GAS
N2 (G)
O2 (G)
H2 (G)
H2O (G)

AQUEOUS
H2O
NA E-1 (AQ)
H E-1 (AQ)
CA E-2 (AQ)
O H E (AQ)
S O4E2 (AQ)
```

```
DEBYE-HUCKEL PARAMETERS FOR AQUEOUS PHASE
A(PHI) =    0.3915      B(PHI) =    0.2522

TEMPERATURE OF SYSTEM    298.15 K

PRESSURE OF SYSTEM   1.000      ATM

EQUILIBRIUM STATE OF SYSTEM AT    298.15 K AND     1.0000 ATM

GAS                   MOLE        GRAM        MOLE FRN    FUGACITY

N2 (G)             0.1000E-02  0.2801E-01  0.7750      0.7750
O2 (G)             0.2500E-03  0.8000E-02  0.1938      0.1938
H2 (G)             0.0000E+00  0.0000E+00  0.0000E+00  0.0000E+00
H2O (G)            0.4028E-04  0.7257E-03  0.3122E-01  0.3122E-01

PHASE  TOTALS      0.1290E-02  0.3674E-01   1.000

AQUEOUS               MOLE        GRAM        MOLALITY    ACTIVITY
H2O                 55.55        1001.        55.51       0.9996
NA E-1 (AQ)        0.2000E-05  0.4598E-04  0.1999E-05  0.1655E-05
H E-1 (AQ)         0.1232E-06  0.1242E-06  0.1231E-06  0.9957E-07
CA E-2 (AQ)        0.1444E-01  0.5788      0.1443E-01  0.4872E-02
O H E (AQ)         0.1232E-06  0.2096E-05  0.1231E-06  0.1026E-06
S O4E2 (AQ)        0.1444E-01   1.387      0.1443E-01  0.4873E-02

PHASE  TOTALS       55.58        1003.        55.54
ION STRENGTH        0.5773E-01

SINGLE                MOLE        GRAM                    ACTIVITY

CA S O6H4          0.9856        169.7                    1.000
NA2 S O4           0.0000E+00  0.0000E+00              0.7040E-14

SYSTEM TOTALS       56.56        1172.
```

FIGURE 7. Input data and resulting output on application of THERMODATA/CHEMIX to calculation of the solubility of gypsum in high-ionic strength solution. (From Turnbull, A. G. and Wadsley, M. W., The CSIRO-SGTE THERMODATA System (Version V), CSIRO Division of Mineral Chemistry, Melbourne, 1987. With permission.)

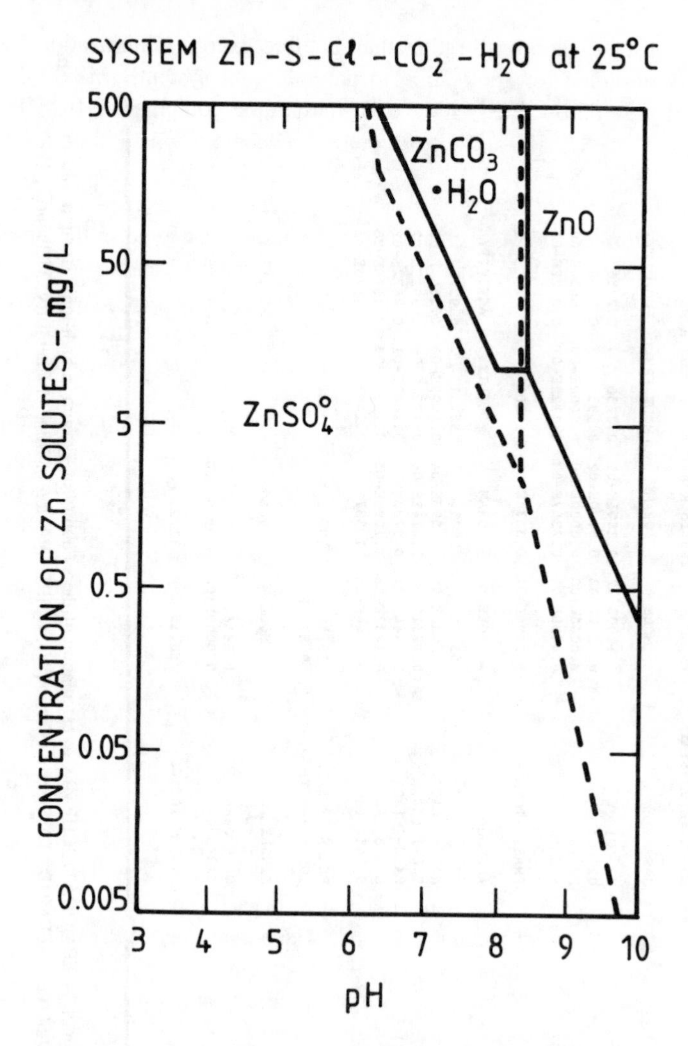

FIGURE 8. Concentration (mg/l) of dominant soluble zinc species plotted against pH for groundwaters exhibiting salinity extremes typical of Australian groundwaters. The solid lines correspond to chloride, sulfate and carbonate concentrations of 1.0, 0.1, and 0.01 *M*, respectively, while the dotted lines correspond to chloride, sulfate and carbonate concentrations of 0.01, 0.01, and 0.001 *M*, respectively. (From Giblin, A. M., Unpublished data, 1987.)

Giblin[34] has applied the Pitzer model within the CHEMIX program of THERMODATA to estimation of activity coefficients for a wide range of ionic species within Australian groundwaters. Activity coefficients estimated in this way have then been used to obtain concentration vs. pH diagrams for trace elements such as copper, zinc, and uranium. An example of such computations for zinc in groundwaters exhibiting order of magnitude differences in salinity is shown in Figure 8. The predominance diagrams were obtained using the SYSTEM program of THERMODATA. This program utilizes a modified version of the routine, STABLE, developed by Froning et al.[111]

B. MINEQL/MICROQL

1. Program Synopsis

As indicated earlier, MINEQL[65] and MICROQL[69] have evolved from ideas and experience gained through the use of the REDEQL series of programs.[63,64] Both programs use an "equilibrium constant" approach rather than a "free energy minimization" approach and are specifically designed for application to natural water systems. MINEQL is written in Fortran IV and may be operated either on a mainframe computer or IBM PC, while MICROQL is written in BASIC and is particularly amenable to operation on a personal computer.

Both MINEQL and MICROQL utilize the concept of "species" and "components". Species are here defined as every chemical entity to be considered in the problem, while a set of components is defined in such a way that every species can be written as the product of a reaction involving only the components, and no component can be written as the product of a reaction involving only the other components. In algebraic terms, the components can be described as a linearly independent set spanning species space. For each component there will be a material balance equation, while associated with each of the species is a mass law equation; these two sets of equations define the chemical equilibrium problem.

When using MICROQL, all species to be considered in the computation must be included in the input data file, while for MINEQL the only input data required is a list of components for the system of interest and associated measured total and predicted free concentrations for all components. All species that are linear combinations of the chosen components and that are included in a stored thermodynamic data base (with associated formation constants) are then considered in the ensuring computations:

- **Type I species** — are the soluble species which correspond to the components.
- **Type II species** — are the complexes (all soluble species which are not components).
- **Type III species** — are precipitated solids which are not allowed to dissolve, even if the amount of these solids becomes negative. Examples are gases at a fixed partial pressure, or simply a solid phase which is specified to be present. In addition, if the concentration (activity) of any soluble species is to be fixed (e.g., pH), this fixed soluble species is included as a Type III species.
- **Type IV species** — are precipitated solids which are subject to dissolution if the amount present becomes less than zero.
- **Type V species** — are dissolved solids which are subject to precipitation if the solubility product is exceeded.
- **Type VI species** — are species which are not to be considered at all (e.g., dissolved solids which are not subject to precipitation).

The extensive thermodynamic data base provided with MINEQL has been compiled from a variety of sources[112-117] and contains stability constants for most of the species expected in natural waters. The facility for modification of stored data or inclusion of additional species and associated formation constants not present in the existing data base is available. The Davies extension of the Debye-Huckel expression is included in MINEQL, enabling computations up to an ionic strength of about 0.7 *M*, but no facility is provided for computations at temperatures other than 25°C. MICROQL does not contain routines for ionic strength or temperature correction; thus, conditional stability constants must be supplied for analysis of systems at conditions other than zero ionic strength or 25°C.

2. Mathematical Formulation of MINEQL/MICROQL

The mathematical framework of these programs stems directly from the definitions of species and components given above and from the material balance and mass law expressions given in Equations 1 and 11. Rather than retain the terminology used in Section II, we will adopt that used by Westall et al.[65,69] in describing MINEQL and MICROQL.

Mass law equations for all species in an aqueous system may be written in the form:

$$C_i = K_i^c \prod_{j=1}^{n} X_j^{a_{ij}} \qquad \text{for} \quad i = 1,\ldots,m \tag{31}$$

where C_i is the concentration of the species i; K_i is the "concentration" formation constant for species i; X_j is the concentration of component j; and a_{ij} is the stoichiometric coefficient of component j in species i. Equation 31 can also be written in logarithmic form:

$$C_i^* = K_i^{c*} + \sum_{j=1}^{n} a_{ij}X_j^* \qquad \text{for} \quad i = 1,\ldots,m \tag{32}$$

where $C_i^* = \log C_i$; $K_i^{c*} = \log K_i^c$; and $X_i^* = \log X_j$. Equation 32 may be written in matrix form, i.e.,

$$C^* = K^{c*} + A \cdot X^* \tag{33}$$

where C^* is a column vector of C_i^*; K^{c*} is a column vector of K_i^{c*}; A is a matrix of stoichiometric coefficients; and X^* is a column vector of X_j^*. Similarly, material balance equations may be written in the general form

$$\sum_{i=1}^{m} a_{ij}C_i - T_j = Y_j \qquad \text{for} \quad j = 1,\ldots,n \tag{34}$$

where T_j is the total (analytical) concentration of component j, and Y_j is the error in the material balance equation for component j. The problem is solved when all $Y_j = 0$. Equation 34 can also be written in matrix form:

$$^tA \cdot C - T = Y \tag{35}$$

where ^tA is the transpose of the stoichiometry matrix A; C is a column vector of C_i; T is a column vector of T_j; and Y is a column vector of Y_j.

As chemical equilibrium problems are normally posed, we are given the total (analytical) concentrations of all components, the stoichiometry and stability constants of all species, and are asked to find the free equilibrium concentration of all components (from which we can easily compute the free concentrations of all species). This problem is solved in the manner described below.[69]

From an initial guess for the concentrations of components, the concentration of each species is computed:

$$C^* = K^{c*} + A \cdot X^* \tag{33}$$

From the concentration of the species, the error, or the remainder in the material balance equations is computed:

$$Y = {}^tA \cdot C - T \tag{35}$$

Then an iterative technique is used to find improved values of X such that the value of the error Y becomes smaller. MINEQL and MICROQL use the Newton-Raphson method in which improved values for X are found from the matrix equation

$$Z \cdot \Delta X = Y \tag{36}$$

where Z (a square matrix) is the Jacobian of Y with respect to X, i.e.,

$$Z_{jk} = \frac{\partial Y_j}{\partial X_k} = \sum_{i=1}^{m} a_{ij}a_{ik}C_i/X_k \quad (37)$$

ΔX (a column vector) is the improvement in X, i.e.,

$$\Delta X = X_{original} - X_{improved} \quad (38)$$

Y (a column vector) is the error in the material balance equations.

Equation 36 can be solved for ΔX by inversion of the Z matrix or by solution of the set of linear equations. In some cases, the computed $X_j^{new} = X_j^{old} - \Delta X_j$ may become less than zero. To avoid this physically meaningless situation, the value of X_j^{new} is set equal to $X_j^{old}/10$ in such cases, and iteration is continued. The iterative procedure is carried out until the error in the material balance equation is small with respect to the terms in the equation. Mathematically, the criterion for convergence may be written

$$\frac{|Y_j|}{\max(Y_j)} < \epsilon \qquad j = 1,\ldots,n \quad (39)$$

where max (Y_j) is the maximum of the absolute values of the set of terms (T_j; a_{ij} C_i, for all species i) of which Y_j is the sum.

In order to improve the efficiency of computation, it is often necessary to transform the basis set $X_1, \ldots, X_n$ such that the resulting Z matrix approaches a diagonal matrix. This transformation is carried out by substitution of one of the C_i, say C_{ii}, for one of the X_j. C_{ii} then becomes basis element X_{n+1}. The Z matrix will approach diagonality if the chosen components are either dominant species within the system of interest or are species of fixed activity. The ability to transform the basis component set is an important facet of MINEQL and MICROQL that aids rapid convergence to the equilibrium solution.

3. Application of MINEQL and MICROQL

Three examples of the application of MINEQL are given below. In Example 1, the equilibrium composition of a solution to which 10^{-3} mol/l of $CaCO_3$ has been added is determined and described in some detail as a demonstration of the method of operation of this model. The results of computations on more complex systems are given in Examples 2 and 3. The distribution of species in the growth medium AQUIL is considered in Example 2, while the changes that occur in trace element speciation on oxidation of sewage are modeled in Example 3.

a. Example 1. Addition of 10^{-3} M $CaCO_3$ to Solution

In this example the aqueous phase and the two solid phases $CaCO_3(s)$ and $Ca(OH)_2(s)$ are considered, but the gas phase is not. The species to be considered in this computation and their respective stability constants are given in Table 4, and a schematic presentation of these data, similar to their representation in MINEQL and MICROQL, is given in Table 5. The entities Ca^{2+}, H^+, and CO_3^{2-} have been chosen as components, and every species can be written as a linear combination of these components. It is assumed initially that no solid phases are present. The solubility of each of the possible solids will be tested after the equilibrium speciation for the soluble species has been computed.

An initial uneducated guess for $[Ca^{2+}]$, $[H^+]$ and $[CO_3^{2-}]$ is made;

$$[Ca^{2+}]^0 = 10^{-3}\,M; \quad [H^+]^0 = 10^{-10}\,M; \quad [CO_3^{2-}]^0 = 10^{-4}\,M$$

and the concentrations of each species is calculated using the expressions in Table 4. The

Table 4
SPECIES CONSIDERED IN THE MINEQL SOLUTION TO THE CARBONATE PROBLEM

Species	Mass law expression	Formation constants
1. Ca^{2+}		
2. H^+		
3. CO_3^{2-}		
4. $CaCO_3$	$[Ca^{2+}][CO_3^{2-}]K = [CaCO_3]$	Log K = 3.0
5. $(CaHCO_3)^+$	$[Ca^{2+}][H^+][CO_3^{2-}]K = [CaHCO_3^+]$	Log K = 11.6
6. $(CaOH)^+$	$[Ca^{2+}][H^+]^{-1}K = [CaOH^+]$	Log K = −12.2
7. HCO_3^-	$[H^+][CO_3^{2-}]K = [HCO_3^-]$	Log K = 10.2
8. H_2CO_3	$[H^+]^2[CO_3^{2-}]K = [H_2CO_3^*]$	Log K = 16.5
9. OH^-	$[H^+]^{-1}K = [OH^-]$	Log K = −14.0
10. $Ca(OH)_2(s)$	$[Ca^{2+}][H^+]^{-2}K = [Ca(OH)_2(s)]$	Log K = −21.9
11. $CaCO_3(s)$	$[Ca^{2+}][CO_3^{2-}]K = [CaCO_3(s)]$	Log K = 8.3

Table 5
SCHEMATIC REPRESENTATION OF DATA FOR MINEQL SOLUTION TO THE CALCIUM CARBONATE PROBLEM BEFORE AND AFTER MODIFICATION FOR THE PRESENCE OF $CaCO_3(s)$

	Initial component set				New component set			
Species	Ca^{2+}	H^+	CO_3^{2-}	Log K	Ca^{2+}	H^+	$CaCO_3(s)$	Log K
1. Ca^{2+}	1	0	0	0.0	1	0	0	0.0
2. H^+	0	1	0	0.0	0	1	0	0.0
3. CO_3^{2-}	0	0	1	0.0	−1	0	1	−8.3
4. $CaCO_3(aq)$	1	0	1	3.0	0	0	1	−5.3
5. $CaHCO_3^+$	1	1	1	11.6	0	1	1	3.3
6. $CaOH^+$	1	−1	0	−12.2	1	−1	0	−12.2
7. HCO_3^-	0	1	1	10.2	−1	1	1	1.9
8. H_2CO_3	0	2	1	16.5	−1	2	1	8.2
9. OH^-	0	−1	0	−14.0	0	−1	0	−14.0
10. $Ca(OH)_2(s)$	1	−2	0	−21.9	1	−2	0	−21.9
11. $CaCO_3(s)$	1	0	1	8.3	0	0	1	0.0

"calculated analytical concentrations" of Ca^{2+}, H^+, and CO_3^{2-}, as determined from the concentration of all species, is then computed:

$$[Ca^{2+}]_T^{calc} = [Ca^{2+}] + [CaCO_3] + [CaHCO_3] + [CaOH^+]$$
$$= 1.11 \times 10^{-3}\ M$$

$$[H]_T^{calc} = [H^+] + [CaHCO_3^+] - [CaOH^+] + [HCO_3^-] + 2[H_2CO_3^*] - [OH^-]$$
$$= 5.62 \times 10^{-5}\ M$$

$$[CO_3^{2-}]_T^{calc} = [CO_3^{2-}] + [CaCO_3] + [CaHCO_3^+] + [HCO_3^-] + [H_2CO_3^*]$$
$$= 3.62 \times 10^{-5}\ M$$

Then the difference function between the imposed analytical concentration $[X]_T$ and the calculated $[X]_T^{calc}$ is evaluated:

$$Y_{Ca^{2+}} = [Ca^{2+}]_T^{calc} - [Ca^{2+}]_T = 1.10 \times 10^{-4}\, M$$

$$Y_{H^+} = [H^+]_T^{calc} - [H^-]_T = 5.62 \times 10^{-5}\, M$$

$$Y_{CO_3^{2-}} = [CO_3^{2-}]_T = 6.37 \times 10^{-4}\, M$$

According to the Newton-Raphson method, the difference functions (Y's) and their derivatives may be used to find improved values for $[Ca^{2+}]$, $[H^+]$, and $[CO_3^{2-}]$ such that the calculated analytical concentrations, $[X]_T^{calc}$, more closely approach the imposed analytical concentrations, $[X]_T$. Thus, for Ca^{2+}:

$$\frac{\partial Y_{Ca^{2+}}}{\partial [Ca^{2+}]}([Ca^{2+}]^0 - [Ca^{2+}]^{calc}) + \frac{\partial Y_{Ca^{2+}}}{\partial [H^+]}([H^+]^0 - [H^+]^{calc})$$

$$\frac{\partial Y_{Ca^{2+}}}{\partial [CO_3^{2-}]}([CO_3^{2-}] - [CO_3^{2-}]^{calc}) = Y_{Ca^{2+}} \qquad (40)$$

The derivatives are readily evaluated; for example,

$$\frac{\partial Y_{Ca^{2+}}}{\partial [Ca^{2+}]} = 1 + \frac{[CaCO_3]}{[Ca^{2+}]} + \frac{[CaHCO_3^+]}{[Ca^{2+}]} + \frac{[CaOH^+]}{[Ca^{2+}]} = 1.11$$

Substitution of values for derivatives in Equation 40 results in a system of simultaneous equations which, on solution, yield

$$[Ca^{2+}]^{calc} = 6.45 \times 10^{-4}\, M, \quad [H^+]^{calc}$$
$$= 8.33 \times 10^{-4}\, M, \quad [CO_3^{2-}]^{calc} = 3.68 \times 10^{-4}\, M$$

The new value for $[H^+]$ is negative and physically meaningless; thus, $[H^+]^{calc}$ is set to $[H^+]^0/10$ and the iterative process continued until the value of each difference function is small compared to the terms in the difference function ($\epsilon = 10^{-4}$ has been chosen here). The full solution to this problem is given in Table 6.

The solubility of the solid phases must now be computed; i.e.,

$$Ca(OH)_2(s){:}\ [Ca^{2+}][H^+]^{-2} \cdot 10^{-21.9} = 10^{-4.24}$$

$$CaCO_3(s){:}\ [Ca^{2+}][CO_3^{2-}] \cdot 10^{8.3} = 10^{+1.78}$$

It is seen that the solubility product of $CaCO_3(s)$ is exceeded, i.e., $[Ca^{2+}]\,[CO_3^{2-}]\,K > 1$; thus, the problem must be modified to include the solid phase $CaCO_3(s)$. The schematic representation of the modified problem is given in Table 5. (Note that, by modification, the basis element CO_3^{2-} has effectively been eliminated from the computation.) The previously calculated values for $[Ca^{2+}]$ and $[H^+]$ are used as initial guesses, and the modified problem is solved as before for the soluble species. The results of these computations are given in Table 6. A check on the mass balance for Ca^{2+} indicates that $8.73 \times 10^{-4}\, M$ $CaCO_3(s)$ is present and the solubility product of $Ca(OH)_2(s)$ is not exceeded.

b. Example 2. Species Distribution in the Growth Medium AQUIL

An essential requirement of studies of trace metal availability or toxicity to phytoplankton is that metal speciation is known as precisely as possible. This has been achieved in the growth medium AQUIL[118] by careful control of contaminants, by avoidance of formation of precipitates and adsorbates, and by computer simulation of the distribution of species.

Table 6
SOLUTIONS TO THE CALCIUM CARBONATE PROBLEM OBTAINED USING MINEQL IN THE ABSENCE AND PRESENCE OF $CaCO_3(s)$

Species	Solution 1: no solid phases −log[x]	Solution 2: $CaCO_3(s)$ present −log[x]
Ca^{2+}	3.16	3.91
H^+	10.41	9.91
CO_3^{2-}	3.36	4.38
$CaCO_3$	3.53	5.30
$CaHCO_3^+$	5.34	6.61
$CaOH^+$	4.95	6.21
HCO_3^-	3.58	4.09
$H_2CO_3^*$	7.69	7.70
OH^-	3.59	4.09
$CaCO_3(s)$	—	3.13
$Ca(OH)_2(s)$	—	—

The results of such a simulation for AQUIL using MINEQL are shown in Table 7. These computations involved 99 complexes and 27 possible solids and were performed at a fixed pH of 8.10 and an ionic strength of 0.5 *M*. The fixed pH condition is achieved by classifying H^+ as a Type III species.

c. Example 3. Trace Element Speciation in Sewage

Morel et al.[119] have investigated the behavior of trace elements in sewage upon dilution and oxidation of the waste and have made extensive use of speciation modeling in these investigations. The model REDEQL[63] (a precursor to MINEQL and essentially identical in structure) was used in these studies. The input to the model includes the analytical concentrations of trace metals found in the sewage of interest (in this case the waste from the Los Angeles County Sanitation District), the analytical concentrations of major metals and ligands, and the measured pH of 7.7 (Table 8, Column 1).

The redox level of the sewage, indicated in this model by $pe = -\log\{e^-\}$, is certain to be a critical determinant of metal speciation. Morel et al.[119] consider that the pe is probably best described as a function of the bacterially mediated reduction of sulfate:

$$SO_4^{2-} + 8H^+ + 8e^- \rightleftharpoons S^{2-} + 4H_2O$$

coupled with the oxidation of organic carbon. The activity of SO_4^{2-} is much greater than that of S^{2-} and effectively constant; thus, the free sulfide activity is a valuable indicator of redox level. Modeling results of trace metal speciation as a function of the negative logarithm of the total sulfide concentration in the sewage (pTOT S^{2-}) are shown in Figure 9. It is seen that over the range pTOT S^{2-} = 3.7 to 5.5, the activity of free sulfide ion is controlled successively by the solubilities of iron, zinc, and copper sulfides; accordingly the pe as computed from the sulfate/sulfide equilibrium varies over three plateaus at pe −4.0, −3.5, and −2.9. Morel et al.[119] report that experimental data show the majority of trace metals to be in the nonfilterable fraction of this sewage, with the exception of nickel, cobalt, and manganese which are found in approximately equal concentrations in the filterable and nonfilterable fractions. The transition range between pe of −4.0 and pe of −3.5 appears to be in reasonable agreement with this information — most metals are in an insoluble

Table 7
CHEMICAL SPECIATION OF THE GROWTH MEDIUM AQUIL OBTAINED USING THE EQUILIBRIUM MODEL MINEQL

	Analytical concentration (*M*)	Computed −log activity	Computed major species (%)
Bromide	8.40×10^{-4}	3.24	Br^-(100)
Borate	4.85×10^{-4}	4.30	H_3BO_3(85); $B(OH)_4^-$(15)
Calcium	1.05×10^{-2}	2.67	Ca^{2+}(87); $CaSO_4$(12)
Carbonate	2.38×10^{-3}	4.97	HCO_3^-(65); $MgHCO_3^+$(17); $MgCO_3$(7); $NaCO_3^-$(3); $CaHCO_3^+$(3); CO_3^{2-}(2)
Chloride	5.59×10^{-1}	0.41	Cl^-(100)
Cobalt	2.50×10^{-9}	11.50	CoY^{2-}(99)
Copper	9.97×10^{-10}	14.41	CuY^{2-}(100)
EDTA(Y)	5.00×10^{-6}	14.68	CaY^{2-}(87); $FeYOH^{2-}$(9); MgY^{2-}(4)
Fluoride	7.14×10^{-5}	4.54	F^-(59); MgF^+(39); CaF^+(2)
Iron	4.51×10^{-7}	20.23	$FeYOH^{2-}$(98); FeY^-(1); $Fe(OH)_2^+$(1)
Magnesium	5.46×10^{-2}	1.96	Mg^{2+}(85); $MgSO_4$(14); $MgCO_3$(1)
Manganese	2.30×10^{-8}	8.89	$MnCl^+$(38); $MnCl_2$(10); $MnCl_3^-$(2); Mn^{2+}(24); MnY^{2-}(22); $MnSO_4$(3)
Molybdate	1.50×10^{-9}	9.45	MoO_4^{2-}(100)
Nitrate	1.00×10^{-4}	1.16	NO_3^-(100)
Phosphate	1.00×10^{-5}	10.25	HPO_4^{2-}(48); $MgHPO_4$((46); $CaHPO_4$(3); H_2PO_4(3)
Potassium	1.03×10^{-2}	2.16	K^+(97); KSO_4^-(3)
Silicate	1.25×10^{-5}	11.26	$HSiO_3^-$(6); H_2SiO_3(94)
Sodium	4.80×10^{-1}	0.49	Na^+(99); $NaSO_4^-$(1)
Strontium	6.38×10^{-5}	4.83	Sr^{2+}(100)
Sulfate	2.88×10^{-2}	2.53	SO_4^{2-}(44); $MgSO_4$(27); $NaSO_4$(24); $CaSO_4$(4); KSO_4^-(1)
Zinc	4.00×10^{-9}	11.50	ZnY^{2-}(100)

From Morel, F. M. M., Rueter, J. G., Anderson, D. M., and Guillard, R. R. L., *J. Phycol.*, 15, 135, 1979. With permission.

sulfide (Zn, Hg, Ag, Cu, Cd, Pb) or oxide (Cr, Fe) form; the sulfides of cobalt and nickel dissolve in this region, cobalt being replaced by a more soluble $CoCO_3$. Significant amounts of the soluble trace elements such as nickel and manganese may be adsorbed to other particulate matter. A summary of the major metal species in this intermediate range, computed at a pe of −3.6, is given in Table 8.

To make the model more realistic, organic ligands should be included. While little definitive analytical information on the identity and concentration of organic compounds in sewage is available, calculations in which a range of likely compounds at realistic concentrations are included indicate that organic complexation is unlikely to dramatically modify the results of inorganic modeling (see Table 8). The only metals that show significant binding with the organic ligands are cobalt and nickel; the reactivity of most other metals is limited due to the extreme insolubility of the solid forms.[119]

C. EQ3/6

1. Program Synopsis

EQ3/6 is a set of related computer codes and data files for use in geochemical modeling of aqueous systems. Development of EQ3/6 was initiated by Wolery[120] in 1975 for the purpose of modeling the reaction of seawater and basalt in midocean ridge hydrothermal systems. Considerable upgrading of the capabilities of EQ3/6 has occurred in recent years

Table 8
TRACE ELEMENT SPECIATION IN LOS ANGELES COUNTY SEWAGE IN THE ABSENCE AND PRESENCE OF SELECTED ORGANIC COMPOUNDS COMPUTED USING THE CODE REDEQL

−Log total concentration	Inorganic model,[a] %	Addition of organics,[a,b] %
Fe, 3.7	$Fe_3O_4(s)$, 100	$Fe_3O_4(s)$, 100
Cr, 4.8	$Cr(OH)_3(s)$, 97	$Cr(OH)_3(s)$, 97
	$Cr(OH)_4^-$, 3	$Cr(OH)_4-$, 3
Cu, 5.0	CuS(s), 100	CuS(s), 100
Cd, 6.5	CdS(s), 100	CdS(s), 100
Pb, 6.0	PbS(s), 100	PbS(s), 100
Zn, 4.5	ZnS(s), 100	ZnS(s), 99
Ag, 6.7	$Ag_2S(s)$, 100	$Ag_2S(s)$, 100
Hg, 8.3	HgS(s), 100	HgS(s), 100
Ni, 5.4	NiS(s), 42	NiS(s), 22
	Ni^{2+}, 2	Ni^{2+}, 2
	$Ni(CN)_4^{2-}$, 56	$Ni(CN)_4^{2-}$, 56
		GLY, 10
		GLU, 10
Co, 6.8	CoS(s), 97	CoS(s), 95
	Co^{2+}, 2	Co^{2+}, 2
		GLU, 2
Mn, 5.6	Mn^{2+}, 50	Mn^{2+}, 50
	$MnHCO_3^+$, 24	$MnHCO_3^+$, 24
	$MnSO_4$,19	$MnSO_4$, 19
	$MnCl^+$, 6	$MnCl^+$, 6

[a] Other inputs to the model: pH = 7.7, pCa = 2.75, pMg = 3.0, pBa = 5.0, pAl = 4.5, pCO_3 = 2.0, pSO_4 = 2.3, pCl = 1.8, pF = 4.0, pNH_4 = 3.3, pCN = 5.0. Ionic strength = 0.01, temp. = 25°C.
[b] Organic ligands added: pAcetate = 3.3, pGlycine = 3.8, pTartrate = 3.6, pGlutamate = 3.7, pSalicylate = 3.8, pPhthalate = 3.9.

From Morel, F. M. M., Westall, J. C., O'Melia, C. R., and Morgan, J. J., *Environ. Sci. Technol.*, 9, 756, 1975. With permission.

through funding provided by the U.S. Department of Energy Radioactive Waste Management Program.

The EQ3/6 package centers around two large computer codes, EQ3NR and EQ6, which are supported by a common thermodynamic data base. EQ3NR is a speciation-solubility code,[121] the function of which is to compute a model of the state of an aqueous solution. Input may consist of analytical measurements, assumptions (such as that the fluid is in equilibrium with specified minerals), or some mixture of measurements and assumptions. The output contains the distribution of aqueous species, their thermodynamic activities, and saturation indices for various solids. The output also includes a calculation of the electrical balance of the fluid, which is considered a useful indicator of the quality and completeness of solution analyses.[121]

EQ6 is a reaction path code which calculates models of changes in aqueous systems as

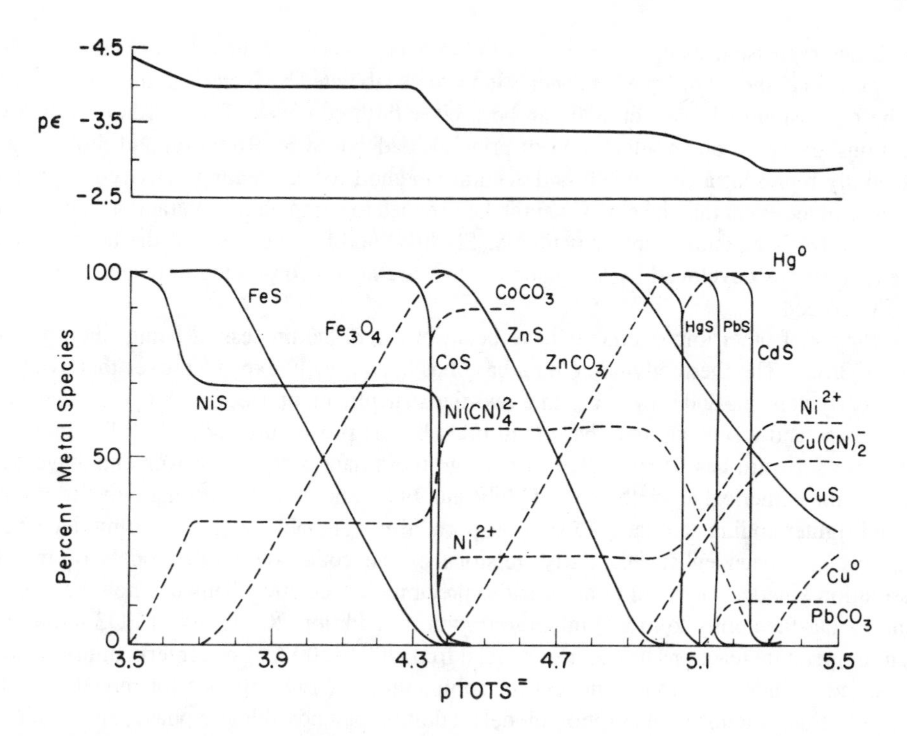

FIGURE 9. Trace element speciation in sewage as a function of total sulfide concentration. (From Morel, F. M. M., Westall, J. C., O'Melia, C. R., and Morgan, J. J., *Environ. Sci. Technol.*, 9, 756, 1975. With permission.)

they proceed toward a state of overall chemical equilibrium.[79,122] These models are divisible into those assuming "instant" equilibration of a system not at equilibrium, reaction-path calculations using arbitrary kinetics to describe mass transfer for reactions that do not achieve instantaneous equilibrium (these calculations essentially represent titrations), and reaction-path calculations using actual kinetic rate laws. EQ6 runs are initialized by entering a description of the starting aqueous fluid (from EQ3NR) and defining the constraints of the reaction path (i.e., identifying the irreversibly reacting materials and specifying the controls on their rate of reaction). EQ6 can also model reaction paths where temperature and/or pressure vary according to a time-dependent equation and can model the time-controlled addition of other reactants such as a gas, aqueous species, or another aqueous solution.

EQ3/6 is currently being supported and developed by the Lawrence Livermore National Laboratory (LLNL) in California. The principal usages of EQ3/6 by LLNL include calculating aqueous speciation and mineral saturation indices from groundwater analysis (EQ3NR), screening analytical data by calculation of apparent charge imbalance (EQ3NR), designing pH buffer compositions for single mineral solubility and dissolution rate experiments at various temperatures (EQ3NR), calculating solubilities of trace element and radionuclide-bearing solids in groundwaters (EQ3NR, EQ6), and simulation of solid/water interaction experiments.

Early versions of EQ3/6 (up to and including version 3230, released in August 1983) used an extended Debye-Huckel expression to account for ionic strength effects, but two alternative sets of equations for computing the activity coefficients of aqueous solute species and the activity of water in high salinity systems have been incorporated in recent versions of the EQ3/6 code. These two sets comprise the equations formulated by Pitzer[27,29] as well as an original set of "hydration theory" equations.[123] The implementation of Pitzer's equations has been undertaken in a very general fashion to allow EQ3/6 to use any of the several data bases regressed, using various conventions and choices of aqueous components. For example, either the set of aqueous species in the data tabulated by Pitzer[29] or the set supported by the regression of Harvie et al.[87] can be used with these codes. The "hydration theory" equations are based on an adaptation of principles advanced by Robinson and Stokes[22] and potentially represent a convenient and accurate method for calculating activity coefficients in brines in the event that the experimental data needed to compute the interaction coefficients used in Pitzer's equations are unavailable. Slightly modified versions of the activity coefficient equations developed by Helgeson et al.[124] are also incorporated in recent versions of the EQ3/6 code.

A variety of other improvements have been, or are in the process of being, incorporated into EQ3/6.[125] The thermodynamic data base has been greatly expanded over that available in the early versions and now exists in a two-tier structure composed of the operational EQ3/6 data base and a separate data base controlled by the processing code MCRT. Much new data base software has been written to manage these data bases. Improvements have been made in the numerical algorithms in EQ3NR and EQ6, and there has been a move to Fortran 77 and tighter coding standards to support portability. The ability of EQ6 to model kinetic data has also been enhanced. Early versions of the code could run models of mineral dissolution kinetics, but could not consider precipitation kinetics. This has now been overcome as has the ability to model mineral growth. In addition, the ability of EQ3/6 to model high-temperature reactions has been expanded from 300 to 500°C. Longer term improvements envisaged include site-mixing models for solid solutions, EQ6 options for modeling redox disequilibrium, inclusion of sorption models, addition of a possible gas phase, and continued expansion and maintenance of necessary data bases.[125]

2. Mathematical Formulation of EQ3/6

The EQ3NR code performs aqueous speciation calculations using an "equilibrium constant" approach in conjunction with either an ion-association model or, at high ionic strengths, a hybrid ion-association-specific interaction model. A subset of aqueous species is classified as the basis set, with this set further classified as strict or auxiliary basis. Except for $O_2(g)$, which is used as a hypothetical aqueous species, the species in the strict basis set correspond one to one with chemical elements. The auxiliary basis set represents the chemical elements in different valence forms. These species can either act as master species if "switched", or as derivative species if "not switched". With these definitions, the aqueous system can be defined by the following mass balance, mass action, and electroneutrality equations:

mass balance:

$$C_{T,i} = C_i + \sum_{j=1}^{n} a_{ij} C_i \qquad \text{for } i = 1,\ldots,m \tag{41}$$

mass action:

$$C_j = K_j^c \prod_{i=1}^{m} C_i^{a_{ij}} \qquad \text{for } j = 1,\ldots,n \tag{42}$$

or

$$\gamma_j C_j = K_j \prod_{i=1}^{m} (\gamma_i C_i)^{a_{ij}} \qquad \text{for } j = 1,\ldots,n \tag{43}$$

electroneutrality:

$$0 = \sum_{i=1}^{m} C_i z_i + \sum_{j=1}^{n} C_j z_j \quad (44)$$

where

C_i = concentration of free element of species i.
C_j = concentration of complex ion j containing a_{ij} moles of i.
$C_{T,i}$ = total (analyzed) concentration of i.
a_{ij} = stoichiometric coefficient of ion i in species j.
γ_i and γ_j = activity coefficients for free elements i and solution complex j, respectively.
K_j^c = "concentration" formation constant of complex j.
K_j = "activity" formation constant of complex j.
z_i and z_j = charges on dissolved element i and complex j, respectively.

Input to EQ3NR include the temperature, solution density, concentration of total dissolved salts, and a redox parameter (either log oxygen fugacity, E_h or pe). The identity of the ion species that will be used in electrical balancing must also be specified. The user has the option of suppressing certain reactions or of altering thermodynamics for selected reactions. He can also designate any aqueous species that should be "basis switched" with the master species. In addition, the total (analytical) concentrations of elements in the aqueous phase must be specified. Once this data has been provided, Equations 41 to 44 are solved iteratively. Early versions of the programs used the method of successive approximations, but this has been replaced with the Newton-Raphson method in recent versions (hence, the designation EQ3NR).

The EQ6 code is implemented subsequent to an EQ3NR computation if there is a requirement to model instantaneous or time-varying precipitation or dissolution reactions or any other mass-transfer phenomenon. Solution reactions with solid or gas phases are handled in much the same way as are ion complexation reactions in the solution itself. For each mineral or gas reacting, a mass action expression can be written similar to Equation 43:

$$K_k = \prod_{i=1}^{m} (\gamma_i C_i)^{a_{ik}} \qquad \text{for} \quad k = 1,\ldots,p \quad (45)$$

γ_i = activity coefficients for dissolved free ions i.
C_i = concentration of dissolved free ions i.
K_k = equilibrium constant for the dissolution of phase k.
a_{ik} = stoichiometric coefficient of ion i in phase k.

The addition of this equation to the set of equations describing the system allows the addition of another unknown to the mass balance equations for each ion i. This additional term is simply the number of moles of i which must be transferred between the solution and the phase k to bring the system to equilibrium.[126]

The reaction path modeling capability of EQ6 is similar in function to the program PATHI or PATHCALC developed by Helgeson et al.,[71-74] but differs in numerical detail. Both EQ6 and Helgeson's model of partial equilibrium mass transfer allow division of the aqueous system into an equilibrium subsystem (aqueous solution plus minerals with which it is in heterogeneous equilibrium) and a reactant subsystem (another aqueous solution or a set of minerals that are not in equilibrium with the solution in the first subsystem). The reactant components are then titrated into the equilibrium subsystem according to a reaction progress variable (ξ) and specified relative rate constants. At any value of ξ, the problem reduces to finding the distribution of phases and species in an aqueous system at fixed temperature, pressure, and known mass for each element present.

Tracing a reaction path involves, in its simplest form, changing the mass constraints ($C_{T,i}$, $i = 1, \ldots, m$) on the closed system (the equilibrium subsystem) so that a series of compositionally neighboring chemical equilibrium problems may be solved. The user specifies a set of q reactants (q may be zero) and a corresponding set of reactant tracking coefficients (ξ), which may be interpreted as relative rate constants.[71] Then, using a power-series format to describe the rate of change in concentration of reactant l, we may write

$$\frac{-dC_l}{d\xi} = \xi_{1l} \pm \xi_{2l}\xi - \xi_{3l}\xi^2 \qquad \text{for} \quad l = 1,\ldots,q \tag{46}$$

where ξ_1, ξ_2, ξ_3 are reactant tracking coefficients for the input of a reactant into the closed system (and may be considered to be relative rate constants of order increasing from one), and ξ is the reaction progress variable. The change in the bulk (elemental) composition of the closed system with reaction progress ξ is given by

$$\frac{dC_{T,i}}{d\xi} = \sum_{l=1}^{q} a_{il}(dC_l/d\xi) \tag{47}$$

Integration of Equations 46 and 47 yields

$$C_{T,i}(\xi) = C_{T,i}(\xi = 0) + \sum_{l=1}^{q} a_{il}(\xi_{1l}\xi + \xi_{2l}\xi^2/2 + \xi_{3l}\xi^3/3) \tag{48}$$

The Newton-Raphson method is again used to solve the set of Equations 41, 44, 45, and 48. Information at previous points can be used to obtain good starting estimates at a new point of reaction progress. The technique is based on use of Taylor's series, as was the original method of Helgeson,[71] and all estimates of derivatives are based on finite-difference calculations. Further details of these procedures are reported by Wolery.[77]

3. *Application of EQ3/6*

a. Example 1. Iron and Uranium Speciation Under Reducing Conditions

In this example, the iron and uranium speciation that results on reducing an oxygenated, calcite- and hematite-saturated solution by adding methane is investigated. The problem is performed in two steps:[126]

- Simulate a starting solution saturated with respect to calcite ($CaCO_3$) and hematite (Fe_2O_3) and with fixed sodium, uranium, sulfate, and chloride concentrations. This solution is open to gas phase with fixed partial pressures of O_2 and CO_2.
- The starting solution, now no longer open to the O_2- and CO_2-containing gas phase, is reduced by the addition of methane to it. During this reduction, the solution maintains saturation with calcite. The system is assumed to contain 0.1 mmol of hematite per kilogram of water, and the solution stays in equilibrium with that hematite until it has entirely dissolved. Finally, it is assumed that the minerals uraninite (UO_2) and pyrite (FeS_2) will precipitate should the solution become supersaturated with respect to them.

The initial solution can be specified by combining fixed input molalities of some dissolved species with mineral equilibria control on others and gas partial-pressure specifications on the remainder. The solution pH will be calculated to satisfy electroneutrality. The information required as input to EQ3 is given in Table 9, and the composition of the starting solution calculated by EQ3 is given in Table 10.

To simulate the behavior of this system on reduction with methane requires only the

Table 9
INPUT DATA REQUIRED BY EQ3 IN CALCULATION OF INITIAL SOLUTION COMPOSITION FOR THE IRON/ URANIUM PROBLEM

Species	Specified by
Sodium	Input at 7.0 E-03 molal
Calcium	Calcite ($CaCO_3$) saturation
Iron	Hematite (Fe_2O_3) saturation
Uranium	Input at 4.0 E-05 molal
Carbonate	Input P(CO_2) of 0.01 atmosphere
Sulfate	Input at 1.0 E-03 molal
Chloride	Input at 5.0 E-03 molal
Hydrogen (pH)	Solution electroneutrality
Oxygen (pe)	Input P(O_2) of 0.2 atmosphere

Table 10
COMPOSITION OF INITIAL SOLUTION COMPUTED BY EQ3 FOR THE IRON/URANIUM PROBLEM

Species or condition		EQ3 results
Sodium	(Log molality)	−2.155
Calcium	(Log molality)	−2.752
Iron	(Log molality)	−16.304
Uranium[a]	(Log molality)	−4.398
Carbonate	(Log molality)	−2.415
Sulfate[a]	(Log molality)	−3.000
Chloride[a]	(Log molality)	−2.301
pH	(Units)	7.312
pe	(Units)	13.289
Ionic Strenth	(Molality)	0.0128
P_{CO_2}[a]	(Atmospheres)	0.010
P_{O_2}[a]	(Atmospheres)	0.200
Dissolved O_2	(Log molality)	−3.599

[a] Values fixed by input.

From Noronha, C. J. and Pearson, F. J., Jr., Report ONWI-472, Office of Nuclear Waste Isolation, U.S. Department of Energy, Washington, D.C., 1983.

starting system description (i.e., speciation results from Table 10), the specification that 1 mol of calcite and 0.1 mmol of hematite are available as reactants, and that the reaction path consists of the addition of a total of 1.2 mmol methane. With this input, the code itself determines which phases saturate or become exhausted, and at what values of reaction progress these system changes occur. The results of this simulation are shown in Figure 10. A number of abrupt changes in solution chemistry occur as solid phases saturate or become exhausted.

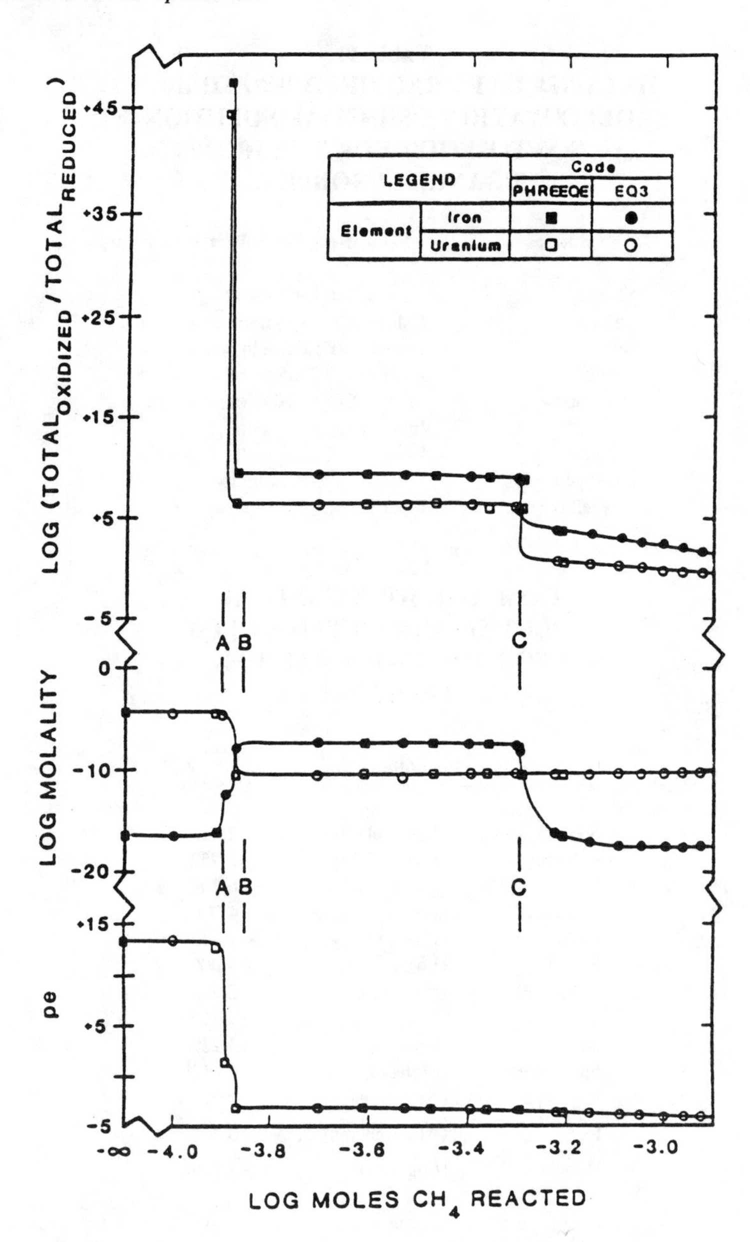

FIGURE 10. Variation in iron and uranium chemistry with reaction progress as simulated by the speciation codes PHREEQE and EQ3/6. The system state at selected points in the reaction path are as follows: (A) uraninite saturates; continuing calcite and hematite saturation; (B) pyrite saturates; saturated calcite, hematite, and uraninite; (C) hematite exhausted; saturated calcite, uraninite, and pyrite. (From Noronha, C. J. and Pearson, F. J., Jr., Report ONWI-472, Office of Nuclear Waste Isolation, U.S. Department of Energy, Washington, D.C., 1983.)

VI. MODEL INTERCOMPARISONS

Two extensive intercomparison studies of the major chemical speciation codes[12,127] and various comparative studies of particular aspects of these codes[13,33,46] have been published in recent years. Such studies give an indication of the aspects of chemical speciation that are adequately modeled by available codes and also highlight areas that are particularly

Table 11
DATA SETS TYPICAL OF RIVER, SEA, AND GROUND WATERS USED BY NORDSTROM ET AL.[12] AND BROYD ET AL.[127] IN MODEL INTERCOMPARISON STUDIES

River water		Seawater		Groundwater	
Species	mg/l	Species	mg/l	Species	mg/l
Na	12.0	Ca	421.931	Ca	62.1
K	1.4	Mg	1321.976	Mg	18.9
Ca	12.2	Na	11019.54	K	129.0
Mg	7.5	K	408.423	Na	3470
Si	8.52	Cl	19805.09	HCO_3	43.8
HCO_3	75.2	So_4	2775.35	SO_4	282
Cl	9.9	Alkal	144.992	Cl	5360
SO_4	7.7	Br	68.872	Al	0.0027
B	0.050	Sr	8.3302	T°C	25
Br	0.006	B	4.5540	pH	7.0
I	0.0018	SiO_2	4.3800		
F	0.10	F	1.4225		
PO_4	0.210	Ba	0.0205		
NO_3	0.898	I	0.06345		
NO_2	0.019	PO_4	0.0614		
NH_4	0.144	NO_3	0.2968		
Fe(II)	0.015	NO_2	0.0205		
Fe(III)	0.0007	NH_4	0.0307		
Mn	0.0044	Fe	0.00205		
Al	0.005	Mn	0.000205		
Zn	0.00040	Al	0.00205		
Cd	0.0001	Zn	0.005014		
Hg	0.00001	Cd	0.000102		
Pb	0.00003	Hg	0.0000307		
Cu	0.0005	Pb	0.0000512		
Co	0.0005	Cu	0.000716		
Ni	0.0018	Co	0.0000512		
Cr	0.0005	Ni	0.001740		
Ag	0.00004	Cr	0.000307		
Mo	0.0005	Ag	0.0000409		
As	0.002	Sb	0.0003377		
H_2S	0.002	Mo	0.00512		
DO	10.94	As	0.000409		
Eh(V)	0.440	Li	0.18523		
DOC	2.5	Rb	0.11973		
T°C	9.5	Cs	0.00409		
pH	8.01	DO	6.6		
Density	1.00	Eh (V)	0.500		
		T°C	25.0		
		pH	8.22		
		Density	1.02336		

sensitive to model formulation. Selected results of model intercomparison studies are presented here.

Nordstrom et al.[12] have performed model intercomparisons using data sets typical of river and sea waters, while Broyd et al.[127] used a data set comprised of selected elements at concentrations typical of a groundwater. The data used in these studies are given in Table 11. Results obtained by Nordstrom et al.[12] using 14 popular speciation codes are presented in Tables 12 to 15. All codes used are the "equilibrium constant" type, except that of

Table 12
p(M) (−LOG MOLALITY) OF SELECTED MAJOR SPECIES IN THE RIVER WATER TEST CASE

Program	EQUIL	EQ3	IONPAIR	MINEQL2[a]	MIRE	MINEQL/ REDEQL2	REDEQL2	SOLMNEQ	WATEQF	WATEQ2	WATSPEC
Ca^{2+}	3.532	3.529	3.522	3.54[b]	3.530	3.54[b]	3.539	3.527	3.525	3.525	3.525
$CaSO_4^0$	5.537	5.578	5.6	5.58[b]	5.533	5.55[b]	5.58	5.545	5.578	5.577	5.545
$CaHCO_3^+$	5.288	5.343	—	5.17[b]	5.409	5.22[b]	5.27	5.518	5.722	5.723	5.714
$CaCO_3^0$	5.681	5.933	6.	5.76[b]	5.732	5.82[b]	5.27	5.959	6.000	6.001	5.992
Mg^{2+}	3.523	3.519	3.518	3.53	3.521	3.54	3.53	3.519	3.519	3.519	3.520
$MgSO_4^0$	5.586	5.750	5.7	5.47	3.593	5.45	5.47	5.726	5.767	5.756	5.600
$MgHCO_3^+$	5.636	5.491	5.6	5.16	5.568	5.21	5.25	6.027	5.495	5.496	5.486
$MgCO_3^0$	5.470	6.133	6.	5.55	5.988	5.61	5.65	5.622	6.156	6.157	6.147
Na^+	3.283	3.282	3.283	3.28	3.284	3.28	3.28	3.283	3.283	3.283	3.283
$NaSO_4^-$	6.447	6.819	7.	6.82	7.280	6.81	6.82	6.617	6.819	6.820	6.793
$NaHCO_3^0$	6.505	—	—	—	6.500	—	—	6.496	6.495	6.496	6.494
K^+	4.446	4.446	4.45	4.45	4.444	4.45	4.45	4.446	4.446	4.446	4.446
KSO_4^-	7.830	7.896	—	7.59	7.819	7.58	7.60	7.942	7.935	7.936	7.910
SO_4^{2-}	4.129	4.122	4.12	4.14	4.128	4.15	4.14	4.122	4.121	4.121	4.164
Cl^-	3.401	3.554	3.554	3.55	3.556	3.55	3.55	3.558	3.554	3.554	3.554
HCO_3^-	2.924	2.920	2.915	2.93	2.919	2.91	2.93	2.918	2.917	2.917	2.917
CO_3^{2-}	5.164	5.346	5.3	5.01	5.168	5.11	5.12	5.328	5.334	5.333	5.333
$B(OH)_3^0$	5.361	—	—	5.37	—	—	—	5.379	5.354	5.355	5.355
$B(OH)_4^-$	6.570	—	—	6.41	—	—	6.42	6.726	6.690	6.711	6.688
Bi^-	—	—	—	7.12	—	—	7.12	—	7.124	7.124	7.120
F^-	5.281	—	—	5.29	—	—	5.29	5.282	5.284	5.284	—
$H_4SiO_4^0$	3.528	—	—	3.53[b]	—	—	3.53[b]	3.523	3.851	3.520	3.520
H^+	7.988	7.989	—	—	—	8.01	—	7.990	7.987	7.987	7.987
OH^-	5.966	6.518	6.	5.94	5.967	—	5.95	6.525	6.502	6.501	6.502
Ionic strength	0.00238	0.00239	0.00240	0.00300	0.0024	—	0.00300	0.00240	0.00240	0.00241	0.00239

[a] Calculated at 25°C.

[b] MINEQL2 and REDEQL2 results were calculated in such a way that the solution was equilibrated with supersaturated phases. P, Fe, Al, and Si species are not exactly comparable to the same species from the other programs because of this equilibrium process, and, of course, all species have been affected to some degree.

From Nordstrom, D. K. et al., in *Chemical Modeling in Aqueous Systems*, ACS Symp. Ser. 93, Jenne, E. A., Ed., American Chemical Society, Washington, D. C., 1979, 857. With permission.

Table 13
p(M) (−LOG MOLALITY) OF SELECTED MINOR SPECIES IN THE RIVER WATER TEST CASE

Program	EQUIL	EQ3	IONPAIR	MINEQL2[a]	MIRE	REDEQL2	SOLMNEQ	WATEQF	WATEQ2	WATSPEC
Cr^{3+}	—	—	—	20.02	—	16.15	—	—	—	—
Mn^{2+}	7.100	8.205	—	7.14	7.138	7.13	7.099	7.119	7.119	—
$MnOH^{+}$	—	11.403	—	9.28	9.804	9.28	—	10.264	10.347	—
$MnSO_4^{0}$	9.166	10.383	—	9.18	9.213	9.18	9.289	9.864	9.864	—
Fe^{2+}	6.711	6.639	—	15.18[b]	6.896	15.18	8.389	12.167	6.579	11.953
$FeOH^{+}$	7.075	7.428	—	15.82[b]	6.855	15.83[b]	26.711	14.257	8.669	13.860
Fe^{3+}	19.04	16.212	—	20.88	—	20.640[b]	25.811	17.635	18.596	17.408
$Fe(OH)_4^{-}$	—	8.072	—	10.44[b]	—	10.35[b]	18.172	6.989	8.340	7.028
Ni^{2+}	—	—	—	7.59	—	7.93	—	—	8.892	—
$NiCO_3^{0}$	—	—	—	—	—	7.76	—	—	7.539	—
Cu^{2+}	10.15	8.763	—	9.67	—	10.42	14.772	—	11.293	—
$Cu(OH)_2^{0}$	—	—	—	13.68	—	13.22	—	—	9.045	—
Ag^{+}	—	9.660	—	—	—	15.51	9.651	—	15.891	—
$AgHS^{0}$	—	—	—	—	—	—	—	—	9.432	—
Zn^{2+}	8.131	8.129	—	8.15	—	8.48	7.129	—	8.504	—
$ZnCO_3^{0}$	—	—	—	—	—	8.51	—	—	8.720	—
$Zn(HS)_2^{0}$	—	—	—	—	—	—	—	—	8.790	—
Cd^{2+}	—	—	—	9.10	—	9.41	—	—	11.612	—
$CdOH^{+}$	—	—	—	10.15	—	11.46	—	—	12.651	—
$CdHS^{+}$	—	—	—	—	—	—	—	—	9.078	—
Hg^{2+}	—	16.875	—	21.07	—	20.15	41.872	—	—	—
Al^{3+}	15.69	13.697	—	17.49[b]	—	17.164[b]	13.598	13.908	15.565	14.778
$Al(OH)_4^{-}$	7.090	6.732	—	8.66	—	8.37[b]	6.765	6.741	8.406	6.733
Pb^{2+}	9.845	14.380	—	12.07	—	11.93	9.843	—	11.749	—
$PbOH^{+}$	—	—	—	11.21	—	11.68	—	—	10.378	—
$PbCO_3^{0}$	—	9.839	—	9.89	—	9.86	—	—	10.025	—
NO_3^{-}	4.839	—	4.84	—	—	—	4.839	4.839	4.839	4.839
NH_4^{+}	5.098	—	—	5.14	—	5.13	5.105	5.106	5.106	10.392
PO_4^{3-}	10.03	—	—	11.41[b]	—	11.40[b]	10.189	10.156	10.156	—
HPO_4^{2-}	5.818	—	—	7.07[b]	5.773	7.16[b]	5.813	5.793	5.793	—
$H_2PO_4^{-}$	6.694	—	—	7.78[b]	6.654	8.06[b]	6.622	6.625	6.625	—
$HAsO_4^{2-}$	—	—	—	—	—	—	13.990	—	7.595	—

Table 13 (continued)
p(M) (−LOG MOLALITY) OF SELECTED MINOR SPECIES IN THE RIVER WATER TEST CASE

Program	EQUIL	EQ3	IONPAIR	MINEQL2[a]	MIRE	REDEQL2	SOLMNEQ	WATEQF	WATEQ2	WATSPEC
$H_2AsO_4^-$	—	—	—	—	—	—	15.046	—	8.885	—
S^{2-}	—	17.140	—	—	—	—	16.489	12.613	12.870	12.615
HS^-	—	7.268	—	—	—	—	7.299	7.288	7.544	7.288
H_2S^0	—	8.076	—	—	—	—	8.071	8.146	8.403	8.146
I^-	—	—	—	7.785	—	7.86	—	—	—	—

[a] Calculated at 25°C.

[b] MINEQL2 and REDEQL2 results were calculated in such a way that the solution was equilibrated with supersaturated phases. P, Ca, Fe, Al, and Si species are not exactly comparable to the same species from the other programs because of this equilibration process, and, of course, all species have been affected to some degree.

From Nordstrom, D. K. et al., in *Chemical Modeling in Aqueous Systems,* ACS Symp. Ser. 93, Jenne, E. A., Ed., American Chemical Society, Washington, D.C., 1979, 857. With permission.

Table 14
p(M) (−LOG MOLALITY) OF SELECTED MAJOR SPECIES IN THE SEAWATER TEST CASE

Program	EQUIL	EQ3	GEOCHEM	K+K[a]	MINEQL2	MIRE	MINEQL/ REDEQL2	REDEQL2	SEAWAT	SIAS	SOLMNEQ	WATEQF	WATEQ2	WATSPEC
Ca^{2+}	2.013	2.027	2.21	2.024	2.12[b]	2.074	2.03	2.11[b]	2. 053	2.058	2.013	2.024	2.025	2.023
$CaSO_4^0$	3.062	2.928	3.12	3.046	3.00[b]	2.744	2.97	3.00[b]	2.764	2.781	3.056	2.940	2.938	2.947
$CaHCO_3^+$	4.261	4.234	4.48	4.504	4.92[b]	4.366	4.26	4.84[b]	4.242	4.285	4.261	4.481	4.449	4.465
$CaCO_3^0$	4.636	4.540	4.59	4.755	5.30[b]	4.734	4.63	5.30[b]	4.715	4.365	4.635	4.649	4.616	4.665
$CaCl^+$	—	—	2.46	—	—	—	—	—	—	—	—	—	—	—
Mg^{2+}	1.304	1.319	1.46	1.328	1.34	1.322	1.33	1.34	1.347	1.357	1.299	1.311	1.312	1.310
$MgSO_4^0$	2.305	2.175	2.47	2.190	2.11	2.299	2.17	2.13	2.008	2.080	2.360	2.235	2.222	2.241
$MgHCO_3^+$	3.777	3.554	3.83	3.405	4.14	3.676	3.60	4.06	3.588	2.684	3.834	2.655	3.624	5.486
$MgCO_3^0$	3.619	3.940	4.14	3.745	4.32	3.684	3.67	4.32	4.121	4.664	3.689	4.045	4.014	4.062
$MgCl^+$	—	—	1.79	—	—	—	—	—	—	—	—	—	—	—
Na^+	0.336	0.320	0.41	0.328	0.34	0.335	0.33	0.34	0.321	0.314	0.336	0.320	0.320	0.323
$NaSO_4^-$	1.964	2.208	1.94	2.286	2.18	2.716	2.20	2.19	2.107	—	1.965	2.178	2.195	2.146
$NaHCO_3^0$	3.873	—	3.40	3.466	—	3.851	3.84	—	—	—	3.860	3.785	3.752	3.807
K^+	1.982	1.983	2.04	2.007	2.01	2.000	1.99	2.01	1.981	1.976	1.982	1.983	1.983	1.985
KSO_4^-	3.839	3.748	3.76	3.802	3.45	3.789	3.67	3.46	3.916	—	3.871	3.768	3.766	3.736
Cl^-	0.257	0.247	0.33	0.25	0.25	0.265	0.25	0.25	0.247	0.253	0.256	0.247	0.247	0.249
SO_4^{2-}	1.906	1.823	2.11	1.804	1.91[b]	1.836	1.84	1.89[b]	2.009	1.723	1.881	1.811	1.808	1.822
HCO_3^-	2.884	2.816	2.87	2.792	3.56	2.815	2.90	3.39	2.808	—	2.885	2.825	2.836	2.845
CO_3^{2-}	4.419	4.371	4.48	4.622	4.91	4.415	4.46	4.89	4.369	5.507	4.484	4.422	4.566	4.443
$B(OH)_3^0$	3.435	—	3.51	—	3.49	—	3.45	3.48	—	—	3.506	3.430	3.431	3.481
$B(OH)^{4-}$	4.226	—	4.23	—	4.06	—	4.20	4.10	—	—	4.234	4.257	4.255	4.015
Br^-	—	—	3.06	—	3.07	—	—	3.07	3.059	3.065	—	3.059	3.059	3.059
F^-	4.228	—	4.36	—	4.39	—	—	4.38	—	4.411	4.379	4.410	4.410	—
MgF^+	4.792	—	4.53	—	4.56	—	—	4.58	—	—	4.464	4.442	4.443	—
Sr^{2+}	4.016	—	4.28	—	4.13[b]	—	—	4.119[b]	—	—	4.045	4.016	4.016	4.016
$H_4SiO_4^0$	4.155	4.159	4.16	—	4.19[b]	—	—	4.17[b]	—	—	4.161	4.145	4.145	4.147
H^+	8.125	8.126	8.15	7.966	—	—	—	—	8.095	—	8.125	8.093	8.093	8.098
OH^-	5.603	5.606	5.65	6.038	5.46	5.552	—	5.54	5.613	—	5.601	5.660	5.660	5.571
Ionic strength	0.663	0.6772	0.5298	—	0.65	0.655	—	0.65	0.6595	—	0.6662	0.6801	0.6799	0.6770

[a] Karpov and Kaz'min (1972) using Gibbs free energy minimization on a similar seawater.

[b] MINEQL2 and REDEQL2 results were calculated in such a way that the solution was equilibrated with supersaturated phases. Si, Al, Fe, Ca, Sr, Ba, Zn, SO_4, and PO_4 species are not exactly comparable to the same species from the other programs because of this process, and, of course, all species have been affected to some degree.

From Nordstrom, D. K. et al,. in *Chemical Modeling in Aqueous Systems,* ACS Symp. Ser. 93, Jenne, E. A., Ed., American Chemical Society, Washington, D.C., 1979, 857. With permission.

Table 15
p(M) (−LOG MOLALITY) OF SELECTED MINOR SPECIES IN THE SEAWATER TEST CASE

Program	EQUIL	EQ3	GEOCHEM	MINEQL2[a]	MIRE	REDEQL2	SIAS	SOLMNEQ	WATEQF	WATEQ2	WATSPEC
Li^+	4.572	—	4.66	—	—	—	—	4.568	4.573	4.573	4.573
Rb^+	—	—	5.87	—	—	—	—	—	—	5.848	—
Cs^+	—	—	8.56	—	—	—	—	—	—	8.506	—
Ba^{2+}	—	—	6.93	8.03[b]	—	8.01[b]	—	6.857	6.821	6.821	6.821
Cr^{3+}	—	—	25.63	28.34	—	24.55	—	—	—	—	—
Mn^{2+}	8.456	9.287	9.33	8.96	8.592	8.94	—	8.478	8.654	8.654	—
$MnCl^+$	—	—	10.26	8.74	9.067	8.75	—	29.376	8.880	8.881	—
Fe^{3+}	18.740	17.466	22.94	20.17[b]	—	20.26[b]	—	—	17.897	17.897	17.714
$Fe(OH)_3^0$	—	—	11.28	17.10[b]	—	24.14[b]	—	—	8.071	8.071	7.805
$Fe(OH)_4^-$	—	7.439	12.41	9.96[b]	—	9.94[b]	—	—	7.664	7.664	7.678
Ni^{2+}	—	—	9.23	7.77	—	7.92	7.834	—	—	8.813	—
$NiCl_2^0$	—	—	9.74	—	—	8.10	—	—	—	9.335	—
$NiCO_3^0$	—	—	7.91	5.75	—	8.16	—	—	—	7.590	—
Cu^{2+}	13.152	9.056	11.21	11.35	—	11.32	9.579	—	—	10.153	—
$Cu(OH)_3^0$	—	—	12.40	3.22	—	9.06	—	—	—	7.985	—
Ag^+	—	14.797	16.79	16.82	—	16.79	—	14.477	—	14.394	—
$AgCl_4^{3-}$	—	9.517	12.48	12.08	—	12.05	—	9.617	—	9.684	—
Zn^{2+}	7.323	7.346	8.35	7.74[b]	—	8.06[b]	7.690	7.321	—	7.547	—
$ZnCO_3^0$	—	—	7.33	—	—	8.95[b]	—	—	—	7.894	—
Cd^{2+}	—	—	11.11	10.59	—	10.70	10.606	—	—	11.257	—
$CdCl^+$	—	—	9.99	9.47	—	9.41	—	—	—	9.371	—
$CdHS^+$	—	—	—	—	—	—	—	—	—	9.078	—
$CdCl_2^0$	—	—	9.94	9.34	—	9.48	—	—	—	9.369	—
Hg^{2+}	—	23.310	—	—	—	—	—	23.426	—	—	—
$HgCl_3^-$	—	9.816	—	—	—	—	—	9.965	—	—	—
Al^{3+}	15.963	16.076	16.37	16.67[b]	—	16.15[b]	—	19.277	15.956	16.091	16.996
$Al(OH)_3^0$	7.600	—	7.72	—	—	7.58[b]	—	—	—	7.665	—
$Al(OH)_4^-$	7.287	7.114	7.25	7.53[b]	—	7.32[b]	—	10.611	7.115	7.259	7.114
Pb^{2+}	10.446	13.746	11.68	—	—	—	—	10.442	—	11.335	—
PbI_2^0	10.380	13.591	11.41	—	—	—	—	0.372	—	11.017	—
$PbCO_3^0$	—	9.602	10.09	—	—	—	—	9.743	—	9.743	—
NO_3^-	5.315	—	5.30	—	—	—	—	—	5.315	5.314	5.315
PO_4^{3-}	9.979	—	10.48	11.81[b]	—	11.76[b]	—	10.126	10.482	10.480	—
HPO_4^{2-}	6.858	—	7.02	8.48[b]	6.673	8.48[b]	—	6.897	6.989	6.988	—
$H_2PO_4^-$	8.474	—	8.49	9.93[b]	8.295	10.06[b]	—	8.449	8.382	8.381	—
AsO_4^{3-}	—	—	—	11.01	—	—	—	—	—	10.007	—
$HAsO_4^{3-}$	—	—	8.45	—	—	—	—	—	—	7.264	—
I^-	—	—	6.30	6.30	—	6.31	—	—	—	6.296	—

[a] Calculated at 25°C.

[b] MINEQL2 and REDEQL2 results were calculated in such a way that the solution was equilibrated with supersaturated phases. Si, Al, Fe, Ca, Sr, Ba, Zn, SO_4, and PO_4 species are not exactly comparable to the same species from the other programs because of this process, and, of course, all species have been affected to some degree.

From Nordstrom, D. K. et al., in *Chemical Modeling in Aqueous Systems*, ACS Symp. Ser. 93, Jenne, E. A., Ed., American Chemical Society, Washington, D.C., 1979, 857. With permission.

Karpov and Kaz'min,[83] which uses the free energy minimization approach. In general, there is better agreement between the major species concentrations than the minor species, and the results for the river water tend to agree better than those of seawater. Nordstrom et al.[12] point out that better agreement would be expected in the river water test case because of the smaller amount of complexing in the dilute solution and the more consistent and reliable activity coefficients which can be obtained at low ionic strength. Broyd et al.[127] similarly found good agreement between major species, but significant scatter between minor species on application of ten different codes to the groundwater data set.

While the results obtained using the various models are, in general, surprisingly consistent, the numerous minor and occasional major differences can be attributed to a range of limitations inherent in speciation modeling. The reliability of the equilibrium constants or free energies used in the computations must rank as the major likely source of error and is an issue which must be considered in attempting to model any system, however simple. Another factor likely to have significant consequences in speciation modeling, particularly when using the equilibrium constant approach, is the number of complexes included in the thermodynamic data base. If one model has significantly more metal-ligand complexes than another model, then it will predict a lower concentration of the free ligand, assuming negligible differences in the data base. For example, the inclusion of the calcium and magnesium chloride complexes by GEOCHEM tends to lower the free calcium, magnesium, and chloride ion concentrations below those of the other models.[12]

The reliability of the thermodynamic data base and the necessity for inclusion of all appropriate species are particularly critical when considering mineral precipitation. This is vividly demonstrated in the results of a study by Broyd et al.[127] on the behavior of aluminium in a groundwater of composition shown in Table 10. All possible mineral phases were first allowed to supersaturate before allowing precipitation. The saturation indices obtained from the various programs before precipitation are shown in Table 16. (The saturation index, SI, is defined as log IAP/K_{sp} where IAP is the ion activity product for the mineral, and K_{sp} is the solubility product). The minerals listed have all been indicated as supersaturated by at least one of the programs, and, where possible, the saturation indices for a particular mineral are recorded for each program whether it is saturated or not. While there is a fair degree of agreement regarding the non-aluminum-containing minerals considered (No. 1 to 12) and the corresponding values of saturation index, the results obtained for the aluminum-containing minerals (No. 13 to 31) are very inconsistent in both minerals considered and computed saturation index.

The next step in this test case was to precipitate minerals such that there were no supersaturates remaining. This was investigated by noting the amounts of particular species in solution before and after precipitation. The species $Al(OH)_4^-$ was chosen by Broyd et al.[127] as a suitable indicator for the aluminum system. Results for this species are shown in Figure 11 and indicate excellent agreement on the concentration of $Al(OH)_4^-$ before precipitation, but a large amount of scatter in the concentration of the same species after precipitation. The reasons for this scatter are relatively obvious. The version of MINEQL used at Gesellschaft for Strahlen and Unweltforschung (GSF) in West Germany has no aluminum supersaturates, thus, no aluminum is precipitated. On the other hand, the version of WATEQ2 used at the RISO National Laboratory (CMCP) in Denmark predicts a large number of highly supersaturated aluminum minerals, and this is reflected in the low concentration of $Al(OH)_4^-$ in the final solution. Broyd et al.[127] note that while the code DISSOL (a reaction path code similar to PATHI, developed by Fritz[128] at the University of Strasbourg) shows no aluminum minerals to be supersaturated initially, the aluminum mineral gibbsite becomes supersaturated during the precipitation of calcite and is precipitated leading to a reduction in the concentration of dissolved $Al(OH)_4^-$. It is apparent that, for this type of problem, the critical factors are the minerals included in the data base and their associated thermodynamic constants. Broyd

Table 16
SATURATION INDICES PRIOR TO PRECIPITATION FOR THE MAJOR MINERALS OF POSSIBLE IMPORTANCE IN A GROUNDWATER OF COMPOSITION SHOWN IN TABLE 10

	Saturation index							
Mineral	WHATIF	CMCP	DISSOL	EQ3/6	WATRA	MINEQL GSF	SOLMNQ UWIST	SOLMNQ AECL
1. Calcite	1.766	1.591	1.492	1.686	1.61	1.53	1.49	1.404
2. Dolomite	3.628	2.426	2.506	3.6	2.51	2.26	2.47	2.324
3. Magnesite	0.278	0.339	0.405	0.318	0.78	−0.16	0.377	0.341
4. Aragonite	1.604	1.452		1.521	1.35		1.454	1.214
5. Tremolite	2.955	9.530		3.97	10.69		1.483	9.129
6. Serpentine	1.665	2.269		2.97	2.41		0.307	1.859
7. Crysotile								4.771
8. Talc	2.065	1.838		3.24	3.34		1.828	0.529
9. Diopside	—	0.705		−0.55	0.89			−0.006
10. Huntite	—	0.015		0.904	0.72		2.365	
11. Antigorite	—			58.6				
12. Hyd. mag.	—	−7.32		−5.81	1.606			
13. 14A clinochlore	—			4.52				
14. Phlogopite	3.06			4.68	−0.39			
15. 7A clinochlore	2.50			1.14				
16. Chlorite	—	4.353			5.65			
17. Diaspore		0.006		−1.78	−0.83	−0.58		
18. Halloysite		9.011			−8.97			
19. Illite		4.737		−3.33	−5.98			
20. Kaolinite		13.10	−2.76	−3.21	−2.64			
21. K mica		10.76			−7.32			
22. Montmoril Ca		10.45			−7.54			
23. Pyrophilite		14.93		−6.55	−8.92			
24. Na saponite				4.49				
25. K saponite				4.23				
26. Ca saponite				4.43				
27. Mg saponite				4.34				

28. H saponite		2.85			
29. 14A penninite		6.85			
30. Chlor Mg				11.41	14.58
31. Phillipsite	−3.39		−3.34	3.73	6.89

From Broyd, T. W., Grant, M. M., and Cross, J. E., Report EUR 10231 EN, Commission of the European Communities, Luxembourg, 1985.

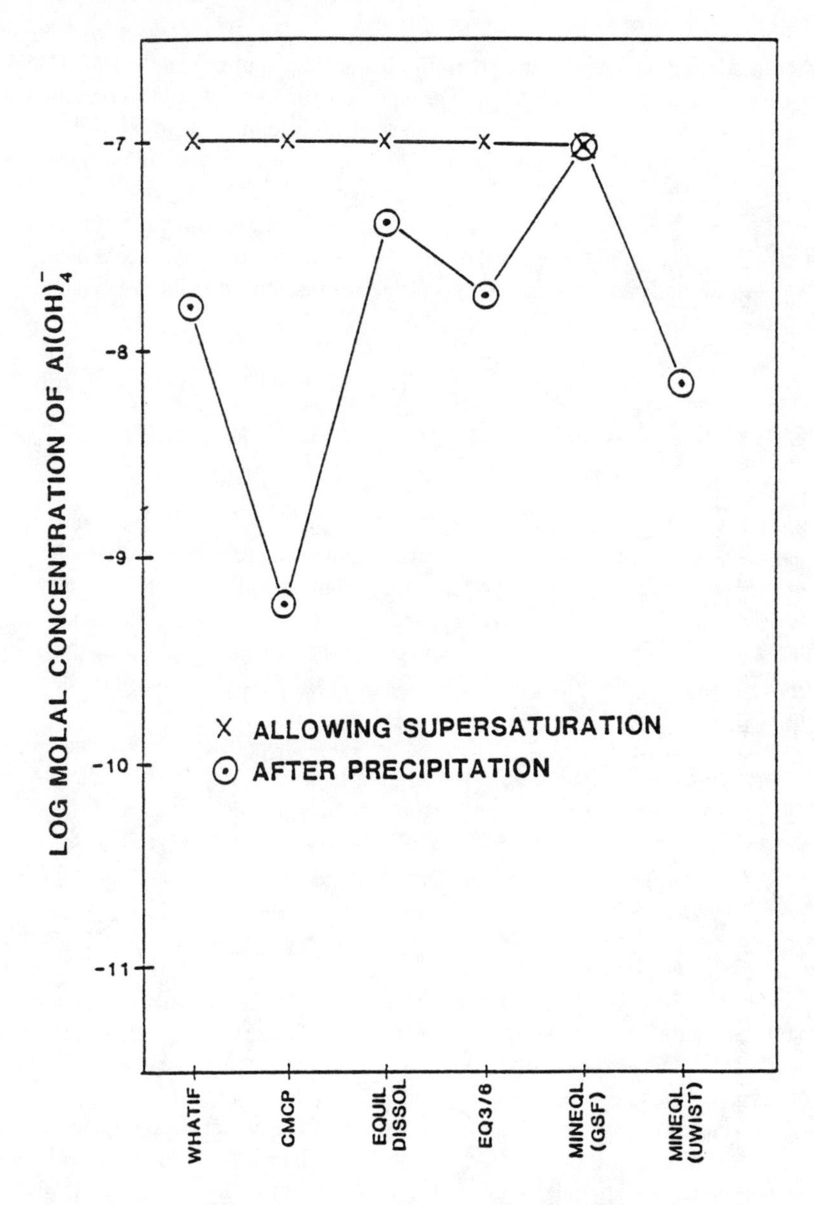

FIGURE 11. Concentration of dissolved $Al(OH)_4^-$ computed using various speciation codes before and after precipitation of supersaturated minerals. The codes used are as follows: WHATIF — a speciation code developed at the RISO National Laboratory (CMCP) in Denmark; CMCP — a RISO derivative of WATEQ2 translated into ALGOL; EQUIL/DISSOL — codes based on PATHI and developed at the Université de Louis Pasteur in Strasbourg; EQ3/6, MINEQL (GSF), and MINEQL (UWIST) — modified versions of MINEQL in use at Gesellschaft für Strahlen und Unweltforschung (GSF) and the University of Wales Institute of Science and Technology (UWIST). (From Broyd, T. W., Grant, M. M., and Cross, J. E., Report EUR 10231 EN, Commission of the European Communities, Luxembourg, 1985.)

et al.[127] consider it likely that, given a consistent data base, all the programs considered would yield comparable results.

Differences in the method of ionic strength correction are unlikely to be a major source of error in computations at the relatively low ionic strengths of the river and seawater test cases considered by Nordstrom et al., but as pointed out in Section III, must be kept in mind when modeling waters of ionic strength greater than approximately 1 *M*. Codes which utilize a Davies or extended Debye-Huckel equation for ionic strength correction will not be suitable for modeling brines or groundwaters of high ionic strength.[13,33,46]

Another source of difference between model results arises because of the variety of ways the available programs handle the carbonate system.[12,127] The concentration of inorganic carbon in a natural water is usually determined from the titration alkalinity. However, the titration alkalinity also contains contributions from species other than carbonate, including various weak acids and bases and inorganic complexes (assuming an ion-association approach is being used);[36] i.e.,

$$\begin{aligned} \text{Alk} = & -[H^+] + [OH^-] + [HCO_3^-] + 2[CO_3^{2-}] + [NH_3] + [HS^-] + 2[S^{2-}] \\ & + [HSiO_3^-] + [B(OH)_4^-] - [H_3PO_4^{2-}] + 2[PO_4^{3-}] + 2[NaCO_3^-] \\ & + 2[CaCO_3] + [CaOH^+] + [CaHPO_4] + \ldots. \end{aligned} \tag{49}$$

and the estimate for inorganic carbon concentration must be corrected for these. Most models correct for the presence of $HSiO_3^-$ and $B(OH)_4^-$, but in some cases, other species shown in Equation 49 may be significant and must be accounted for. Some models, such as MINEQL, avoid the ambiguity of the correction procedure by accepting only analytically determined values of total inorganic carbon concentration rather than values deduced from alkalinity measurements.

Different assumptions made about the redox state of aqueous systems is another possible reason for variability in results of model intercomparison studies.[12] The distribution of species of multiple oxidation state will be dependent on the redox potential assumed to be appropriate for the system of interest. The redox potential may be that measured using a platinum electrode or may be determined by the presence of a dominant couple such as the Fe^{3+}/Fe^{2+} couple, the SO_4^{2-}/S^{2-} couple, the O_2/H_2O couple, the O_2/H_2O_2 couple, or the NO_3^-/NH_4^+ couple. Nordstrom et al.[12] consider the wide range of values for iron, manganese, chromium, and arsenic species reported in Tables 13 and 15 to be attributable, in part, to differences in redox assumptions. In order to examine how different redox controls affect the distribution of species, while other factors are kept constant, Nordstrom et al.[12] computed the distribution of the different oxidation states of iron and arsenic, using the WATEQ2 code, for several imposed redox potentials and fixed total concentrations of iron and arsenic. The results of these computations for both the river and seawater cases defined in Table 11 are shown in Table 17. The range of concentrations obtained for iron and arsenic is several orders of magnitude and includes most of the values listed in Tables 13 and 15. Nordstrom et al.[12] conclude that if chemical models are to be applied to interpretation of natural water chemistry involving redox aspects, then individual redox elements such as total ferrous and total ferric concentrations should be analyzed separately when possible, rather than assuming that they can be distributed according to some other equilibrium. Given the slow rates of many redox processes (e.g., oxidation of Fe(II) and Mn(II) in waters of pH <7), it is unlikely that homogenous redox equilibrium is often obtained in real systems.

As is clear from Section III, differences in the method of temperature correction of thermodynamic parameters is another possible source of variation in any intercomparison exercise. Some codes, such as REDEQL2 and MINEQL, have no facility for temperature correction, while others, such as EQ3/6, incorporate sophisticated procedures for correction of thermodynamic data.

VII. ADSORPTION MODELING

In natural waters, trace elements will be present as soluble and precipitated species and,

Table 17
VARIATIONS IN THE DISTRIBUTION OF SELECTED REDOX SPECIES CAUSED BY CHANGING THE IMPOSED REDOX POTENTIAL

Imposed redox potential	Implied E_h (V)	Fe^{2+}	Fe^{3+}	H_2AsO_3	$HAsO_4$
River Water Test Case					
Pt E_h	0.440	11.78	17.25	25.80	7.60
O_2/H_2O	0.783	17.89	17.25	38.03	7.60
O_2/H_2O_2	0.141	6.81	17.61	15.14	7.60
NO_3^-/NO_2^-	0.445	11.87	17.25	25.98	7.60
NO_3^-/NH_4^+	0.328	9.78	17.25	21.81	7.60
S^{2-}/SO_4^{2-}	−0.532	6.56	29.36	7.59	24.04
Fe^{2+}/Fe^{3+}	0.073	6.58	18.50	12.72	6.60
Seawater Test Case					
Pt E_h	0.550	13.94	17.90	29.17	7.26
O_2/H_2O	0.731	17.85	17.90	36.98	7.26
O_2/H_2O_2	0.133	7.94	18.09	16.76	7.26
NO_3^-/NO_2^-	0.390	12.08	17.90	25.45	7.26
NO_3^-/NH_4^+	0.269	10.04	17.90	21.36	7.26

From Nordstrom, D. K., et al., in *Chemical Modeling in Aqueous Systems*, ACS Symp. Ser. 93, Jenne, E. A., Ed., Amerian Chemical Society, Washington, D.C., 1979, 857. With permission.

in some systems, as species adsorbed to solid inorganic or organic phases. This adsorbed component cannot be ignored in any realistic attempt to model the distribution of trace element species. The most common inorganic adsorbents are hydrous metal oxides (as homogeneous phases or coatings), clays, and sands, while the most common organic adsorbents consist of plant and animal remains and humic coatings on mineral surfaces. Among the inorganic adsorbents, hydrous metal oxides have the highest affinity for ions because of their charged, reactive hydroxyl surface sites combined with their high surface area. The oxides of iron, manganese, aluminum, and silicon are of major importance in natural waters, and considerable effort has been devoted to the development of models that satisfactorily describe the interaction of trace elements with these adsorbents. Adsorption of trace elements by organic coatings on mineral surfaces has also been shown to be of significance in aquatic systems containing high concentrations of humic substances,[128] though mathematical description of this interaction is still in the early stages of development.

A variety of empirical approaches have been frequently used to model trace element binding by solid surfaces, including partition coefficients, isotherm equations (such as those of Langmuir and Freundlich), and Kurbatov plots.[129,130] Such descriptions are necessarily limited to particular conditions and are not easily extrapolated to other conditions of reactant concentration, pH, or ionic strength. A more fundamental, mechanistic approach which uses the surface ionization (or complexation) concept first introduced in studies of ion binding by proteins[131] in conjunction with classical Gouy-Chapman double-layer theory and its extensions (Stern-Grahame) is now widely accepted in modeling the interaction of solute ions with surfaces.[132]

In the surface complexation approach, adsorption is considered to take place at defined coordination sites (present in finite numbers), with adsorption reactions described quantitatively by mass law expressions.[90] For example, surface groups of metal oxides exposed to water may be represented by a general hydrolyzed species $\equiv SOH^0$ and acid-base reactions may be written by considering the loss or gain of a proton, i.e.,

$$\equiv SOH_2^+ \rightleftharpoons \equiv SOH^0 + H_s^+; \qquad K_{a1}^{int} \tag{50}$$

$$\equiv SOH^0 \rightleftharpoons SO^- + H_s^+; \qquad K_{a2}^{int} \tag{51}$$

where H_s^+ denotes protons in the surface plane and K_{1a}^{int}, K_{2a}^{int} are intrinsic acidity constants. Protons at the surface are distinguished from protons in bulk solution because the electrical potential difference (ψ_0) between these regions results in a different chemical potential of the proton. Electrostatic energy is required to move ions through the interfacial potential gradient. The proton activity at the surface is related to the bulk solution activity according to the Boltzmann distribution:

$$\{H_s^+\} = \{H^+\} \exp(-F\psi_0/RT) \tag{52}$$

where ψ_0 represents the surface potential (compared to a reference potential of zero in bulk solution) and R, T, and F are the gas constant, absolute temperature, and Faraday constant, respectively.

Cation and anion binding can be characterized quantitatively by considering the formation of surface complexes; for example,

$$\equiv SOH^0 + M^{2+} \rightleftharpoons \equiv SOM^+ + H_s^+; \qquad K_M^{int} \tag{53}$$

$$\equiv SOH^0 + A^{2-} + H_s^+ \rightleftharpoons \equiv SA^- + H_2O; \qquad K_A^{int} \tag{54}$$

A variety of other possible surface complexes may be envisaged including ternary complexes of the Type I (surface-metal-ligand) and Type II (surface-ligand-metal) forms. The choice of surface species (or possibly multiple surface species) for a particular situation will be dictated by the available experimental data. A mole balance equation for total surface sites may thus be written:

$$TOT(\equiv SOH) = N_s = [\equiv SOH^0] + [\equiv SOH_2^+] + [\equiv SO^-] + [\equiv SOM^+] + [\equiv SA^-] \tag{55}$$

In this surface complexation approach, surface charge is considered to develop as a result of acid-base and complexation reactions. In the absence of specifically adsorbed cations and anions, the surface charge σ is computed as

$$\sigma = F/AS([\equiv SOH_2^+] - [\equiv SO^-] = F(\Gamma_H - \Gamma_{OH}) \tag{56}$$

where A is the specific surface area, S is the solid concentration, and Γ_H, Γ_{OH} are the adsorption densities of H^+ and OH^- ions. In the presence of cationic or anionic surface complexes of the form shown in Equations 53 and 54, the surface charge becomes

$$\sigma = F/AS([\equiv SOH_2^+] + [\equiv SOM^+] - [\equiv SO^-] - [\equiv SA^-]) \tag{57}$$

The mass law, mole balance, and surface charge expressions given in Equations 50 to 57 in conjunction with a set of equations representing the constraints imposed by the model chosen for the electrical double-layer structure constitute the surface complexation approach for adsorption. These equations are readily incorporated in solution-phase speciation models of the equilibrium constant form, such as REDEQL, MINEQL, GEOCHEM, MINTEQ, and MICROQL, by including components for surface sites and surface charge and making other modifications to enable coulombic correction of stability constants.[133-135]

The variety of different adsorption models based on the surface complexation approach are differentiated by the geometric location assigned to adsorbed ions and the associated description of electrostatic interactions at and near the surface. In the basic two-layer model of ion adsorption on oxide surfaces, specifically adsorbed ions are all considered to be part of the solid surface (one layer), the resulting electrostatic charge of which is balanced by an adjacent diffuse layer in solution (the other layer) containing nonspecifically adsorbed counterions.[90] This view of charge distribution at an oxide/water interface is shown in Figure 12. The surface potential ψ_0 (volts) is related to the surface charge σ according to the Gouy-Chapman theory (for a symmetrical electrolyte with valence z at 25°C) by

$$\sigma = 0.1174 I^{1/2} \sinh(zF\psi_0/2RT) \qquad (58)$$

where I is the molar electrolyte concentration. This diffuse-layer model of the electrical double layer[136,137] may be simplified to a constant capacitance model[138,139] under conditions of high ionic strength and/or low potential (<25 mV). In this case, Equation 58 reduces to

$$\sigma = C\psi_0 \qquad (59)$$

where C is a constant with dimensions of a capacitance. Though the capacitance can be estimated from the Gouy-Chapman theory, it is often taken as an adjustable parameter and the model applied to systems of all ionic strengths.

While the basic two-layer surface complexation models accurately predict the variation in oxide acid-base properties as a function of ionic strength and the effects of pH and ionic strength on specific ion adsorption, there are a number of issues that are inadequately accounted for and that have led to the development of more elaborate models of interactions at the oxide/water interface. First, electrokinetic potential measurements indicate that the diffuse-layer charge adjacent to the oxide surface is much lower than the surface charge measured by acid-base titration[140] (though considerable debate exists concerning the significance of this difference).[90] A variety of closely related models have been developed to resolve the issue of charge on oxide surfaces. Both the variable surface charge-variable surface potential model of Bowden et al.[141] (and essentially equivalent to the Stern model for specific adsorption at an electrified interface[142]), and the triple-layer model of Yates et al.[143] consider separate planes of adsorption for different solutes and include additional surface capacitances as fitting parameters (see Figure 12). In applying the triple-layer model, Davis et al.[144] used additional fitting parameters by considering surface reactions with background electrolyte ions. Like the compact layer capacitances, these reactions serve the purpose of reducing surface charge, thus enabling satisfactory modeling of zeta potential data. Bowden et al.[145,146] have also developed a model (termed the objective model by Sposito[70]) in which adsorption is considered to occur in four planes in the interfacial region. Chemical reactions are not employed in this model, though the binding constants used can be related to the intrinsic equilibrium constants of the surface complexation approach. Values for the parameters used to describe adsorption in the four planes are obtained by optimization (curve-fitting) methods. Interestingly, Westall and Hohl[147] applied five of the common electrostatic models for the oxide/water interface to acid-base titration data for two oxides and found all of the models to be equally capable of fitting the data. Based on these and other comparative exercises, Dzombak and Morel[90] recommend the use of the simple diffuse layer description of electrostatics at the oxide/water interface and suggest that the "charge problem" be overcome by specifying the location of the shear plane as an empirical function of solution conditions.

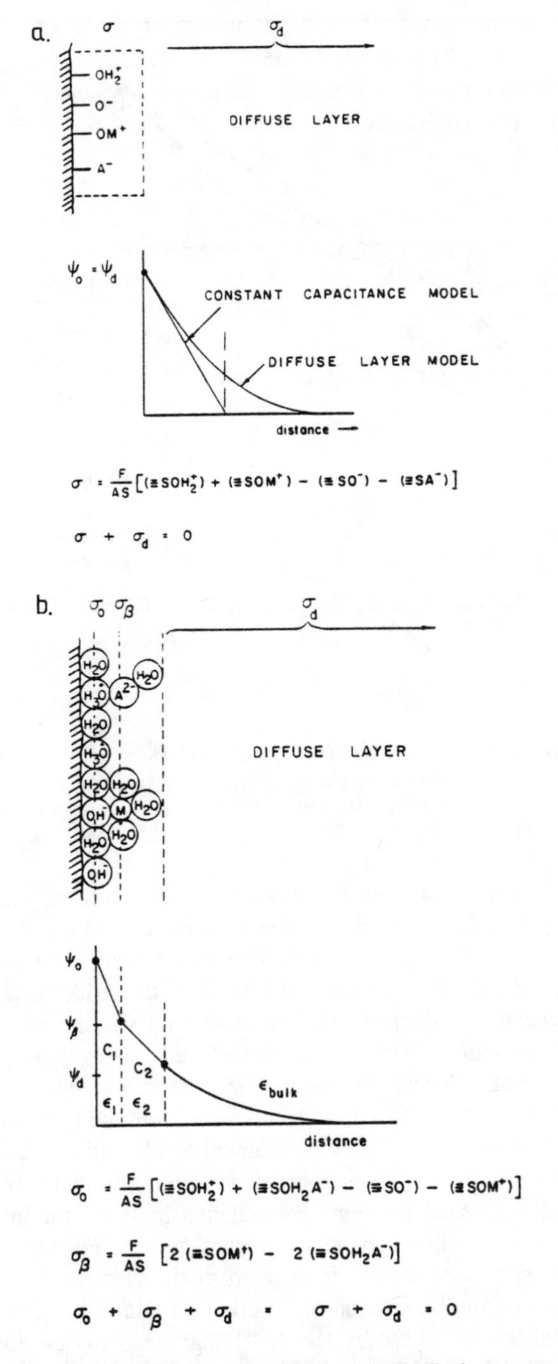

FIGURE 12. Schematic representation of charge distribution on an oxide surface and potential decay away from the surface used in (a) basic (two-layer) surface complexation models, and (b) the triple-layer model. (From Dzombak, D. A. and Morel, F. M. M., *Am. Soc. Civ. Eng. J. Hydraul.*, 113, 430, 1987. With permission.)

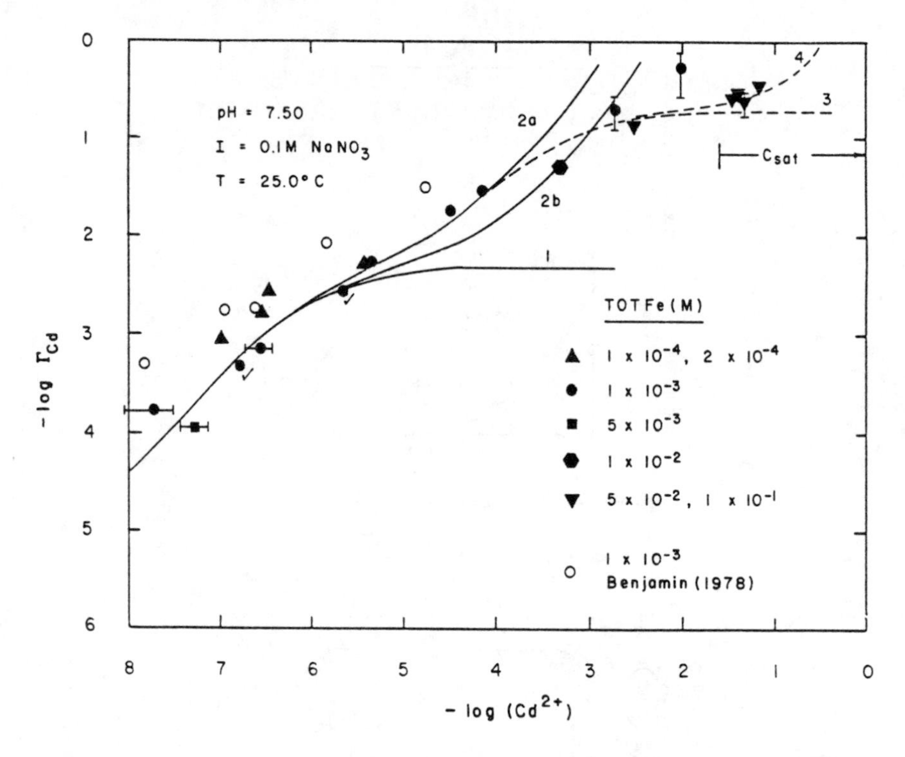

FIGURE 13. Isotherm for equilibrium sorption of Cd^{2+} on hydrous ferric oxide at a pH of 7.50. C_{sat} represents Cd^{2+} concentrations at which bulk solution precipitation of $Cd(OH)_2(s)$ is expected based on published solubility products. The various fitted curves are described in Table 18. (From Dzombak, D. A. and Morel, F. M. M., *J. Colloid Interface Sci.*, 112, 588, 1986. With permission.)

A second reason for extension of the basic two-layer surface complexation models relates to the inabilty of these models to predict cation adsorption over a wide range of adsorbate concentrations.[90] This issue has resulted in two major refinements of the simple model. In the multiple-site-type model of Benjamin and Leckie,[148] the oxide surface is considered to consist of several chemically distinct types of sites with varying affinities for adsorbate cations and can be implemented with any of the surface complexation models above. Two-site types generally appear adequate to model cation adsorption data,[90] though the effect of surface heterogeneity on metal adsorption is an issue of continuing debate.[149] In the surface precipitation model of Farley et al.,[150] the mechanism of adsorption is assumed to shift from surface complex formation to surface precipitation at high adsorbate concentration.

The various models discussed above are being increasingly used in interpretation of trace element speciation under conditions typical of natural waters. For example, the basic two-layer model with constant capacitance description of the electrical double layer has been used by Goldberg[151] to describe arsenate adsorption on aluminum and iron oxides, while the triple-layer model has been used by Hsi and Langmuir[152] to describe the adsorption of uranyl ion onto ferric oxyhydroxides, by LaFlamme and Murray[153] to interpret the effect of carbonate alkalinity on thorium speciation in the presence of goethite, and by Hayes and Leckie[154] to rationalize the effects of ionic strength variations on the adsorption of cadmium and lead to goethite. Dzombak and Morel[155] have used the surface precipitation model described by Farley et al.[150] in conjunction with a two-site, two-layer surface complexation model to satisfactorily model the adsorption of cadmium to amorphous ferric oxyhydroxide for cadmium concentrations spanning eight orders of magnitude. As can be seen from Figure 13 and the associated information in Table 18, a one-site surface complexation model provides a reasonable fit to the data at low sorbate/sorbent ratios, but fails badly at high cadmium

Table 18
MODELS AND PARAMETER VALUES USED TO FIT THE CADMIUM-HYDROUS FERRIC OXIDE KINETICS AND EQUILIBRIUM SORPTION DATA

(1) One-site surface complexation model:

$$X_{T1} = 0.005 \cdot TOTFe$$
$$K_1^{cond} = 7.04 \times 10^5; = 10^{5.85}$$
$$f_1 = 2.82 \times 10^4 h^{-1} M^{-1}$$
$$b_1 = 0.04\ h^{-1}$$

(2a) Surface precipitation model, one-site type:

$$X_{T1} = 0.005 \cdot TOTFe$$
$$K_{s1}^{app} = K_1^{cond} \cdot \{H^+\}^2 \cdot K_{SP(Fe)} = 10^{-6.55}$$
$$\log K_{SP(Cd)} = 12.5; \log K_{SP(Fe)} = 2.6$$

(2b) Same as model 2a, but with $\log K_{SP(Cd)} = 13.0$

(3) Two-site surface complexation model (X_{T1}, K_1^{cond}, f_1, and b_1 as above):

$$X_{T2} = 0.2 \cdot TOTFe$$
$$K_2^{cond} = 1.58 \times 10^3 = 10^{3.20}$$
$$f_2 = 110.6\ h^{-1} M^{-1}$$
$$b_2 = 0.07\ h^{-1}$$

(4) Two-site model with surface precipitation on Type II sites:

$$X_{T1} = 0.005 \cdot TOTFe; K_1^{app} = K_1^{cond} \cdot \{H^+\}^2 = 10^{-9.15}$$
$$X_{T2} = 0.2 \cdot TOTFe; K_{s2}^{app} = 10^{-9.2}$$
$$\log K_{SP(Cd)} = 14.9; \log K_{Sp(Fe)} = 2.6$$

From Dzombak, D. A. and Morel, F. M. M., *J. Colloid Interface Sci.*, 112, 588, 1986. With permission.

concentrations. Under these conditions, a two-site model with surface precipitation at one-site type must be invoked.[155]

As noted earlier, organic matter, particularly humic substances, present in natural waters may affect the adsorption of trace elements to solid surfaces. The major effects are likely to result from the change in surface properties that occur due to the adsorption of organic matter. The primary functional moieties of humic substances are carboxyl and phenolic groupings, and the adsorptive behavior of these substances is, in general terms, similar to that of simple organic acids.[156] Thus, pH edges for the binding of natural organic matter of oxides indicate that adsorption is greatest at low pH. As with other anions, protons are consumed by reactions of humic substances with oxide surfaces.[157] However, while the adsorption of simple organic acids to oxide surfaces has been modeled successfully with surface complexation models,[156] data for humic substance adsorption to oxide surfaces is not readily amenable to surface complexation modeling because of the lack of information on functional group concentrations and because of our limited understanding of the mode of interaction of these ill-defined substances with the solid surface.

Davis[158] used an empirical relationship based on experimental data to describe the ratio of adsorbed to soluble humic material as a function of pH. Cation binding was then modeled using complexation constants for cation binding to the adsorbed and soluble humic material and to the "bare" oxide surface. This approach provided a reasonable model for a given set of conditions, but certainly has no general applicability. Considerable advances in our ability to mathematically describe metal-humate interactions have been made in recent years.[159,160] Discrete ligand models in particular enable incorporation of humic substances in speciation models of the equilibrium constant type, though extensive research into issues such as the nature of ternary surface complexes involving humic substances is required before reliable models can be constructed of systems in which both natural organic compounds and surfaces exert significant influences on trace element speciation.

VIII. INCORPORATION OF KINETICS IN SPECIATION MODELS

While thermodynamic considerations are invaluable in that they identify the direction in which chemical reactions will proceed, a variety of reactions proceed toward the ultimate equilibrium state relatively slowly, and kinetic information is required to satisfactorily describe the system of interest. Processes in natural waters involving trace elements for which a kinetic treatment may be necessary include the adsorption of trace metals to surfaces, the precipitation and dissolution of minerals, and certain ligand exchange reactions. While such time-dependent processes can, in some cases, be satisfactorily ignored by construction of partial equilibrium models,[36] the time scale of interest in other cases necessitates their inclusion in any realistic attempt at modeling trace element speciation. Kinetics modeling, however, is inherently more difficult than that of thermodynamics because time-dependent processes are path dependent.[161] This dependence on reaction path renders the generalization of the mathematical description of kinetics difficult, and accounts, in part, for the omission of kinetics routines from most speciation codes to date.

Substantial effort in recent years has been spent on elucidation of rate laws associated with the kinetics of precipitation and dissolution, and attempts have been made to incorporate general descriptions of these processes in speciation codes. Considerable progress in this direction has been made by the developers of the EQ3/6 code at the Lawrence Livermore National Laboratory with mineral dissolution kinetics included in early versions of the code and precipitation kinetics incorporated more recently.[162] While the precipitation or dissolution of many minerals involves only major species and has little direct impact on trace element speciation, this is not always the case. For example, the solution phase concentration of Al^{3+} in some systems (e.g., groundwaters) may be controlled by the rates of dissolution of alkali feldspar, $KAlSi_3O_8$, and precipitation of gibbsite, $Al(OH)_3$; i.e.,

$$KAlSi_3O_8(s) + 4H + 4H_2O \rightleftharpoons K^+ + Al^{3+} + 3H_4SiO_4$$

$$Al^{3+} + 3OH^- \rightleftharpoons Al(OH)_3(s)$$

Since trace elements are typically undersaturated with respect to formation of solid phases in most natural waters, the time dependency of sorption processes is likely to be more critical in influencing trace element speciation than are precipitation or dissolution kinetics. Particular interest in the kinetics of trace element sorption reactions at the mineral-water interface has arisen in recent years with the inclusion of adsorption routines in solute transport codes.

Adsorption of trace elements to mineral surfaces (oxides and clays) is generally characterized by two-step kinetics consisting of a rapid initial uptake (seconds to minutes) followed by a slower step (hours to weeks).[163] The first step is attributed to adsorption onto easily accessible surface sites,[164] while the second step appears to be related to solid-state diffusion of the adsorbate into the bulk adsorbent. Two-step kinetics have been reported for adsorption of trace metals to a variety of substrates including manganese dioxide,[165] amorphous aluminum oxyhydroxide,[166] titanium dioxide, amorphous and crystalline iron oxides,[163,167] soils,[168] and a calcareous sand.[169] Results of cadmium uptake experiments on a calcareous sand are shown in Figure 14 and clearly exhibit the two-stage process described above. When surface sites are in large excess, i.e., when the ratio of adsorbate/adsorbent is low, the contribution

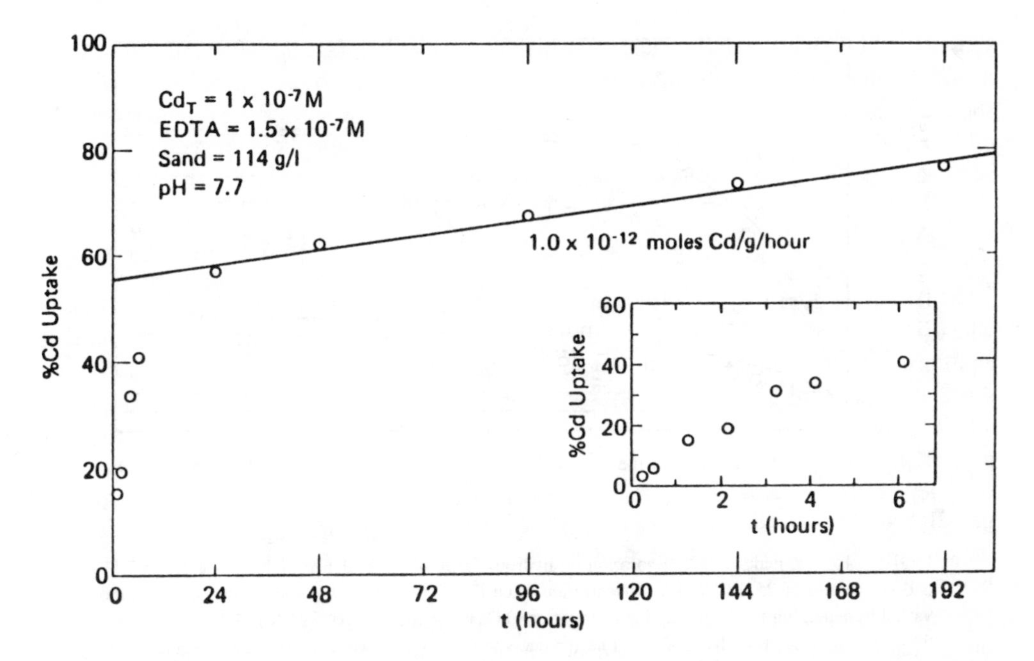

FIGURE 14. Rate of Cd(II) sorption by 114 g/l of a calcareous sand suspended in an artificial groundwater at pH 7.7 with a slight excess of EDTA. The line represents the least squares linear fit of data for t <24 h. (From Fuller, C. C. and Davis, J. A., *Geochim. Cosmochim. Acta*, 51, 1491, 1987. With permission.)

of the slow adsorption step to total adsorption is small and equilibrium is achieved rapidly. However, at high adsorbate/adsorbent ratios, adsorption kinetics can slow considerably, and the second step can become much more important. It should be noted that for natural waters, rapid adsorption equilibrium is a good assumption in most circumstances, but for treatment processes and other applications with high adsorbate concentrations, the validity of assuming equilibrium must be carefully examined.

Molecular diffusion theory (Fick's first law) has been widely used to model the adsorptive process, with liquid-phase diffusion of solute molecules through the solid liquid interfacial boundary layer used to describe the rapid first step and solid-phase diffusion used to describe the subsequent slower step. A variety of empirical expressions based on diffusion principles have been constructed and are useful for any given system, but a new set of empirical constants must be obtained once reactants and conditions change.[170]

Other general rate equations simply describe the kinetics of the adsorption process as a bimolecular reversible chemical reaction between solute and sorbent molecules. Dzombak and Morel[155] have used this approach in modeling the uptake of cadmium on amorphous iron oxide; i.e., expressing the adsorption of cadmium as a surface complexation reaction:

$$\equiv FeOH_2^+ + Cd^{2+} + H_2O \rightleftharpoons \equiv FeOCdOH_2^+ + 2H^+ \tag{60}$$

At constant pH and electrolyte concentration, the mass law expression for Equation 60 may be written

$$K^{cond} = \frac{[XC]}{[X][C]} = \frac{f}{b} \tag{61}$$

where $X = \equiv FeOH_2^+$, $C = Cd^{2+}$, $XC = \equiv FeOCdOH_2^+$, and f and b are forward and reverse rate constants, respectively. The rate expression for the surface complexation reaction is

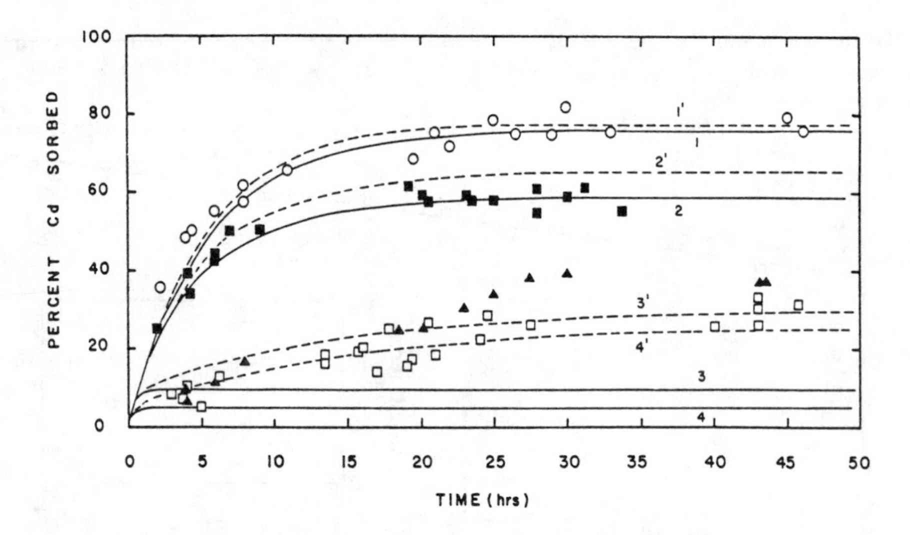

FIGURE 15. Kinetics of Cd^{2+} sorption on hydrous ferric oxide in 0.1 *M* $NaNO_3$ at a pH of 7.50 and temperture of 25.0°C. Solid concentration fixed (TOTFe = 10^{-3} M) and sorbate/sorbent ratio varied by changing total cadmium concentration: TOTCd = 6.5×10^{-7} *M* (○), 5×10^{-6} *M* (■), 5×10^{-3} *M* (△), 1×10^{-4} *M* (□). Fits for one-site (——) and two-site (---) rate expressions are also shown; parameter values are given in Table 18. (From Dzombak, D. A. and Morel, F. M. M., *J. Colloid Interface Sci.*, 112, 588, 1986. With permission.)

$$d[XC]/dt = f[X][C] - b[XC] \tag{62}$$

After including the appropriate mole balance quations for X_T and C_T, Equation 62 may be rewritten as

$$d[XC]/dt = f[XC]^2 - [XC]\{fX_T + fC_T + b\} + fX_TC_T \tag{63}$$

Dzombak and Morel[155] determined by trial and error values of f and b that provided reasonable fits of the experimental data subject to the constraints that $K^{cond} = f/b$. As shown in Figure 15, a one-site rate expression adequately fits the kinetic data at low sorbate/sorbent ratios, but in accord with the equilibrium modeling results discussed earlier, two-site expressions are essential at the higher loadings. Parameter values (f,b) used in modeling this kinetic data are given in Table 18.

IX. TRANSPORT MODELING

As noted earlier, an important extension of the application of chemical speciation codes is to the modeling of pollutants transport in both surface and subsurface waters. While transport modeling in general involves solution of the convective-diffusion equation, the time scales of interest and problems encounted are very different for surface compared to subsurface waters.

A. Surface Waters

While a variety of models have been constructed which consider a limited set of chemical reactions in conjunction with dispersion,[171,172]only a few workers have combined solution of the convective-dispersion equation with a general chemical speciation code. A good example of such a code is RIVEQL, developed by Chapman et al.,[94,94,173,174] in which a convolution integral technique is used to solve a modified form of the one-dimensional convective-diffusion equation to provide values of analytical concentrations of selected reactants at any point along a river at any time. From these concentrations, the chemical submodel, which uses the program MINEQL, calculates the equilibrium distribution of chemical species.

Chapman et al.[95] have satisfactorily modeled the effect of increase in pH on transport of soluble and colloidal forms of heavy metals in a stream contaminated with acid mine drainage using RIVEQL. The increase in pH was obtained experimentally by injection of the weak bases sodium acetate and potassium bicarbonate to the stream, and samples were taken for chemical analysis at sites downstream of the base injection point at regular time intervals. The convective-diffusion equation used in the model contained source terms, but no sink terms implying that any precipitates formed do not sediment, but are constrained to remain suspended as stable colloids and to disperse in a similar manner to solutes, and that adsorptive loss to stationary reactive surfaces (such as sediments and plants) is unimportant; i.e.,

$$\frac{\partial \mathbf{m}}{\partial t} = \frac{\partial}{\partial y}\left(D^* \frac{\partial \mathbf{m}}{\partial y}\right) - \frac{\partial \mathbf{m}}{\partial y} + f \tag{64}$$

where $\mathbf{m}$ is a vector of molar flow rates for each of the chemical components; t is time; y, the time of travel, is a measure of the distance x downstream from a fixed reference point in the river, such that $dy = dx/u$; D^* is the velocity-reduced dispersion coefficient; and f is the source vector for input of chemical components. The velocity-reduced dispersion coefficient $D^* = D/U^2$, where D is the dispersion coefficient and U is the average velocity over the river cross-section.

The semianalytical convolution integral method provides a solution to Equation 64 very efficiently and does not suffer the numerical instabilities which can affect finite-difference solutions of the convective-diffusion equation.[95] Once the molar flow-rate vector $\mathbf{m}$ has been obtained, it can be converted to a concentration vector $\mathbf{c}$ using the approximate relation

$$\mathbf{m} = \nu\mathbf{c} \tag{65}$$

where ν is the volumetric flow rate of the stream.

Measured concentrations of the soluble forms of major and trace elements, pH, and conductivity, together with lines representing the corresponding RIVEQL calculations for each parameter are shown in Figures 16a and 16b. In general, the model results describe the experimental values extremely well, confirming the validity of the modeling approach taken. In the case of sodium acetate addition (Figure 16a), RIVEQL predicts that the solids $Fe(OH)_3$ and $Al(OH)_3$ should form, and, indeed, marked decreases in the concentrations of soluble iron and aluminum are observed. Bicarbonate ion is a stronger base than the acetate ion; thus, a similar concentration of bicarbonate causes the stream pH to be higher (Figure 16b). At this higher pH, RIVEQL predicts that $Fe(OH)_3$, $Al(OH)_3$, $Cu_2(OH)_2CO_3$, $ZnCO_3$, and $ZnSiO_3$ should form. Decreases in the analytically determined concentrations of soluble Fe, Al, Cu, and Zn at the sampling points substantiate this prediction.

In cases where the effects of sedimentation and adsorption to stationary surfaces are significant, a more detailed description incorporating these phenomena becomes necessary.[94] Such a model has been applied successfully to the prediction of change in trace element speciation in an acidic stream on injection of strong base (NaOH). Although the model essentially assumes equilibrium, a pseudokinetic component has been added in order to deal with redissolution of precipitates and dissociation of surface species. Due to the added complexities, the run times of the modified version of RIVEQL were an order of magnitude greater than those for the simpler model described above.

B. Subsurface Waters

Modeling of solute transport in subsurface waters also requires solution of the convective-

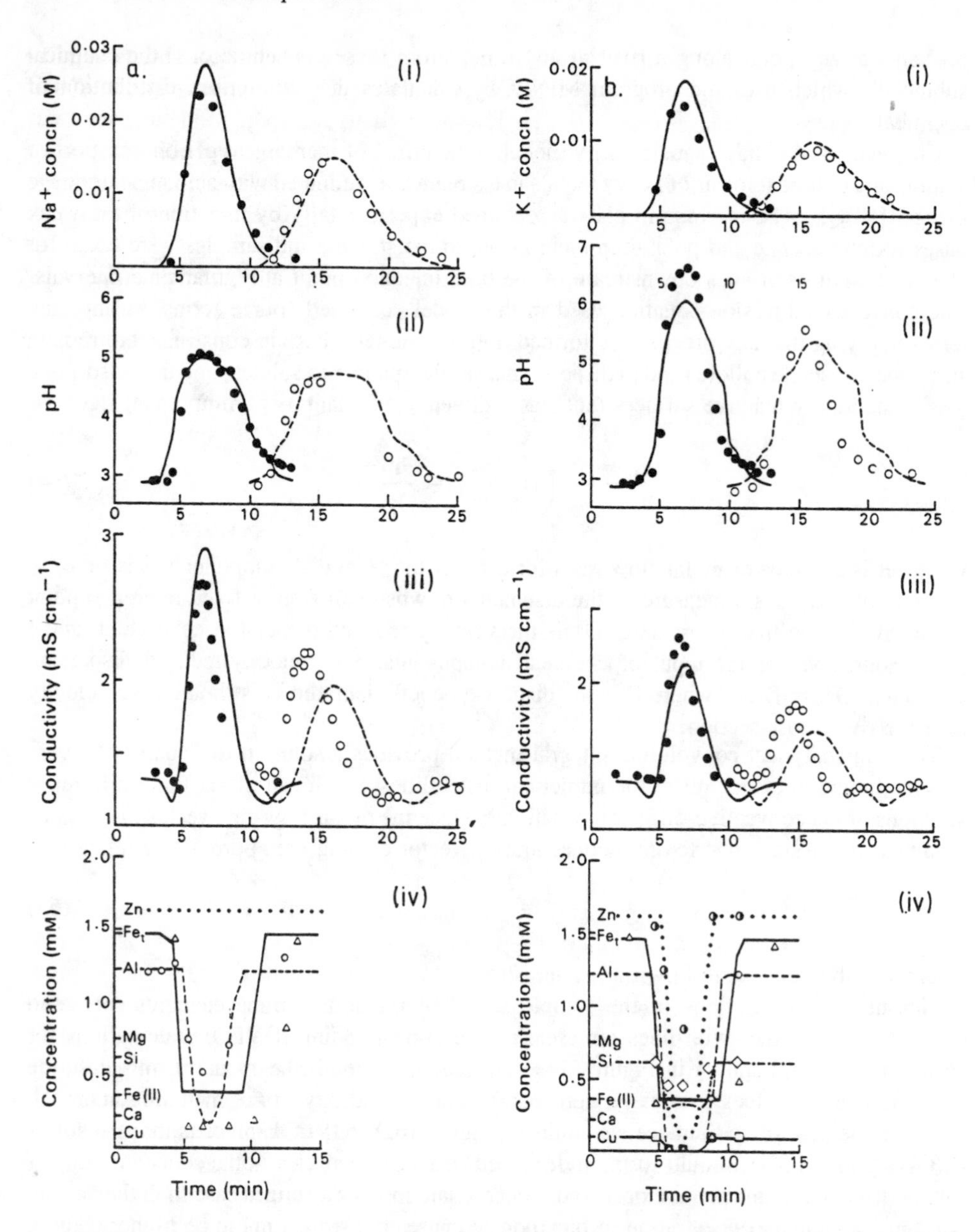

FIGURE 16. Na^+ (a) and K^+ (b) concentration (i), pH (ii), and conductivity (iii) as a function of time for experiments in which the pH of a stream is altered by addition of sodium acetate (a) and potassium bicarbonate (b). Observed values (points) and model simulations (lines) are shown for sampling stations at 41 m (——) and 93 m (---) downstream of the base injection point. (iv) Measured concentrations of components in soluble forms observed at the 41-m station are shown for Fe_t (△), Al (○), Zn (◑), Si (◇), and Cu (□). Simulated values are shown as lines. Where the concentration of an element did not change with time, that concentration is indicated on the vertical axis. (From Chapman, B. M., James, R. O., Jung, R. F., and Washington, H. G., *Aust. J. Mar. Freshwater Res.*, 33, 617, 1982. With permission.)

diffusion equation, but given the high solids to fluid ratio compared to surface waters, much greater attention must be given to solute-substrate interaction. In many modeling studies, aqueous-phase species distribution has been ignored entirely, and elements have simply been partitioned into either aqueous or sorbed components using a distribution coefficient (K_d) approach.[175,176] Although the convenience and computational ease of the K_d approach is beyond dispute, it has had limited success in modeling observed behavior, and its validity as a means of developing reliable predictions of the behavior of contaminants in actual subsurface systems is questionable.[177] The K_d approach to transport modeling does not account for changes in contaminant concentration due to geochemical processes such as complexation or changes in pH, and the K_d value provides neither information about specific chemical reactions occurring in the system nor insight into the physical or chemical factors controlling these reactions.

A number of subsurface solute transport codes have now been developed which incorporate chemical speciation in solution of the mass transport equations. These equations have the following form for one-dimensional flow in a homogeneous, isotropic porous medium:

$$\frac{\partial C_i}{\partial t} + \frac{\rho \partial \overline{C}_i}{\theta \partial t} = D_L \frac{\partial^2 C_i}{\partial x^2} - \nu \frac{\partial C_i}{\partial x} + S_{ci} \qquad i = 1,\ldots,n_{aq} \tag{66}$$

where

C_i = aqueous phase concentration of species i
$\bar{C}_i$ = sorbed phase concentration of species i
n_{aq} = number of aqueous species to be considered
S_{ci} = source-sink term for species i due to aqueous phase complexation
θ = medium porosity
ρ = bulk density of medium
D_L = coefficient of hydrodynamic dispersion
ν = average pore velocity
t = time
x = distance

Jennings et al.[178] suggest that there are two approaches to incorporation of chemical speciation in solution of mass transport equations. The first approach is to insert all of the interaction chemistry directly into the transport equations and to reduce the problem to one strongly nonlinear equation set. The second approach is to pose the equilibrium interaction chemistry independently of the mass transport equations, leading to a set of algebraic equations for the chemistry coupled to a set of differential equations for the mass transport which require sequential solution. While the first approach has been used by a number of workers,[178-183] the method appears to be expensive and unwieldy, except for a small number of solutes, and more attention is now being focused on the second approach.[184]

The technique of separating the algebraic equation set for the equilibrium chemistry from the differential equation set for mass transport has been used with some success in modeling the movement of trace elements and other pollutants in subsurface waters,[97-100,185,186] though it cannot be claimed that a general code for reliable prediction of trace element movement in subsurface waters is as yet available. Grove and Wood[185] coupled mineral dissolution and precipitation and ion exchange for three reactive solutes to a one-dimensional transport model which neglected transverse dispersion, while Schulz and Reardon[186] presented a combined two-dimensional mixing cell-analytical solution model which described multiple reactive solute transport in unidirectional groundwater flow regimes. The chemical equilibria for four solutes undergoing ion exchange and mineral precipitation and dissolution was calculated using a mixing cell model. The mixing cell model was coupled to and solved sequentially with an analytical transport model. This model is limited in its application to groundwater sites containing one-dimensional flow, although both transverse and longitudinal dispersion were considered.

More elaborate chemical speciation routines have been included in a limited number of models in which the equilibrium chemistry and mass transport equations are coupled and solved sequentially. Thus, MINEQL is coupled with the one-dimensional mass transport code FEAP in the model FIESTA.[97] The transport equation for each chemical component is solved using the Galerkin finite-element method subject to concentration or flux boundary conditions and appropriate initial conditions.

A maximum of six components can be transported simultaneously using FIESTA with the possibility that each component (as well as each aqueous complex) may interact with the solid phase through adsorption. Either 1:1 ion exchange or concentration Langmuir adsorption can be specified, but pH-dependent surface complexation models are not included. FIESTA has been used to simulate the movement of strontium in 1- and 5-m columns containing a homogeneous solid and found to give reliable results, provided sufficiently small time and space increments are used for calculations.[184]

The surface complexation model MICROQL has been coupled to the mass transport code ISOQUAD in the model TRANQL developed by Cedeberg et al.[98,99,177] The mass transport equations are solved using the Galerkin finite-element method with an implicit time-stepping scheme, and the algebraic chemical equation set (MICROQL) is solved using the Newton-Raphson method. A solution to a given problem is found by iterating between the two equation sets. TRANQL has been applied to the simulation of cadmium movement in laboratory column studies and the results compared to those obtained using a K_d approach.[98,177] Comparative studies were conducted in the presence of varying amounts of EDTA, a nonadsorbing (thus conservative) complexing agent for cadmium. As shown in Figure 17, significant differences in the cadmium mobility were predicted by the two approaches, with the largest differences occurring in cases where the cadmium was not entirely complexed by EDTA. TRANQL has also been used with some success to simulate the movement of cadmium in two-dimensional field-scale experiments.[177] Grover and Freyberg[99] and Cedeberg[177] have suggested a number of improvements that could be made to TRANQL including (1) the incorporation of more varied and sophisticated chemical equilibrium reaction sets (e.g., to include precipitation/dissolution reactions), and (2) the incorporation of non-equilibrium geochemical processes.

Selected precipitation/dissolution reactions have been included in the geochemical transport code DYNAMIX (DYNamic MIXing) developed by Narasimhan et al.[100] As noted earlier, DYNAMIX couples the chemical speciation algorithm, PHREEQE, with a modified form of the transport algorithm, TRUMP, specifically designed to handle the simultaneous migration of several chemical constituents under transient saturated or unsaturated flow conditions. Narasimhan et al.[100] investigated the transport of pollutants away from a uranium mill tailings pile and divided the overall problem of simulating the evolution and migration of the contaminant plume into three subproblems that were solved in sequential stages. These were the infiltration problem, the aquifer-tailings interaction problem, and the plume migration problem. Reasonable agreement was obtained between numerical and field data, and Narasimhan et al.[100] consider that, with some improvement (e.g., the replacement of the explicit method used in solution of the transport equations with a more efficient implicit scheme, such as that of Crank-Nicolson), the DYNAMIX algorithm has the potential of becoming a viable and practical tool in quantifying and predicting the transport of trace metals and other chemical species in groundwater systems.

All transport models discussed above incorporate a simplification known as the "local equilibrium assumption" in relation to sorption reactions at the sediment-waters interface. Chemical reactions are considered to occur so rapidly compared to the bulk fluid flow rate that they can be assumed to occur instantaneously.[2] While such an assumption may be valid in some cases, Valocchi[187] has shown that its validity is dependent on a range of transport system parameters such as seepage velocity, dispersion coefficients, and boundary conditions. Kirkner et al.[188] have facilitated investigation of the validity of the local equilibrium

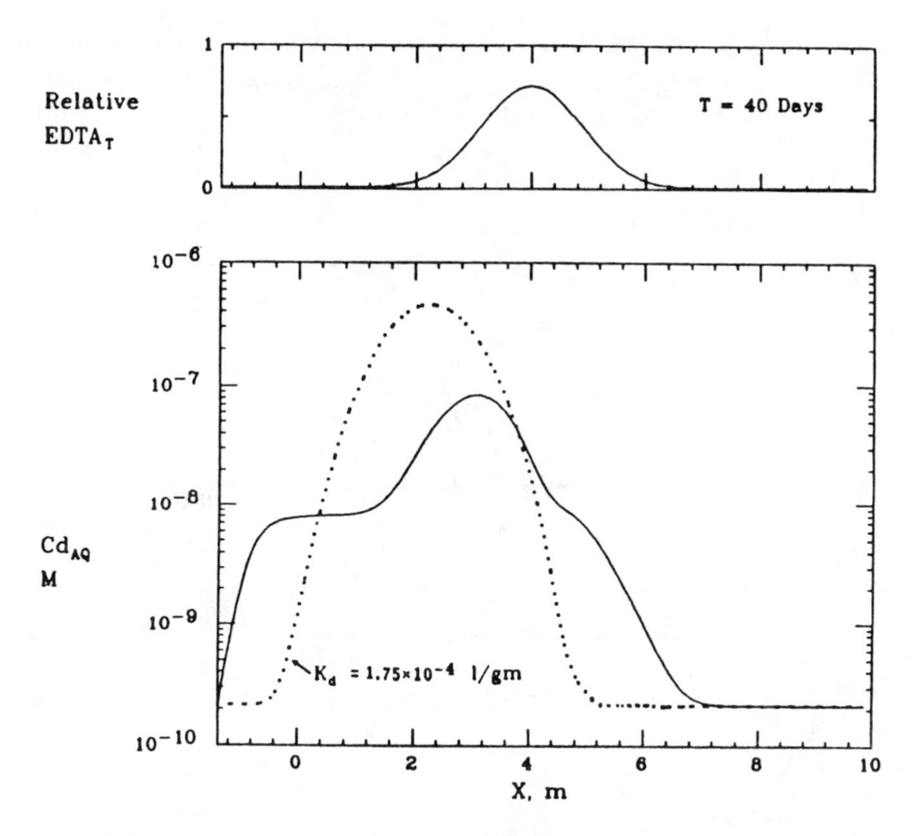

FIGURE 17. Comparison of one-dimensional TRANQL and K_d simulations for movement of cadmium through a column under conditions typical of those found in groundwater systems following pulse inputs of both EDTA and cadmium at T = 0 d and X = 0 m. EDTA (peak concentration) = 2.51×10^{-6} *M* and $Cd_T = 10^{-6}$ *M*. A mean velocity of 0.1 m/d, porosity of 0.4, dispersivity of 0.6 m, and bulk density of 1760 g/l were used in these simulations. (From Cedeberg, G. A., A Groundwater Mass-Transport and Equilibrium Chemistry Model for Multicomponent Systems, Ph.D. thesis, Stanford University, Stanford, CA, 1985.)

assumption by inclusion of a dynamic form of the Langmuir isotherm in the transport equation (Equation 66). Both equilibrium- and time-dependent sorption reactions can be considered using the formulation presented by Jennings et al.[188] These workers compared experimental and simulated results for the one-dimensional transport of nickel and cadmium through a dunal sand under both adsorptive and desorptive conditions. Reasonable agreement was found for the adsorptive case assuming local equilibrium, but the results for the desorptive case clearly indicated that the rate of desorption could not be considered rapid compared to the fluid flow rate (Figure 18). Simulation using the dynamic form of the Langmuir isotherm (the "kinetic" formulation) provided much better agreement between theory and experiment than obtained with the equilibrium formulation.

X. CONCLUSIONS

In this chapter an overview has been presented of the mathematical modeling of chemical speciation with an emphasis on application to trace elements in natural waters. An attempt has been made to provide essential mathematical concepts with some rigor, in addition to providing information of use to the applications-oriented reader. While not a classical analytical technique, mathematical modeling clearly has a place in extending the information that can be obtained on trace element species distribution by other methods and will be of

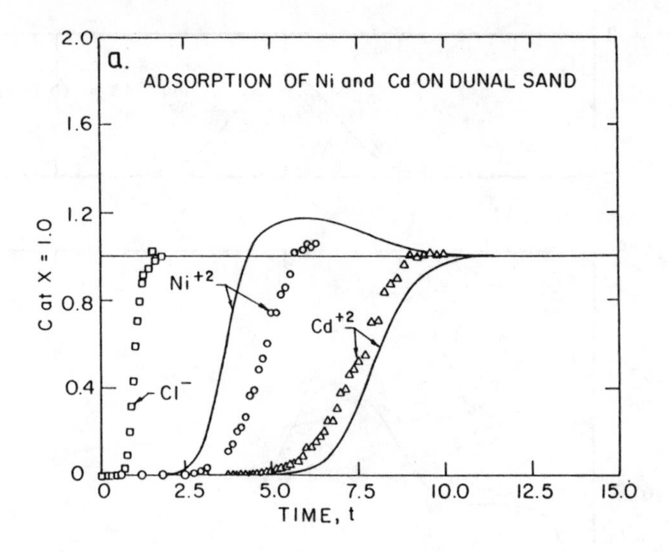

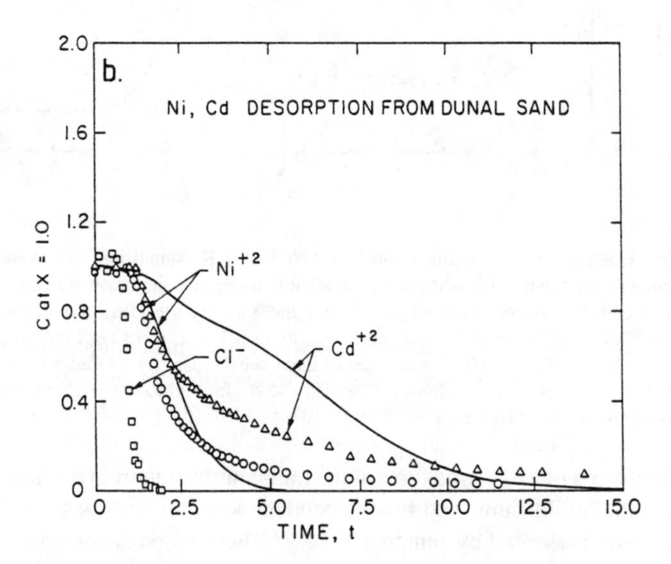

FIGURE 18. Comparison of measured and simulated results from column experiments in which either the local equilibrium assumption (LEA) is made or a "kinetic form" of the transport equation is used; (a) adsorption of Ni and Cd on dunal sand — LEA used in simulation, (b) desorption of Ni and Cd from dunal sand — LEA used in simulation, (c) desorption of Ni and Cd from dunal sand — kinetic form used in simulation. (From Kirkner, D. J., Jennings, A. A., and Theis, T. L., *J. Hydrol.*, 76, 107, 1985. With permission.)

particular use in systems for which determination of concentrations of all species of interest is impossible because of sensitivity constraints or other analytical difficulties. In addition, the mathematical modeling of reliable species concentration data may provide valuable thermodynamic parameters such as stability constants and free energies of formation for particular species.

A wide range of computer codes is available which may be used in modeling various aspects of trace element behavior in natural waters, though considerable attention must be given to selection of the code most appropriate to the task at hand. Thus, a code incorporating

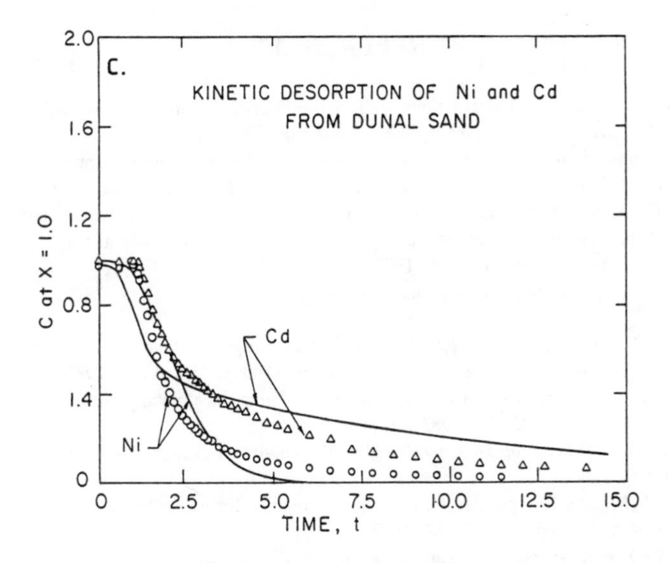

FIGURE 18 (continued)

virial equations, such as those of Pitzer for the computation of activity coefficients, should be used in preference to a code using a modified Debye-Huckel expression in modeling trace element behavior under high-ionic strength conditions typical of many groundwater and hydrothermal fluids. Similarly, a code to be used in modeling trace element transport in subsurface waters must contain an adequate sorption routine. Irrespective of the code being used, an essential prerequisite to the production of reliable information is the careful validation for completeness and correctness of the supplied thermodynamic data.

A variety of areas remain which require considerable development. Foremost of these is the incorporation of kinetic aspects in speciation codes. While all available speciation models, of necessity, are structured on the assumption that equilibrium conditions hold, the concentration of some trace element species in natural waters may be far from equilibrium and may approach such a state relatively slowly. Considerable progress has been made in modeling the dynamics of mineral precipitation and dissolution, but models of the time dependence of redox reactions or sorption/desorption processes tend to be reactant or concentration specific and not particularly amenable to general description. Similarly, our inability to adequately describe the thermodynamic properties of naturally occurring organic materials and surfaces or the mode of interaction of these complex entities with trace elements severely limits our ability to reliably model natural water processes, though steady progress in this direction is being made.

While general models such as THERMODATA/CHEMIX, MINEQL, and EQ3/6 have been and will continue to be used extensively, the proliferation of situation- or process-specific models is likely to continue. One explanation for the ever-expanding library of speciation codes may be the inaccessibility and inadequate documentation of many of the large, general routines, though as noted by Nordstrom et al.,[12] the satisfaction that researchers apparently obtain from programming may also account for the growth in this area.

REFERENCES

1. **Sunda, W. G.,** Trace metal interactions with marine algae, in *Marine Photosynthesis,* Alberte, R. and Barber, R. T., Eds., Oxford University Press, New York, in press.
2. **Rubin, J.,** Transport of reacting solutes in porous media: relation between mathematical nature of problem formulation and chemical nature of reactions, *Water Resour. Res.,* 19, 1231, 1983.
3. **Heinrich, C. A. and Eadington, P. J.,** Thermodynamic predictions of the hydrothermal chemistry of arsenic, and their significance for the paragenetic sequence of some cassiterite-arsenopyrite-base metal sulfide deposits, *Econ. Geol.,* 81, 511, 1986.
4. **Florence, T. M. and Batley, G. E.,** Chemical speciation in natural waters, *CRC Crit. Rev. Anal. Chem.,* 9, 219, 1980.
5. **Turner, D. R.,** Relationships between biological availability and chemical measurements, in *Metal Ions in Biological Systems,* Vol. 18, Sigel, H., Ed., Marcel Dekker, New York, 1984.
6. **Luoma, S. N. and Davis, J. A.,** Requirements for modeling trace metal partitioning in oxidized estuarine sediments, *Mar. Chem.,* 12, 159, 1983.
7. **Denbigh, K. G.,** *The Principles of Chemical Equilibrium,* 4th ed., Cambridge University Press, Cambridge, England, 1981.
8. **Morel, F. M. M. and Morgan, J. J.,** A numerical method for computing equilibria in aqueous chemical systems, *Environ. Sci. Technol.,* 6, 58, 1972.
9. **Zeleznik, F. J. and Gordon, S.,** Calculation of complex chemical equilibria, *Ind. Eng. Chem.,* 60, 27, 1968.
10. **Van Zeggeren, F. and Storey, S. H.,** *The Computation of Chemical Equilibria,* Cambridge University Press, Cambridge, England, 1970.
11. **Smith, W. R. and Missen, R. W.,** *Chemical Reaction Equilibrium Analysis,* Wiley-Interscience, New York, 1982.
12. **Nordstrom, D. K., Plummer, L. N., Wigley, T. M. L., Wolery, T. J., Ball, J. W., Jenne, E. A., Bassett, R. L., Crerar, D. A., Florence, T. M., Fritz, B., Hoffman, M., Holdren, G. R., Lafon, G. M., Mattigod, S. V., McDuff, R. E., Morel, F., Reddy, M. M., Sposito, G., and Thrailkill, J.,** A comparison of computerized chemical models for equilibrium calculations in aqueous systems, in *Chemical Modeling in Aqueous Systems,* ACS Symp. Ser. 93, Jenne, E. A., Ed., American Chemical Society, Washington, D.C., 1979, 857.
13. **Nordstrom, D. K. and Ball, J. W.,** Chemical models, computer programs and metal complexation in natural waters, in *Complexation of Trace Metals in Natural Waters,* Kramer, C. J. M. and Duinker, J. C., Eds., Martinus Nijhoff, The Hague, 1984, 149.
14. **Brinkley, S. R.,** Calculation of the equilibrium composition of systems of many constituents, *J. Chem. Phys.,* 15, 107, 1947.
15. **White, W. B., Johnson, S. M., and Dantzig, G. B.,** Chemical equilibrium in complex mixtures, *J. Chem. Phys.,* 28, 751, 1958.
16. **Shapiro, N. Z. and Shapley, L. S.,** Mass Action Laws and the Gibbs Free Energy Function, RAND Corporation Memo 3935-PR, AD-605 919, RAND, Santa Monica, CA, 1964.
17. **Clasen, R. J.,** The Numerical Solution of the Chemical Equilibrium Problem, RAND Corporation Memo 4345-PR, AD-609 904, RAND, Santa Monica, CA, 1965.
18. **Walsh, G. R.,** *Methods of Optimization,* John Wiley & Sons, New York, 1975.
19. **Ralston, A. and Rabinowitz, P.,** *A First Course in Numerical Analysis,* 2nd ed., McGraw-Hill, New York, 1978.
20. **Wigley, T. M. L.,** WATSPEC: a computer program for determining the equilibrium speciation of aqueous solutions, *Br. Geomorphol. Res. Group Tech. Bull.,* 20, 1977.
21. **Wigley, T. M. L.,** Ion pairing and water quality measurements, *Can. J. Earth Sci.,* 8, 468, 1971.
22. **Robinson, R. A. and Stokes, R. H.,** *Electrolyte Solutions,* 2nd ed. (revised), Butterworths, London, 1970.
23. **Bronsted, J. N.,** Studies on solubility. IV. The principle of the specific interaction of ions, *J. Am. Chem. Soc.,* 44, 877, 1922.
24. **Guggenheim, E. A.,** The specific thermodynamic properties of aqueous solutions of strong electrolyes, *Philos. Mag.,* 19, 588, 1935.
25. **Mayer, J. E.,** The theory of ionic solutions, *J. Chem. Phys.,* 18, 1426, 1950.
26. **Scatchard, G.,** The excess free energy and related properties of solutions containing electrolytes, *J. Am. Chem. Soc.,* 90, 3124, 1968.
27. **Pitzer, K. S.,** Thermodynamics of electrolytes. I. Theoretical basis and general equations, *J. Phys. Chem.,* 77, 268, 1973.
28. **Pitzer, K. S.,** Electrolyte theory-improvements since Debye and Huckel, *Acc. Chem. Res.,* 10, 371, 1977.
29. **Pitzer, K. S.,** Theory: ion interaction approach, in *Activity Coefficients in Electrolyte Solutions,* Vol. 1, Pytkowicz, R. M., Ed., CRC Press, Boca Raton, FL, 1979, 157.

30. **Whitfield, M.,** An improved specific interaction model for seawater at 25°C and 1 atmosphere pressure, *Mar. Chem.*, 3, 197, 1975.
31. **Whitfield, M.,** Activity coefficients in natural waters, in *Activity Coefficients in Electrolyte Solutions*, Vol. II, Pytkowicz, R. M., Ed., CRC Press, Boca Raton, FL, 1979, 153.
32. **Millero, F. J. and Schreiber, D. R.,** Use of the ion pairing model to estimate activity coefficients of the ionic components of natural waters, *Am. J. Sci.*, 282, 1508, 1982.
33. **Harvie, C. E. and Weare, J. H.,** The prediction of mineral solubilities in natural waters: the Na-K-Mg-Ca-Cl-SO_4-H_2O system from zero to high concentration at 25°C, *Geochim. Cosmochim. Acta*, 44, 981, 1980.
34. **Giblin, A. M.,** unpublished data, 1987.
35. **Garrels, R. M. and Christ, C. L.,** *Solutions, Minerals and Equilibria*, Freemen, Cooper, San Francisco, 1965.
36. **Morel, F. M. M.,** *Introduction to Aquatic Chemistry*, John Wiley & Sons, New York, 1983.
37. **Klotz, I. and Rosenberg, R. M.,** *Chemical Thermodynamics*, Benjamin, New York, 1972.
38. **Khodakovskiy, I. L., Ryzhenko, B. N., and Naumov, G. B.,** Thermodynamics of aqueous electrolyte solutions at elevated temperatures (temperature dependence of the heat capacity of ions in aqueous solutions), *Geokhimiya*, 14, 86, 1968.
39. **DeBethune, A. J.,** Irreversible thermodynamics in electrochemistry, *J. Electrochem. Soc.*, 107, 829, 1960.
40. **Criss, C. M. and Cobble, J. W.,** The thermodynamic properties of high temperature aqueous solutions. V. The calculation of ionic heat capacities up to 200°. Entropies and heat capacities above 200°, *J. Am. Chem. Soc.*, 86, 5390, 1964.
41. **Helgeson, H. C.,** Thermodynamics of complex dissociation in aqueous solution at elevated temperatures, *J. Phys. Chem.*, 71, 3121, 1967.
42. **Barner, H. E. and Scheuerman, R. V.,** *Handbook of Thermochemical Data for Compounds and Aqueous Species*, John Wiley & Sons, New York, 1978.
43. **Lowson, R. T.,** Aluminium corrosion studies. I. Potential-pH-temperature diagrams for aluminium, *Aust. J. Chem.*, 27, 105, 1974.
44. **Jenne, E. A.,** Groundwater Modelling: A Review, PNL-3574, Pacific Northwest Laboratory, Richland, WA, 1981.
45. **Kincaid, C. T., Morrey, J. R., and Rogers, J. E.,** Geohydrochemical Codes for Solute Migration, Volume I: Process Description and Computer Code Selection, EPRI EA-3417, Vol. 1, Electric Power Research Institute, Palo Alto, CA, February, 1984.
46. **Kincaid, C. T., Morrey, J. R., Yabusaki, S. B., Felmy, A. R., and Rogers, J. E.,** Geohydrochemical Models for Solute Migration, Volume 2: Preliminary Evaluation of Selected Codes, EPRI EA-3417, Vol. 2, Electric Power Research Institute, Palo Alto, CA, November, 1984.
47. **Garrels, R. M. and Thompson, M. E.,** A chemical model for sea water at 25°C and one atmosphere total pressure, *Am. J. Sci.*, 260, 57, 1962.
48. **Sillen, L. G.,** High-speed computers as a supplement to graphical methods. I. Functional behaviour of the error square sum, *Acta Chem. Scand.*, 16, 159, 1962.
49. **Ingri, N., Kakolwicz, W., Sillen, L. G., and Warnquist, G.,** High-speed computers as a supplement of graphical methods. V. HALTAFALL, a general program for calculating the composition of equilibrium mixtures, *Talanta*, 14, 1261, 1967.
50. **Dryssen, D., Jagner, D., and Wengelin, F.,** *Computer Calculation of Ionic Equilibria and Titration Procedures*, John Wiley & Sons, New York, 1968.
51. **Perrin, D. D.,** Multiple equilibria in assemblages of metal ions and complexing species: a model for biological systems, *Nature (London)*, 206, 170, 1965.
52. **Perrin, D. D. and Sayce, I. G.,** Computer calculation of equilibrium concentrations in mixtures of metal ions and complexing species, *Talanta*, 14, 833, 1967.
53. **Ginzburg, G.,** Calculation of all equilibrium concentrations in a system of competing complexation, *Talanta*, 23, 149, 1976.
54. **Legget, D. J.,** Machine computation of equilibrium concentrations — some practical considerations, *Talanta*, 24, 535, 1977.
55. **Fardy, J. J. and Sylva, R. N.,** SIAS, A Computer Program for the Generalized Calculation of Speciation in Mixed Metal-Ligand Aqueous Systems, AAEC/E445, Lucas Heights Research Laboratories, Australian Atomic Energy Commission, Australia, 1978.
56. **Barnes, I. and Clarke, F. E.,** Chemical properties of groundwater and their encrustation effects on wells, *U.S. Geol. Surv. Prof. Papers*, 498-D, 1969.
57. **Truesdell, A. H. and Jones, B. F.,** WATEQ, a computer program for calculating chemical equilibria of natural waters, *U.S. Geol. Surv. J. Res.*, 2, 233, 1974.
58. **Plummer, L. N., Jones, B. F., and Truesdell, A. H.,** WATEQF — a Fortran IV version of WATEQ, a computer program for calculating chemical equilibrium of natural waters, *U.S. Geol. Surv. Water Resour. Invest. Rep.*, 76-13, 1976.

59. **Ball, J. W., Jenne, E. A., and Nordstrom, D. K.,** WATEQ2, a computerized chemical model for trace and major element speciation and mineral equilibria of natural waters, in *Chemical Modelling in Aqueous Systems,* ACS Symp. Ser. 93, Jenne, E. A., Ed., American Chemical Society, Washington, D.C., 1979, 815.
60. **Ball, J. W., Nordstrom, D. K., and Jenne, E. A.,** Additional and revised thermochemical data and computer code for WATEQ2 — a computerized chemical model for trace and major element speciation and mineral equilibria of natural waters, *U.S. Geol. Surv. Water Resour. Invest.,* 78-116, 1980.
61. **Ball, J. W., Jenne, E. A., and Cantrell, M. W.,** WATEQ3 — a geochemical model with uranium added, *U.S. Geol. Surv. Open-File Rep.,* 81-1183, 1981.
62. **Kharaka, Y. K. and Barnes, I.,** SOLMNEQ: solution — mineral equilibrium computations, *U.S. Geol. Surv. Water Resour. Invest.,* 73-002, 1973.
63. **Morel, F. M. M. and Morgan, J. J.,** A numerical method for computing equilibria in aqueous systems, *Environ. Sci. Technol.,* 6, 58, 1972.
64. **McDuff, R. E. and Morel, F. M. M.,** Description and Use of the Chemical Equilibrium Program REDEQL2, Tech. Rep. EQ-73-02, W. M. Keck Laboratory, California Institute of Technology, Pasadena, CA, 1973.
65. **Westall, J. C., Zachary, J. L., and Morel, F. M. M.,** MINEQL, a Computer Program for the Calculation of Chemical Equilibrium Composition of Aqueous Systems, Tech. Note 18, Department of Civil Engineering, Massachusetts Institute of Technology, Cambridge, 1976.
66. **Harriss, D. K., Ingle, S. E., Magnuson, V. R., and Taylor, D. K.,** Programmer's Manual for REDEQL.UMD, University of Minnesota, Duluth, 1984.
67. **Sposito, G. and Mattigod, S. V.,** GEOCHEM: A Computer Program for the Calculation of Chemical Equilibria in Soil Solutions and Other Natural Water Systems, University of California, Riverside, 1980.
68. **Felmy, A. R., Girvin, D., and Jenne, E. A.,** MINTEQ: A Computer Program for Calculating Aqueous Geochemical Equilibria, U.S. Environmental Protection Agency, Washington, D.C., 1983.
69. **Westall, J. C.,** MICROQL, A Chemical Equilibrium Program in BASIC, Swiss Federal Institute of Technology (EAWAG), Zurich, 1979.
70. **Tripathi, V. S.,** An algorithm and a FORTRAN program (CHEMEQUIL-2) for calculation of complex equilibria, *Talanta,* 33, 1015, 1986.
71. **Helgeson, H. C.,** Evaluation of irreversible reactions in geochemical processes involving minerals and aqueous solutions. I. Thermodynamic relations, *Geochim. Cosmochim. Acta,* 32, 853, 1968.
72. **Helgeson, H. C., Garrels, R. M., and Mackenzie, F. T.,** Evaluation of irreversible reactions in geochemical processes involving minerals and aqueous solutions. II. Applications, *Geochim. Cosmochim. Acta,* 33, 455, 1969.
73. **Helgeson, H. C., Brown, T. H., Nigrini, A., and Jones, T. A.,** Calculation of mass transfer in geochemical processes involving aqueous solutions, *Geochim. Cosmochim. Acta,* 34, 569, 1970.
74. **Helgeson, H. C.,** A chemical and thermodynamic model of ore deposition in hydrothermal systems, *Miner. Soc. Am. Spec. Paper,* 3, 155, 1970.
75. **Miller, D. G., Piwinskii, A. J., and Yamauchi, R.,** The Use of Geochemical-Equilibirum Computer Calculations to Estimate Precipitation from Geothermal Brines, UCRL-52197, Lawrence Livermore National Laboratory, Livermore, CA, 1977.
76. **Mercer, J. W., Faust, C. R., Miller, W. J., and Pearson, F. J.,** Review of simulation techniques for aquifer thermal energy storage (ATES), *Adv. Hydrosci.,* 13, 1, 1982.
77. **Wolery, T. J.,** Calculation of Chemical Equilibrium Between Aqueous Solution and Minerals: The EQ3/6 Software Package, UCRL-52658, Lawrence Livermore National Laboratory, Livermore, CA, 1979.
78. **Reed, M. H.,** Calculation of multicomponent chemical equilibria and reaction processes involving minerals, gases and an aqueous phase, *Geochim. Cosmochim. Acta,* 46, 513, 1982.
79. **Wolery, T. J., Isherwood, D. J., Jackson, K. J., Delany, J. M., and Purgdomenech, I.,** EQ3/6: Status and Applications, Lawrence Livermore National Laboratory, Livermore, CA, 1984.
80. **Parkhurst, D. L., Thorstenson, D. C., and Plummer, L. N.** PHREEQE — a computer program for geochemical calculations, *U.S. Geol. Surv. Water Resour. Invest. Rep.,* 80-96, 1980.
81. **Plummer, L. N., Parkhurst, D. L., and Thorstenson, D. C.,** Development of reaction models for groundwater systems, *Geochim. Cosmochim. Acta,* 47, 665, 1983.
82. **Deland, E. C.,** CHEMIST — The Rand Chemical Equilibrium Program, RAND Corporation Memo 5404-PR, AD-664 045, RAND, Santa Monica, CA, 1967.
83. **Karpov, I. K. and Kaz'min, L. A.,** Calculation of geochemical equilibria in heterogeneous multicomponent systems, *Geochem. Int.,* 9, 252, 1972.
84. **Karpov, I. K., Kaz'min, L. A., and Kashik, S. A.,** Optimal programming for computer calculation of irreversible evolution in geochemical systems, *Geochem. Int.,* 10, 464, 1973.
85. **Erikkson, G.,** An algorithm for the computation of aqueous multi-component, multiphase equilibria, *Anal. Chim. Acta,* 112, 375, 1979.

86. **Gautem, R. and Seider, W. D.,** Computation of phase and chemical equilibrium. I. Local and constrained minima in Gibbs free energy. II. Phasesplitting. III. Electrolytic solutions, *Am. Inst. Chem. Eng. J.*, 25, 991, 1979.
87. **Harvie, C. E. and Weare, J. H.,** The prediction of mineral solubilities in natural waters: the Na-K-Mg-Ca-Cl-SO_4-H_2O system from zero to high concentration at 25° C, *Geochim. Cosmochim. Acta,* 44, 981, 1980.
88. **Harvie, C. E., Greenberg, J. P., and Weare, J. H.,** A chemical equilibrium algorithm for highly non-ideal multiphase systems: free energy minimization, *Geochim. Cosmochim. Acta,* 51, 1045, 1987.
89. **Turnbull, A. G. and Wadsley, M. W.,** The CSIRO-SGTE THERMODATA System (Version V), CSIRO Division of Mineral Chemistry, Melbourne, 1987.
90. **Dzombak, D. A. and Morel, F. M. M.,** Adsorption of inorganic pollutants in aquatic systems, *Am. Soc. Civ. Eng. J. Hydraul.*, 113, 430, 1987.
91. **Faughnam, J.,** The SURFEQL/MINEQL Manual, W. M. Keck Laboratory, California Institute of Technology, Pasadena, 1981.
92. **Bourg, A. C. M.,** ADSORB, a chemical equilibria computer program accounting for adsorption processes in aquatic systems, *Environ. Technol. Lett.*, 3, 305, 1982.
93. **Bourg, A. C. M. and Mouvet, C.,** A heterogeneous complexation model of the adsorption of trace metals on natural particulate matter, in *Complexation of Trace Metals in Natural Waters,* Kramer, C. J. M. and Duinker, J. C., Eds., Martinus Nijhoff, The Hague, 1984, 267.
94. **Chapman, B. M.,** Numerical simulation of the transport and speciation of non-conservative chemical reactants in rivers, *Water Resour. Res.*, 18, 155, 1982.
95. **Chapman, B. M., James, R. O., Jung, R. F., and Washington, H. G.,** Modelling the transport of reacting chemical contaminants in natural streams, *Aust. J. Mar. Freshwater Res.*, 33, 617, 1982.
96. **Felmy, A. R., Brown, S. M., Onishi, Y., Yabusaki, S. B., and Argo, R. S.,** MEXAMS — The Metals Exposure Analysis Modelling System, Battelle Memorial Institute Pacific Northwest Laboratories, Richland, WA, 1985.
97. **Theis, T. L., Kirkner, D. J., and Jennings, A. A.,** Multi-Solute Subsurface Transport Modeling for Energy Solid Wastes, Department of Civil Engineering, University of Notre Dame, Notre Dame, IN, 1982.
98. **Cedeberg, G. A., Street, R. L., and Leckie, J. O.,** A groundwater mass transport and equilibrium chemistry model for multicomponent systems, *Water Resour. Res.*, 21, 1095, 1985.
99. **Grover, D. A., and Freyberg, D. L.,** User's Guide to TRANQL, A groundwater Mass Transport and Equilibrium Chemistry Model for Multicomponent systems, Department of Civil Engineering, Stanford University, Stanford, CA, 1986.
100. **Narasimhan, T. N., White, A. F., and Tokunaga, T.,** Groundwater contamination from an inactive uranium mill tailings pile. II. Application of a dynamic mixing model, *Water Resour. Res.*, 22, 1820, 1986.
101. **Ingri, N. and Sillen, L. G.,** High speed computers as a supplement of graphical methods. IV. An ALGOL version of LETAGROP VRID, *Arkiv. Kemi,* 23, 97, 1965.
102. **Sabatini, A., Vaca, A., and Gans, P.,** MINIQUAD — a general computer programme for the computation of formation constants from potentiometric data, *Talanta,* 21, 53, 1974.
103. **Westall, J. C. and Morel, F. M. M.,** FITEQL — An Algorithm for Determination of Metal-Ligand Complex Formation Constants, Tech. Note 19, Ralph M. Parsons Laboratory, Massachusetts Institute of Technology, Cambridge, 1977.
104. **Westall, J. C.,** FITEQL — A Program for the Determination of Chemical Equilibrium Constants from Experimental Data, User's Guide, Version 1.2, Chemistry Department, Oregon State University, Corvallis, 1982.
105. **Turnbull, A. G. and Wadsley, M. W.,** Thermodynamic modelling of metallurgical processes by the CSIRO-SGTE THERMODATA system, in Proc. Symp. Extractive Metallurgy, Australasian Institute of Mining and Metallurgy, Melbourne, November 12 to 14, 1984, 79.
106. **Barin, I. and Knacke, O.,** *Thermochemical Properties of Inorganic Substances,* Springer-Verlag, Berlin, 1973.
107. **Barin, I., Knacke, O., and Kubaschewski, O.,** *Thermochemical Properties of Inorganic Substances* (Suppl.), Springer-Verlag, Berlin, 1977.
108. **Robie, R. A., Hemingway, B. S., and Fisher, J. R.,** Thermodynamic properties of minerals and related substances, *U.S. Geol. Surv. Bull.*, 1452, 1978.
109. **Stull, D. R., Westrum, E. F., and Sinke, G. C.,** *The Chemical Thermodynamics of Organic Compounds,* John Wiley & Sons, New York, 1969.
110. **Stull, D. R. and Prophet, H.,** JANAF Thermochemical Tables, 2nd ed., NSRDS NBS 37, U.S. Government Printing Office, Washington, D.C., 1971.
111. **Froning, M. H., Shanley, M. E., and Verink, E.D., Jr.** An improved method for calculation of potential-pH diagrams of metal ion-water systems by computer, *Corros. Sci.*, 16, 371, 1976.

112. **Sillen, L. G. and Martell, A. E.,** *Stability Constants of Metal-Ion Complexes,* Special Publ. 17, Chemical Society, London, 1964.
113. **Sillen, L. G. and Martell, A. E.,** *Stability Constants of Metal-Ion Complexes* (Suppl. 1), Special Publ. 25, Chemical Society, London, 1971.
114. **Martell, A. E. and Smith, R. M.,** *Critical Stability Constants,* Vol. 1, Plenum Press, New York, 1974.
115. **Smith, R. M. and Martell, A. E.,** *Critical Stability Constants,* Vol. 2, Plenum Press, New York, 1975.
116. **Martell, A. E. and Smith, R. M.,** *Critical Stability Constants,* Vol. 3, Plenum Press, New York, 1976.
117. **Smith, R. M. and Martell, A. E.,** *Critical Stability Constants,* Vol. 4, Plenum Press, New York, 1976.
118. **Morel, F. M. M., Rueter, J. G., Anderson, D. M., and Guillard, R. R. L.,** AQUIL: a chemically defined phytoplankton culture medium for trace metal studies, *J. Phycol.,* 15, 135, 1979.
119. **Morel, F. M. M., Westall, J. C., O'Melia, C. R., and Morgan, J. J.,** Fate of trace metals in Los Angeles County wastewater discharge, *Environ. Sci. Technol.,* 9, 756, 1975.
120. **Wolery, T. J.,** Some Chemical Aspects of Hydrothermal Processes at Mid-Ocean Ridges: A Theoretical Study, Ph.D. thesis, Northwestern University, Evanston, IL, 1978.
121. **Wolery, T. J.,** EQ3NR, A Computer Program for Geochemical Aqueous Speciation — Solubility Calculations: User's Guide and Documentation, UCRL-5344, Lawrence Livermore National Laboratory, Livermore, CA, 1983.
122. **Wolery, T. J.,** EQ6, A Computer Program for Reaction-Path Modelling of Aqueous Geochemical Systems: User's Guide and Documentation, Lawrence Livermore National Laboratory, Livermore, CA, 1984.
123. **Jackson, K. J. and Wolery, T. J.,** Models for activity coefficients in brines, in Collected Abstracts for the Workshop on Geochemical Modelling, Fallen Leaf Lake, California, September 14 to 17, 1986.
124. **Helgeson, H. C., Kirkham, D., and Flowers, G. C.,** Theoretical prediction of the thermodynamic behaviour of aqueous electrolytes at high pressures and temperatures. IV, *Am. J. Sci.,* 281, 1249, 1981.
125. **Wolery, T. J.,** EQ3/6 — status and future directions, in Collected Abstracts for the Workshop on Geochemical Modelling, Fallen Leaf Lake, California, September 14 to 17, 1986.
126. **Noronha, C. J. and Pearson, F. J., Jr.,** EQ3/EQ6: A Geochemical Speciation and Reaction Path Code Package Suitable for Nuclear Waste Performance Assessment. ONWI-472, Office of Nuclear Waste Isolation, U.S. Department of Energy, Washington, D. C., 1983.
127. **Broyd, T. W., Grant, M. M., and Cross, J. E.,** A Report on Intercomparison Studies of Computer Programs Which Respectively Model: (i) Radionuclide Migration, (ii) Equilibrium Chemistry of Groundwater, EUR 10231 EN, Commission of the European Communities, Luxembourg, 1985.
128. **Davis, J. A.,** Complexation of trace metals by absorbed natural organic matter, *Geochim. Cosmochim. Acta,* 48, 679, 1984.
129. **Kurbatov, M. H., Wood, G. B., and Kurbatov, J. D.,** Isothermal adsorption of cobalt from dilute solutions, *J. Phys. Chem.,* 55, 1170, 1951.
130. **Kinniburgh, D. G. and Jackson, M. L.,** Concentration and pH dependence of calcium and zinc adsorption by iron hydrous oxide gel, *Soil Sci. Soc. Am. J.,* 46, 56, 1982.
131. **Tanford, C.,** Multiple equilibria, in *Physical Chemistry of Macromolecules,* John Wiley & Sons, New York, 1961, 526.
132. **James, R. O. and Parks, G. A.** Characterization of aqueous colloids by their electrical double-layer and intrinsic surface chemical properties, in *Surface and Colloid Science,* Vol. 12, Matijevic, E., Ed., Plenum Press, New York, 1982, 119.
133. **Sposito, G.,** Chemical models of inorganic pollutants in soils, *CRC Crit. Rev. Environ. Control,* 15, 1, 1985.
134. **Morel, F. M. M., Yeasted, J. G., and Westall, J. C.,** Adsorption models: a mathematical analysis in the framework of general equilibrium calculations, in *Adsorption of Inorganics at Solid-Liquid Interfaces,* Anderson, M. A. and Rubin, A. J., Eds., Ann Arbor Science, Ann Arbor, MI, 1981, 263.
135. **Westall, J.,** MICROQL, II. Computation of Adsorption Equilibria in BASIC, Swiss Federal Institute of Technology (EAWAG), Duebendorf, Switzerland, 1979.
136. **Huang, C. P. and Stumm, W.,** Specific adsorption of cations on hydrous γ-Al_2O_3, *J. Colloid Interface Sci.,* 43, 409, 1973.
137. **Huang, C. P.,** The surface acidity of hydrous solids, in *Adsorption of Inorganics at Solid-Liquid Interfaces,* Anderson, M. A. and Rubin, A. J., Eds., Ann Arbor Science, Ann Arbor, MI, 1981, 183.
138. **Stumm, W., Hohl, H., and Dalang, F.,** Interactions of metal ions with hydrous oxide surfaces, *Croat. Chem. Acta,* 48, 491, 1976.
139. **Schindler, P. W., Furst, B., Dick, R., and Wolf, P. U.,** Ligand properties of surface silanol groups, *J. Colloid Interface Sci.,* 55, 469, 1976.
140. **Breeuwsma, A. and Lyklema, J.,** Physical and chemical adsorption of ions in the electrical double layer on hematite (α-Fe_2O_3), *J. Colloid Interface Sci.,* 43, 437, 1973.
141. **Bowden, J. W., Posner, A. M., and Quirk, J. P.,** Ionic adsorption on variable charge mineral surfaces. Theoretical charge development and titration curves, *Aust. J. Soil Res.,* 15, 121, 1977.

142. **Overbeek, J. T. G.,** Electrochemistry of the double layer, in *Colloid Science,* Vol. 1, Kruyt, H. R., Ed., Elsevier, Amsterdam, 1952, 115.
143. **Yates, D. E., Levine, S., and Healy, T. W.,** Site-binding model of the electrical double layer at the oxide/water interface, *J. Chem. Soc. Faraday Trans. I,* 70, 1807, 1974.
144. **Davis, J. A., James, R. O., and Leckie, J. O.,** Surface ionization and complexation at the oxide/water interface, *J. Colloid Interface Sci.,* 63, 480, 1978.
145. **Barrow, N. J., Bowden, J. W., Posner, A. M., and Quirk, J. P.,** An objective method for fitting models of ion adsorption on variable charge surfaces, *Aust. J. Soil Res.,* 18, 34, 1980.
146. **Barrow, N. J., Bowden, J. W., Posner, A. M., and Quirk, J. P.,** Describing the adsorption of copper, zinc and lead on a variable charge mineral surface, *Aust. J. Soil Res.,* 19, 309, 1981.
147. **Westall, J. and Hohl, H.,** A comparison of electrostatic models for the oxide/solution interface, *Adv. Colloid Interface Sci.,* 12, 265, 1980.
148. **Benjamin, M. M. and Leckie, J. O.,** Multiple-site adsorption of Cd, Cu, Zn and Pb on amorphous iron oxyhydroxide, *J. Colloid Interface Sci.,* 79, 209, 1981.
149. **Van Reimsdijk, W. H., De Wit, J. C. M., Koopal, L. K., and Bolt, G. H.,** Metal ion adsorption on heterogeneous surfaces: adsorption models, *J. Colloid Interface Sci.,* 116, 511, 1987.
150. **Farley, K. J., Dzombak, D. A., and Morel, F. M. M.,** A surface precipitation model for the sorption of cations on metal oxides, *J. Colloid Interface Sci.,* 106, 226, 1985.
151. **Goldberg, S.,** Chemical modelling of arsenate adsorption on aluminium and iron oxide minerals, *Soil Sci. Soc. Am. J.,* 50, 1154, 1986.
152. **Hsi, C.-K. D. and Langmuir, D.,** Adsorption of uranyl onto ferric oxyhydroxides: application of the surface complexation site-binding model, *Geochim. Cosmochim. Acta,* 49, 1931, 1985.
153. **LaFlamme, B. D. and Murray, J. D.,** Solid/solution interaction: the effect of carbonate alkalinity on adsorbed thorium, *Geochim. Cosmochim. Acta,* 51, 243, 1987.
154. **Hayes, K. F. and Leckie, J. O.,** Modelling ionic strength effects on cation adsorption at hydrous oxide/ solution interfaces, *J. Colloid Interface Sci.,* 115, 564, 1987.
155. **Dzombak, D. A. and Morel, F. M. M.,** Sorption of cadmium on hydrous ferric oxide at high sorbate/ sorbent ratios: equilibrium, kinetics, and modelling, *J. Colloid Interface Sci.,* 112, 588, 1986.
156. **Kummert, R. and Stumm, W.,** The surface complexation of organic acids on hydrous γ-Al_2O_3, *J. Colloid Interface Sci.,* 75, 373, 1980.
157. **Davis, J. A.,** Adsorption of natural dissolved organic matter at the oxide/water interface, *Geochim. Cosmochim. Acta,* 46, 2381, 1982.
158. **Davis, J. A.,** Complexation of trace metals by adsorbed natural organic matter, *Geochim. Cosmochim. Acta,* 48, 679, 1984.
159. **Cabaniss, S. E., Shuman, M. S., and Collins, B. J.,** Metal-organic binding: a comparison of models, in *Complexation of Trace Metals in Natural Waters,* Kramer, C. J. M. and Duinker, J. C., Eds., Martinus Nijhoff, The Hague, 1984, 165.
160. **Fish, W., Dzombak, D. A., and Morel, F. M. M.,** Metal-humate interactions. II. Application and comparison of models, *Environ. Sci. Technol.,* 20, 676, 1986.
161. **Lasaga, A. C.,** Rate laws of chemical reactions, in *Kinetics of Geochemical Processes,* Vol. 8, Reviews in Mineralogy, Lasage, A. C. and Kirkpatrick, R. J., Eds., Mineralogical Society of America, Washington, D.C., 1981.
162. **Delany, J. M., Puigdomenech, I., and Wolery, T. J.,** Precipitation Kinetics Option for the EQ6 Geochemical Reaction Path Code, Lawrence Livermore National Laboratory, Livermore, CA, 1986.
163. **Benjamin, M. M.,** Effects of Competing Metals and Complexing Ligands on Trace Metal Adsorption at the Oxide/Solution Interface, Ph.D. thesis, Stanford University, Stanford, CA, 1978.
164. **Yates, D. E.,** The Structure of the Oxide/Aqueous Electrolyte Interface, Ph.D. thesis, University of Melbourne, Parkville, Australia, 1975.
165. **Loganathan, P. and Burau, R.,** Sorption of heavy metal ions by a hydrous manganese oxide, *Geochim. Cosmochim. Acta.,* 37, 1277, 1973.
166. **Anderson, M. A., Ferguson, J. F., and Gavis, J.,** Arsenate adsorption on amosphous aluminium hydroxide, *J. Colloid Interface Sci.,* 54, 391, 1976.
167. **Davis, J. A.,** Adsorption of Trace Metals and Complexing Ligands at the Oxide/Water Interface, Ph.D. thesis, Stanford University, Stanford, CA, 1977.
168. **Barrow, N. J.,** Testing a mechanistic model. II. The effects of time and temperature on the reaction of zinc with a soil, *J. Soil Sci.,* 37, 277, 1986.
169. **Fuller, C. C. and Davis, J. A.,** Processes and Kinetics of Cd^{2+} sorption by a calcareous aquifer sand, *Geochim. Cosmochim. Acta,* 51, 1491, 1987.
170. **Hendricks, D. W. and Kuratti, L. G.,** Derivation of an empirical sorption rate equation by analysis of experimental data, *Water Res.,* 16, 829, 1982.
171. **Knowles, G. and Wakeford, A. C.,** A mathematical deterministic river quality model. I. Formulation and description, *Water Res.,* 12, 1149, 1978.

172. **Somlyody, L.,** An effort for modelling the transport of micropollutants in rivers, in Modelling Water Quality Hydrology Cycle Symp., Proc. Baden Symp., *IAHS AISH Publ.* 125, 39, 1978.
173. **Chapman, B. M.,** Dispersion of soluble pollutants in non-uniform rivers. I. Theory, *J. Hydrol.*, 40, 139, 1979.
174. **Chapman, B. M.,** Dispersion of soluble pollutants in non-uniform rivers. II. Application to experimental results, *J. Hydrol.*, 40, 153, 1979.
175. **Anderson, M. P.,** Using models to simulate the movement of contaminants through groundwater flow systems, *CRC Crit. Rev. Environ. Control,* 9, 97, 1979.
176. **Valocchi, A. J.,** Describing the transport of ion-exchanging contaminants using an effective K_d approach, *Water Resour. Res.*, 20, 499, 1984.
177. **Cedeberg, G. A.,** TRANQL: A Groundwater Mass-Transport and Equilibrium Chemistry Model for Multicomponent Systems, Ph.D. thesis, Stanford University, Stanford, CA, 1985.
178. **Jennings, A. A., Kirkner, D. J., and Theis, T. L.,** Multicomponent equilibrium chemistry in groundwater quality models, *Water Resour. Res.*, 18, 1089, 1982.
179. **Rubin, J. and James, R. V.,** Dispersion-affected transport of reacting solutes in saturated porous media: Galerkin method applied to equilibrium-controlled exchange in unidirectional steady water flow, *Water Resour. Res.*, 9, 1332, 1973.
180. **Valocchi, A. J., Street, R. L., and Roberts, P. V.,** Transport of ion-exchanging solutes in groundwater: chromatographic theory and field simulation, *Water Resour. Res.*, 17, 1517, 1981.
181. **Miller, C. W. and Benson, L. V.,** Simulation of solute transport in a chemically reactive heterogeneous system: model development and application, *Water Resour. Res.*, 19, 381, 1983.
182. **Gureghian, A. B., Ward, D. S., and Cleary, R. W.,** Simultaneous transport of water and reacting solutes through multilayered soils under transient unsaturated flow conditions, *J. Hydrol*, 41, 253, 1979.
183. **Pandey, R. S., Singh, S. R., and Sinha, B. K.,** An extrapolated Crank-Nicolson method for solving the convection-dispersion equation with non-linear adsorption, *J. Hydrol.*, 56, 277, 1982.
184. **Morrey, J. R., Kincaid, C. T., Hostetler, C. J., Yabusaki, S. B., and Vail, L. W.,** Geohydrochemical models for Solute Migration, Volume 3: Evaluation of Selected computer codes, EPRI EA-3417, Vol. 3, Electric Power Research Institute, Palo Alto, CA, March, 1986.
185. **Grove, D. B. and Wood, W. W.,** Prediction and field verification of subsurface water quality and changes during artificial recharge, Lubbock, Texas, *Ground Water,* 17, 250, 1979.
186. **Schulz, H. D. and Reardon, E. J.,** A combined mixing cell/analytical model to describe two-dimensional reactive solute transport for unidirectional groundwater flow, *Water Resour. Res.*, 19, 493, 1983.
187. **Valocchi, A. J.,** Validity of the local equilibrium assumption for modelling sorbing solute transport through homogeneous soils, *Water Resour. Res.*, 21, 808, 1985.
188. **Kirkner, D. J., Jennings, A. A., and Theis, T. L.,** Multisolute mass transport with chemical interaction kinetics, *J. Hydrol,* 76, 107, 1985.

Chapter 6

APPLICATIONS OF HIGH-PERFORMANCE LIQUID CHROMATOGRAPHY TO TRACE ELEMENT SPECIATION STUDIES

Graeme E. Batley and Gary K.-C. Low

TABLE OF CONTENTS

I. INTRODUCTION

In recent years, high-performance liquid chromatography (HPLC) has rapidly become the fastest growing area of analytical chemistry, and it is not surprising, therefore, that it has found application in trace element speciation studies. In studies of natural waters or biological fluids, the ability of HPLC to effect separations in solution at ambient temperatures makes it potentially more suited to speciation studies than, for example, gas chromatography (GC), where separations are limited to species which are volatile at elevated temperatures. At these temperatures both decomposition and *in situ* formation of new species is a possibility. A further advantage of HPLC separations, unlike GC, where only the stationary phase is variable, is that both the stationary and mobile phases can be varied to enhance a separation. In HPLC the mobile phase plays an important role in the separation, whereas in GC it is only a carrier. In addition, HPLC is compatible with a wider range of instrumental detection techniques including spectrometry, fluorimetry, voltammetry, and refractive index and conductivity measurements.

Traditionally HPLC has found its greatest application to the analysis of organic compounds; however, the use of reversed-phase and ion-exchange columns has made possible the separations of inorganic cations and anions in addition to those of metal chelates and organometallic species.

II. PRACTICAL HPLC

A. Column Technology

1. Reversed-Phase Separations

Reversed-phase HPLC separations, which utilize a nonpolar stationary phase and a polar mobile phase, are the most versatile and popular separation mode and commonly involve octadecylsilane (C_{18}) phases, but can also involve C_2, C_8, phenyl, cyano, amino, or diol, as well as polymeric phases. The resolving power of reversed-phase columns varies greatly from column to column, even with columns from the same supplier, due to differences in the silica support substrate and in the means of producing the alkyl-bonded phase.[1]

In columns having monomeric alkyl-bonded stationary phases, less than 50% of the available silanol groups on the silica surface are reacted,[2] leaving highly acidic groups which can themselves have a high retention capacity particularly for basic compounds. End capping of these unreacted surface groups with a short-chain alkylsilane is often carried out with many commercial columns; however, despite this, a substantial percentage of silanol groups still remain. Small di- or trialkylamine compounds are often included in the mobile solvent to minimize the strong interaction between basic compounds and the ionized form of the silanol groups above pH 4.[1]

The use of a polymeric support rather than the conventional silica base on which to bind the C_{18} or C_8 chains shows considerable promise.[3] Styrene-divinylbenzene copolymers or polyacrylamide-based columns are stable over a wider pH range than the narrow pH 2 to 8 range which silica columns can withstand.

In trace element speciation studies, the separation of neutral hydrophilic molecules is a rare requirement, and more often the concern is for anionic or cationic species, including organometallic species, free metal ions or metal complexes. Such compounds will usually elute from a reversed-phase column with the solvent front unless some charge-neutralizing species is present in the mobile phase, as in the so-called paired-ion HPLC. Here the addition of an anionic species such as *n*-pentane sulfonate will facilitate the separation of cationic species such as those of Cr(III) on a C_{18} column,[4] while organoarsenic anions, for example, can be separated with the addition to the mobile phase of a cationic species such as *t*-butylammonium phosphate.[5] The mechanism for this separation process is the subject of debate, but is generally believed to involve the *in situ* modification of the reversed-phase column into a dynamic ion-exchange system. Reversed-phase columns have, however, been used to separate dissolved organic matter in natural waters, and metal associations have been implied on the basis of simultaneous monitoring of heavy metals and organic carbon in the eluent fractions.[6,7]

2. Normal-Phase Separations

The use of normal-phase columns in speciation studies is limited to the separation of stable, neutral metal chelates. The volatile nonpolar solvents that are required may be desirable in interfacing with some detection systems. Silica and alumina have been used successfully as stationary phases for the separation of dithiocarbamate, dithizonate, thiooxinate, thionate, and β-diketonate metal complexes, which all exhibit good solubility in nonpolar solvents.[8]

These same stationary phases have also recently been used in the reversed-phase mode, giving greatly improved separations with reversed-phase mobile solvents instead of the conventional normal-phase eluents.[1]

3. Ion Chromatography

Ion-exchange stationary phases are capable of separating free ions, complex ions, and neutral species simultaneously, and hence have by far the widest application to speciation studies.[9] They comprise both strong and weak, anion and cation exchange functional groups grafted onto either silica or styrene-divinylbenzene polymer substrates. In general, the improvement in polymeric supports in recent years makes them preferable to silica, in addition to their obvious advantage with respect to their usable pH range. Mass transfer to the former is slower, although this can be overcome by the use of elevated temperatures.

Polymeric ion-exchange resins may be either micro- or macroporous. The former swell and form a gel when hydrated, and shrink if the resin becomes dehydrated. Macroporous resins tend to be more highly cross-linked, with a large number of hard polymeric microspheres in each bead. They generally undergo less swelling and shrinking than do microporous resins.

Low-capacity ion-exchange columns have been advocated for unsuppressed ion chromatography, particularly for use with conductivity detection where very dilute eluents are required for the detection of low concentrations of sample ions. For most speciation applications this requirement does not exist, and optimum separations have generally been obtained using the conventional high-capacity resins.

Cation-exchange preconcentration columns have also been used for the trace enrichment of heavy metals from natural waters.[10] A small cartridge (3 × 3 mm) of 13 μm Aminex-5 (Bio-Rad,® Richmond, CA) has been used in the sample loop of an injector, with preconcentrated species subsequently separated on a larger high-pressure anion exchange column and detected via post-column reaction with pyridylazoresorcinol. Significant differences were observed for the same metal ions, e.g., Cu(II), concentrated from acidified and unacidified solutions, in keeping with the expected differences in their speciation.[11] Acidification is likely to dissociate weak complexes and release metals associated with colloidal species. The system has not, however, been fully exploited in speciation studies.

4. Gel Permeation Chromatography

Separations of metal species on the basis of molecular size have been successfully achieved using open-column techniques[12-16] (Chapter 3). A range of size-exclusion HPLC columns is now commercially available in both analytical and preparative column sizes[17] and they have been applied to separate metal ions and high-molecular-weight complexes in aqueous solution.[18] Hausler and Taylor[19,20] used a 100-Å μ-Styragel column (Waters, Milford, MA) to resolve metal species in pyridine, chloroform, and tetrahydrofuran extracts of solvent-refined coal, using metal-specific, inductively coupled plasma emission (ICP) detection. No specific applications of these HPLC columns to metal speciation have yet been reported, principally because the concentration ranges generally of interest in environmental or biological samples are too low to be detected. Indeed, even the open-column separations of Steinberg[12] required some preconcentration. Other techniques of size classification, e.g., ultrafiltration or dialysis, are more easily performed (Chapter 3).

B. Pumps and Hardware

In conventional HPLC, pumps, connecting tubing, and fittings are generally of stainless steel construction for both strength and corrosion resistance, although Teflon® connecting tubing is occasionally used. Particularly in studies of heavy metals, sample contamination from metal components of the system must be avoided. Teflon® or glass-lined stainless steel tubing is preferred with other fittings made from Teflon,® polypropylene or Kel-F.

Although pumps constructed from inert components are commercially available, they have been found to be difficult to operate, and to give irreproducible flow rates,[21] making them poorly suited to gradient elution. Cassidy and Elchuk[10] overcame this problem by using an inert pump (CMP-1 Laboratory Data Control, FL) and an inert sampling valve (Durham Chemical Corp., CA) in their enrichment of ionic trace metals from natural freshwaters on a strong-acid cation-exchange column. Conventional stainless steel pumps were used to control the citrate or oxalate eluents. No major blank problems were observed using this system and isocratic elution.

Chloride ions are particularly corrosive to stainless steel components. In HPLC studies of trace metals in seawater, inert hardware is essential. Mackey[22,23] used a CMP-3K pump (Laboratory Data Control) for studies of Amberlite® XAD resins for the preconcentration of trace metal species from seawater. Bank et al.[24] used a tungsten HPLC pump in conjunction with a thermospray sample introduction system for atomic absorption spectrometry (AAS) and incorporated an ion-exchange column at the point of solvent delivery, prior to the injection valve, to remove metal contaminant originating from pump components.

C. Detection Systems

The traditional UV-visible absorbance detector, which has found wide application in the HPLC analysis of organics, is seldom used in trace element speciation, principally because of the low molar absorptivities of these species. Post-column reactions with liquids forming highly absorbing complexes have been used to separate different metals,[10,25] but not for speciation. The general lack of selectivity of absorbance detection makes it prone to interferences from other absorbing species in the sample. A further limitation of post-column reactors is the dilution and potential dispersion of the eluting peaks due to reagent addition which will make it more difficult to resolve closely eluting species.

This can be minimized and the efficiency of mixing, and hence reaction with the added reagent, enhanced by several means. The use of pulseless gas pressure mixing was found to be far superior[26] to the pulsed delivery of a second pump. A helium-pressurized reservoir or a syringe pump was recommended. Mixing via either an annular membrane reactor or a screen-tee reactor was highly efficient (Figure 1). The former consists of a hollow fiber tube containing a coiled nylon filament, and the latter, screen disks within a tee mixer. In another novel mixing procedure reported recently,[27] the sample-containing eluent is passed through a hollow fiber membrane bundle immersed in a reagent solution, e.g., ammonia, when it is necessary to change the pH of the eluent to enable detection (Figure 2). A number of novel miniaturized post-column reactors have recently been described for use with microbore HPLC columns.[28,29] The latter ensure a dramatic reduction in band broadening.

For nonabsorbing species, indirect UV absorbance measurements can be made, where,

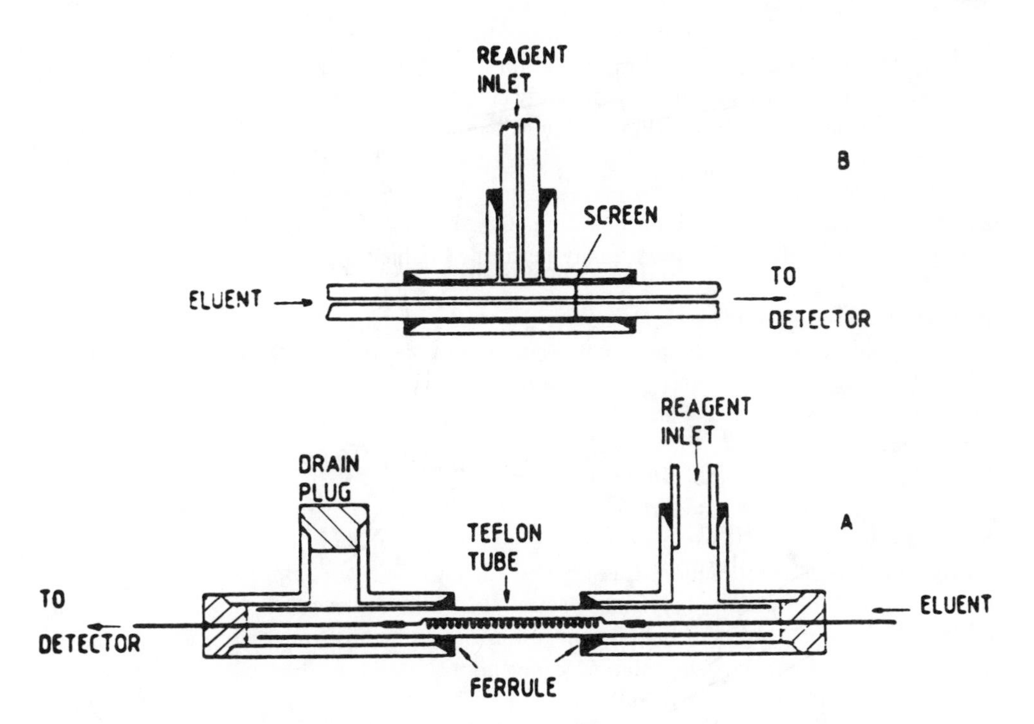

FIGURE 1. Membrane (A) and screen-tee (B) mixers for HPLC post-column reaction detection. (From Cassidy, R. M., Elchuk, S., and Dasgupta, P. K., *Anal. Chem.*, 59, 85, 1987. With permission.)

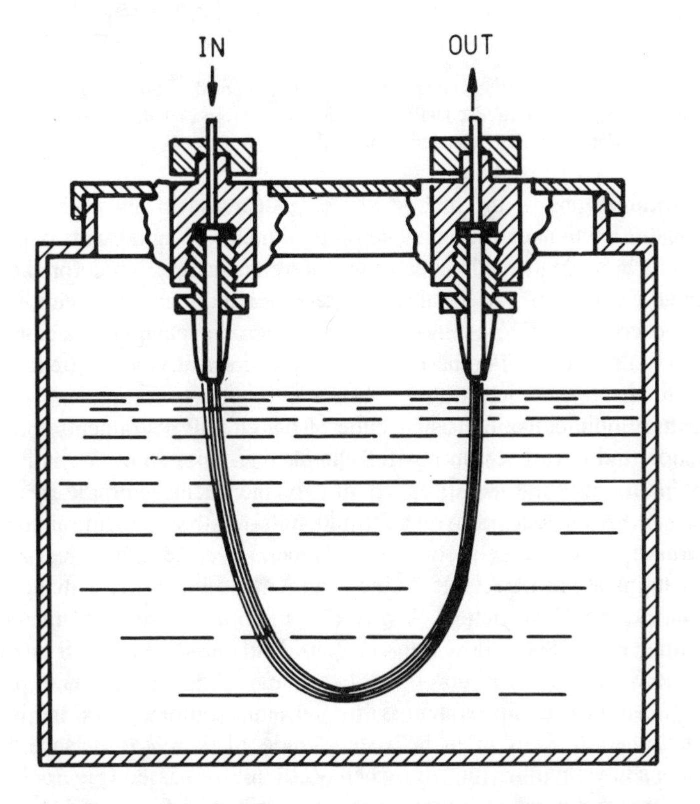

FIGURE 2. Hollow-fiber membrane post-column reactor filled with 2 *M* NH_3 for raising eluate pH. (From Davis, J. C. and Peterson, D. P., *Anal. Chem.*, 57, 768, 1985. With permission.)

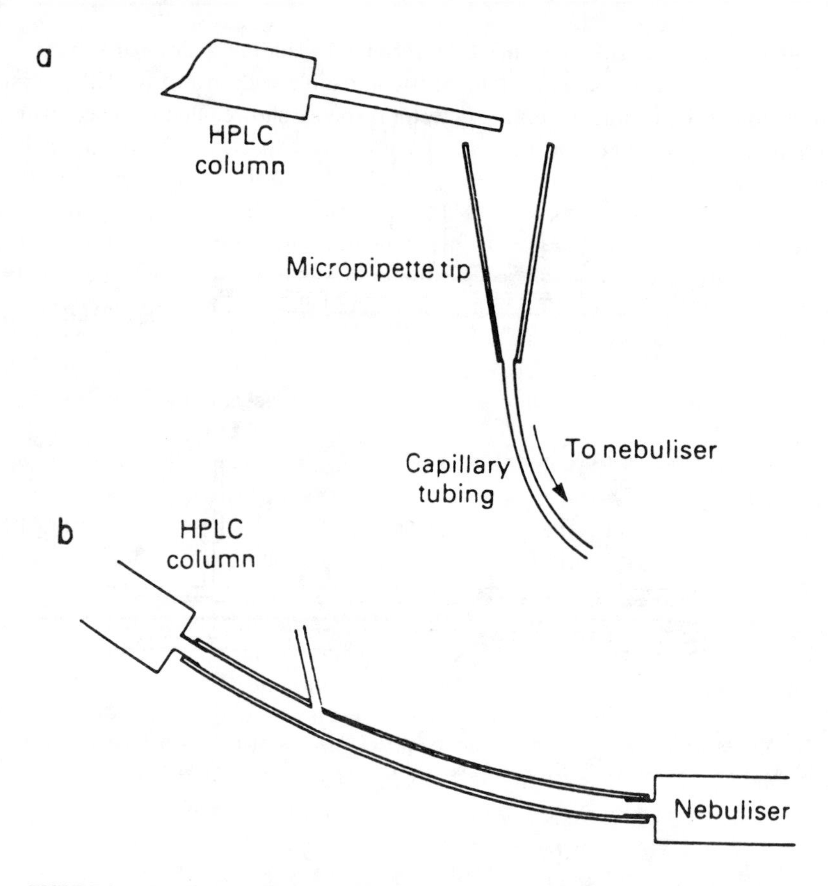

FIGURE 3. Nebulizer coupling systems for atomic absorption detection of HPLC effluents. (a) Funnel collection; (b) direct coupling with air intake (From Ebdon, L., Hill, S. J., and Jones, P., *Analyst,* 110, 515, 1985. With permission.)

using light-absorbing ions in the mobile phase, a decrease in absorbance is seen at the detector during the elution of the nonabsorbing solute.[30] Eluants such as benzylamine,[30] picolinic acid,[3] and benzyltrimethylammonium ion[31] have been used for the detection of cations. Recently, Sherman and Danielson[32] reported indirect fluorescence detection by monitoring the decrease in fluorescence at the emission wavelength of a fluorescent counterion, cerium (III) in the mobile phase. Selectivity and sensitivity are greatly enhanced by comparison with absorbance detection.

Atomic spectroscopic detection, using either AAS or ICP instruments, permit element-specific detection and have been successfully applied in a range of speciation studies.[33-37] The major problem with these detectors lies in their interfacing with the HPLC system.

The options for AAS detection involve introduction of either a liquid sample into a flame or graphite furnace, or a gaseous product, e.g., a metal hydride, into a quartz tube furnace, for subsequent atomization. Direct interfacing of an AAS spectrometer with the HPLC system is possibly the least sensitive option. A major problem is the apparent incompatibility of flow rates, with typical HPLC flow rates of 1 to 2 ml min^{-1} and AAS nebulizer uptake rates nearer 6 to 8 ml min^{-1}. Ebdon et al.[38] overcame this problem using direct volume nebulization, where the column effluent is first introduced directly into the injection loop, allowing the nebulizer to draw in air between aliquots of sample to balance the flow from the column. The spray chamber runs dry when no sample is passed (Figure 3a). Sensitivity is compatible with direct uptake from a funnel, as described by Slavin and Schmidt[29] (Figure 3b). This system directs successive, approximately 100-μl drops of eluent to the nebulizer. The resultant signal trace returns to the baseline between each drop.

Other alternatives are to either balance the flows by adding additional solvent,[40] or to simply reduce the nebulizer uptake so that it is compatible with that of the HPLC system.[41] An examination of the factors governing detector response showed that with decreasing flow rates, the peak heights decrease, but not to the extent expected.[42] Fewer metal ions are reaching the nebulizer per unit time at the lower flow rate, but a larger proportion are being nebulized. At the same time, with water as the mobile phase, quenching of the flame, which reduces sensitivity at high flow rates, is less evident at low flows, the net result being that the fall-off in sensitivity with decreasing flow is less than expected, enabling reasonable sensitivities to be achieved for 2-ml min^{-1} flows and below.[42] For 1-ml min^{-1} flows in an aqueous mobile phase, less than one third of the sample has been shown to be nebulized. This process is more efficient for organic solvents, which are both more readily vaporized and generally increase the temperature of the flame,[42] with resultant greater sensitivity being achieved.

Under the above low-flow conditions, the burner is effectively starved for liquid. This will lead to a reduced-pressure region post-column which can result in bubble formation. Adjustment of the nebulizer to apply slight back pressure[43] eliminates aspiration. Lower signal to noise ratios, smaller droplet sizes, and increased transport efficiencies were evident with lower flows and greater back pressures.

Flame atomization is limited by background interferences with many solvent mixtures and particularly where the mix is changing, as in gradient elution. Such problems can be overcome using graphite furnace atomization. Increased sensitivities are also obtained because of the greater atom concentrations achieved. Continuous signal recording is not possible, however, and a system of incremented measuring, as described by Brinckman and co-workers,[5,44] must be used. Column effluent is directed into the base of a 50 μl Teflon® well sampler (Figure 4). Samples of preselected volumes (10 to 50 μl) are then automatically sampled from the well at fixed time intervals and injected into the graphite furnace (Figure 4). The well overflow is directed to waste. An option is to split the HPLC effluent stream and collect 100- to 500-μl aliquots in cups in a carosel sample holder for auto-sampling as before. The incremental signal (Figure 5) can be optimized by changes in flow rate and sampling interval. Typically flow rates of 0.5 to 1.0 ml min^{-1} and sampling intervals of 30 to 45 s are used.[44] The signal output is quantified either by calculating the area under the curves obtained by connecting the incremental signal maxima, or by summing the heights of the signals corresponding to a particular peak. The latter method is easier and was generally preferred.[44] To avoid losses of volatile species in furnace methods, ashing temperatures are critical. Interferences caused by the eluent can also present problems, although generally less than encountered in flame methods.

A limitation of the incremented sampling method is that only broad peaks can be analyzed or low flow rates used. Vickrey et al.[45] examined a peak storage procedure where the peak-containing eluate was diverted via a switching valve to a Teflon® capillary tube. The stored eluate was then displaced by a syringe pump to the graphite cup of a Zeeman® AAS instrument. This system (Figure 6) permitted more frequent incremental measurements to be made over a given eluate volume, giving better precision over a sharply eluting peak.

Koizumi et al.[46] found it difficult to get the same absorbance response for different organolead species using the conventional graphite furnace and experimented with an alternative design to achieve the higher gas temperatures necessary to bring about complete dissociation. Even double-chamber variants of the standard cuvette gave incomplete recoveries of tetraethyl- and tetramethyllead, whereas a furnace where sample is vaporized in a tantalum cup and then passed through porous graphite into the graphite tube absorption cell (Figure 7) allowed temperatures of 2800°C to be achieved and permitted 100% recovery of both of the above compounds. The furnace also reduced chemical interferences of salts such as $MgCl_2$. Solvent interference effects were overcome by utilizing the Zeeman® AAS. The

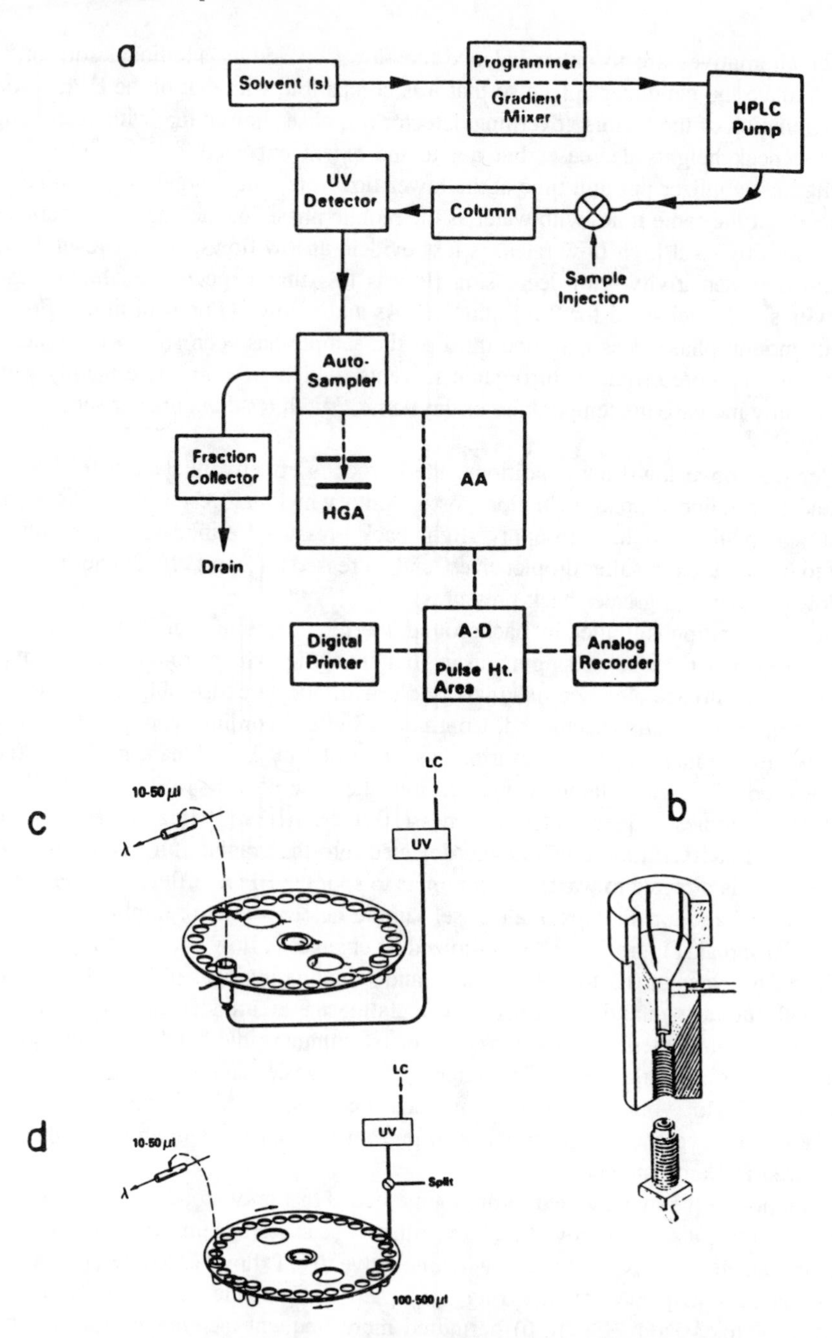

FIGURE 4. Coupling of HPLC to a graphite furnace atomic absorption system. (a) Block diagram of HPLC system; (b) Teflon® well sampler; (c) periodic stream sampling using well sampler; (d) Stream-splitting and carousel cup collection. (From Brinckman, F. E., Blair, W. R., Jewett, H. L., and Iverson, W. P., *J. Chromatogr. Sci.*, 15, 493, 1977. With permission.)

solvent flow rate was 0.67 ml min^{-1}, with 10 µl from each 250 µl of column effluent being intermittently introduced into the furnace and measured while the flow was stopped.

Another approach to the direct detection of HPLC effluents using AAS, by Hill et al.,[47,48] utilizes a rotating platinum spiral collection system for directly introducing the eluate into a flame (Figure 8). Spirals of platinum wire are mounted at 45° to each other in a rotating

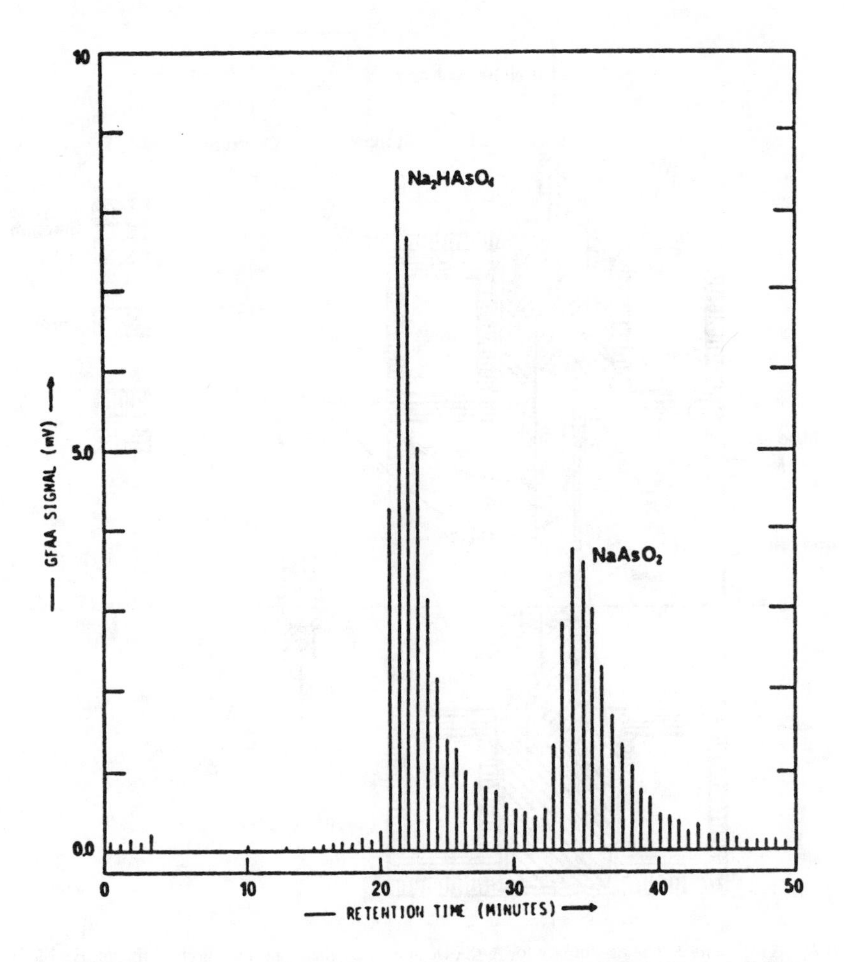

FIGURE 5. Detection of As(III) and As(V) (100 ng each) by the HPLC-GFAA technique. Using an anion-exchange column, acetate buffer mobile phase, 0.15 ml min^{-1} flow, with 20-μl injections every 40 s. (From Brinckman, F. E., Jewett, K. L., Iverson, W. P., Irgolic, K. L., Ehrhardt, K. C., and Stockton, R. A., *J. Chromatogr.*, 191, 31, 1980. With permission.)

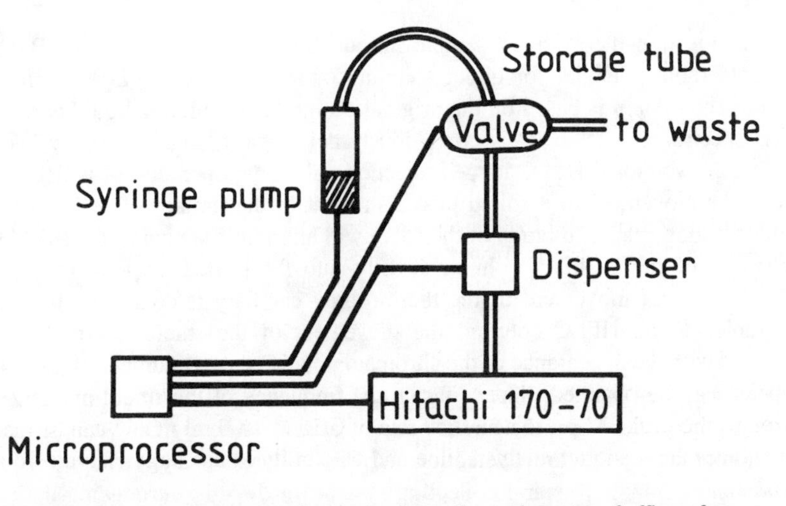

FIGURE 6. Schematic diagram of apparatus for on-line peak storage of effluent from an HPLC column for graphite furnace analysis. (From Vickrey, T. M., Howell, H. E., and Paradise, M. T., *Anal. Chem.*, 51, 1880, 1983. With permission.)

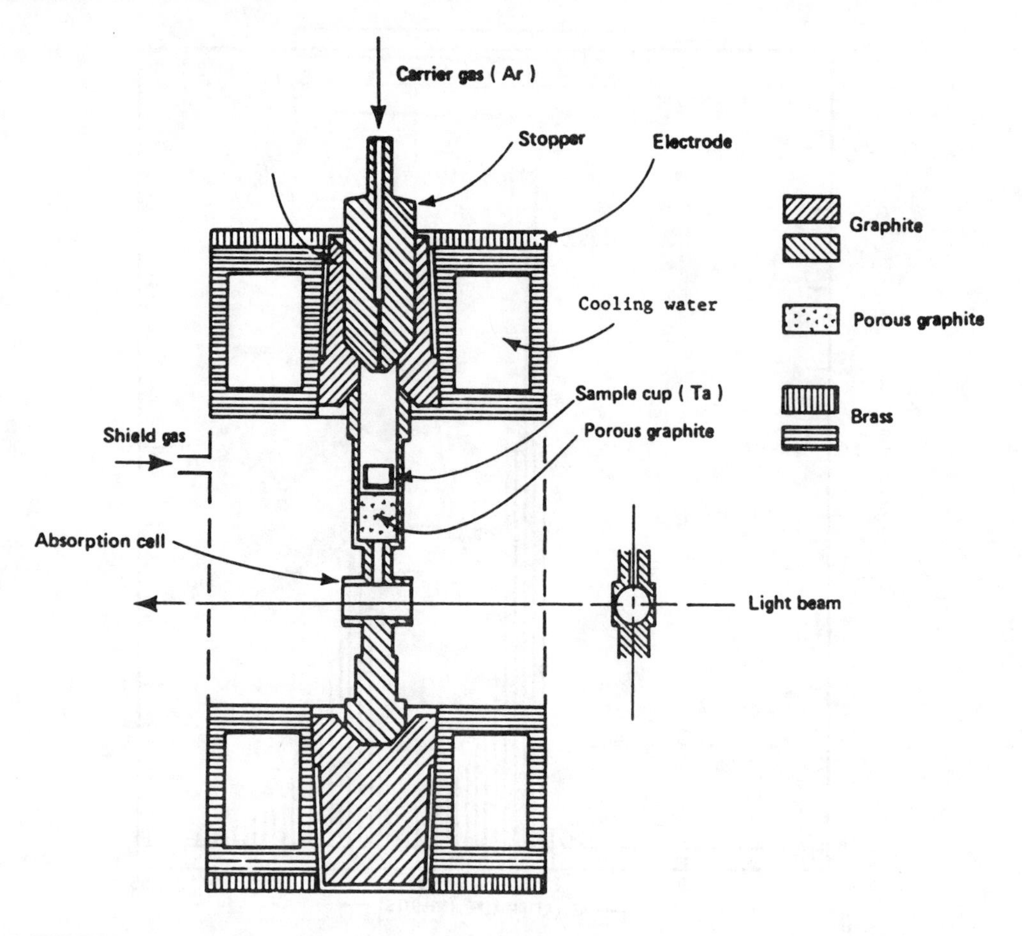

FIGURE 7. A high-temperature gas furnace for AAS detection of alkyllead species. (From Koizumi, H., McLaughlin, R. D., and Hadeishi, T., *Anal. Chem.*, 51, 387, 1979. With permission.)

disk. A 2-mm-I.D. column is used to permit flow rates below 1 ml min^{-1}. A microburner is used to drive off excess solvent from the loaded spirals before flame atomization. In many respects this is similar to the Brinckman system,[44] in that sequential segmented measurements are obtained.

The flow-injection thermospray sample introduction system described by Bank et al.[24] offers considerable potential for direct coupling of an HPLC to a graphite furnace AAS instrument. The system is based on the original design for coupling a liquid chromatograph to a mass spectrometer. Solvent is continuously delivered through a fused silica capillary column in a conventional HPLC looped injection valve, the operation of which is synchronized with the movement of a shield placed in front of the furnace. The sample is loaded in the injector loop, then activation of the valve will align the opening of the shield to permit the protrusion of the capillary column directly into the heated graphite furnace (100 to 1000°C). The in-out movement of the thermospray capillary is controlled by a solenoid. When coupled to an HPLC column, the movement of the shield has to be accurately synchronized with the appearance of the chromatographic peak. A number of discrete signals on the peak may be obtained, depending on the frequency of the shield movement during the elution of the peak. At present, a flow rate of 0.15 to 1.00 ml min^{-1} can be maintained. The advantages are a shorter analysis time and the ability to use larger sample volumes (up to 100 μl).

For elements such as As, Se, Sb, and Pb, which form volatile hydrides, post-column

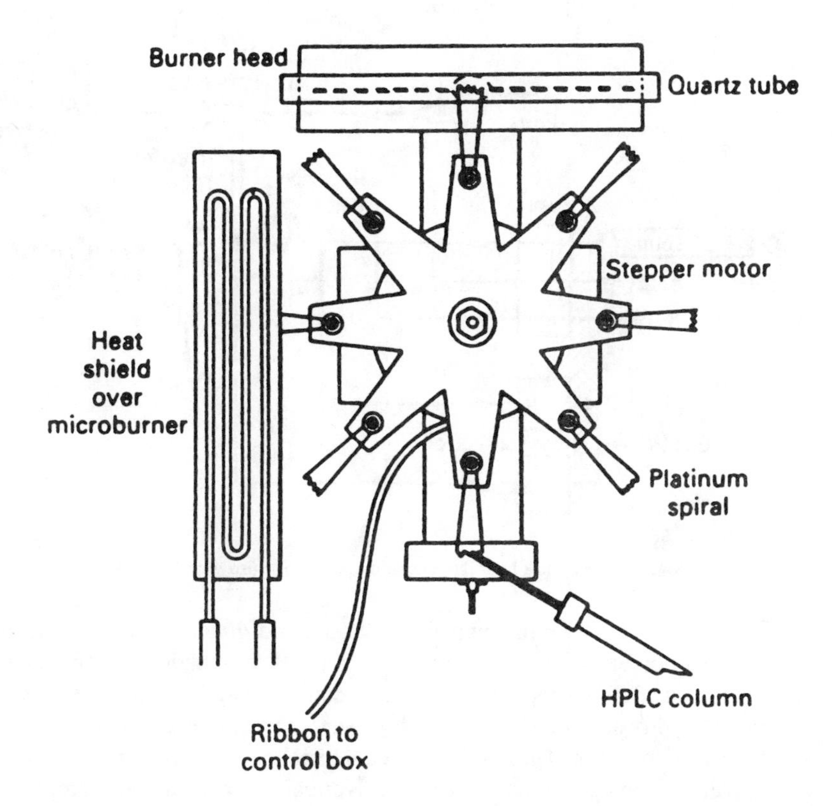

FIGURE 8. Rotating platinum-wire collection of HPLC column effluent with detection by flame AAS. (From Hill, S., Ebdon, L., and Jones, P., *Anal. Proc.*, 23, 6, 1986. With permission.)

hydride generation followed by AAS detection, offers a sensitive method of detection.[36,37,49-51] Several commercial hydride-generation attachments are now available consisting essentially of a peristaltic pump capable of mixing the sample sequentially with acid and sodium borohydride. Reaction occurs in a mixing coil with the gaseous hydrides being separated from the solution phase in a gas-liquid separator. The gas is then directed to a quartz-tube furnace aligned in the light path of the atomic absorption spectrometer. The furnace is heated either electrically or by the conventional air-acetylene burner. These systems were designed for continuous sample aspiration, but are easily adapted to receive the effluent of an HPLC column (Figure 9). It is essential that the hydrogen, produced from the borohydride, mix efficiently with the eluting peak. This necessitates reducing the borohydride concentration below the 3% w/v generally recommended for continuous aspiration to 1.5% w/v, to provide both good mixing and reproducible results.[52] The mixing coil diameter and length is also critical for complete reaction. The quartz furnace used in this technique is a further source of problems. With continued use, the interior of the furnace slowly vitrifies becoming opaque and brittle, with a gradual decrease in signal sensitivity. Presiliconizing of the furnace interior and flushing of the entire system with water at the end of each series of measurements slowed this deterioration.[52]

The use of an inductively coupled plasma emission spectrometer as an HPLC detector has also been fully investigated.[34,35,53-60] With aqueous eluents there is very little peak broadening and detection limits are satisfactory if conventional nebulizers are used.[54] Organic solvents in the mobile phase do, however, create problems.[55] With conventional nebulizers a limiting aspiration rate is reached beyond which plasma stability is adversely affected.[54] Methanol, acetone, tetrahydrofuran, and acetonitrile are particularly bad in this regard,

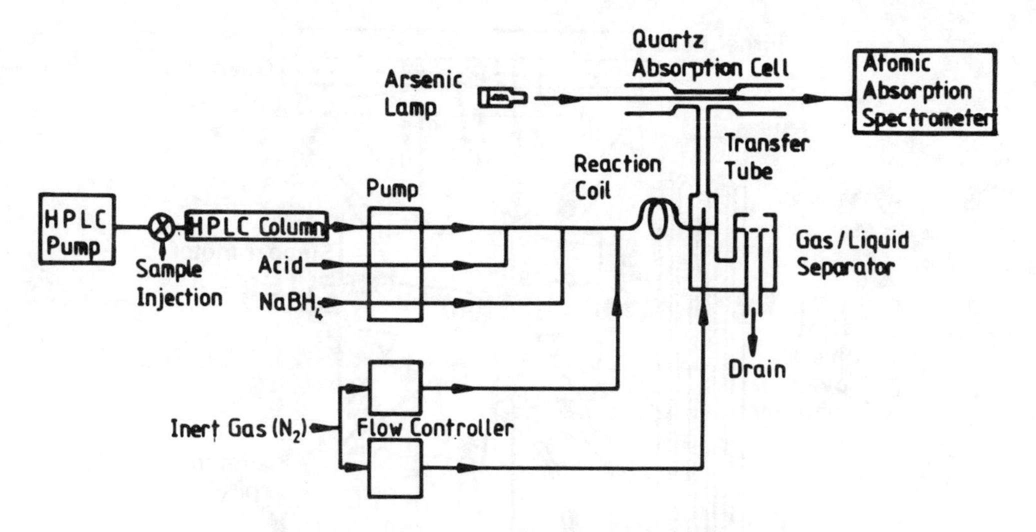

FIGURE 9. Block diagram of a hydride-generation system for use with HPLC systems.

whereas ethanol, higher alcohols, methylisobutylketone, and dimethylsulfoxide are preferable. In addition, carbon emission may interfere with the determination of some elements. Uden et al.[56] noted that while reversed-phase solvents could generally be used with a standard ceramic nebulizer, hydrocarbon and halocarbon solvents virtually extinguished the plasma due mainly to carbon deposition. These solvents could be tolerated using a modified nebulizer system to eliminate carbon deposition (Figure 10). Normal-phase separations are, however, rarely employed in metal speciation studies.

An investigation of the properties of the nebulizer/spray chamber system as an interface for HPLC-ICP revealed some interesting facts concerning the effects of transport mechanisms.[57] Using a spray chamber located external to the plasma box, i.e., with sample transport mainly as an aerosol, peak height was found to be virtually independent of flow rate, and peak broadening was less than that obtained with the conventional internal spray chamber, where liquid transport predominates. In the latter case, peak heights decrease rapidly with increasing flow rates.

Currently, ICP detection using conventional nebulizers is limited to concentrations above about 100 $\mu g\ l^{-1}$ for 100-μl injections, although this can be reduced to 10 $\mu g\ l^{-1}$ by use of a large injector loop (up to 1000 μl) or a concentrator column.[58] The major limitations with ICP are associated with inefficient sample introduction procedures. Current on-line nebulizers, result in only 1 to 20% of the original sample being detected, mainly because the majority of droplets are too large to be easily atomized and are removed by the spray chamber. In attempts to improve this situation, Fassel and co-workers[60-62] have developed a total injection microconcentric nebulizer (Figure 11) which yields 100% nebulization and transport efficiency. Detection limits were better or comparable to those with the conventional cross-flow nebulizer, and stable plasma operation was obtained for a variety of mobile phases, including up to 100% methanol and acetonitrile at flow rates from 2 to 200 $\mu l\ min^{-1}$. The increased loading of the plasma with the direct injection nebulizer (DIN) does, however, cause some changes in plasma characteristics, particularly in the excitation temperatures obtained in the axial channel, which affects the achievable detection limits. For the study of As, Se, and Cr speciation, detection limits using the continuous-flow DIN were one to two orders of magnitude better than those for conventional HPLC-ICP.[62]

The use of supercritical fluids (SCF) as mobile phases has been studied as another means of overcoming the problems of ICP sample introduction.[63] At elevated pressures, SCFs have physical properties intermediate between liquids and gases. Their densities and solvating

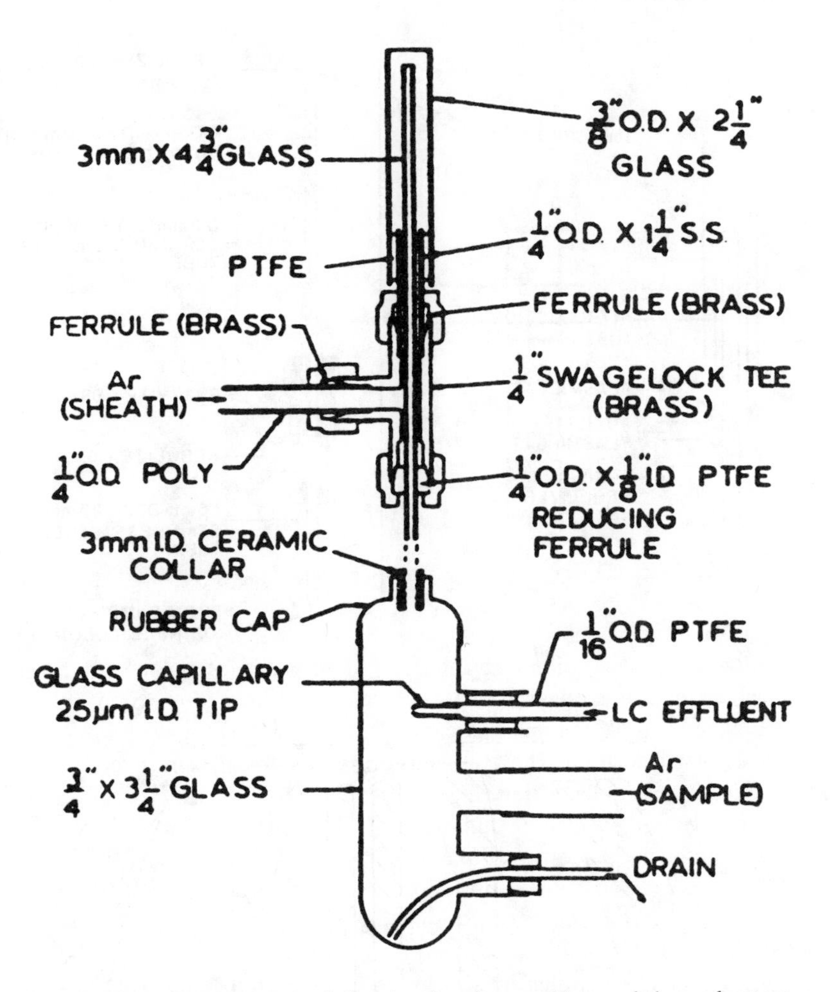

FIGURE 10. Modified aerosol-nebulizer interface for use with normal-phase solvent systems. (From Uden, P. C., Quimby, B. D., Barnes, R. M., and Elliott, W. G., *Anal. Chim. Acta,* 101, 99, 1978. With permission.)

powers are similar to liquids, but their diffusivities and viscosities are more similar to gases, thus allowing rapid mass transfer of solutes. The compressibility above their critical temperature is large, and small changes in pressure result in large changes in their density and solvating power. Thus, variable solvating power can be achieved by pressure manipulation.[64] In the SCF-based sample introduction system, the sample, extracted or injected into the SCF, is passed through a 1-m × 250-µm-I.D. capillary tube to a point just below the plasma. In the atmospheric pressure plasma, the SCF transforms to a gas, dispersing the sample in an easily atomized form with virtually 100% transport efficiency. The main disadvantages appear to be the effects of the SCF on excitation conditions and background emission, which are similar to those observed when organic solvents are used with conventional nebulizer systems.[63]

The state-of-the-art HPLC detection system is the ICP-mass spectrometer.[58] A description of such a system, using a new ultrasonic nebulizer able to tolerate a 20% methanol/water solvent mixture, has been published by Thompson and Hook.[58] The possibilities for this technique in speciation studies have yet to be fully explored.

Atomic fluorescence detection has been used by Mackey.[22,23] A variable nebulizer on the detector was adjusted to match the flow rate through the column. The potential advantages over AAS detection include greater sensitivity, increased linear range, and simultaneous

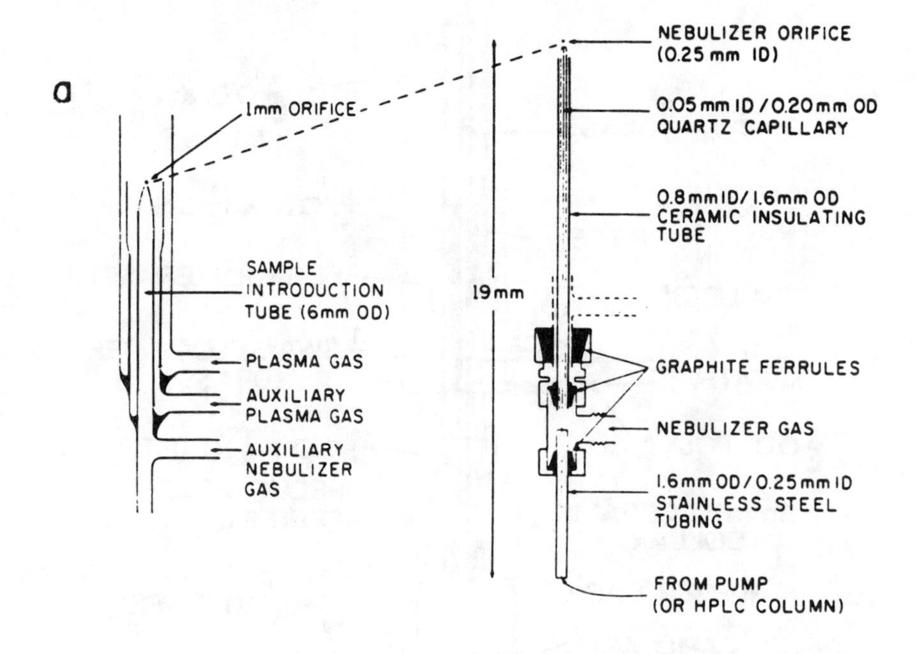

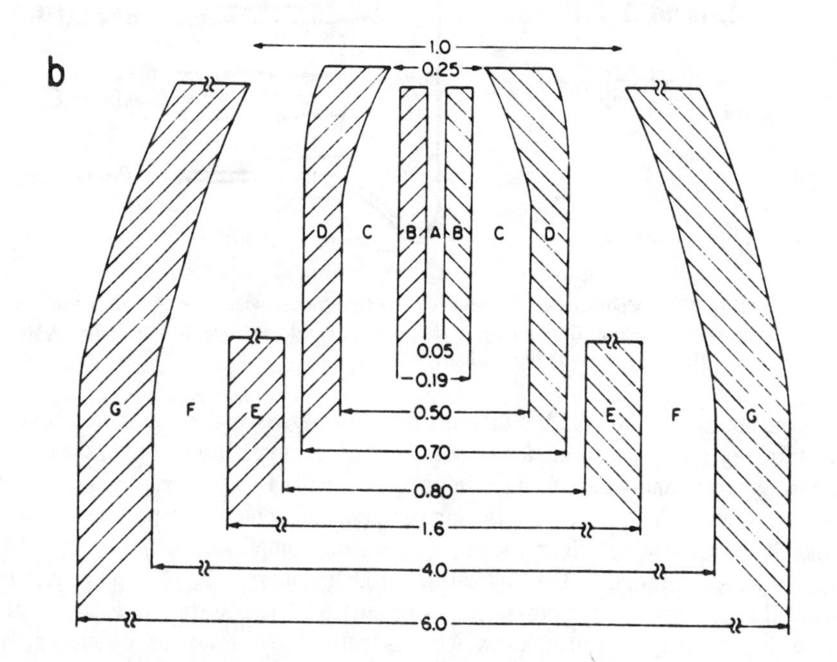

FIGURE 11. Microconcentric nebulizer for use with HPLC-ICP. (a) Nebulizer and torch; (b) magnified view of nebulizer tip: (A) liquid sample; (B) fused silica inner capillary tube; (C) nebulizer argon gas flow; (D) fused silica outer capillary tube; (E) ceramic insulating tube; (F) auxilliary argon gas; (G) normal sample introduction tube. (From Lawrence, K. E., Rice, G. W., and Fassel, V. A., *Anal. Chem.*, 56, 289, 1984. With permission.)

multielement compatability. The low popularity of this technique is most probably due to the limited availability of commercial instrumentation. Demers and Allemand[65] claimed that atomic fluorescence gives less spectral and background interferences than ICP.

Electrochemical (EC) detection is also being increasingly used in studies of metal speciation using HPLC. It has the advantage of being highly selective and relatively inexpensive

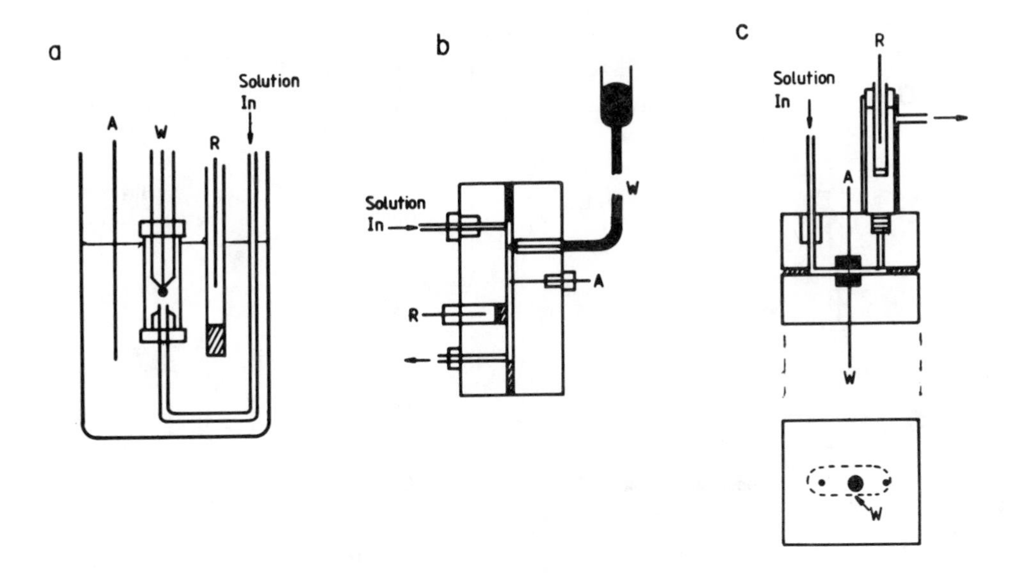

FIGURE 12. Types of electrochemical detectors. (a,b) Dropping mercury electrodes; (c) glassy carbon electrode. W = working electrode; R = reference electrode; A = auxilliary electrode.

compared, for example, to AAS detection. The range of EC detector systems (Figure 12) has been described in a number of recent comprehensive reviews.[66-68] The most widely used commercial EC detectors utilize glassy carbon working electrodes; however, although these are useful for oxidative detection of organic molecules, their limited cathodic range makes them of little value for metal speciation studies where reductive processes are usually involved. The higher hydrogen overpotential of mercury extends this cathodic range. The use of a mercury film on glassy carbon or gold electrodes, or of a dropping mercury electrode, is therefore preferred for reductive detection. The dropping electrode has the advantage that the periodic renewal of the mercury surface eliminates the problems of electrode fouling and high residual currents often observed at solid electrodes.

Because it is also reducible, oxygen poses a particular problem in reductive EC detection and must be removed. Refluxing is often recommended to deoxygenate the mobile phase;[69,70] however, ultrasonication and purging with nitrogen, helium, or argon is usually adequate. In trace metal studies, it is often desirable to replace potentially contaminating stainless steel tubing and fittings with Teflon® components; however, the high permeability of Teflon® to oxygen is itself a problem, the ultimate solution to which is to enclose the entire chromatograph in a box which can be purged with nitrogen.[71]

Additional detector selectivity can be obtained using dual working electrodes capable of measurements at differing potentials.[72] Their use in either series or parallel operation allows either reductions at two selected potentials (parallel operation) or oxidation of products from reduction at the first electrode (series operation). As yet, dual electrode systems have not been applied to speciation studies. This electrode offers a particular advantage for analytes having high redox potentials where the products of oxidation or reduction, if electroactive, can be sensed at the series electrode.[73]

The actual measurement process in EC detection utilizes either a coulometric or amperometric measurement. In the former, the analyte is fully electrolyzed, requiring a high-surface-area electrode such as reticulated glassy carbon. In amperometric detection, the instantaneous current due to diffusion and reaction of some fraction of the analyte is measured. Normally a direct current signal is measured; however, added selectivity, but not necessarily sensitivity, can be obtained using a superimposed differential waveform (e.g.,

Table 1
COMMON ARSENIC SPECIES

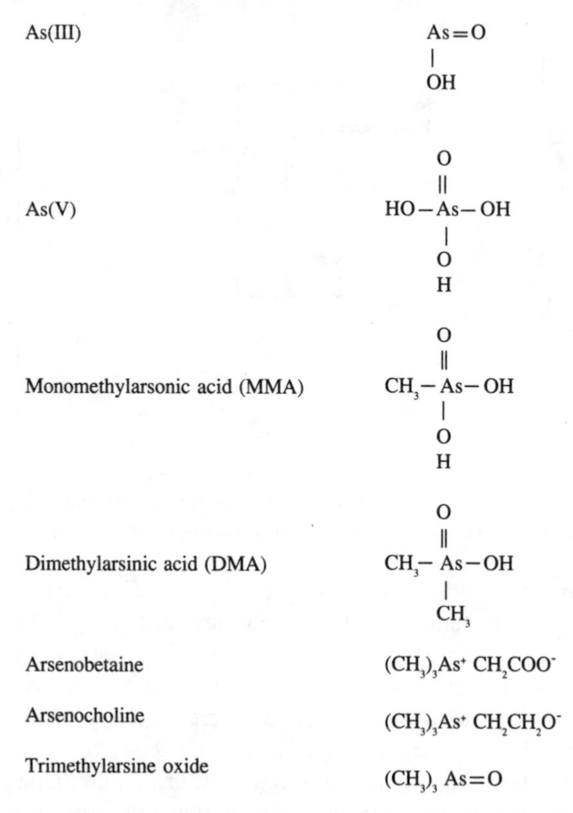

Species	Structure
As(III)	$As{=}O$, with OH bonded to As
As(V)	$HO{-}As({=}O)({-}OH){-}OH$
Monomethylarsonic acid (MMA)	$CH_3{-}As({=}O)({-}OH){-}OH$
Dimethylarsinic acid (DMA)	$CH_3{-}As({=}O)({-}OH){-}CH_3$
Arsenobetaine	$(CH_3)_3As^+ CH_2COO^-$
Arsenocholine	$(CH_3)_3As^+ CH_2CH_2O^-$
Trimethylarsine oxide	$(CH_3)_3 As{=}O$

differential pulse or square wave voltammetry).[68] The application of differential pulse detection to organometal speciation has been described by MacCrehan.[74]

II. SPECIATION STUDIES

A. Arsenic

1. Occurrence

Of the elements subjected to speciation analysis by HPLC, arsenic has received by far the greatest attention. The commonly encountered species of arsenic are shown in Table 1. In oxygenated natural waters, arsenic is generally present as the arsenate ion $HAsO_4^{2-}$, although arsenic (III) as H_3AsO_3, is significant under less-oxidizing conditions.[75] In sediments, methanogenic bacteria can promote the formation of trimethyl-arsine[76] with monomethylarsonic acid (MMA) and dimethylarsinic acid (DMA) being intermediates in this process. Studies by Andreae[77,78] have indicated the presence of the methylated acids in both freshwater and seawater, resulting presumably from phytoplankton or heterotrophs closely associated with the primary producers.

Marine algae have been shown to contain arsenic phospholipids and arsenolipids, where because of its chemical similarity, arsenic has replaced phosphorus. This occurs specifically in phosphate-deficient waters.[79] Arsenocholine and arsenobetaine are related organoarsenic species to these, and their presence has been confirmed in marine crustacea[80,81] and fish.[82]

Animals metabolize inorganic arsenic to the methyl acids.[83,84] The principal product is DMA, although MMA is an intermediate. Analyses of human blood and urine have detected these species, together with arsenobetaine specifically associated with a dietary intake of fish, shrimp, or lobster. Arsenocholine, which is also present in these biota, has been shown to be rapidly hydrolyzed and excreted as arsenobetaine.[85]

Until recently, arsenic was also widely used in pesticides, in particular, DMA (cacodylic acid), which was used as a silvicide.[86] In oil shale waste waters, a range of organoarsenic acids, including phenylarsonic acid, have been identified,[87] presumably resulting from reactions, during retorting at temperatures near 600°C, of inorganic arsenic minerals and organic arsenic present in the algal precursors in the raw shale.

2. Analysis

The large differences in acid dissociation constants of the common arsenic species (Table 1) make them ideally suited to separations by ion exchange. Open-column separations using both anion- and cation-exchange resins have been reported by Henry and Thorpe,[88] Yamamoto et al.,[89] and others.[90-92]

Many attempts to separate arsenic species by HPLC have utilized strong anion-exchange columns.[93-95] Many studies used silica-based columns which greatly restricted the pH range over which they could be applied, ideally between pH 3 and 7. Brinckman and co-workers,[5] using a LiChrosorb® SAX column (Alltech Associates, Inc.) were able to separate As(III), DMA, and MMA, but found arsenite too strongly adsorbed using 0.05 *M* sodium dihydrogenphosphate, pH 4.5, as eluent. Increasing the phosphate concentration reduced retention times; however, the sensitivity of the graphite furnace AAS detection system, was reduced. Replacement of phosphate by an acetate buffer of the same pH reduced this problem. The order of elution was As(III), DMA, and MMA, with strong retention still of As(V). The peaks for As(III) and DMA were poorly resolved. It was noted also that the resolution of all peaks diminished after 1 month of operation unless the system was well flushed with methanol after each run and kept filled with methanol when not in use.

Woolsen and co-workers[93,96] used a resin-based anion-exchange column, Bio-Rad® Aminex A-27, to extend the usable eluent pH range. They were able to successfully elute As(V) by changing the eluent from water to 0.2 *M* ammonium carbonate after the elution of As(III), DMA, and MMA. Tye et al.[37] loaded arsenic from 1.8×10^{-5} *M* sulfuric acid onto a 10-cm × 5-mm column of a pellicular silica-based anion-exchange resin (Zipax,® 40 μm) for subsequent separation on a 20-cm × 5-mm column of SAX-10 (5 μm), a polystyrene-based strong anion-exchange resin, using 0.1 *M* ammonium carbonate as eluent. Good separation of As(III) DMA, MMA, and As(V) was obtained.

Ion chromatography using a Dionex® system (Dionex Corp., Sunnyvale, CA) has been explored by several workers.[97] Tan and Dutrizak[97] separated only As(III) and As(V). Ricci et al.,[95] using the Dionex® 3-mm × 500-mm anion separator column, separated all except As(III) and DMA using 0.0014 *M* $NaHCO_3$/0.0019 *M* Na_2CO_3/0.001 *M* $Na_2B_4O_7$ eluent, and needed a separate analysis using a lower ionic strength eluent, 0.005 *M* $Na_2B_4O_7$, to resolve As(III) and DMA. Their optimum separation required gradient elution from an initial borax buffer to a final carbonate buffer. The order of elution was DMA, As(III), MMA, and As(V) with respective detection limits of 6.5, 4.0, 3.0, and 20 $\mu g\ l^{-1}$, using the hydride generation-AAS system. A similar gradient separation has been reported by Spall et al.,[98] who used an 8-mm × 10-cm Aminex A-27 Radial-Pak column and a gradient from water to 0.5 *M* $(NH_4)_2CO_3$. Gradient elution is not a preferred technique for large numbers of samples because of the need to reequilibrate the columns after each run.

Morita et al.[94,99] examined both anion and cation columns for the separation of arsenic species. A column (25 cm × 3 mm I.D.) packed with Nagel-N$(CH_3)_2$ gave a good separation of arsenobetaine, As(III), MMA, DMA, and As(V) in under 15 min, using 0.05 *M* phosphate buffer (pH 7.2). Using a cation-exchange packing (Nagel-SO_3H-10), the elution order As(III), As(V) > MMA > DMA > arsenobetaine was followed. This may only partially be the result of charge effects, as all except arsenobetaine are negatively charged and are not expected to be retained by the sulfonate groups of the same charge on the resin, and most probably also involves the silica backbone of the column packing.

A number of studies have investigated reversed-phase columns, using ion pairing to facilitate separation. Brinckman et al.[5] used a 10-μm μ-Bondapak®-C_{18} column (20 cm × 4 mm I.D.) with 0.005 *M* *t*-butylammonium phosphate, pH 7.3, in 95:5 water-methanol. Under these conditions, As(III) and DMA were poorly separated. An Alltech LiChrosorb® RP-18 (25 cm × 4.6 mm) column, using saturated *t*-heptylammonium acetate in 75:25 water-methanol (pH 7.6), separated these species, but As(V) could only be eluted with 100% methanol. Studies by other workers have in general confirmed the unsuitability of reversed-phase ion-pairing techniques.[100-102] Separation of selected organoarsenic compounds can, however, be quite readily accomplished by this technique. Francesconi et al.,[80] for example, separated MMA, DMA, arsenobetaine, and arsenocholine using a Hamilton® PRP-1 column (10-μm macroporous styrene-divinylbenzene copolymer) and a mobile phase comprising 0.05 *M* sodium heptanesulfonate in 2.5% aqueous acetic acid.

All of the studies so far discussed have been concerned with the determination of arsenic species in aqueous solutions. Analyses in biological matrices or seawater, which are high in chloride, have, however, been shown to cause problems. Low et al.[51] attempted separations in chloride-containing solution using a Hamilton® PRP-X100 column (2.5 cm × 4.1 mm I.D.), a low capacity, strong-base anion-exchange phase on a macroporous styrene-divinylbenzene copolymer. Because only a small percentage of the surface is bonded to anion-exchange functional groups, the column possesses some reversed-phase characteristics;[103] however, unlike silica-based columns, it is able to function over a wide pH range. Resolution of standard arsenic compounds on this column was achieved using mobile phases of ammonium carbonate or ammonium phosphate adjusted to between pH 6 and 8.2 with ammonia or phosphoric acid, with the ionic strength adjusted using sodium sulfate. In this pH range, chloride was found to interfere with the elution of MMA, resulting in peak splitting (Figure 13). This occurred with a range of eluents in the pH range 6 to 8. The extent of peak splitting was regulated by chloride concentration. It was postulated that this arose from an interplay between the elution of MMA in the mobile solvent, in the slug of the injection solvent, and on the column surface perturbed by chloride. A model for this behavior incorporating *in situ* microenvironmental changes in the column surface was proposed. The removal of chloride from the sample before injection onto the column proved to be a difficult problem, and no adsorbent could be found which did so without also removing one of the arsenic species.

From a practical standpoint, analyses of urine and seawater for arsenic species were undertaken using the above system[52] with integration of peaks, including the split peak, to obtain the MMA concentration. Detection limits of 1 μg l^{-1} As(III), 10 μg l^{-1} As(V), 4 μg l^{-1} MMA, and 4 μg l^{-1} DMA were reported for a 100-μl injection.

To resolve arsenobetaine in combination with the four common arsenic metabolites, Low et al.[104] experimented with an anion column and a reversed-phase C_{18} column in series and utilized a column-switching procedure to optimize the separations (Figure 14).

The most useful detection system for the common arsenic species is the use of AAS to detect hydrides formed by reaction with sodium borohydride and acid.[49] The optimum conditions for hydride generation in the effluent from an HPLC column differ from those recommended using continuous aspiration. With 3% w/v sodium borohydride and above, similar responses are obtained for As(III), MMA, and DMA, with As(V) somewhat less,[44] using continuous aspiration. With the column procedure under these conditions, the higher volumes of hydrogen so produced did not mix efficiently with the eluting sample, resulting in poor reproducibility, and a lower borohydride concentration 1.5% w/v was preferred.[52] Experiments with the size of the mixing coil using the Varian® VGA-76 hydride generator for HPLC detection showed that more efficient mixing of eluent and hydrogen was achieved

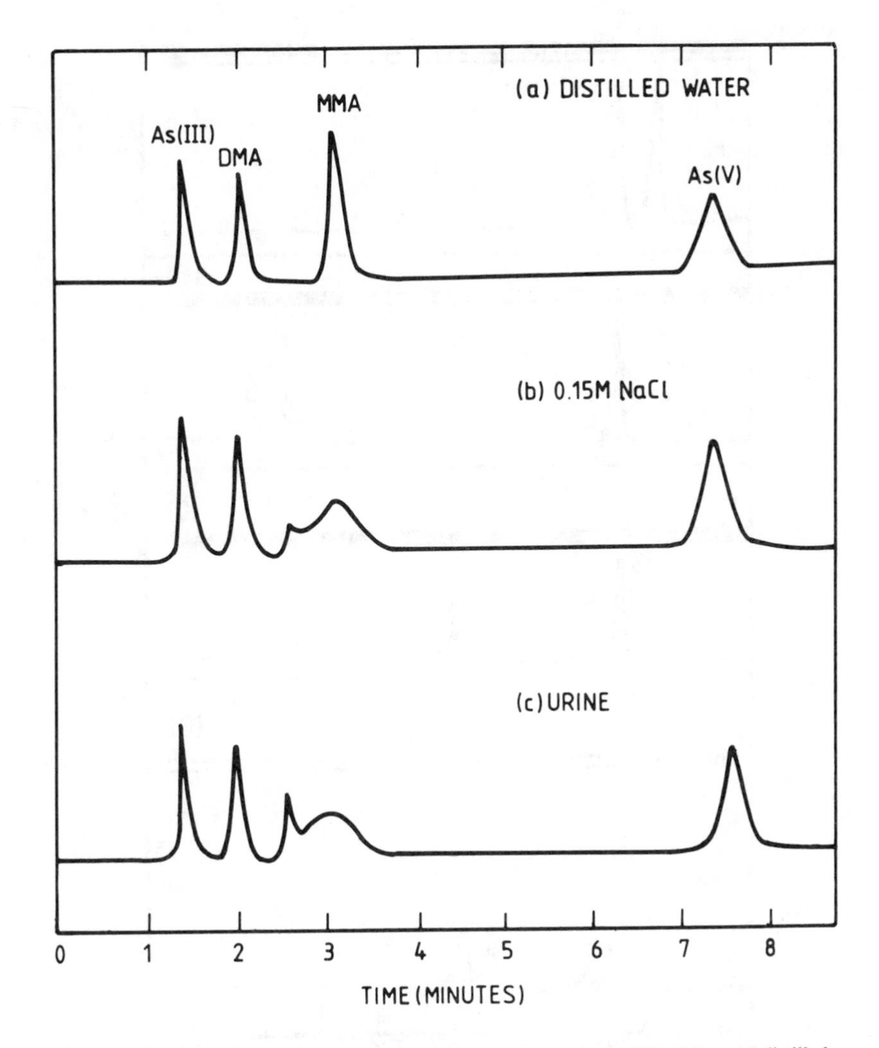

FIGURE 13. Separation of standard mixtures of arsenic species by HPLC from (a) distilled water, (b) 0.15 *M* NaCl, and (c) urine, using an anion-exchange column with ammonium carbonate eluent and hydride-AAS detection.

using a larger diameter coil than the standard 0.5-mm size. The optimum coil design comprised a 90-cm × 0.5-mm-diameter Teflon® coil in advance of the 0.5-mm diameter autoanalyzer tube. The hydride system has the disadvantage that it is sensitive only to those arsenic species which form hydrides and thus will not detect arsenobetaine or arsenocholine. Coupling of the column effluent to an ICP spectrometer will overcome this problem; however, the sensitivity of ICP detection is much lower. Based on 100-μl injections, detection limits for As(III), DMA, MMA, and As(V) have been reported as 3 to 7 mg l^{-1}, 2 to 6 mg l^{-1}, 4 to 6 mg l^{-1}, and 3 to 13 mg l^{-1}, respectively.[94,102] The lower figures are those of Morita et al.[99] These are at least two orders of magnitude above those for the hydride-AAS system.[52] The hydride system can of course be coupled with an ICP detector to achieve a similar increase in sensitivity, but there would appear to be no particular advantages to be gained by this over AAS detection.

A particular idiosyncracy of the ICP method, when applied to arsenic speciation in salt-containing matrices, was the appearance of a spurious peak at the arsenic wavelength, which was not in fact due to an arsenic compound.[105] This only occurred when analyzing an HPLC column effluent and resulted from a disturbance in the emission background of the plasma

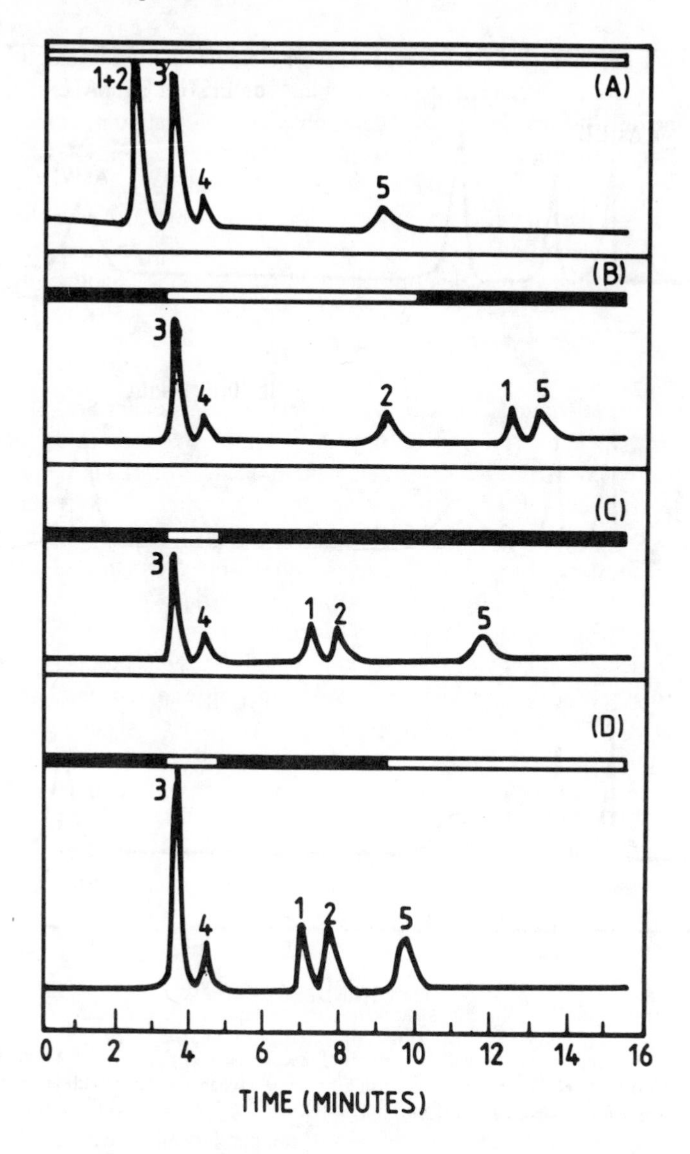

FIGURE 14. Separation of arsenic species by HPLC using column switching between a C_{18} and an anion-exchange column and ICP detection. (Open bar areas) anion column only; (closed bar areas) anion column followed by C_{18} column. (From Low, G. K.-C., Batley, G. E., and Buchanan, S. J., *J. Chromatogr.*, 368, 423, 1986. With permission.)

when an easily ionizable element, e.g., Na or K, in the sample matrix was eluted from the HPLC column as a concentrated slug.

The use of graphite-furnace AAS detection[5,44] is unsuited to continuous analysis and, as discussed earlier, the effluent has to be collected and injected repeatedly into the furnace to obtain an elution profile. The system used by Brinckman et al.[5] for arsenic speciation has a programmable furnace which uses a measurement cycle of 10-s drying at 75°C, 10-s ashing at 300°C, and 7-s atomization at 2500°C, with detection using the 193.7-mm arsenic line. An ashing cycle was eliminated to minimize volatilization. The signal clusters so obtained (Figure 4), converted to numerical values by summation of the peak heights, yielded detection limits near 5 $\mu g\ l^{-1}$, comparable to those for the hydride-AAS system. This method has been applied to drinking waters, soil digests, and oil shale waste waters.[5,86]

Tan and Dutrizac[97] analyzed As(III) and As(V) by a combination of electrochemical and conductivity detection, after ion chromatographic separation. Pentavalent arsenic is not electroactive and only the trivalent form is detectable. A gold, platinum, or mercury electrode is suitable.[97,106] In most previous electrochemical studies,[106,107] arsenic(V) has been detected after reduction to As(III) using hydrazine or a similar reductant. The peak current is pH dependent and is greatest for the free arsenite ion (pKa = 9.2). Using the Dionex® ion chromatograph,[97] an electrolyte of 1.8 m*M* H_2SO_4 was selected and a detection limit of 5 μg l^{-1} obtained. Arsenic(V) was determined to 22 μg l^{-1} using conductivity detection.

B. Selenium

1. Occurrence

In natural waters, selenium is present in inorganic forms as either Se(VI), selenate (pKa = 1.92), or Se(IV), selenite (pKa_1 = 2.46, pKa_2 = 7.31). In oxygenated seawater, despite thermodynamic predictions that Se(VI) should be dominant,[108] Se(IV) is the principal species detected, with increasing selenite concentrations at greater depths.[109] Regenerative cycles to account for this behavior have been postulated.[110] In anoxic water, analyses revealed an organic selenide, most of which was associated with dissolved amino acids, and it has been proposed that this selenium might exist as seleno-amino acids.[111]

In sediments, biomethylation and reduction of organic sediments may lead to the formation of dimethylselenide and dimethyldiselenide, both volatile gaseous products.[112] Evidence for the production of these species from inorganic selenide by fungi, plants, and animals has also been presented.[113] Certain plants biosynthesize organoselenium compounds such as Se-methylselenomethionine, Se-methylselenocysteine, and selenocystathionine.[114,115] Organic forms of dietary selenium include such compounds together with selenocysteine and selenomethionine.

Selenium detoxification in the body is believed to occur by metabolic conversion of selenite to the trimethylselenium ion $(CH_3)_3Se^+$,[116,117] and this compound has been detected in urine samples.

2. Analysis

The applications of HPLC to the speciation of selenium have to date been restricted to the separation of inorganic forms only. Urasa and Ferede[59] separated Se(IV) and Se(VI) by ion chromatography using a Dionex® HPLC-AS4 analytical column, a mobile phase of 3.0 m*M* $NaHCO_3$/2.4 m*M* Na_2CO_3, and ICP detection. The shapes of the eluted peaks were ionic strength dependent. Peak areas were species independent, unlike conductivity detection where the ratio for Se(IV) to Se(VI) was 0.6. Chakraborti et al.,[118] using a similar Dionex® system and a 1-ml injection loop, were able to detect 20 ng of each selenium species, or 0.2 ng if an anion separator preconcentration column was used.

McCarthy et al.,[102] using a 25-cm Nucleosil®-$NH(CH_3)_2$ anion-exchange column, separated Se(IV) and Se(VI), with the former eluted by a mobile phase consisting of 0.002 *M* ammonium dihydrogenphosphate and 0.05 *M* ammonium acetate (pH 4.6), while selenite was strongly retained and could only be eluted with a change to 0.08 *M* ammonium dihydrogenphosphate adjusted to pH 6.9 with ammonia. A change of solvent is unnecessarily complex, and it is possible to separate both species on an anion-exchange column using a single eluant, 5 m*M* sodium citrate pH 6.[119] Concommitant separation of arsenic species is also possible, although the detection is more complex. Silica-based columns should not be used because they are degraded by citrate buffers.

Detection of both arsenic and selenium species via ICP requires either sequential detection or a fixed wavelength polychromator system.[102] Nevertheless, the detection limits without preconcentration are generally poor (1.4 mg l^{-1} for Se(IV) and 0.9 mg l^{-1} for Se(VI)). Detection via hydride formation, as described for arsenic, is also possible, but only Se(IV) gives complete recoveries. Those for Se(VI) are greater at pH values below 3, with negligible reaction at pH 9.6.[120] In analyses using open-column chromatographic separation of selenium anions from interfering elements,[50,111] heating with dilute HCl has been used to convert Se(VI) to Se(IV); however, given the poor reproducibility of this procedure, photoreduction[121] might be a better alternative.

Selenium, unlike arsenic, is more amenable to electrochemical detection. Although similarly only the lower oxidation state is electroactive, reduction of Se(VI) is more readily accomplished and is amenable to an on-line system.[121]

The application of HPLC to the analysis of organoselenium species has not yet been reported.

C. Tin

1. Occurrence

Tin is present in natural waters in concentrations below 50 $\mu g\ l^{-1}$ and is generally accepted as being principally Sn(IV) on the basis of thermodynamic equilibria. However, it is possible that kinetic control of steady-state speciation may indeed favor Sn(II), especially in polluted waters, and this is not always readily measurable.[122]

The major source of tin compounds in the aquatic environment is as organotin compounds, particularly trialkyltins, which are extensively used as biocides, algicides, fungicides, and molluscides, in marine antifouling paints, and agriculture. Dialkyl species are used as heat and light stabilizers in polyvinylchloride manufacture and as catalysts for the curing of silicones and production of polyurethanes.[123] Photodegradation and biologically mediated methylation of the products in sediments can lead to the presence in both waters and sediments of a range of alkylated tin species.[123,124]

2. Analysis

Although gas chromatography has been frequently employed for the detection of organotin species in volatile stannanes, HPLC separations offer a reliable direct speciation alternative. Brinckman and co-workers[122] reported the first such application, resolving triphenyl-, tributyl-, and tripropyltin using a LiChrosorb® C_2 reversed-phase 10-μm silica column and a mobile phase of 100% methanol. Detection utilized the graphite furnace atomic absorption system described earlier for arsenic.[44] For the organotin species studied, this form of detection was, as expected, far more sensitive than UV detection.

The same detection system was applied by Vickrey et al.[125] using a LiChrosorb® 10-μm C_{18} column and a mobile phase of 97.5:2.5 methanol-water and a flow rate of 0.1 ml min^{-1}. Responses were reported for $SnCl_4$, $SnPh_4$, $SnBu_4$, $SnBu_2Cl_2$, $SnMe_3Cl$, $SnMe_2Cl_2$, $(SnBu_3)_2SO_4$, $SnPr_2Cl_2$, $Sn(C_6H_{11})_2Br$, and $SnPh_3Cl$, but no HPLC retention data were published.

Burns et al.[41,126] reported separations of methyl- and ethyltin isomers using a C_{18} column (Spherisorb® S5W, Phase Separations, Ltd.) and mobile phases of 60:40 acetone:pentane (methyltins) and 70:30 acetone-pentane (ethyltins) (Figure 15). Both direct flame atomization and hydride-generation quartz tube atomization have been employed with AA detection. The former gave poor detection limits near 0.2 to 0.7 mg l^{-1}; however, with hydride generation, the limit was 0.2 to 0.3 $\mu g\ l^{-1}$ for a 50-μl injection. As found for arsenic speciation, the detector response to the hydrides produced will vary depending on the nature of the species.

Ebdon et al.[38] also employed a silica column (Whatman® Partisil 10 SCX) with 70:30 methanol-water containing 0.1 *M* ammonium acetate as mobile phase coupled with a discrete volume nebulization system as discussed earlier. Using a double-slotted quartz tube atom trap and a 5-cm slot burner, a detection limit of 200 $\mu g\ l^{-1}$ was obtained. Separation of Sn(II), Sn(IV), and tributyltin chloride was achieved.

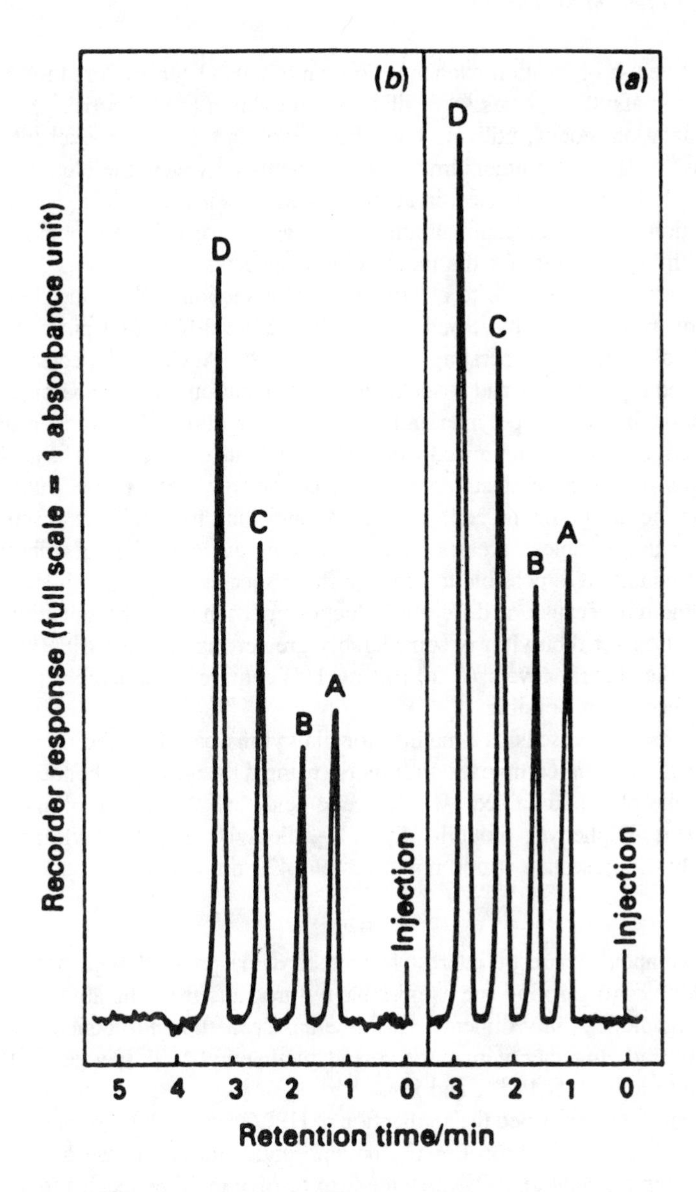

FIGURE 15. Separation of methyltin and ethyltin compounds on a C_{18} column using hydride-AAS detection. (a) A, Me_4Sn; B, Me_3SnCl; C, Me_2SnCl_2; D, $MeSnCl_3$. (b) A, Et_4Sn; B, Et_3SnCl; C, Et_2SnCl_2; D, $EtSnCl_3$. (From Burns, D. T., Glockling, F., and Harriott, M., *J. Chromatogr.*, 200, 305, 1980. With permission.)

The application of an ion-pairing reagent with a reversed-phase column system has been reported by Krull and Panaro[127] to provide separations of ionizable methyltins. Using 0.003 *M* hexanesulfonic acid, 0.003 *M* KF, and 2.5% acetic acid, all in 0.02 *M* sulfuric acid, and a Hamilton® PRP-1 column, mono-, di-, and trimethyltin were resolved. Both direct ICP detection and ICP detection after hydride generation were examined. As found for arsenic, the detection limits in the absence of hydride formation are poor (250 $\mu g\ l^{-1}$) and an order of magnitude below that in the hydride forms. The presence of fluoride presumably results in an initial exchange of all chloride present, preventing the formation of mixed ligand complexes that occur with chloride and the ion-pairing reagent only present. Optimum hydride conditions required 0.1% $NaBH_4$, 1 to 4% KOH at 0.5 to 7.5 ml min^{-1}, mixing with the eluting sample at 2.5 to 3.0 ml min^{-1}. Tin is detected at 303.4 nm.

The alternative use of a cation-exchange column has also been explored for the separation of the protonated alkyltin species.[128] A silica-based column (e.g., Partisil® SCX) offers the secondary separation modes, utilizing both the silanol and the cation-exchange sites. With a mobile phase of 0.01 *M* ammonium acetate in methanol-water, the elution order was Bu < Pr < Et < Me for R_3Sn^+ species, in agreement with their pKa values. Using the graphite furnace detection system, detection limits as low as 25 μg l^{-1} were obtained for 200-μl injections, with higher limits for the more volatile species.

Tin species are also amenable to electrochemical detection.[71,73,74] MacCrehan[74] reported the analysis of tributyl-, triethyl-, and trimethyltin using differential pulse detection at an amalgamated gold electrode, operating at −1.10 V vs. Ag/AgCl. The species were separated using the procedure of Jewett and Brinckman.[128] Conventional DC amperometric detection was less sensitive because of polymerization reactions involving the radical products, which coated the electrode. This is minimized with differential pulse detection. The need to carefully exclude oxygen in reductive electrochemical detection, however, places quite stringent requirements on the design of the cell, ancillary tubing, and hardware, as discussed earlier.

In natural water samples, greatest interest lies in the detection of tributyltin and its degradation products. It is possible to separate these species, all of which are electroactive, by HPLC, although because of differing reduction mechanisms, the sensitivity of electrochemical detection for dibutyltin is considerably greater than for the tributyl species.[129,130] Nevertheless, the natural levels (2 to 100 ng l^{-1}) require considerable preconcentration before their detection is possible.

Langseth[131] reported a sensitive method for dialkyltins based on the normal-phase separation of their fluorescent complexes with morin using a cyanopropyl-bonded silica column. A typical mobile phase contained 3% v/v acetic acid, 2% v/v methanol and 0.0015% v/v morin in toluene. Diphenyl-, dibutyl-, diproply-, diethyl-, and dimethyltin were separated and detected by fluorescence at 560 nm emission, 420 nm excitation.

D. Lead

Alkyllead compounds are of interest from their widespread though declining usage as antiknock additives to gasoline and as possible by-products from the abiotic or biotic methylation of inorganic lead in sediments. Gasoline has been shown to contain both tetraethyl- and tetramethylead, together with three mixed methyethyl lead species.[132] Photolytically these degrade to R_3Pb^+, R_2Pb^{2+}, and Pb^{2+}.[133]

Several papers have described the application of HPLC[46,125,132,133] to organolead speciation. Vickrey et al.[125] used a LiChrosorb® C_{18} column and a mobile phase of 80:20 methanol-water, with a step gradient to 100% methanol, to resolve the five species found in gasoline. Iodine was added to the eluent prior to graphite-furnace analysis to improve the sensitivity by reducing the lead volatility.

Similar separations were employed by Koizumi et al.,[46] Messman and Rains,[132] and Botra et al.[134] with minor differences in mobile phase compositions and differing detection systems (Figure 16). Because of differences in the atomization conditions for different species and their high volatility, maximum signals can only be obtained in atomic absorption detection systems by using the closed-system furnace discussed earlier.[46] This system utilizes 10 μl from each 250 μl of column effluent being sequentially analyzed. Using a Zeeman® AAS instrument, the typical gasoline concentrations (≈20 mg l^{-1}) are readily detected without the interferences from magnesium and calcium chlorides encountered in conventional furnaces.

Direct introduction of an acetonitrile/H_2O effluent, at a flow rate of 3 ml min^{-1}, into the aspiration uptake of the nebulizer of an air/acetylene flame in an AAS instrument yielded detection limits of 0.5 mg l^{-1} for a 20-μl injection.[132] (Figure 16). Direct nebulization into an ICP spectrometer yielded similar limits.[135] Although it is generally accepted that the

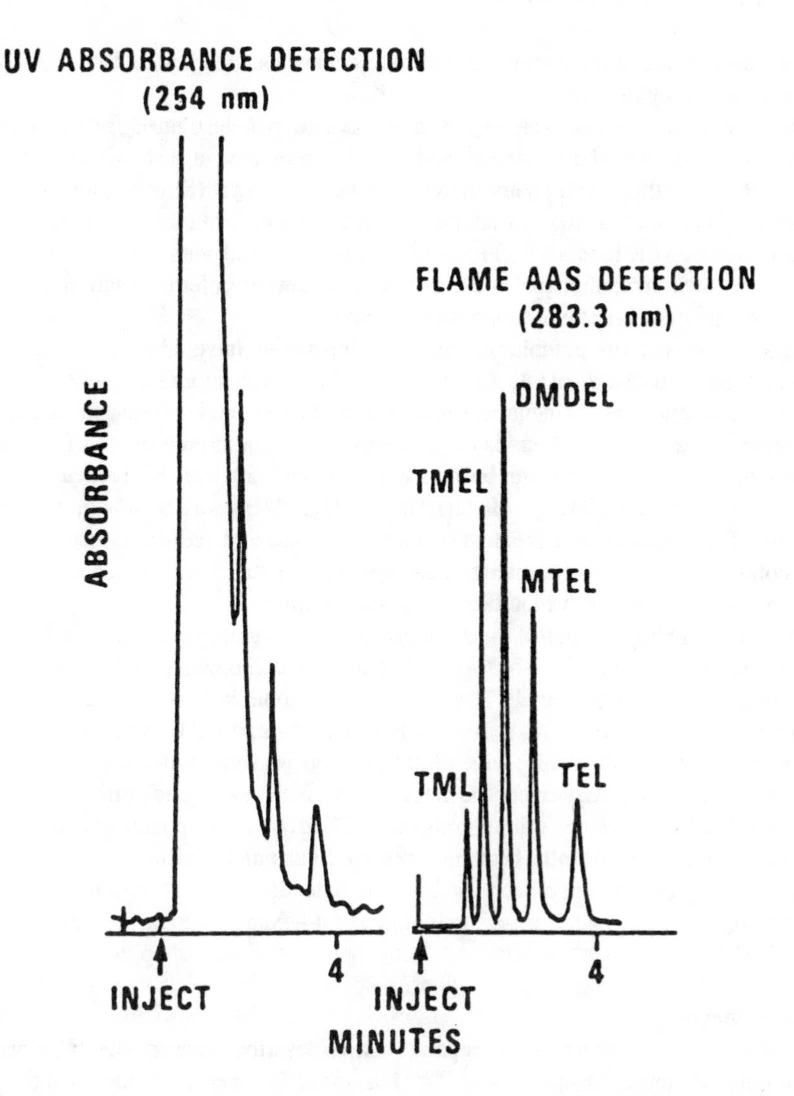

FIGURE 16. Comparison of flame AAS and UV detection of tetramethyl-, trimethylethyl-, dimethyldiethyl-, methyltriethyl-, and tetraethyllead, after separation on a C_{18} column. Eluent 70:30 acetonitrile-water at 3 ml min^{-1}. (From Messman, J. D. and Rains, T. C., *Anal. Chem.*, 53, 1632, 1981. With permission.)

emission sensitivity is independent of chemical form, the sensitivity for tetramethyl lead was found to be five times that for tetraethyllead. This must relate either to more efficient transport of the more volatile species to the plasma or most likely to more efficient dissociation.

The above means of detection have been shown to be superior to UV detection because of their selectivity. Electrochemical detection is also a possibility, as the stripping voltametric detection of these species has been observed;[136] however, no applications to species separated by HPLC have yet been reported.

E. Mercury

The important aspect of the environmental chemistry of mercury is its biomethylation in sediments to mono- and dimethylmercury. These species are more toxic than inorganic mercury, and through their high lipid solubility can be concentrated in higher aquatic organisms. Bioaccumulation factors of 10^3 to 10^4 have been reported for mercury in fish, principally as methylmercury.[25]

Both normal- and reversed-phase columns have been investigated for the separation of organomercury salts by HPLC. Gast and Kraak[25] separated a test mixture of diphenyl-, propyl-, ethyl-, methyl-, and phenylmercury using a silica gel (SI 60) column and a mobile phase of 10% butanol in *n*-hexane saturated with tetramethylammonium chloride (≈0.01%). A reversed-phase (LiChrosorb® RP8) column was also used with 0.05 *M* aqueous sodium chloride plus 30% methanol (pH 3.5). For detection, post-column reaction with dithizone gave UV-detectable complexes absorbing at 480 nm.

Separations based on precolumn complex formation have also been employed.[13,138] Langseth[137] showed that at pH 4, dithizone complexes with inorganic and organic mercury salts, after extraction into toluene, evaporation to dryness, and solution in methanol, could be separated on a Waters® Nova-pak C_{18} column. Using a mobile phase of 2:1 tetrahydrofuran-methanol, 0.05 *M* acetate buffer, pH 4.0, and 50 μ*M* EDTA, an elution order MeHg(HDz) > EtHg(HDz) > PhHg(HDz) > $Hg(HDz)_2$ was obtained, where H_2Dz is dithizone. The addition of EDTA prevented exchange and reduction reactions involving metal components of the chromatographic system and the metal chelates. This technique has been applied to water, vegetable, and urine samples.

Neutral complexes between 2-mercaptoethanol and organomercurials were separated by MacCrehan et al.[71] using NH_2- or C_{18}-reversed-phase columns. A mobile phase of 0.05 *M* ammonium acetate, pH 5.5, in 40% methanol-water containing 0.5 m*M* 2-mercaptoethanol satisfactorily separated Hg^{2+}, CH_3Hg^+, $CH_3CH_3Hg^+$, and $C_6H_5Hg^+$, in that order. Electrochemical detection is suitable, with all of these species being reducible at potentials below −0.65 V vs. Ag/AgCl. Amperometric detection at a mercury-coated gold electrode at −0.90 V vs. Ag/AgCl yielded detection limits near 5 μg l^{-1}. Optimized conditions for these separations have more recently been reported by Evans and McKee.[138]

Atomic absorption has also been used for the detection of ethyl- and methylmercuric ion down to 1 μg l^{-1},[139] using a Corasil® I column and *n*-hexane as eluent, and a quartz vaporizer tube at 300°C.

F. Other Metals

Paired-ion reversed-phase HPLC separations, as described earlier, have also been applied to cadmium and chromium speciation.[101,140] Nisamaneepong et al.[100] examined the separation of cadmium NTA and EDTA complexes using tetrabutylammonium phosphate as the ion-pair reagent in 95:5 water-methanol. Krull et al.[140] investigated Cr(III) and Cr(VI) separation. Both studies utilized ICP detection, but neither examined natural samples. This is particularly important in the case of chromium, where it is assumed that Cr(III), being cationic, will not be retained in the presence of the positively charged counterion, while anionic Cr(VI) will be. The reverse will be true using an anionic counterion (e.g., *n*-pentane sulfonate or camphor sulfonate) (Figure 17). As discussed previously,[11] $Cr(OH)_4^-$ is a likely form of Cr(III) in natural waters, and a study of the behavior of this species could be profitably pursued.

The separation of inorganic cobalt(II) and cyanocobalamin has been described by Koizumi et al.[141] Using an anion-exchange column and a mobile phase of 2 *M* ammonium acetate, excellent resolution was obtained using Zeeman® AAS detection, where 10 μl from each 3 ml of eluent (flow rate 1 ml min^{-1}) was measured.

Many studies have exploited the HPLC separation of inorganic metal ions by either post-, pre-, or on-column formation of metal chelates.[10,21] Uden et al.,[142] for example, examined the separation of a range of diethyldithiocarbamate and β-diketone complexes. The use of precolumns to enrich metal species enabled detection at microgram-per-liter concentrations and below,[11] but this could only be achieved by the use of metal-free pumps and associated equipment. This procedure has yet to be applied to speciation studies.

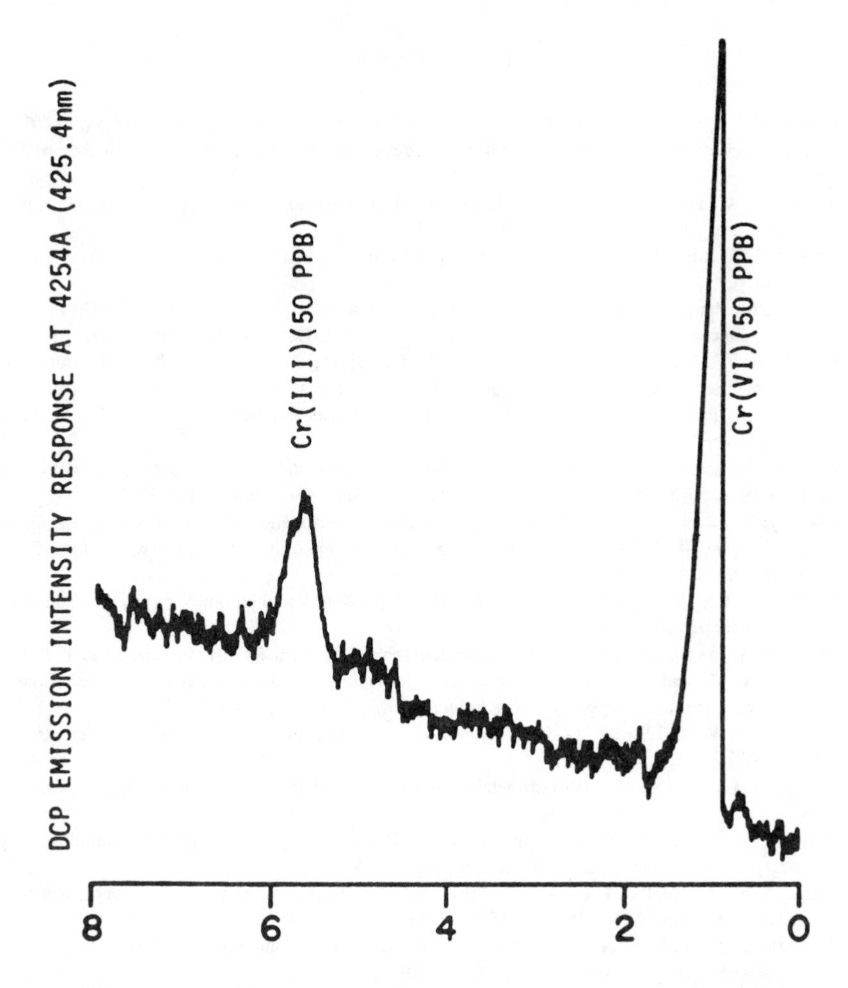

FIGURE 17. Separation of Cr(III) and Cr(VI) on a C_{18} column using camphor sulfonate (PIC A) as eluent; flow rate 1.5 ml min^{-1}. Detection by ICP at 425.4 nm. (From Krull, I. S., Bushee, D., Savage, R. N., Schleicher, R. G., and Smith, S. B., *Anal. Lett.*, 15, 267, 1982. With permission.)

Mueller and Lovett[142] developed a preconcentration procedure for diethyldithiocarbamate complexes by salt extraction into acetrontrile for injection into the HPLC system. Schwedt[144] used ammonium pyrrolidinedithiocarbamate in preference to diethyldithiocarbamate to determine Cr(III) and Cr(VI). The separation of Mn(II) and Mn(III), based on reversed-phase HPLC of their oxinate complexes, was described by Hoffmann and Schwedt.[145] Both pre- and on-column derivitization can be used, but the latter was preferred, as oxidation is more likely with the precolumn procedure. Citrate buffers were avoided because of the possible oxidation of Mn(II).

G. Anion Speciation

The use of ion-exchange HPLC (ion chromatography) for the analysis of anions has experienced a rapid growth in recent years,[9,30] and as already shown, it has found wide application in speciation studies. In studies of anion speciation, besides those of arsenic and selenium already discussed, readers should be aware that the technique has been applied to speciation of nitrogen, sulfur, halide, and phosphorus anions in aqueous samples.[9] An example is the separation of sulfite, sulfate, and thiosulfate, using a step gradient from 4.8 m*M* $NaHCO_3$/4.7 m*M* Na_2CO_3 to 7.2 m*M* $NaHCO_3$/9.1 m*M* Na_2CO_3 on a Dionex® anion separator column.[146] These analyses will not, however, be discussed further in this chapter.

REFERENCES

1. **Low, G.K.-C.,** Stationary phases for HPLC — some present problems and filtered trends, in *HPLC in the Clinical Laboratory,* Clinical Biochemist Monogr., Australian Association of Clinical Biochemists, Sydney, 1986, 12.
2. **Sander, L. C. and Wise, S. A.,** Synthesis and characterization of polymeric C_{18} stationary phases for liquid chromatography, *Anal. Chem.,* 56, 504, 1984.
3. **Benson, J. R. and Wood, D. J.,** Polymeric columns for liquid chromatography, *J. Chromatogr. Sci.,* 22, 386, 1984.
4. **Krull, I. S., Panaro, K. W., and Gershman, L. L.,** Trace analysis and speciation for Cr(VI) and Cr(III) via HPLC-direct current plasma emission spectroscopy (HPLC-DCP), *J. Chromatogr. Sci.,* 21, 460, 1983.
5. **Brinckman, F. E., Jewett, K. L., Iverson, W. P., Irgolic, K. J., Ehrhardt, K. C., and Stockton, R. A.,** Graphite furnace atomic absorption spectrophotometers as automated element-specific detectors for high-pressure liquid chromatography. The determination of arsenate, methylarsonic acid and dimethylarsinic acid, *J. Chromatogr.,* 191, 31, 1980.
6. **Mills, G. L. and Quinn, J. G.,** Isolation of dissolved organic matter and copper-organic complexes from estuarine waters using reverse-phase liquid chromatography, *Mar. Chem.,* 10, 93, 1981.
7. **Mackey, D. J.,** Metal-organic complexes in seawater-an investigation of naturally occurring complexes of Cu, Zn, Fe, Mg, Ni, Cr, Mn and Cd using high-performance liquid chromatography with atomic fluorescence detection, *Mar. Chem.,* 13, 169, 1983.
8. **Steinbrech, B.,** Thin layer chromatography and high pressure liquid chromatography of metal chelates, *J. Liq. Chromatogr.,* 10, 1, 1987.
9. **Smith, F. C. and Chang, R. C.,** Ion chromatography, *CRC Crit. Rev. Anal. Chem.,* 9, 197, 1980.
10. **Cassidy, R. M. and Elchuk, S.,** Trace enrichment methods for the determination of metal ions by high performance liquid chromatography, *J. Chromatogr. Sci.,* 18, 217, 1980.
11. **Florence, T. M. and Batley, G. E.,** Chemical speciation in natural waters, *CRC Crit. Rev. Anal. Chem.,* 9, 219, 1980.
12. **Steinberg, G.,** Species of dissolved metals derived from oligotrophic hard water, *Water Res.,* 14, 1239, 1980.
13. **Bender, M. R., Matson, W. R., and Jordan, A. R.,** On the significance of metal complexing agents in secondary sewage effluents, *Environ. Sci. Technol.,* 4, 520, 1976.
14. **Sugai, S. F. and Healy, M. L.,** Voltammetric studies of the organic association of copper and lead in two Canadian inlets, *Mar. Chem.,* 6, 291, 1978.
15. **Sterritt, R. M. and Lester, J. N.,** Speciation of copper and manganese in effluents from the activated sludge process, *Environ. Pollut. Ser. A,* 27, 37, 1982.
16. **De Mora, S. J. and Harrison, R. M.,** The physicochemical speciation of lead in tapwater, in *Proc. 4th Int. Symp. on Heavy Metals in the Environment,* Vol. 2, CEP Consultants, Edinburgh, 1983, 1207.
17. **Pfannkoch, E., Lu, K. C., Regnier, F. E., and Barth, H. G.,** Characterization of some commercial high performance size-exclusion chromatography columns for water-soluble polymers, *J. Chromatogr. Sci.,* 18, 430, 1980.
18. **Yoza, N. and Ohashi, S.,** The application of atomic absorption method as a flow detector to gel chromatography, *Anal. Lett.,* 6, 595, 1973.
19. **Hausler, D. W. and Taylor, L. T.,** Non-aqueous on-line simultaneous determination of metals by size exclusion chromatography with inductively coupled plasma atomic emission spectrometric detection, *Anal. Chem.,* 53, 1223, 1981.
20. **Hausler, D. W. and Taylor, L. T.,** Size exclusion chromatography of organically bound metals and coal-derived materials with inductively coupled plasma atomic emission spectrometric detection, *Anal. Chem.,* 53, 1227, 1981.
21. **Cassidy, R. M. and Elchuk, S.,** Dynamically coated columns for the separation of metal ions and anions by ion chromatography, *Anal. Chem.,* 53, 1227, 1981.
22. **Mackey, D. J.,** The adsorption of simple trace metal cations on Amberlite XAD-1 and XAD-2. A study using a multichannel nondispersive atomic fluorescence detector with quantitation by batch measurements, *J. Chromatogr.,* 236, 8, 1982.
23. **Mackey, D. J.,** Amberlite XAD-2 and XAD-4 as cation exchange resins of low capacity. An investigation using an atomic fluorescence detector directly coupled to a liquid chromatograph, *J. Chromatogr.,* 237, 79, 1982.
24. **Bank, P. C., De Loos-Vollebregt, M. T. C., and De Galen, L.,** Thermospray sample introduction in atomic absorption spectrometry, Paper, Colloq. Spectroscopicum Int., Toronto, 1987.
25. **Gast, C. H. and Kraak, J. C.,** Phase systems and post-column dithizone reaction detection for the analysis of organomercurials by HPLC, *Int. J. Environ. Anal. Chem.,* 6, 297 1979.

26. **Cassidy, R. M., Elchuk, S., and Dasgupta, P. K.,** Performance of annular membrane and screen-tee reactors for postcolumn-reaction detection of metal ions separated by liquid chromatography, *Anal. Chem.*, 59, 85, 1987.
27. **Davis, J. C. and Peterson, D. P.,** Hollow fiber postcolumn reactor for liquid chromatography, *Anal. Chem.*, 57, 768, 1985.
28. **Lillig, B. and Engelhardt, H.,** Fundamentals of reaction systems, in *Reaction Detection in Liquid Chromatography,* Vol. 34, Chromatographic Science Ser., Krull, I. S., Ed., Marcel Dekker, New York, 1986, 1.
29. **Schlabach, T. D. and Weinberger, R.,** Solution chemistry for post-column reaction, in *Reaction Detection in Liquid Chromatography,* Vol. 34, Chromatographic Science Ser., Krull, I. S., Ed., Marcel Dekker, New York, 1986, 63.
30. **Foley, R. C. L. and Haddad, P. R.,** Conductivity and indirect u.v. absorption detection of inorganic cations in non-suppressed ion chromatography using aromatic bases as effluents. I. Principles of operation, *J. Chromatogr.*, 366, 13, 1986.
31. **McAleese, D. L.,** Indirect photometric chromatography of cations and amines in a polymer-based column, *Anal. Chem.*, 59, 541, 1987.
32. **Sherman, J. H. and Danielson, N. D.,** Indirect cationic chromatography with fluorometric detection, *Anal. Chem.*, 59, 1413, 1987.
33. **Iverson, D. G., Anderson, M. A., Holm, T. R., and Stanforth, R. R.,** An evaluation of column chromatography and flameless atomic absorption spectrophotometry for arsenic speciation as applied to aquatic systems, *Environ. Sci. Technol.*, 13, 1491, 1979.
34. **Bushee, D. S., Krull, I. S., Demko, P. R., and Smith, S. B.,** Trace analysis and speciation for arsenic anions by HPLC-hydride generation-inductively coupled plasma emission spectroscopy, *J. Liq. Chromatogr.*, 7, 861, 1984.
35. **Spall, W. D., Lynn, J. G., Andersen, J. L., Valdez, J. G., and Gurley, L. R.,** High-performance liquid chromatographic separation of biologically important arsenic species utilizing on-line inductively coupled argon plasma atomic emission spectrometric detection, *Anal. Chem.*, 58, 1340, 1986.
36. **Donard, O. F. X., Rapsomanikis, S., and Weber, J. H.,** Speciation of inorganic tin and alkyltin compounds by atomic absorption spectrometry using electrothermal quartz furnace after hydride generation, *Anal. Chem.*, 58, 772, 1986.
37. **Tye, C. T., Haswell, S. J., O'Neill, P., and Bancroft, K. C. C.,** High-performance liquid chromatography with hydride generation/atomic absorption spectrometry for the determination of arsenic species with application to some water samples, *Anal. Chim. Acta,* 169, 195, 1985.
38. **Ebdon, L., Hill, S. J., and Jones, P.,** Speciation of tin in natural waters using coupled high-performance liquid chromatography-flame atomic-absorption spectrometry, *Analyst,* 110, 515, 1985.
39. **Slavin, W. and Schmidt, G. J.,** Atomic absorption detection for liquid chromatography using metal labeling, *J. Chromatogr. Sci.*, 17, 610, 1979.
40. **Yoza, N. and Ohashi, S.,** The application of atomic absorption method as a flow detector to gel chromatography, *Anal. Lett.*, 6, 595, 1973.
41. **Burns, D. T., Glocking, F., and Harriott, M.,** Investigation of the determination of tin tetraalkyls and alkyltin chlorides by atomic-absorption spectrometry after separation by gas-liquid or high-performance liquid-liquid chromatography, *Analyst,* 106, 921, 1981.
42. **Jones, D. R., Tung, H. C., and Manahan, S. E.,** Mobile phase effects in atomic absorption detectors for high speed liquid chromatography, *Anal. Chem.*, 48, 7, 1976.
43. **Koropchak, J. A. and Coleman, E. N.,** Investigations of nebulizer parameters for on-line flame atomic absorption detection of liquid chromatographic effluents, *Anal. Chem.*, 52, 1252, 1980.
44. **Brinckman, F. E., Blair, W. R., Jewett, H. L., and Iverson, W. P.,** Application of a liquid chromatograph coupled with a flameless atomic absorption detector for speciation of trace organometallic compounds, *J. Chromatogr. Sci.*, 15, 493, 1977.
45. **Vickrey, T. M., Howell, H. E., and Paradise, M. T.,** Liquid chromatogram peak storage and analysis by atomic absorption spectrometry, *Anal. Chem.*, 51, 1880, 1983.
46. **Koizumi, H., McLaughlin, R. D., and Hadeishi, T.,** High gas temperature furnace for species determination of organometallic compounds with a high pressure liquid chromatograph and a Zeeman atomic absorption spectrometer, *Anal. Chem.*, 51, 387, 1979.
47. **Hill, S., Ebdon, L., and Jones, P.,** Novel approaches to directly coupled high-performance liquid chromatography-flame atomic absorption spectrometry for trace metal speciation, *Anal. Proc.*, 23, 6, 1986.
48. **Ebdon, L., Hill, S., and Jones, P.,** Interface system for directly coupled high-performance liquid chromatography-flame atomic absorption spectrometry, *J. Anal. At. Spectrom.*, 2, 205, 1987.
49. **Gunn, A. W.,** An automated hydride generation atomic absorption spectrometric method for the determination of total arsenic in raw and potable waters, WRC Tech. Rep. TR191, 1983.

50. **Rodin, D. R. and Tallman, D. E.,** Determination of inorganic selenium species in groundwater containing organic interferences by ion chromatography and hydride generation/atomic absorption spectrometry, *Anal. Chem.,* 59, 307, 1982.
51. **Low, G. K.-C., Batley, G. E., and Buchanan, S. J.,** Interference of chloride in the speciation of arsenic by ion chromatography, *Chromatographia,* 22, 292, 1987.
52. **Batley, G. E., Low, G. K.-C. and Smith, W. J.,** Unpublished results, 1986.
53. **Gast, C. H., Kraak, J. C., Poppe, H., and Maessen, F. J. M. J.,** Capabilities of on-line element-specific detection in high-performance liquid chromatography using an inductively coupled argon plasma emission source detector, *J. Chromatogr.,* 185, 549, 1979.
54. **Krull, I. S. and Jordan, S.,** Interfacing GL and HPLC with plasma emission spectrometry, *Am. Lab.,* 21, 1980.
55. **Boorn, A. W. and Browner, R. F.,** Effects of organic solvents in inductively coupled plasma atomic emission spectrometry, *Anal. Chem.,* 57, 140, 1982.
56. **Uden, P. C., Quimby, B. D., Barnes, R. M., and Elliott, W. G.,** Interfaced d.c. argon-plasma emission spectroscopic detection for high-pressure liquid chromatography of metal compounds, *Anal. Chim. Acta,* 101, 99, 1978.
57. **Whaley, B. S., Snable, K. R., and Browner, R. F.,** Spray chamber placement and mobile phase flow rate effects in liquid chromatography/inductively coupled plasma atomic emission spectrometry, *Anal. Chem.,* 54, 162, 1982.
58. **Thompson, J. J. and Hook, R. S.,** Inductively coupled plasma mass spectrometric detection for multielement flow injection analysis and elemental speciation by reversed-phase liquid chromatography, *Anal. Chem.,* 58, 2541, 1986.
59. **Urasa, I. T. and Ferede, F.,** Use of direct current plasma as an element selective detection for simultaneous ion chromatography of arsenic(III) and arsenic(IV) in the presence of other common anions, *Anal. Chem.,* 59, 1563, 1987.
60. **Lawrence, K. E., Rice, G. W., and Fassel, V. A.,** Direct liquid sample introduction for flow injection analysis and liquid chromatography with inductively coupled argon plasma spectrometric detection, *Anal. Chem.,* 56, 289, 1984.
61. **LaFreniere, K. E., Rice, G. W., and Fassel, V.A.,** Flow injection analysis with inductively coupled plasma-atomic emission spectroscopy: critical comparison of conventional pneumatic, ultrasonic and direct injection nebulization, *Spectrochim. Acta.,* 400, 1495, 1985.
62. **LaFreniere, K. E., Fassel, V. A., and Eckels, D. E.,** Elemental speciation via high performance liquid chromatography combined with inductively coupled plasma atomic emission spectroscopic detection: application of a direct injection nebulizer, *Anal. Chem.,* 59, 879, 1987.
63. **Olsiek, J. W. and Olsiek, S. V.,** Supercritical fluid-based sample introduction for inductively coupled plasma atomic spectrometry, *Anal. Chem.,* 59, 796, 1987.
64. **Peaden, P. A. and Lee, M. L.,** Theoretical treatment of resolving power in open tubular column supercritical fluid chromatography, *J. Liq. Chromatogr.,* 5, 179, 1982.
65. **Demers, D. R. and Allemend, C. D.,** Atomic fluorescence spectrometry with an inductively coupled plasma as atomization cell and pulsed hollow cathode lamps for excitation, *Anal. Chem.,* 53, 1915, 1981.
66. **Stulik, K. and Pacukova, V.,** Electrochemical detection in high-performance liquid chromatography, *CRC Crit. Rev. Anal. Chem.,* 14, 297, 1985.
67. **White, P. C.,** Recent developments in detection techniques for high-performance liquid chromatography. I. Spectroscopic and electrochemical detection. A review, *Analyst,* 109, 677, 1984.
68. **Krstulovic, A. M., Colin, H., and Guiochon, G. A.,** Electrochemical detectors for liquid chromatography, *Adv. Chromatogr.,* 24, 83, 1985.
69. **Lewis, J. Y., Zodda, J. P., Deutsch, E., and Heineman, W. R.,** Determination of pertechnetate by liquid chromatography with reductive electrochemical detection, *Anal. Chem.,* 55, 708, 1983.
70. **Bratin, K. and Kissinger, P. T.,** Glassy-carbon amperometric transducers as electrochemical detectors in liquid chromatography. The influence of oxygen, *Talanta,* 29, 365, 1982.
71. **MacCrehan, W. A., Durst, R. A., and Bellama, J. M.,** Electrochemical detection in liquid chromatography: application to organometallic speciation, *Anal. Lett.,* 10, 1175, 1977.
72. **Roston, D. A., Shoup, R. E. and Kissinger, P. T.,** Liquid chromatography/electrochemistry: thin-layer multiple electrode detection, *Anal. Chem.,* 57, 1417A, 1982.
73. **MacCrehan, W. A. and Durst, R. A.,** Dual-electrode, liquid chromatographic detector for the determination of analytes with high redox potentials, *Anal. Chem.,* 53, 1700, 1981.
74. **MacCrehan, W. A.,** Differential pulse detection in liquid chromatography and its application to the determination of organometal cations, *Anal. Chem.,* 53, 74, 1981.
75. **Ferguson, J. F. and Gavis, J.,** A review of the arsenic cycle in natural waters, *Water Res.,* 6, 1259, 1972.
76. **McBride, B. C. and Wolfe, R. S.,** Biosynthesis of dimethylarsine by methanobacterium, *Biochemistry,* 10, 4312, 1971.

77. **Andreae, M. O.,** Determination of arsenic species in natural waters, *Anal. Chem.,* 49, 820, 1977.
78. **Andreae, M. O.,** Distribution and speciation of arsenic in natural waters and some marine algae, *Deep-Sea Res.,* 25, 391, 1978.
79. **Benson, A. A., Cooney, R. V., and Herrera-Lasso, J. M.,** Arsenic metabolism in algae and higher plants, *J. Plant Nutr.,* 3, 285, 1981.
80. **Francesconi, K. A., Micks, P., Stockton, R. A., and Irgolic, K. J.,** Quantitative determination of arsenobetaine, the major water-soluble arsenical in three species of crab, using high pressure liquid chromatography and an inductively coupled argon plasma emission spectrometer as the arsenic-specific detector, *Chemosphere,* 14, 1443, 1985.
81. **Norin, H., Ryhage, R., Christakopoulas, A., and Sandstrom, M.,** New evidence for the presence of arsenocholine in shrimps *(Panadales borealis)* by use of pyrolysis gas chromatography, — atomic absorption spectrometry/mass spectrometry, *Chemosphere,* 12, 299, 1983.
82. **Luten, J. B., Riekwel-Booy, G., and Rauchbaar, A. B.,** Occurence of arsenic in plaice *(Pleuronectes platessa),* nature of organo-arsenic compound present and its excretion by man, *Environ. Health Perspect.,* 45, 165, 1982.
83. **Crecelius, E. A.,** Changes in the chemical speciation, arsenic following ingestion by man, *Environ. Health Perspect.,* 19, 149, 1977.
84. **Foa, V., Colombi, A., Maroni, M., Buratti, M., and Calzaferri, G.,** The speciation of the chemical forms of arsenic in the biological monitoring of exposure to inorganic arsenic, *Sci. Total Environ.,* 34, 241, 1984.
85. **Marafante, E., Vahter, M., and Dencker, L.,** Metabolism of arsenocholine in mice, rats and rabbits, *Sci. Total Environ.,* 34, 220, 1984.
86. **Tarrant, R. F.,** Arsenic levels in urine of forest workers applying silvicides, *Arch. Environ. Health,* 24, 277, 1972.
87. **Fish, R. H., Brinckman, F. C., and Jewett, K. L.,** Fingerprinting inorganic arsenic and organoarsenic compounds in in situ oil shale retort and process waters using a liquid chromatograph coupled with an atomic absorption spectrometer as detector, *Environ. Sci. Technol.,* 16, 174, 1982.
88. **Henry, F. J. and Thorpe, T. M.,** Determination of arsenic(III), arsenic(V), monomethylarsonate, and dimethylarsinite by differential pulse polarography after separation by ion exchange chromatography, *Anal. Chem.,* 52, 80, 1980.
89. **Yamamoto, M.,** Determination of arsenate, methanearsonate and dimethylarsenite in water and sediment extracts, *Soil Sci. Soc. Am. Proc.,* 39, 859, 1975.
90. **Pacey, G. E. and Ford, J. A.,** Arsenic speciation by ion-exchange separation and graphite-furnace atomic-absorption spectrophotometry, *Talanta,* 28, 935, 1981.
91. **Grabinski, A. A.,** Determination of arsenic(III), arsenic(V), monomethylarsonate, and dimethylarsinate by ion-exchange chromatography with flameless atomic absorption spectrometric detection, *Anal. Chem.,* 53, 966, 1981.
92. **Dietz, E. A. and Perez, M. E.,** Purification and analysis methods for methylarsonic acid and hydroxydimethylarsine oxide, *Anal. Chem.,* 48, 1088, 1976.
93. **Woolsen, E. A. and Aharonson, N.,** Separation and detection of arsenical pesticide residues and some of their metabolites by high pressure liquid chromatography-graphite atomic absorption spectrometry, *J. Assoc. Off. Anal. Chem.,* 63, 523, 1980.
94. **Morita, M., Uehiro, T., and Fuwa, K.,** Speciation and elemental analysis of mixtures by high performance liquid chromatography with inductively coupled argon plasma emission spectrometric detection, *Anal. Chem.,* 52, 349, 1980.
95. **Ricci, G. R., Shephard, L. S., Colovos, G., and Hesler, N. E.,** Ion chromatography with atomic absorption spectrometric detection for determination of organic and inorganic arsenic species, *Anal. Chem.,* 53, 610, 1981.
96. **Pyles, R. A. and Woolsen, E. A.,** Quantitation and characterization of arsenic compounds in vegetables grown in arsenic acid treated soil, *J. Agric. Food Chem.,* 30, 866, 1982.
97. **Tan, L. K. and Dutrizac, J. E.,** Simultaneous determination of arsenic(III) and arsenic(V) in metallurgical processing media by ion chromatography with electrochemical and conductivity detectors, *Anal. Chem.,* 58, 1383, 1986.
98. **Spall, W. D., Lynn, J. G., Andersen, J. L., Valdez, J. E., and Gurley, L. R.,** High-performance liquid chromatographic separation of biologically important arsenic species utilizing on-line inductively coupled argon plasma atomic emission spectrometric detection, *Anal. Chem.,* 88, 1340, 1986.
99. **Morita, M., Uehiro, T., and Fuwa, K.,** Determination of arsenic compounds in biological samples by liquid chromatography with inductively coupled argon plasma-atomic emission spectrometric detection, *Anal. Chem.,* 53, 1806, 1981.
100. **Stockton, R. A. and Irgolic, K. J.,** The Hitachi graphite furnace-Zeeman atomic absorption spectrometer as an automated, element-specific detector for high pressure liquid chromatography: The separation of arsenobetaine, arsenocholine and arsenite/arsenate, *Int. J. Environ. Anal. Chem.,* 6, 313, 1979.

101. **Nisamaneepong, W., Ibrahim, M., Gilbert, T. W., and Caruso, J. A.,** Speciation of arsenic and cadmium compounds by reversed-phase ion-pair LC with single-wavelength inductively coupled plasma detection, *J. Chromatogr. Sci.*, 22, 473, 1984.
102. **McCarthy, J. P., Caruso, J. A., and Fricke, F. L.,** Speciation of arsenic and selenium via anion-exchange HPLC with sequential plasma emission detection, *J. Chromatogr. Sci.*, 21, 389, 1983.
103. **Lee, D. P.,** A new anion exchange phase for ion chromatography, *J. Chromatogr. Sci.*, 22, 327, 1984.
104. **Low, G. K.-C., Batley, G. E., and Buchanan, S. J.,** Application of column-switching in HPLC to arsenic speciation analysis with inductively-coupled argon plasma spectrometric detection, *J. Chromatogr.*, 368, 423, 1986.
105. **Low, G. K.-C., Batley, G. E., and Buchanan, S. J.,** An interference effect in the use of inductively-coupled argon plasma spectrometric detection for high performance liquid chromatography, *Anal. Chim. Acta,* in press.
106. **Lown, J. A., Koile, R., and Johnson, D. C.,** Amperometric flow-through wire detector: a practical design with high sensitivity, *Anal. Chim. Acta,* 116, 33, 1980.
107. **Lown, J. A. and Johnson, D. C.,** Anodic detection of arsenic(III) in a flow-through platinum electrode for flow-injection analysis, *Anal. Chim. Acta,* 116, 41, 1980.
108. **Cutter, G. A. and Bruland, K. W.,** The marine biogeochemistry of selenium: a reevaluation, *Limnol. Oceanogr.*, 29, 1179, 1984.
109. **Sugimura, Y., Suzuki, Y., and Miyake, Y.,** The content of selenium and its chemical form in seawater, *J. Oceanogr. Soc. Jpn.*, 32, 235, 1977.
110. **Measures, C. I. and Burton, J. D.,** The vertical distribution and oxidation states of dissolved selenium in the northeast Atlantic ocean and their relationship to biological processes, *Earth Planet. Sci. Lett.*, 46, 315, 1980.
111. **Cutter, G. A.,** Determination of selenium speciation in biogenic particles and sediments, *Anal. Chem.*, 57, 295, 1985.
112. **Ridley, W. P., Dizikes, L. J., and Wood, J. M.,** Biomethylation of toxic elements in the environment, *Science,* 197, 329, 1977.
113. **Chau, Y. K., Wong, P. T. S., and Goulding, P. D.,** Gas-chromatography — atomic absorption method for the determination of dimethylselenide and dimethyldiselenide, *Anal. Chem.*, 47, 2279, 1975.
114. **Fleming, R. W. and Alexander, M.,** Dimethylselenide and dimethyltelluride formation by a strain of *Penicillium, Appl. Microbiol.*, 24, 424, 1972.
115. **Shrift, A. and Virupaksha, T. K.,** Seleno-amino acids in selenium-accumulating plants, *Biochim. Biophys. Acta,* 100, 65, 1965.
116. **Byard, J. L.,** Trimethylselenide a urinary metabolite of selenite, *Arch. Biochem. Biophys.*, 130, 556, 1969.
117. **Oyamada, N. and Ishizaki, M.,** Determination of trimethylselenonium ions in human urine by graphite furnace atomic absorption spectrometry, *Jpn. J. Ind. Health,* 24, 320, 1982.
118. **Chakraborti, D., Hillman, D. C. J., Irgolic, K. J., and Zingaro, R. A.,** Hitachi Zeeman graphite furnace atomic absorption spectrometer as a selenium-specific detector for ion chromatography. Separation and determination of selenite and selenate, *J. Chromatogr.*, 249, 81, 1982.
119. **Batley, G. E.,** Unpublished results, 1986.
120. **Fodor, P. and Barnes, R. M.,** Determination of some hydride-forming elements in urine by resin complexation and inductively coupled plasma atomic spectroscopy, *Spectrochim. Acta.*, 38B, 229, 1983.
121. **Batley, G. E.,** Differential pulse polarographic determination of selenium species in contaminated waters, *Anal. Chim. Acta,* 187, 109, 1986.
122. **Brinckman, F. E., Jackson, J. A., Blair, W. R., Olson, G. J., and Iverson, W. P.,** Ultratrace speciation and biogenesis of methyltin transport species in estuarine waters, in *Trace Metals in Sea Water,* Wong, C. S., Boyle, E., Bruland, K. W., Burton, J. D., and Golberg, E. D., Eds., Plenum Press, New York, 1984, 31.
123. **Blunden, S. J. and Chapman, A. H.,** The environmental degradation of organotin compounds — a review, *Environ. Technol. Lett.*, 3, 267, 1982.
124. **Seligman, R. F., Valkirs, A. O., and Lee, R. F.,** Degradation of tributyltin in San Diego Bay, California, waters, *Environ. Sci. Technol.*, 10, 1229, 1986.
125. **Vickrey, T. M., Howell, H. E., Harrison, G. V., and Ramelow, G. J.,** Post column digestion methods for liquid chromatography-graphite furnace atomic absorption speciation of organolead and organotin compounds, *Anal. Chem.*, 52, 1743, 1980.
126. **Burns, D. T., Glockling, F., and Harriott, M.,** Comparative assessment of gas-liquid chromatography and high performance liquid chromatography for the separation of tin tetraalkyls and alkyltin halides, *J. Chromatogr.*, 200, 305, 1980.
127. **Krull, I. S. and Panaro, K. W.,** Trace analysis and speciation for methylated organotins by HPLC-hydride generation-direct current plasma emission spectroscopy (HPLC-HY-DCP), *Appl. Spectrosc.*, 39, 960, 1985.

128. **Jewett, K. L. and Brinckman, F. E.,** Speciation of trace di- and triorganotins in water by ion-exchange HPLC-GFAA, *J. Chromatogr. Sci.,* 19, 583, 1981.
129. **Payne, E. and Batley, G. E.,** Unpublished results, 1987.
130. **Haseba, K., Yamamoto, Y., and Kambara, T.,** Differential pulse-polargraphic determination of organotin compounds coated on fishing nets, *Fresenius Z. Anal. Chem.,* 310, 634, 1982.
131. **Langseth, W.,** Determination of diphenyltin and dialkyltin homologues by HPLC with morin in the effluent, *Talanta,* 31, 975, 1984.
132. **Messman, J. D. and Rains, T. C.,** Determination of tetraalkyllead compounds in gasoline by liquid chromatography-atomic absorption spectrometry, *Anal. Chem.,* 53, 1632, 1981.
133. **Chau, Y. K.,** Occurrence and speciation of organometallic compounds in freshwater systems, *Sci. Total Environ.,* 49, 305, 1986.
134. **Botre, C., Cacace, F., and Cozzani, R.,** Direct combination of high pressure liquid chromatography and atomic absorption for the analysis of metallorganic compounds, *Anal. Lett.,* 9, 825, 1976.
135. **Ibrahim, M., Gilbert, T. W., and Caruso, J. A.,** Determination of tetraalkyllead by high performance liquid chromatography with ICP detection, *J. Chromatogr. Sci.,* 22, 111, 1984.
136. **Hodges, D. J. and Noden, F. G.,** The determination of alkyl lead species in natural waters by polarographic techniques, in *Heavy Metals in the Environment,* Proc. Int. Conf. CEP Consultants, Perry, R., Ed., CEP Consultants, Edinburgh, 1979, 408.
137. **Langseth, W.,** Determination of organic and inorganic mercury compounds by reverse phase high-performance liquid chromatography after extraction of the compounds as their dithizonates, *Anal. Chim. Acta,* 185, 249, 1986.
138. **Evans, O. and McKee, G. D.,** Optimisation of high performance liquid chromatographic separations with reductive amperometric electrochemical detection: speciation of inorganic and organomercury, *Analyst,* 112, 983, 1987.
139. **Funasaka, W., Hanai, T., and Fujimura, K.,** High speed liquid chromatographic separation of phthalic esters, carbohydrates, TCA organic acids and organic mercury compounds, *J. Chromatogr. Sci.,* 12, 517, 1974.
140. **Krull, I. S., Bushee, D., Savage, R. N., Schleicher, R. G., and Smith, S. B.,** Speciation of Cr(III) and Cr(VI) via reversed phase HPLC with inductively coupled plasma emission spectroscopic detection (HPLC-ICP), *Anal. Lett.,* 15, 267, 1982.
141. **Koizumi, H., Hadeishi, T., and McLaughlin, R.,** Speciation of organometallic compounds by Zeeman atomic absorption spectrometry with liquid chromatography, *Anal. Chem.,* 50, 1700, 1978.
142. **Uden, P. C., Quimby, B. D., Barnes, R. M., and Elliott, W. G.,** Interfaced D.C. argon-plasma emission spectroscopic detection for high-pressure liquid chromatography of metal compounds, *Anal. Chim. Acta,* 101, 99, 1978.
143. **Mueller, B. J. and Lovett, R. J.,** Salt-induced phase separation for the determination of metals via their diethyldithiocarbamate complexes by high performance liquid chromatography, *Anal. Chem.,* 59, 1405, 1987.
144. **Schwedt, G.,** Application of high pressure liquid chromatography in inorganic analysis. IV. Determination of chromium(III) and chromium(VI) ions in waste water as dithiocarbamate complexes, *Fresenius Z. Anal. Chem.,* 295, 282, 1979.
145. **Hoffmann, B. W. and Schwedt, G.,** Application of HPLC to inorganic analysis. VII. Comparison between pre-column and on-column derivitization and separation of different metal oxinates; quantitative determination of manganese(III) besides manganese(III) ions, *J. High Resolut. Chromatogr. Commun.,* 5, 439, 1982.
146. **Sunden, T., Lindgren, M., Cedergren, A., and Siemer, D. D.,** Separation of sulfite, sulfate, and thiosulfate by ion chromatography with gradient elution, *Anal. Chem.,* 55, 2, 1983.

Chapter 7

APPLICATIONS OF GAS CHROMATOGRAPHY TO TRACE ELEMENT SPECIATION

Y. K. Chau and P. T. S. Wong

TABLE OF CONTENTS

I. GENERAL INTRODUCTION

As many scientific investigations are now conducted at the molecular level, ultrasensitivity in the associated chemical analysis as well as the knowledge of the chemical forms of the molecule are required. Such information is important and often vital in toxicological, environmental, and geochemical studies. As a result of these challenges, a new generation of analytical chemistry has evolved, employing the tandem combination of two or even three techniques, to form a new analytical system.

Chromatographic techniques using either the gas or liquid mode are a useful means of separating mixtures of metal species, with measurement of ultratrace concentrations achievable using a range of element-specific detection systems. Gas chromatography (GC) has now been applied to the analysis of a wide range of metal and organometallic species; however, since only volatile species can be separated, it is necessary in many instances to derivatize the metal species to volatile forms prior to GC separation.

The separation of volatile chelates of inorganic trace metals has been the subject of extensive research.[1,2] In particular, metal β-diketonates are readily separable by gas chromatography. This type of derivatization, while separating different elements, is of little value in speciation research, except for the separation of different valency states, and will therefore not be discussed further in this chapter.

In speciation studies, both hydridization[3,4] and alkylation[5] are commonly used to convert both inorganic and organometallic species to volatile products. These can then be recovered by purge and trap methods for separation by selective volatilization into a suitable detection system, such as an atomic absorption spectrometer.[5-9] Alternatively, solvent extraction[10-12] or solid absorbants[12] can be used to preconcentrate these derivatives prior to a conventional GC separation and detection. This chapter will discuss the methodology that has been developed and applied to the GC separation of trace element species.

II. INSTRUMENTATION

A. Introduction

A range of sophisticated chromatographic instrumentation is now commercially available. Improvements over recent years in column design and in detector sensitivities and selectivities, coupled with good software for data analysis, has enabled the attainment of detection limits in the picogram range for a range of organometals and metal complexes. Despite these advances, good sensitivities can also be obtained using a basic system of a column and oven, coupled with an atomic absorption spectrometer as an element-specific detector.[5] The oven temperature should ideally be programmable to enable the reproducible separation of more complex mixtures having a wide range of boiling points, although good results have been achieved using controlled heating to volatilize trapped hydride or alkylated species.[3,13]

B. GC Columns

The earliest approaches to the GC separation of organometallic or inorganic species involved trapping them as their hydride derivatives on a suitable packed column immersed in liquid nitrogen.[14,15] The trapped compounds were then volatilized and separated by controlled thermal desorption. Using packing materials such as glass beads or glass wool,[2,15-17] poor separations generally resulted, with badly tailing peaks.

In later studies, improved separations were obtained using more conventional GC packings contained in short glass U-tube columns (typically 30 cm × 6 mm I.D.). Nonpolar or low-polarity inert stationary phases, such as silicone oils (OV-1, OV-3, or OV-17) on inert supports such as Chromosorb® W, were successfully applied to the separation of derivatized species of arsenic, antimony, and tin.[2,18,19] Sharp, well-separated peaks were obtained by controlled desorption, using a heating coil around the outside of the U-tube trap. Such a system has been widely used in conjunction with atomic absorption spectrometric detection.

In addition to the above homemade chromatography systems, commercial gas chromatographs have found increasing application in recent years, especially using electron capture (EC) or flame photometric (FP) detection. Separation of complex mixtures can be optimized by programming of the column temperature, a simple process with the microprocessor-controlled instrumentation of today.

The modern gas chromatographer has the choice between traditional packed columns and capillary or megabore columns, and all three types have been used for speciation studies. The superior resolution achievable by capillary and megabore columns can be an advantage, particularly where the detection system lacks selectivity, e.g., with EC detection. Advocates of packed columns prefer their ease of handling and convenience in assembly. Furthermore, in some environmental samples, oils, fats, protein debris, and other organic residues present can deposit on the column, degrading performance. Packed columns, however, can be more easily cleaned up by replacing part of the packing material or even repacking the entire column.

Table 1 lists a selection of GC columns, detectors, and derivatization methods that have been applied to the analysis of organometallic and inorganic species of Sn, Pb, Se, As, Hg, Ge, and Sb.

C. Detection Systems

1. General

For the detection of picogram concentrations of organometals in environmental samples, detection systems with inherently greater sensitivity and selectivity than the traditional first-generation thermal conductivity or flame ionization GC detection are required. Research, particularly in the last decade, has led to the development of a range of highly selective detectors based on atomic spectrometric detection, together with the application of electron capture and mass spectrometric detection.

Atomic spectrometry is a sensitive highly element-selective technique, applicable to a wide range of metals and nonmetals. When used in conjunction with a gas chromatographic separation technique, it can detect elements in the presence of complex organic and biological matrices, without the necessity for cumbersome chemical separations.

Of the three basic areas of atomic spectrometry, namely, atomic absorption (AAS), atomic emission (AES), and atomic fluorescence (AFS) spectrometry, AAS has been the most widely used due to its inherent simplicity in operation and the ready availability of instruments. It is also the most selective of the three techniques because of the selective absorption by the analyte of the spectral lines generated by the same element in the spectral source. This absorption mechanism is a unique feature of AAS which further reduces the possibility of spectral interferences. Since the applications of various plasma techniques such as direct current plasma (DCP), microwave-induced plasma (MIP), and inductively coupled plasma (ICP) in atomic emission spectrometry in recent years, higher excitation temperatures, hence, higher sensitivities, have been achieved. The old technique of emission spectrometry has been resurrected in analytical use. Atomic fluorescence spectrometry (AFS), using a specific line source for fluorescence excitation, is also highly selective. Both emission and fluorescence operations offer multielement detection capabilities, although the actual multielement operation is still very much limited by the high cost of the instrumentation. AFS application is still at the developmental stage and is limited by the availability of commercial fluorescence instruments. Gas chromatography-atomic spectrometry systems have been reviewed by several authors.[6-9]

2. Electron Capture Detection

In electron capture (EC) detection, the effluent from the GC column passes through an ionization chamber subject to a constant flux of β-electrons from a ^3H or ^{63}Ni source. Molecules able to capture free electrons to form negative molecular ions can be detected by the increasing current flow for a particular applied voltage in the ionization chamber, in the presence of an inert gas such as nitrogen, helium, or argon. The EC detector is particularly sensitive to compounds containing halogen atoms or polar functional groups, i.e., having a high electron affinity. This places a limitation on the solvents that can be used for the extraction of organometals. Thus, hexane, pentane, ethyl acetate, and benzene are acceptable, whereas dichloromethane or similar halogenated solvents are not, and if used, must be removed by evaporation and the solute redissolved in a more suitable solvent. Alternatively, a split injection system can be used to permit diversion of the rapidly eluting solvent peak.

Not surprisingly, EC detection has found wide application to the speciation of organometallic compounds of tin,[20,21] lead,[34] mercury,[52-54] selenium,[61,62] and arsenic.[66] Alkyl- and aryltins have been determined as their hydrides in pentane, hexane, or ethyl acetate/ hexane. Alkylleads were phenylated after extraction as their dithizonates and the products measured in hexane.[34] Successful selenium speciation relied on the EC detection of the piazselenols formed from selenium(IV) by reaction with reagents such as 1,2-diamino-3,5-dibromobenzene.[61,62] These were extracted into toluene, with other valency states being determined after oxidation or reduction to the tetravalent form.

Andreae,[16] used the EC detector for mono-, di-, and trimethylarsines, finding detection limits of 0.4, 0.2, and 15 ng, respectively, for optimum pulse interval, gas flow rates, and temperature. The linear range of peak areas extended over two orders of magnitude. Both inorganic arsenic, and the organoarsenic species monomethylarsonic acid (MMA) and dimethylarsinic acid (DMA) have been detected after hydridization[16] or as diethyldithiocarbamates extracted respectively into toluene, benzene, and hexane.[66] Drying of the gas stream is important for GC detection, so in the hydridization method, a dry ice-cooled water trap was included before the GC column.

Possibly the earliest applications of EC detection in speciation studies were to the detection of organomercurials,[74] with the separation and detection of methylmercury salts as chlorides,[75] or of other alkyl- and arylmercurials as dithizonate derivatives.[76] The EC detector was not, however, applicable to underivatized dialkylmercury compounds which could be more effectively detected by emission spectrometric techniques.

The main limitations of EC detectors are their lack of specificity and their sensitivities to contamination and changes in operating conditions. It has been noted that, depending on the type and amount of solute, the detector response may or may not be flow sensitive, concentration sensitive, and temperature sensitive, and this should always be verified for each analytical application. During routine use it is recommended that measurements of samples and standards be performed daily. Frequent cleaning of the detector is also required because of fouling by contaminants from bleeding columns. The EC detector is strongly affected by oxygen and concentrations in the carrier gas should be controlled to below 5 ppm to avoid interferences.[77]

Problems of EC detector poisoning observed with organomercury salts[76] led Longbottom[55] to develop an inexpensive UV absorbance detector. This system required prior conversion of the organomercurials to the elemental state using a combustion furnace and also a flame. It was less sensitive than EC detection for methylmercury, but gave good results for several dialkylmercurials.

3. *Atomic Absorption Detection*

The important process in atomic absorption spectrometry is the production of ground-state atoms for absorption. Two types of atomizers are commonly used for this purpose with either a flame or an electrically heated furnace. Flame atomization is the most convenient and popular technique.

The interfacing of a flame AAS and GC was first reported by Kolb et al.[78] who used an air-acetylene flame to determine tetraalkyllead compounds in gasoline. The coupling of the two instruments is relatively simple and easily performed. The column effluent can be directly ducted through a transfer tube into the nebulizer of the burner system. The flame can be operated in the usual manner. It should be noted, however, that most spray chambers of commercial AAS instruments are coated with an inert plastic (e.g., Penton in Perkin Elmer® nebulizers), and it is necessary to install a glass lining inside the nebulizer chamber to avoid adsorption and contamination of organometallic compounds, notably for organolead and organotin analyses.[48] Sensitivity is generally low (microgram detection) with flame atomization because of the dilution of analytes by the large volume of fuel gases, resulting in a relatively short residence time of the analyte atoms in the flame. Flame AAS detection, however, has the advantage of real-time and continuous operation, giving signals in the form of chromatographic peaks as the compounds elute from the column.

The GC-flame AAS combination has since been applied to tetraalkylleads in atmospheric samples,[45,48] gasoline,[43,48] and to the sequential determination of As, Ge, Se, and Sn after hydride generation,[79] by adjusting the chromatogram programs to allow for manual changes of lamps and monochromator.

Three orders of magnitude greater sensitivities (nanogram level) can be achieved by the use of furnace devices. There are two types of furnace devices in common use today: (1) custom-made quartz or ceramic tubes, electrically heated by resistance wire; and (2) commercial graphite furnaces. In serving as a GC detector, the furnace has to be maintained at about 1000°C during operation for atomization of most of the organometallic compounds. Chau et al.[63] first interfaced an electrically heated, open-ended quartz tube similar to that designed by Chu et al.[80] for the determination of methyl derivatives of selenium compounds (Figure 1). Interfacing of the GC to the quartz furnace was achieved using a stainless steel transfer tube (2 mm O.D.) from the column outlet.[63] For compounds such as methylmercury, for example, easily decomposed on a heated steel surface, a Teflon® tube of similar dimension may be used instead.[81] Other materials, such as nickel[18] and glass,[44] have been used as transfer tubes.

A double-compartment furnace[63] was used for the precombustion of hydrocarbons or solvents which, when present in the sample, cause interferences for elements such as selenium, the effective absorption lines of which are in the ultraviolet region. The hydrogen generated a small flame jet inside the furnace, which gave improved sensitivity. The flow rate of hydrogen was not critical as long as the flame was not visible at the ends of the furnace. The flow of air introduced to the precombustion tube was to support the combustion of the contaminants.

In general, introduction of hydrogen into the furnace was found necessary to enhance the atomization of certain organometals such as alkyllead and alkyltin compounds.[5] The mechanism of enhancement has been studied and explained by the formation of metal hydrides prior to atomization.[82] The length and diameter of the furnace tube are important parameters for increasing the sensitivity for absorption. They must be empirically optimized so as to increase the residence time of the atoms in the light path. While the light path for absorption is maximized, minimum volume inside the tube must be maintained so as to contain the highest density of ground-state atoms. Also, the diameter of the furnace tube must not be so narrow that the entrance of the spectral beam is obstructed. For stabines, hydrogen combustion within the quartz tube was shown to provide an adequate atomization temperature without the need for external heating (Figure 2).[73]

Donard and co-workers,[29] experimenting with furnace design, found a 110- × 12-mm-I.D. furnace to be optimum, allowing a residence time for alkyltin hydrides of 0.3 s. Their furnace was electrically heated via a coiled, doubled strand of Nichrome® heating wire which also covered a gas premixing chamber (20 × 10 mm) inlet in the center of the furnace.

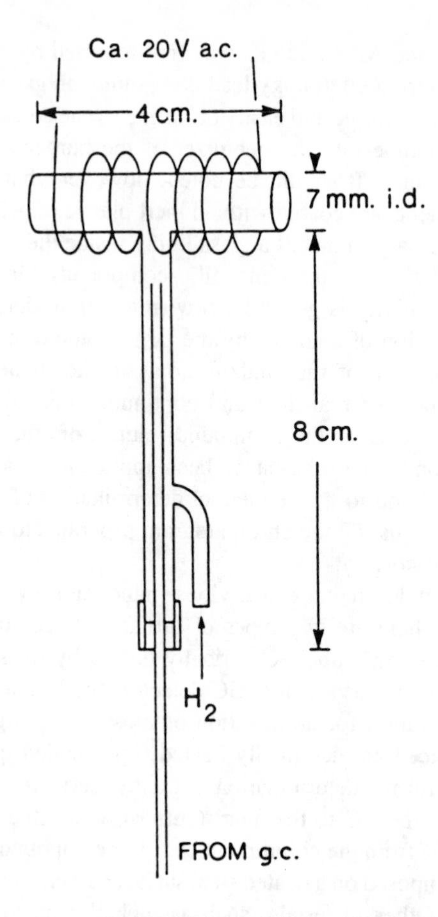

FIGURE 1. Electrically heated quartz tube furnace. (From Chau, Y. K., Wong, P. T. S., and Goulden, P. D., *Anal. Chim. Acta*, 85, 421, 1976. With permission.)

Premixing of the helium carrier gas (400 ml min^{-1}) with hydrogen (1200 ml min^{-1} and then oxygen (90 ml min^{-1}) before the AAS beam path increased its signal to noise ratio. The furnace, in operation, had a stable flame burning on either end, assisted by an inert argon gas blanket (5 l min^{-1}) supplied via the AAS burner head. The eluting hydrides were supplied to the furnace via Teflon® transfer lines (2.5 mm I.D.) inserted in 3-mm-I.D. Tygon tubing which was also electrically heated via a Nichrome® wire coil around the outside. A typical output using this furnace system is shown in Figure 3 for organotin detection after hydridization and trapping of the products in a U-tube packed with 3% SP-2100 on Chromosorb® G.

Other furnace modifications included a quartz tube[7] or a ceramic tube[83] placed over a flame in different configurations. These devices increase the residence time of the atoms in the absorption path. The optimum design of several studied by Ebdon et al.[83] achieved the lowest absolute detection yet reported of 17 pg of lead. This cell consisted of a recrystallized alumina tube, 110 mm long, 10 mm O.D., and 6.25 mm I.D., with a 4.0-mm hole (Figure 4) at right angles to the air-acetylene burner head. The GC effluent is passed along a glass-lined interface tube (0.76 mm I.D.) to a T-piece into which an auxilliary flow of hydrogen is introduced. A hydrogen diffusion flame is burned in the end of the T-piece, with the flame aligned with the center of the hole in the ceramic tube. The hydrogen flame prevents the appearance of large solvent peaks caused by uncorrected molecular absorption. The separation between the hydrogen burner and the ceramic tube (0.25 mm) was critical.

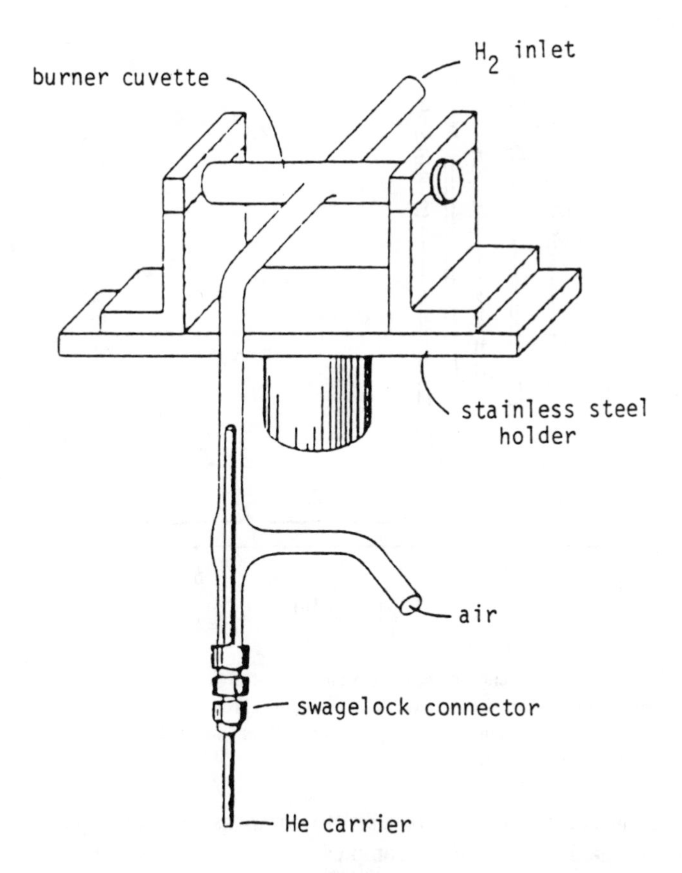

FIGURE 2. Quartz tube burner for AAS detection. (From Andreae, M. O., *Anal. Chem.*, 49, 820, 1977. With permission.)

Higher atomization temperatures (2000°C) are possible using graphite furnaces, and better sensitivity is anticipated. Commercially available graphite furnaces are operated on drying, ashing, and atomizing cycles for discrete sampling and therefore must be modified in order to be compatible with the continuous effluent from the gas chromatograph. Segar[84] first interfaced a GC to a commercial graphite furnace with a tungsten transfer line connected to an enlarged hole in the graphite furnace tube. Parris[85] investigated the effects of varying the atomization temperature, the inner surface of the furnace, and addition of hydrogen to the carrier gas for the determination of As, Se, and Sn. They achieved the best results with standard graphite tubes at an atomization temperature of 1800°C with hydrogen added to the effluent. Radziuk et al.,[49] using a tantalum adapter friction-fitted snugly into the sampling hole of the graphite tube to accommodate a stainless steel transfer tube from the GC, were able to improve the interface and to achieve better sensitivity for alkyllead compounds (Figure 5). Shaikh and Tallman,[69] by using a graphite tube to introduce the hydrides to a graphite furnace, speciated organoarsenic compounds in natural waters.

Commercial graphite furnaces have been used by several workers, taking advantage of the inert gas purge port of the device for introduction of the GC effluent.[51,67,71,85] The major disadvantage of graphite furnace operation is the necessity of maintaining the furnace temperature at ca. 2000°C during the course of chromatography, at which temperature the average life of a graphite tube is shortened to 10 to 15 h or less.[49] The use of graphite furnaces can therefore be quite costly. On the other hand, a custom-made quartz furnace tube is simple and inexpensive and can last over a period of 1 month on continuous 8-h daily use.

In all the combinations using atomic absorption spectrometry as detector for environmental

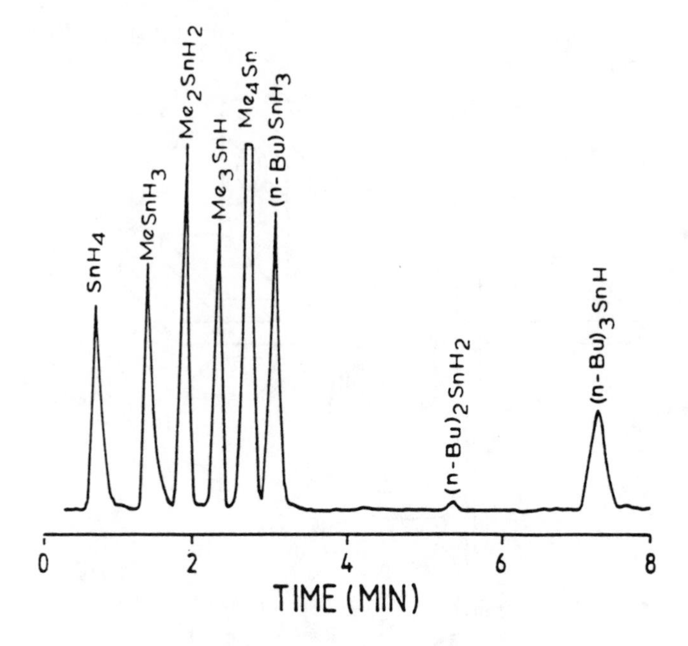

FIGURE 3. Detection of nanogram amounts of organotin species using hydridization GC-quartz furnace AAS detection. (From Donard, O. F. X., Rapsomanikis, S., and Weber, J. H., *Anal. Chem.*, 58, 772, 1986. With permission.)

samples, background correction must be used to compensate for the possible nonspecific absorbance due to organic matter in the samples.

At present, GC-AAS systems are mostly custom built and have not been commercially available. The lack of interest on the part of instrument manufacturers is most likely due to: (1) the high cost of the component instruments involved, (2) the ease of custom construction with existing or even obsolete instruments in most laboratories, and (3) a lack of confidence in marketing. Nonetheless, these systems have been successfully set up in many laboratories interested in the speciation of organometallic compounds, as shown in Table 1.

4. Atomic Emission Detection

The flame photometric (FP) detector, introduced by Grant in 1958, was the first GC detector based on atomic emission.[86] The detector uses a hydrogen-air burner with a photomultiplier located so that emission from the upper portion of the flame is measured. The use of band-pass filters enables the monitoring of selected emission wavelengths. The FP detector is now available as an optional detector for most commercial GC instruments.

The first application of this detector to organometal analysis was described by Aue and Hill,[87] for the detection of tetraethyllead, tetraethyltin, and ferrocene at 405.8, 485.0, and 373.5 nm, respectively, with the minimum detectable amounts being 40, 5, and 2 ng, respectively. Braman and Tomkins[19] found that for tetramethyltin analysis, the red fluorescence at 610 nm due to the molecular SnH emission band was more sensitive than the SnO band, at 485 nm, resulting in a detection limit near 1 ng Sn. They experimented with gas composition, carrier gas flow, and burner size. The signal response to Sn increased as the ratio of hydrogen to air within the burner increased. Flame stability, however, decreased with the hydrogen-rich mixture, and an optimum hydrogen to air ratio of 2:3 and a helium carrier gas to air ratio of 1:2 was selected as a compromise. Because of the diffuse nature of the SnH emission, best sensitivity was obtained with a small burner (8 mm outside collar diameter) (Figure 6). The detector used a photomultiplier module with a maximum response

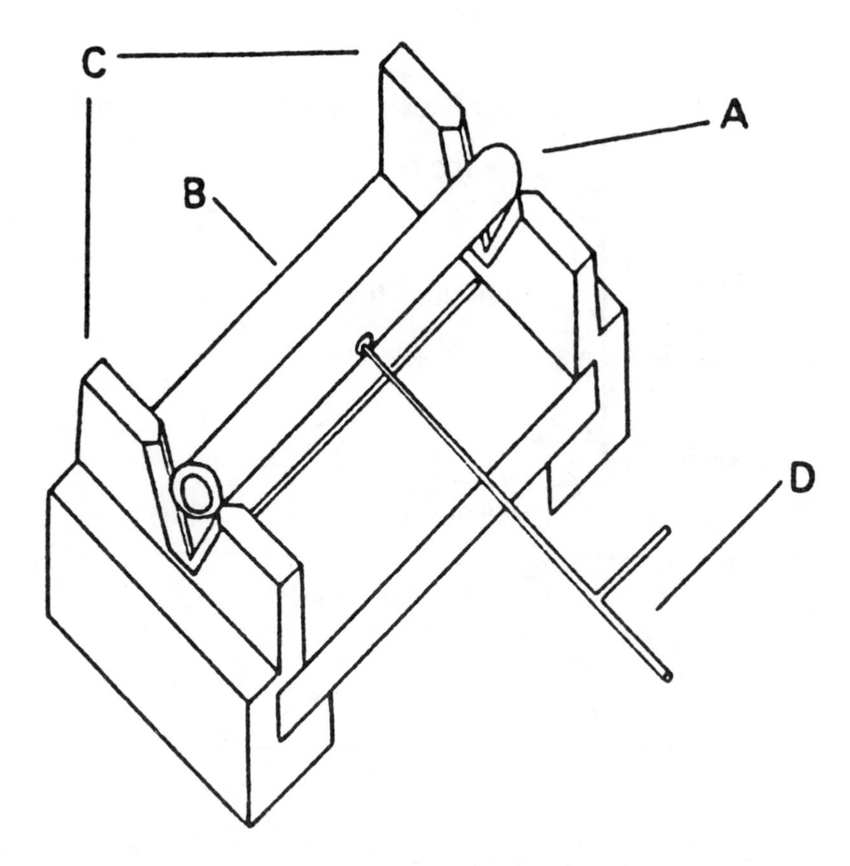

FIGURE 4. Flame-heated ceramic tube atom cell. (A) Ceramic tube; (B) air-acetylene burner lead; (C) stainless steel knife-edge support; and (D) glass-lined T-piece. (From Ebdon, L., Ward, R. W., and Leathard, D. A., *Analyst*, 107, 129, 1982. With permission.)

near 500 nm, and a 610-nm band-pass filter for wavelength selectivity. Later studies[13,23] used commercial detectors, modified to detect tin (600- to 2000-nm band-pass filter) in a hydrogen rich flame,[88] generally with gas flows of 110 ml min^{-1} H_2, 70 ml min^{-1} air, and 20 ml min^{-1} N_2 as carrier gas. An increase in sensitivity of almost two orders of magnitude was obtained in the absence of interference filters, using a dumbbell-shaped quartz tube (3 mm I.D. at its center above the flame.[88] This tube gave reproducibly good peak shapes and increased response for tin compounds, while discriminating against other FP-detectable species. The presence of silicon or phosphorus compounds resulted in depressed, tailing peaks; however, detector response could be restored by the injection of fluorine-containing compounds. Using this system, detection limits near 0.2 pg were obtained for tetraalkyltins.[88] Other applications of the FP detector, principally to tin analysis, are listed in Table 1.

In general, improvements in emission detection have resulted from improvements in the excitation techniques. Braman et al.,[14] using a simple direct current, electrical-discharge excitation emission detector coupled to a cold trap, were able to speciate inorganic arsenic and methylarsenic compounds. More versatile excitation techniques have included the microwave-induced plasma (MIP), the direct current plasma (DCP), and the inductively-coupled plasma (ICP), and these have been comprehensively reviewed elsewhere.[9,89]

The particular advantages of plasma emission detectors are their high sensitivity and selectivity, features common to most spectroscopic detectors, enabling incompletely resolved GC peaks to be tolerated. With the exception of the EC detector, plasma emission detectors offer considerably greater sensitivity than most conventional GC detectors. Generally higher

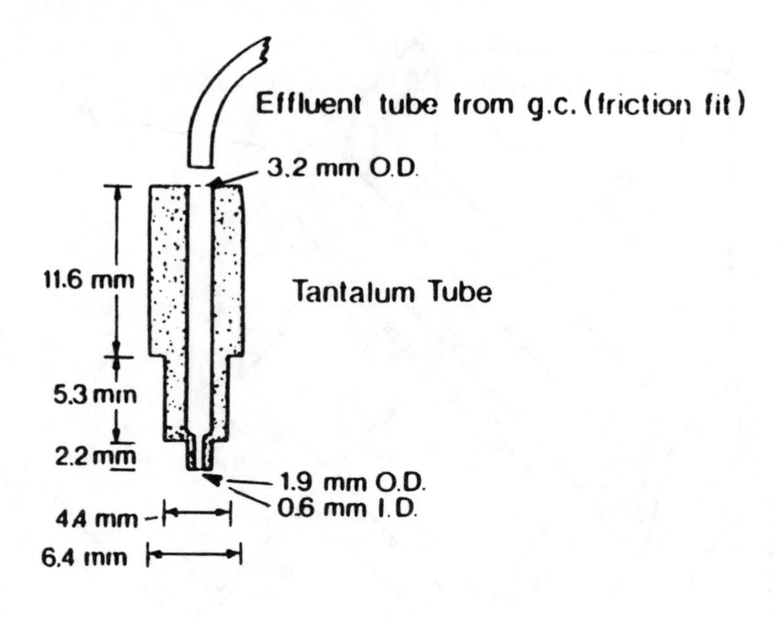

A

6.4 mm

Graphite Tube

HGA – 2100

B

FIGURE 5. Tantalum tube connector (A) for interfacing a GC column with a graphite furnace (B). (From Radziuk, B., Thomassen, Y., Van Loon, J. C., and Chau, Y. K., *Anal. Chim. Acta,* 105, 255, 1979. With permission.)

temperatures can be achieved with plasma techniques, resulting potentially in more efficient atomization.

By far the most widely used plasma detector for speciation studies is the MIP. It was first used as a GC detector for the analysis of halogens and sulfur in organic compounds.[90] Selective determination of phosphorus and iodine in pesticide analyses were subsequently achieved by using an atmospheric argon plasma or a helium plasma.[91] Mixed argon-helium plasmas were also used in the determination of halogens and phosphorus[92] and later applied to the determination of sulfur in various compounds.[93] MIP has since been used for the detection of metal chelates of Cr, Ga, Fe, Sc, V, Cu, and Al,[94-96] as well as for alkylarsenic acids,[68] tetraalkyllead,[50] and the hydride forming elements, As, Ge, Se, Sb, and Sn.[97] Talmi

Table 1
GC METHODS FOR METAL SPECIATION

Element	Species	Derivatization method	Column	Detector[a]	Ref.
Sn	Bu_3Sn^+, Bu_2Sn^{2+}	Hydridization	2.5 m × 2 mm, 2% OV-17 or 10% OV-1 on Chromosorb® W	EC	10
	Ph_4Sn, Ph_2Sn^{2+}, Sn(IV)	Hydridization	0.7 m × 2 mm, 30% OV-17 on Gas-Chrom® Q	EC	20
	Bu_3Sn^+, Bu_2Sn^{2+}, Pr_3Sn^+, Ph_3Sn^+	Hydridization	2 m × 2 mm, 2% OV-17 on Gas-Chrom® Q	EC	21
	Methyltins	Hydridization	1.8 m × 2 mm, 3% SP-2401 and 10% SP-2100 on Supelcoport	FP	22
	Bu_3Sn^+	Methylation	20 m SE-54 (capillary)	FP	12
	Bu_3Sn^+, Bu_2Sn^{2+}, $BuSn^{3+}$	Methylation	1.2 m × 2 mm, 3% OV-17 on Supelcoport or 3% OV-1 on Chromosorb® W (HP)	FP	11
	Butyltins, phenyltins, cyclohexyltins, Sn(IV)	Ethylation	30 m Pluronic® L 64 (capillary)	FP	23
	Butyltins	Hydridization	1.8 m × 2 mm, 1.5% OV-101 on Chromosorb® G (HP)	FP	4,18
	Bu_4Sn, Pr_4Sn, Et_4Sn	—	0.15 m × 0.2 cm, 3% OV-101 on Chromosorb® W	FP	24
	Butyltins, methyltins	Butylation, pentylation	2 m × 2 mm, 3% OV-225 on Chromosorb® W	FP, AAS	25—28
	Butyltins, methyltins, Sn(IV)	Hydridization	3% SP-2100 on Chromosorb® W	AAS	29
	Methyltins, ethyltins	Hydridization	2 m × 6 mm, 3% SE-30 on Chromosorb® G	AAS	30
	Methyltins, Sn(IV)	Hydridization	30 cm × 6 mm, 15% OV-3 on Chromosorb® W	AAS, FES	31,19
	Methyltins	Hydridization	45 cm × 6 mm, 10% SP-2100 on Chromosorb® W	AAS	32
	Methyltins, Sn(IV)	Hydridization	6 cm × 6 mm, glass wool	AAS	33
Pb	Methyl- and ethylleads	Phenylation	30 m DB-1 (capillary)	EC	34
	Methyl- and ethylleads, Pb(II)	Butylation	1.8 m × 6 mm, 3% OV-1 on Chromosorb® W	AAS	35
	Methyl- and ethylleads	Butylation	1.5 m × 2 mm, 10% OV-101 on Chromosorb® W	AAS	36—38
	Trialkylleads	Butylation	12.5 m SP-2100 (capillary)	MIP	39
	Di-, tri-, and tetraalkylleads	Propylation	1 m × 2 mm, 10% OV-101 on Chromosorb®	AAS	40
	Methyl- and ethylleads, Pb(II)	Ethylation, methylation	10% SP-400 on Chromosorb® W	AAS, AAS	41 42
	Tetraalkylleads	—	0.9 m × 4 mm, 10% PEG 20 *M* on Porasil® C	AAS	43
	Di-, tri-, and tetraalkylleads	—	1 m × 2 mm, 3% OV-101 on Gas-Chrom® Q	AAS	44—49
	Tetraalkylleads	—	2.4 m × 2 mm, 20% TCP on Chromosorb® W	AAS	51

Table 1 (continued)
GC METHODS FOR METAL SPECIATION

Element	Species	Derivatization method	Column	Detector[a]	Ref.
Hg	Hg(II), alkylmercury salts	—	45 cm, 10% DEGS on Anakrom SD	EC	52
			2 m × 4 mm, 20% Carbowax® 20 *m* on Chromosorb® W (HP)	EC	53
	Alkyl-, alkylmercury-salts, Hg(II)	Methylation	1.2 m × 2 mm, 1.5% OV-17 + 1.95% QF1 on Chromosorb® W (HP)	EC, UV	54,55
	Dialkylmercurys	—	1.8 m × 2 mm, 5% DC-200 + 3% QF1 or Gas-Chrom® Q	FI	56
	$MeHg^+$, Me_2Hg	—	1 m × 0.5 mm 4% FFAP on Gas-Chrom® Q on Chromosorb® 121	MIP	57,58
	$MeHg^+$, Me_2Hg	—	1 m × 3 mm, 15% DEGS on Chromosorb® W or	MIP	59,60
		—	3 m × 3 mm, 30% OV-17 on Uniport® HP	DCP	59,60
Se	Se(IV), total Se	5-nitropiazselenol	1.8 m × 2 mm, 3% 10C on Chromosorb® W	EC	61
	Se(−II,0), Se(IV), Se(VI)	4,6 dibromopiazselenol	1 m × 3 mm 15% SE-30 on Chromosorb® W	EC	62
	Me_2Se, Me_2Se_2	—	1.8 m × 6 mm, OV-1 on Chromosorb® W	AAS	63
	Me_2Se, Me_2Se_2, Et_2Se	—	2 m × 2 mm, 10% polymetaphenylether on Chromosorb® W	AAS	64,65
As	As(III)	Diethyldithiocarbamates	5% OV-17 on Anakrom AS	EC	66
	As(III) MMA,[b] DMA, Me_3AsO	Hydridization	6 m × 4.8 mm, 16.5% silicone oil on Chromosorb® W	EC,AAS, FI	3,14,16
	As(III)	Phenylation	—		67
	MMA, DMA, alkylarsines	Hydridization	1.8 m × 0.5 mm, 5% Carbowax,® 20 *M* on Chromosorb® 101	MIP	68
	MMA, DMA, inorganic As	Hydridization	40 × 1.2 cm glass beads	AAS	69
Ge	MeGeCl, Me_2GeCl_2, $Ge(OH)_4$, Sb	Hydridization	Glass beads	AES	70
	Inorganic Ge	Hydridization	Silanized glass wool	AAS	71
	Methylgermaniums, inorganic Ge	Hydridization	22 cm, 15% OV-3 on Chromosorb® W	AAS	72
Sb	Sb(III), Sb(V), $MeSb(OH)_2O$, $Me_2Sb(OH)O$	Hydridization	30 cm × 6 mm, 15% OV-3 on Chromosorb® W	AAS	73

[a] EC = electron capture; FP = flame photometry; FI = flame ionization; FES = flame emission spectrometry; MIP = microwave induced plasma; DCP = direct current plasma; AAS = atomic absorption spectrometry; UV = ultraviolet absorbance.

[b] MMA = monomethylarsonic acid; DMA = dimethylarsinic acid.

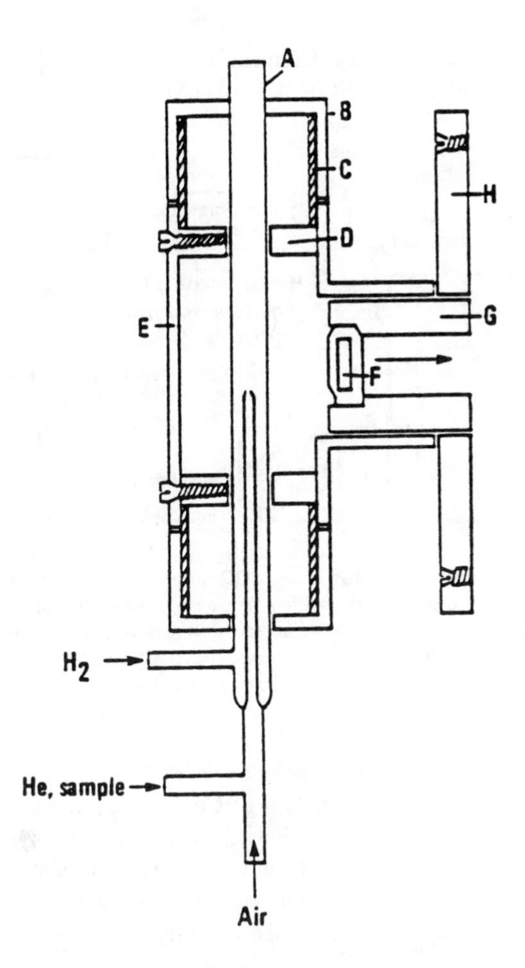

FIGURE 6. Quartz burner and housing for flame photometric detection. (A) Quartz burner (8 mm); (B) PVC cap; (C) PVC tubing; (D) mounting ring; (E) PVC T-joint (32 mm); (F) filter and holder; (G) PVC coupling. (From Braman, R. S. and Tompkins, M. A., *Anal. Chem.*, 51, 12, 1979. With permission.)

et al.[56,57] reported absolute MIP detection limits of 0.2 and 2 pg, respectively, for CH_3HgCl and $(CH_3)_2Hg$ in benzene extracts, compared with reported limits between 1 and 400 pg using EC detection.

The development of a new cavity design by Beenakker[98,99] in 1976 enabled a helium plasma to be produced at atmospheric pressure rather than the usual reduced pressure, with a resulting improvement in detection limits. The Beenakker cavity was easier to tune than earlier cavities and, in general, resulted in a more efficient transfer of microwave power to the plasma, leading to improvements in detection limits. The atmospheric pressure helium MIP has since been applied to both organomercury[100] and organolead detection.[39]

Sample introduction in GC-MIP is facilitated because the carrier gas and plasma gas are the same. Care should be taken with the GC detector interface to achieve a virtually zero dead volume connection and to ensure that the interface is at the same temperature as the column.[89] A typical system is shown in Figure 7.

Problems experienced with the MIP systems included deposition of carbon from solvents on the quartz emission tube and extinction of the plasma by eluting solvents.[98] Molecular oxygen and nitrogen may have to be employed as scavenger gases to prevent the deposits. It has also been reported that in MIP argon plasma detection, atomic emission intensities could vary with molecular structure.[89,92]

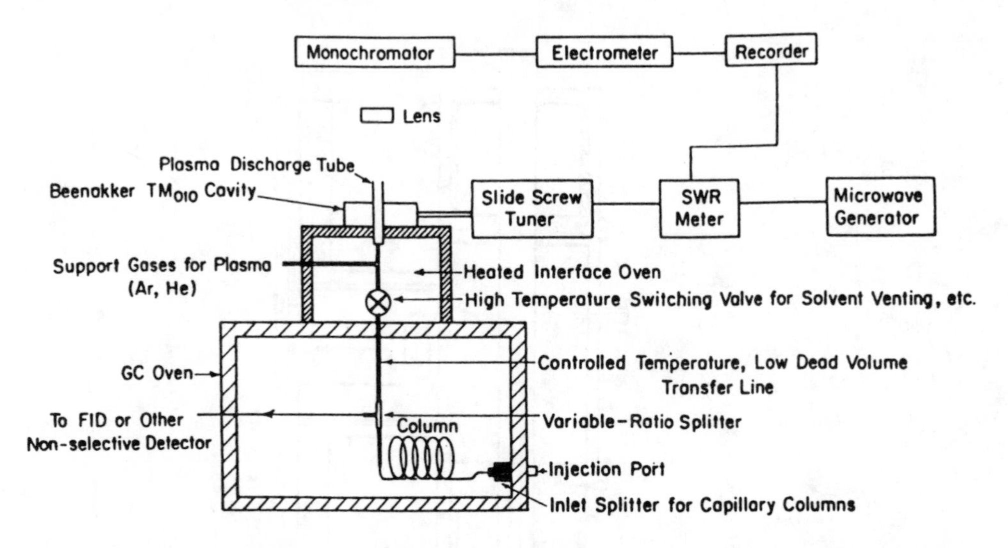

FIGURE 7. Schematic of interfaced GC-MIP system. GC effluent may be split to either MIP detector on nonselective GC detector (FI, TC), or both. Heated interface oven is maintained at same temperature as the GC oven and contains a high-temperature switching valve in order to vent large quantities of solvent to the atmosphere. All fittings should be of zero dead volume with narrow-bore connecting tube (glass-lined) between the GC column, switching valve, and MIP discharge tube. (From Krull, I. S. and Jordan, S., *Am. Lab.*, 12, 21, 1980. With permission.)

There have been relatively few applications of DCP. Lloyd et al.[101] reported its use as a metal-selective detector for the determination of Cu, Ni, Pb, and Cr, while Uden et al.[102] applied DCP to the analysis of cyclopentadienylmanganesetricarbonyl. The use of a heated sheath gas prevented diffusion of sample from the plasma and increased the residence time of the sample within the discharge region, thus increasing the sensitivity. Background emission was also reduced by minimizing the quantity of entrained air into the emission volume.

Panaro et al.[60] used GC-DCP for methylmercury analysis. The effluent from a low-cost, isothermal, packed-column GC was carried directly into the DCP plume by a 6-mm-O.D. × 1-mm-I.D. quartz jet tube wrapped in heating tape to maintain a 190°C temperature compared to the column temperature of 150°C. Advantages claimed for DCP over MIP detection are a tolerance to larger sample volumes, better plasma stability, and better reproducibility and sensitivity.

Similarly, there have been few publications on the use of GC-ICP systems for organometals detection. The first GC-ICP system was reported by Windsor and Denton[103] using ICP as a multielement detector for the simultaneous analysis of Br, C, Cl, F, H, I, Fe, Pb, and Sn. High detection limits were obtained for elements with relatively intense atomic lines (Fe, Pb, Sn, C, H, I, and Si) which were comparable to that of flame photometric detectors and other MIP detectors. However, detection limits observed for Br, Cl, and F were in the microgram range. Duebelbeis et al.[104] examined the performance of ICP detection for a range of organometals, including tetraethyltin, ferrocene, and tetraethyllead, finding detection limits of 25, 15, and 6 pg, respectively. For organoselenium compounds in coal gasification products, the limits were near 100 pg. Their system used a heated glass-lined stainless steel capillary transfer line to deliver the column effluent to within 1 cm of the base of the plasma region (Figure 8).

The higher gas temperature of the ICP detectors compared to the MIP means they are better able to tolerate organic solvents, which could be an advantage in some separations. The major limitations to their adoption are high capital and running costs.

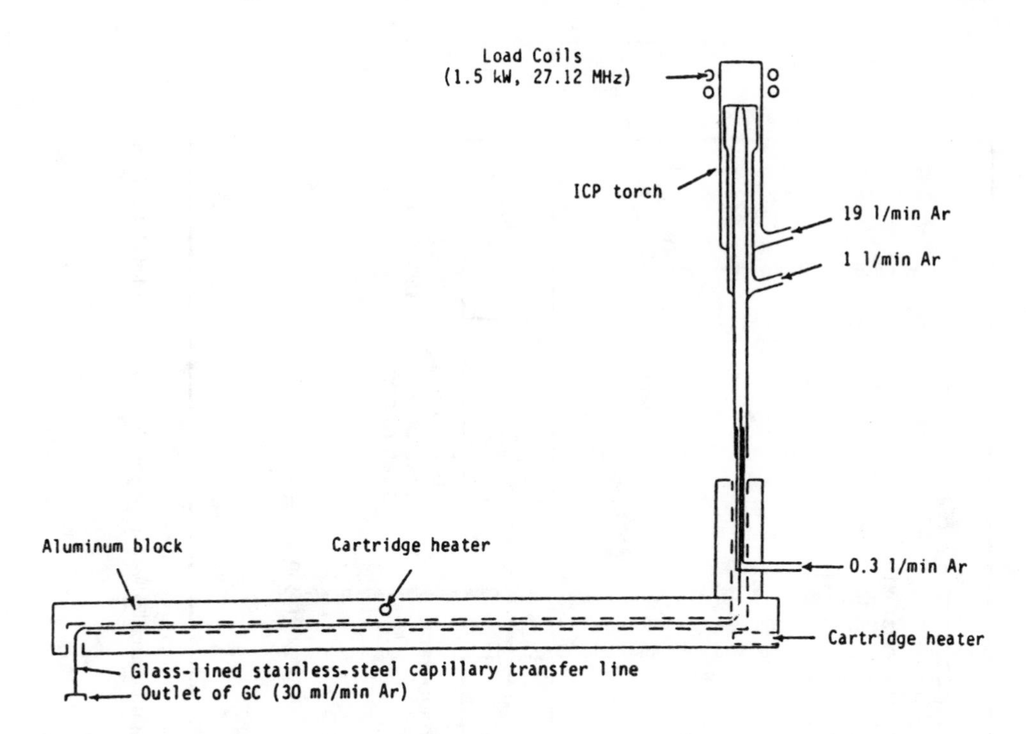

FIGURE 8. A GC-ICP interface. (From Duelbelbeis, D. O., Kapila, S., Yates, D. E., and Manahan, S. E., *J. Chromatogr.*, 351, 465, 1986. With permission.)

5. Atomic Fluorescence Detection

Nondispersive atomic fluorescence spectrometry (AFS) was first used as a GC detector by Radziuk et al.[105] in the determination of tetraalkyllead compounds. In theory, atomic fluorescence has a high sensitivity and multielement detection capability. It has not been well received probably because of the lack of intense and stable light sources for excitation and, most importantly, the lack of appropriate commercial instruments. Sensitivity is not particularly different from AAS when used with either graphite cups or a quartz furnace. The technique is, however, still in its infancy.

6. Mass Spectrometric Detection

Mass spectrometric (MS) detection provides unambiguous identification of metal species separated by GC through analysis of the characteristic mass spectra for each peak in a mass fragmentogram. Because of its poor sensitivity and relatively high cost it is unlikely to challenge any of the more conventional detectors and has therefore not been widely adopted. Meinema et al.[11] examined its application to the analysis of butyltin compounds, and in later studies, both Mueller[12] and other workers[4,25,41] used GC-MS to verify peaks obtained for tin and arsenic derivatives using GC-FPD.

III. SAMPLE PREPARATION

A. Extraction and Preconcentration

Samples for introduction into the GC system should be either in a gaseous form or dissolved in a volatile nonaqueous solvent. Volatile organometallic compounds, such as tetraalkylleads, tetraalkyltins, and methylselenides, can be readily collected in a cryogenic trap or trapped at room temperature on a suitable solid adsorbent such as glass beads, activated charcoal, or GC stationary phases, e.g., 3% OV-1 on Chromosorb® W,[8,45-49] the latter proving most effective (Figure 9). These molecular organometals are highly lipophilic and are therefore

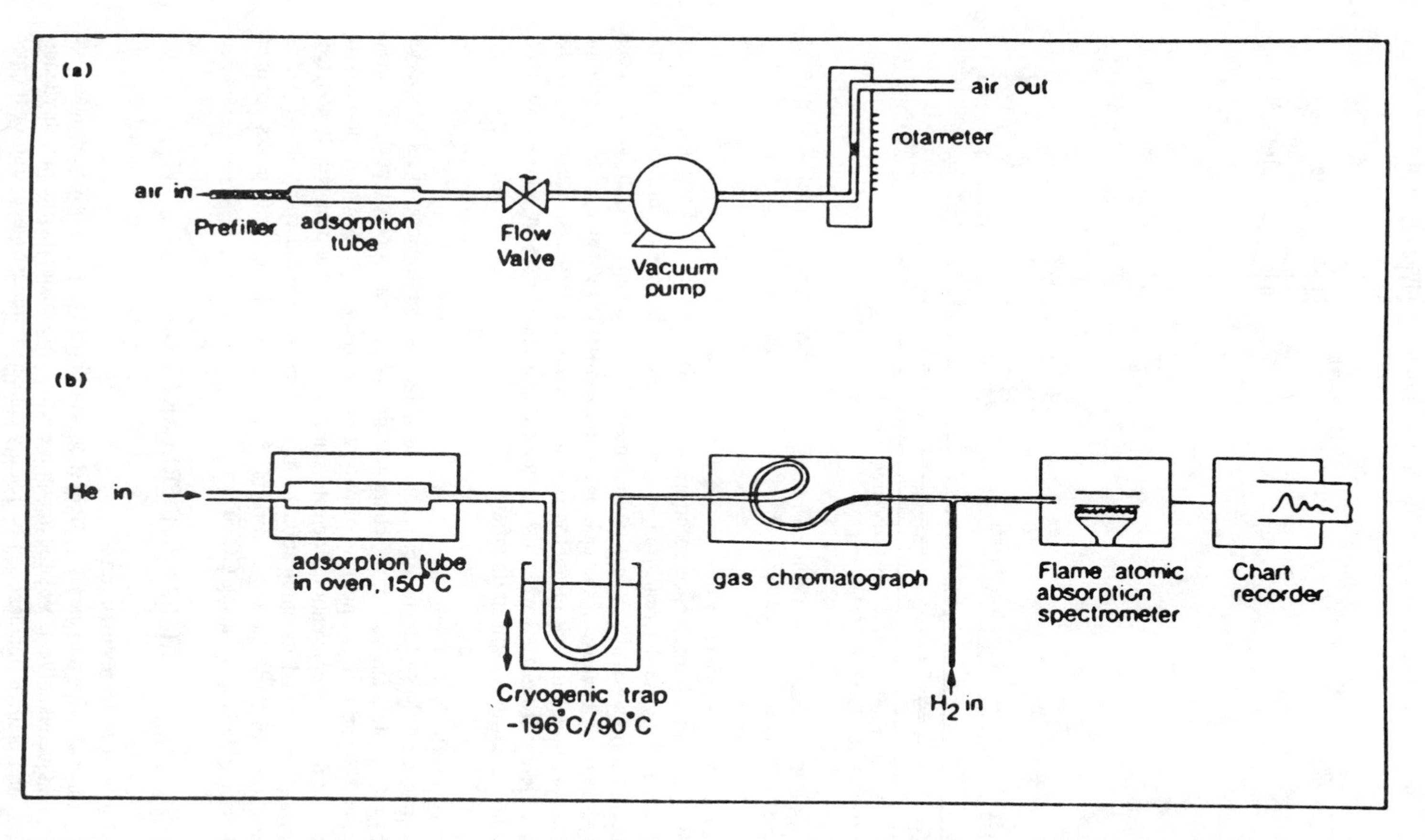

FIGURE 9. Sample collection system (a) and two-stage thermal desorption GC-AAS system (b) for tetraalkyllead in compounds in air. (From Hewitt, C. N. and Harrison, R. M., *Anal. Chim. Acta*, 167, 277, 1985. With permission.)

readily extractable into nonaqueous solvents for direct injection. The extraction procedures are also essential to preconcentrate organometallic species which are present in many environmental samples at ultratrace concentrations (nanogram per liter in natural waters).

For ionic organometallic species such as tri- and dialkylleads or tri- and dialkyltins, which behave like solvated cations, extraction recoveries increase with the degree of alkylation. Tributyltin, for example, can be quantitatively extracted from aqueous solutions using benzene, hexane, chloroform, or dichloromethane, while the recovery was 80 to 90% for dibutyltin, with negligible extraction of monobutyltin.[11] The addition of chelating agents has been practiced to achieve more to complete recoveries. For butyltin species, tropolone or oxine addition resulted in quantitative recoveries of all species.[2] Mueller[23] used tropolone-coated Sep-Pak C_{18} cartridges (Millipore-Waters,® Milford, MA), with subsequent extraction with diethylether. For ionic alkyllead compounds in waters, sediments, or biological tissues, sodium diethyldithiocarbamate addition[35] facilitated extraction into benzene prior to derivatization by butylation. Mercury extractions have been facilitated by the formation of cysteine or quaternary amine complexes,[57] while selenium(IV) can be extracted as a piazselenol.[61]

Dissolution of biological samples without breaking down the compounds for species determination is another challenging area in analytical chemistry. The conventional acid digestion techniques often convert organic and organometallic compounds to their elemental or inorganic forms. Two methods have recently been reported to circumvent such effects. Chau et al.[37] used a tissue solubilizer, tetramethylammonium hydroxide (TMAH), to digest fish, clams, and macrophytes for species determination of alkyllead compounds. The tissues dissolved at near-ambient temperature without the need for violent heating or oxidation. Forsyth and Marshall[34] hydrolyzed egg homogenates and tissues in an enzyme mixture of lipases and proteases to release alkyllead species. Both techniques were effective in releasing alkylleads in their authentic forms from biological tissues and have generally proved useful in dealing with other biological samples.

B. Derivatization Techniques

Derivatization is necessary to transform many organometallic species into volatile forms amenable to GC separation. It is important that the derivatization reactions do not alter the authentic structure of the compound, such as the positions and number of the metal-carbon bonds, or change the characteristics of the compounds. In other words, the main metal-carbon skeleton must remain, no matter what other functionalities are added to the molecule, so that the original identity of the analyte can still be recognized.

Hydridization and alkylation are the two most common derivatization methods; however, other procedures, such as complex formation or oxidation and reduction, have been used to achieve volatile products. Some of these, such as the formation of piazselenol derivatives of selenium, have already been discussed.

1. Hydridization

A range of elements including As, Sb, Bi, Sn, Pb, Se, Te, and Ge are easily converted to volatile covalent hydrides by reaction with sodium borohydride. The reactions are spontaneous and quantitative, even at ultratrace levels. The hydrides may be trapped cryogenically under liquid nitrogen, for later separation and detection, or they may be extracted directly into a nonaqueous solvent for GC analysis.

A major source of errors in the borohydride method is the loss of hydrides by irreversible adsorption to the internal walls of the apparatus.[2] To minimize this, only inert materials such as Teflon® or Pyrex® should be used, and the internal surfaces of all glassware should be silanized.[2,18]

An alternative approach has been to use a purge and trap procedure with desorption into

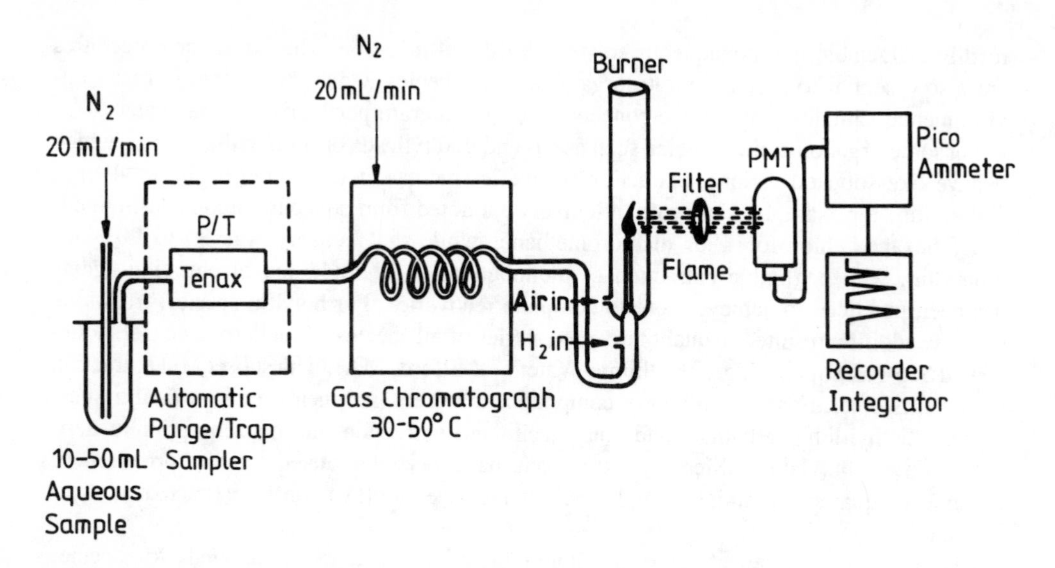

FIGURE 10. A typical purge/trap GC-FPD system.

a gas chromatograph for separation and detection.[13,22,79] Brinckman and co-workers[13,22] used a trap containing the polymeric adsorbent Tenax®-GC (Alltech Associates, IL) to collect organotin hydrides sparged from water (Figure 10). The desorption step involved rapid heating of the trap to 250°C for 5 min, while the carrier gas was directed into a GC column. An additional heating to 300°C for 3 min was used to vent any residue not removed by the first heating. Talmi and Bostick[68] trapped arsenic hydrides in cold toluene at −5°C prior to GC column separation.

Solvent extraction of organotin hydrides has been used in several studies.[4,10,18,20,21] Hattori et al.[21] reacted ethanolic sodium borohydride with a hexane extract of butyltins to form the hydrides. Unreacted borohydride was removed by water addition, with the hydrides partitioning into the hexane layer. Sonderquist and Crosby[20] used lithium aluminum hydride in diethylether as the hydridizing reagent for phenyltins in hexane. Matthias and co-workers[4] found that rather than hydridizing butyltins after solvent extraction, a simultaneous hydridization and extraction resulted in a 50% increase in mono- and dibutyltin and a threefold improvement in tributyltin response (Figure 11). One disadvantage is the possible rearrangement of the alkyl groups observed in the analysis of both methylarsenic[68,107] and methyltin species.[28] While the cause of alkyl rearrangements during the hydridization reaction is not yet fully understood, it is suggested that regulating the borohydride concentration by using borohydride tablets[68] or by adding the reagent uniformly with a peristaltic pump[107] may minimize the rearrangement reactions.

The removal of analytes in the volatile gaseous form from a sample is not only a unique feature as an analytical method, it has also provided an extremely convenient means of interfacing a high-performance liquid chromatograph (HPLC) to an atomic spectrometric furnace, as demonstrated by the conversion to arsines for arsenic[108] and to stannanes for alkyltin speciation.[30] A detailed discussion of such HPLC-AAS systems is given in Chapter 5.

Typical reactions for the hydride derivatization of Sn(IV) and methyltin species are as follows:[97]

$$CH_3SnCl_3 + BH_4^- \rightarrow CH_3SnH_3 \quad \text{(bp 0°C)} \tag{1}$$

$$(CH_3)_2SnCl_2 + BH_4^- \rightarrow (CH_3)_2SnH_2 \quad \text{(35°C)} \tag{2}$$

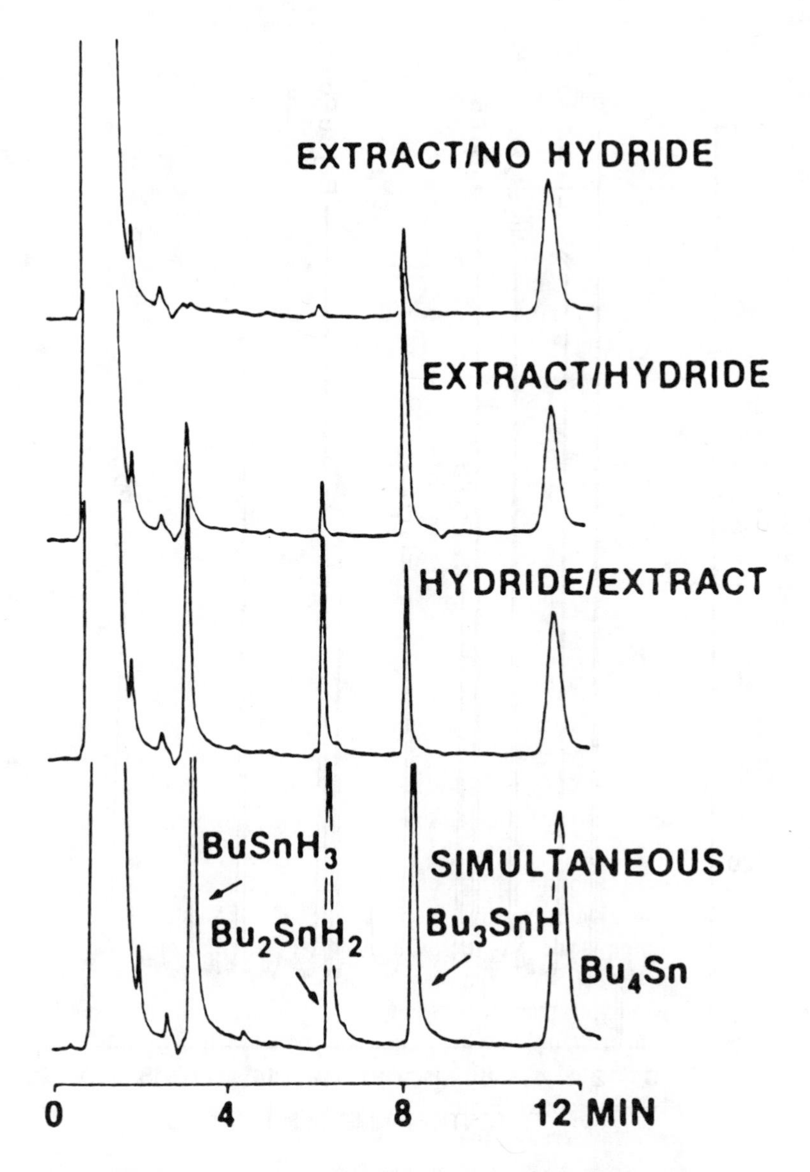

FIGURE 11. Chromatograms for butyltin species in Baltimore Harbor microlayer samples detected using GC-FPD after simultaneous hydridization/extraction. (From Matthius, C. L., Bellama, J. M., Olson, G. J., and Brinckman, F. E., *Environ. Sci. Technol.*, 20, 609, 1986. With permission.)

$$(CH_3)_3SnCl + BH_4^- \rightarrow (CH_3)_3SnH \quad (59°C) \tag{3}$$

$$Sn(IV) + BH_4^- \rightarrow SnH_4 \quad (-52°C) \tag{4}$$

The hydridization reactions result in dramatic changes in boiling points, without alteration of the authenticity of the carbon-metal bonds in the molecules. In the case of arsenic, the respective boiling points of arsine, methylarsine, and dimethylarsine on −62, 2, and 55°C.[106]

2. Alkylation

Another technique for changing the volatility of a metal or organometallic compound is

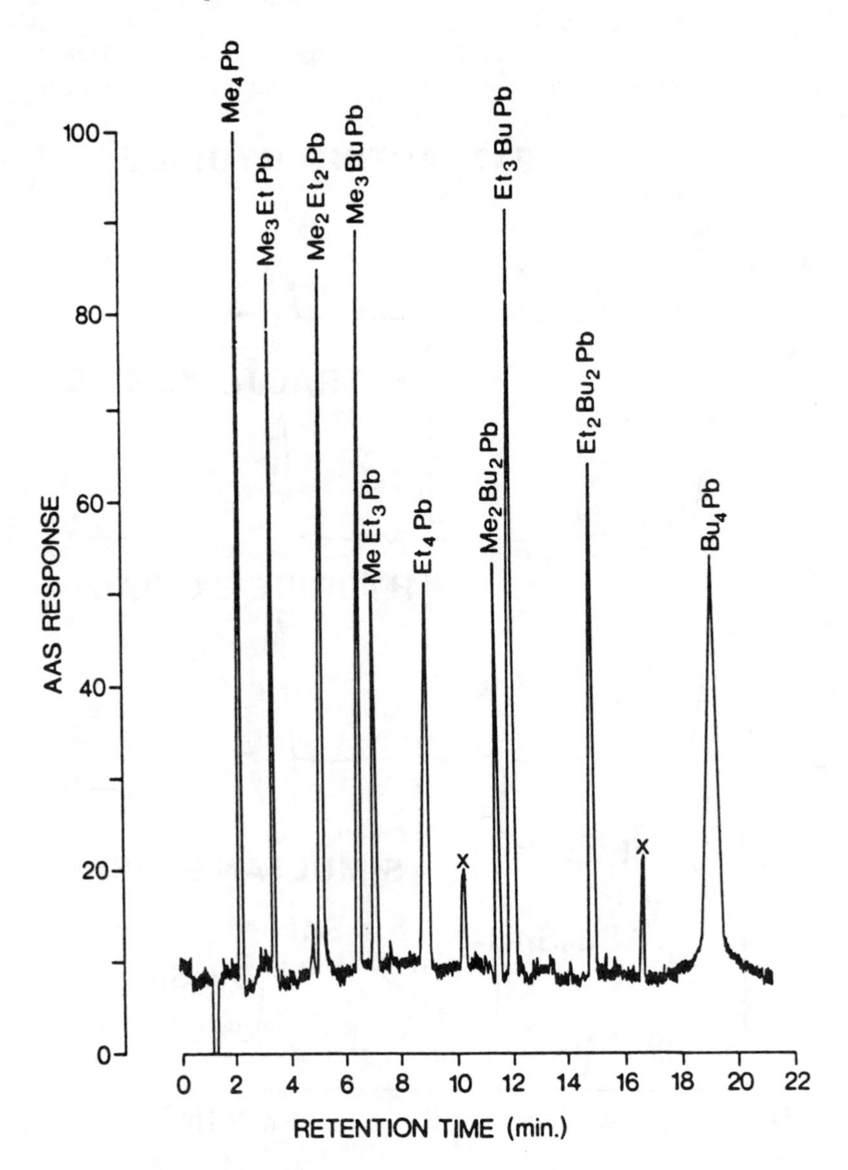

FIGURE 12. GC-AAS of alkyllead compounds after butylation. (From Chau, Y. K., Wong, P. T. S., and Kramar, O., *Anal. Chim. Acta,* 146, 211, 1983. With permission.)

by alkylation. Metals of Group IVA, notably, Ge, Sn, Pb, and their ionic alkylated derivatives, $R_nM^{(4-n)+}$, can be alkylated by a Grignard reagent, RMgCl, to the tetraalkyl-substituted derivatives which are more volatile and stable for separation by GC. For example, inorganic Pb(II) has been methylated to tetramethyllead by reaction with methylmagnesium chloride for detection by furnace AAS.[42] Phenylation has also been used as a volatilization technique in the determination of As, Sb, Tl, Se, Te, Hg, Bi, and Sn after extraction as their dithiocarbamate complexes.[109] For ionic organometallic compounds such as dialkyllead, R_2Pb^{2+}, and trialkyllead, R_3Pb^+ (R = Me, Et), alkylation with a butyl-Grignard reagent can convert them to the tetraalkyl-butyllead, $R_nPbBu_{(4-n)}$, which can be conveniently separated by GC. Ten alkyllead species, including Pb(II), have been separated and determined in one chromatogram at subnanogram levels (Figure 12).[35] Other alkylation techniques applied to the speciation of organometallic compounds include methylation,[11,12] ethylation,[23] butylation,[28] and pentylation[25] for alkyltin speciation, and ethylation,[41] propylation,[40] butylation,[35-37] and phenylation[34] for alkyllead analysis. Again, the derivatization results only in the formation of more volatile compounds without alteration of the integrity of the metal-carbon bonds, thus retaining the original identity of the molecule.

For example, the butylation of diethyllead, triethyllead, and Pb(II) results in the formation of the more volatile tetra-substituted ethylbutyllead species, $Et_nPbBu_{(4-n)}$, which are readily separated by GC.[35]

$$Et_2Pb^{2+} + 2\ BuMgCl \rightarrow Et_2PbBu_2 + 2\ Cl^- + 2\ Mg^{2+} \quad (5)$$

$$Et_3Pb^+ + BuMgCl \rightarrow Et_3PbBu + Mg^{2+} + Cl^- \quad (6)$$

$$Pb^{2+} + 4\ BuMgCl \rightarrow PbBu_4 + 4\ Mg^{2+} + 4\ Cl^- + 2\ e \quad (7)$$

Alkylation reactions need to be carried out in aprotic solvents. The reactions, like hydridization, are spontaneous and, with shaking, are essentially complete in 2 to 3 min. Special precautions against water are not necessary,[35] although Grignard reagents are unstable in the presence of moisture. Drying the reaction products with anhydrous sodium sulfate increases their storage life. Standards are stable for several months at room temperature. Grignard reagents are highly reactive and should be handled with care, with storage in the dark being preferred.

There are advantages and disadvantages of both derivatization methods. The hydridization technique, using trapping, is more sensitive because it is a total sampling technique in which the analyte is completely driven out of the sample and determined as a hydride, whereas in the alkylation method, only an aliquot of the sample is injected to the GC-AAS, as in the hydride method using solvent extraction. On the other hand, alkylation is relatively free from contamination and interferences arising from molecular rearrangements. However, both techniques are widely used.

IV. CONCLUSIONS

The coupling of GC separations with highly selective and sensitive detection systems forms a powerful analytical system capable of the identification and determination of different chemical species that conventional analytical methods cannot achieve. For example, for the first time it has been possible to determine nanogram quantities of alkyllead and alkyltin in environmental samples and to investigate their transport and degradation pathways. The future will see further improvements in sensitivity and specificity and in methods of sample preparation. Such techniques have opened up a new horizon in ultratrace analysis for applications in many areas of research.

REFERENCES

1. **Uden, P. C.,** Inorganic gas chromatography, *J. Chromatogr.*, 313, 3, 1984.
2. **Moshier, R. W. and Sievers, R. E.,** *Gas Chromatography of Metal Chelates,* Pergamon Press, London, 1965.
3. **Andreae, M. O.,** The determination of the chemical species of some of the "hydride elements" (arsenic, antimony, tin and germanium) in seawater, methodology and results, in *Trace Metals in Seawater,* Wong, C. S., Boyle, E., Bruland, K. W., Burton, J. D., and Goldberg, E. D., Eds., Plenum Press, New York, 1983, 1.

4. **Matthias, C. L., Bellama, J. M., Olson, G. J., and Brinckman, F. E.,** Comprehensive method for determination of aquatic butyltin and butylmethyltin species at ultratrace levels using simultaneous hydridization/extraction with gas chromatography-flame photometric detection, *Environ. Sci. Technol.*, 20, 609, 1986.
5. **Chau, Y. K.,** Occurrence and speciation of organometallic compounds in freshwater systems, *Sci. Total Environ.*, 49, 305, 1986.
6. **Fernandes, F. J.,** Metal speciation using atomic absorption as a chromatography detector, *At. Absorpt. Newsl.*, 16, 33, 1977.
7. **Van Loon, J. C.,** Metal speciation by chromatography/atomic spectrometry, *Anal. Chem.*, 51, 1139, 1979.
8. **De Jonghe, W. R. A. and Adams, F. C.,** Measurements of organic lead in air, *Talanta*, 29, 1057, 1982.
9. **Ebdon, L., Hill, S. J., and Ward, R. W.,** Directly coupled chromatography- atomic spectroscopy. I. Directly coupled gas chromatography-atomic spectrometry. A Review, *Analyst*, 111, 1113, 1986.
10. **Tsuda, J., Nakanishi, H., Morita, T., and Takebayashi, J.,** Simultaneous gas chromatographic determination of dibutyltin and tributyltin compounds in biological and sediment samples, *J. Assoc. Off. Anal. Chem.*, 69, 981, 1986.
11. **Meinema, H. A., Burger-Wiersma, T., Versluis-de Hann, G., and Giver, E. C.,** Determination of trace amounts of butyltin compounds in aqueous systems by gas chromatography/mass spectrometry, *Environ. Sci. Technol.*, 12, 288, 1978.
12. **Mueller, M. D.,** Tributyltin detection at trace levels in water and sediments using GC with flame-photometric detection and GC-MS, *Fresenius Z. Anal. Chem.*, 317, 32, 1984.
13. **Brickman, F. E., Jackson, J. A., Blair, W. R., Olsen, G. J., and Iverson, W. P.,** Ultratrace speciation and biogenesis of methyltin transport species in estuarine waters, in *Trace Metals in Seawater*, Wong, C. S., Boyle, E., Bruland, K. W., Burton, J. G., and Goldberg, E. D., Eds., Plenum Press, New York, 1983, 39.
14. **Braman, R. S., Johnson, D. L., Foreback, C. C., Ammons, J. M., and Bricker, J. L.,** Separation and determination of nanogram amounts of inorganic arsenic and methylarsenic compounds, *Anal. Chem.*, 49, 621, 1977.
15. **Hodge, V. F., Scidel, S. L., and Goldberg, E. D.,** Determination of tin(IV) and organotin compounds in natural waters, crustal sediments and macro algae by atomic absorption spectrometry, *Anal. Chem.*, 51, 1256, 1979.
16. **Andreae, M. O.,** Determination of arsenic in natural waters, *Anal. Chem.*, 49, 820, 1977.
17. **Braman, R. S. and Foreback, C. C.,** Methylated forms of arsenic in the environment, *Science*, 182, 1247, 1973.
18. **Valikers, A. O., Seligman, P. F., Olson, G. J., Brinckman, F. E., Matthias, C. L., and Bellama, J. M.,** Di- and tributyltin species in marine and estuarine waters. Inter-laboratory comparison of two ultratrace analytical methods employing hydride generation and atomic absorption or flame photometric detection, *Analyst*, 112, 17, 1987.
19. **Braman, R. S. and Tompkins, M. A.,** Separation and determination of nanogram amounts of inorganic tin and methyltin compounds in the environment, *Anal. Chem.*, 51, 12, 1979.
20. **Soderquist, C. J. and Crosby, D. G.,** Determination of triphenyltin hydroxide and its degradation products in water, *Anal. Chem.*, 50, 1435, 1978.
21. **Hattori, Y., Kobayashi, A., Takemoto, S., Takami, K., Kuge, Y., Sugimai, A., and Nakamoto, M.,** Determination of trialkyl, dialkyl and triphenyltin compounds in environmental water and sediments, *J. Chromatogr.*, 315, 314, 1984.
22. **Jackson, J. A., Blair, W. R., Brinckman, F. E., and Iverson, W. P.,** Gas chromatographic speciation of methylstannanes in the Chesapeake Bay using purge and trap sampling with a tin-selective detector, *Environ. Sci. Technol.*, 16, 110, 1982.
23. **Mueller, M. D.,** Comprehensive trace level determination of organotin compounds in environmental samples using high resolution gas chromatography with flame photometric detection, *Anal. Chem.*, 59, 617, 1987.
24. **Kapila, S. and Vogt, C. R.,** Some aspects of organotin analysis by gas chromatography-flame photometry, *J. Chromatogr. Sci.*, 18, 144, 1980.
25. **Maguire, R. J. and Huneault, H.,** Determination of butyltin species in water by gas chromatography with flame photometric detection, *J. Chromatogr.*, 209, 458, 1981.
26. **Maguire, R. J. and Tkacz, R. J.,** Analysis of butyltin compounds by gas chromatography; comparison of flame photometric and atomic absorption spectrophotometric detectors, *J. Chromatogr.*, 268, 99, 1983.
27. **Maguire, R. J., Tkacz, K. J., Chau, Y. K., Bengert, G. A., and Wong, P. T. S.,** Occurrence of organotin compounds in water and sediment in Canada, *Chemosphere*, 15, 253, 1986.
28. **Chau, Y. K., Wong, P. T. S., and Bengert, G. A.,** Determination of methyltin(IV) and tin(IV) species in water by gas chromatography-atomic absorption spectrophotometry, *Anal. Chem.*, 54, 246, 1982.
29. **Donard, O. F. X., Rapsomanikis, S., and Weber, J. H.,** Speciation of inorganic tin and alkyltin compounds by atomic absorption spectrometry using electrothermal quartz furnace after hydride generation, *Anal. Chem.*, 58, 772, 1986.

30. **Burns, D. T., Glockling, F., and Harriott, M.,** Investigation of the determination of tin tetraalkyls and alkyltin chlorides by atomic-absorption spectrometry after separation by gas-liquid or high performance liquid-liquid chromatography, *Analyst,* 106, 921, 1981.
31. **Andreae, M. O. and Byrd, J. T.,** Determination of tin and methyltin species by hydride generation and detection with graphite-furnace atomic absorption or flame emission spectrometry, *Anal. Chim. Acta,* 156, 147, 1984.
32. **Donard, O. F. X. and Weber, J. H.,** Behaviour of methyltins under simulated estuarine conditions, *Environ. Sci. Technol.,* 19, 1104, 1985.
33. **Hodge, V. F., Seidel, S. L., and Goldberg, E. D.,** Determination of tin (IV) and organotin compounds in natural waters, coastal sediments and macro algae by atomic absorption spectrometry, *Anal. Chem.,* 51, 1256, 1979.
34. **Forsyth, D. S. and Marshall, W. D.,** Determination of alkyllead salts in water and whole eggs by capillary column gas chromatography with electron capture detection, *Anal. Chem.,* 55, 2132, 1983.
35. **Chau, Y. K., Wong, P. T. S., and Kramar, O.,** The determination of dialkyllead, trialkyllead, tetraalkyllead and lead(II) ions in water by chelation/extraction and gas chromatography/atomic absorption spectrometry, *Anal. Chim. Acta,* 146, 211, 1983.
36. **Chakraborti, D., De Jonghe, W. R. A., Van Mol, W. E., Van Cleuvenbergen, R. J. A., and Adams, F. C.,** Determination of ionic alkyllead compounds in water by gas chromatography/atomic absorption spectrometry, *Anal. Chem.,* 56, 2692, 1984.
37. **Chau, Y. K., Wong, P. T. S., Bengert, G. A., and Dunn, J. L.,** Determination of dialkyllead, trialkyllead, tetraalkyllead and lead(II) compounds in sediment and biological samples, *Anal. Chem.,* 56, 271, 1984.
38. **Chau, Y. K., Wong, P. T. S., Bengert, G. A., Dunn, J. L., and Glen, B.,** Occurrence of alkyllead compounds in the Detroit and St. Clair Rivers, *J. Great Lakes Res.,* 11, 313, 1985.
39. **Estes, S. A., Uden, P. C., and Barnes, R. M.,** Determination of *n*-butylated trialkyllead compounds by gas chromatography with microwave plasma detection, *Anal. Chem.,* 54, 2402, 1982.
40. **Radojevic, M., Allen, A., Rapsomanikis, S., and Harrison, R.,** Propylation technique for the simultaneous determination of tetraalkyllead and ionic alkyllead species by gas chromatography/atomic absorption spectrometry, *Anal. Chem.,* 58, 658, 1986.
41. **Rapsomaniskis, S., Donard, O. F. X., and Weber, J. H.,** Speciation of lead and methyllead ions in water by chromatography/atomic absorption spectrometry after ethylation with sodium tetraethylate, *Anal. Chem.,* 58, 35, 1986.
42. **Brueggemeyer, T. W. and Caruso, J. A.,** Determination of lead in aqueous samples as the tetramethyllead derivative by atomic absorption spectrometry, *Anal. Chem.,* 54, 872, 1982.
43. **Coker, D. T.,** Determination of individual and total lead alkyls in gasoline by a simple rapid gas chromatography/atomic absorption spectrometry technique, *Anal. Chem.,* 47, 386, 1975.
44. **De Jonghe, W., Chakraborti, D., and Adams, F.,** Graphite furnace atomic absorption spectrometry as a metal specific detection system for tetraalkyllead compounds separated by gas chromatography, *Anal. Chim. Acta,* 115, 89, 1980.
45. **Hewitt, C. N. and Harrison, R. M.,** A sensitive, specific method for the determination of tetraalkyllead compounds in air by gas chromatography/atomic absorption spectrometry, *Anal. Chim. Acta,* 167, 277, 1985.
46. **Harrison, R. M. and Radojevic, M.,** Determination of tetraalkyl and ionic alkyllead compounds in environmental samples by butylation and gas chromatography-atomic absorption, *Environ. Technol. Lett.,* 6, 129, 1985.
47. **Chau, Y. K., Wong, P. T. S., and Goulden, P. D.,** Gas chromatography-atomic absorption spectrometry for the determination of tetraalkyllead compounds, *Anal. Chim. Acta,* 85, 421, 1976.
48. **Chau, Y. K., Wong, P. T. S., and Saitoh, H.,** Determination of tetraalkyllead compounds in the atmosphere, *J. Chromatogr. Sci.,* 14, 162, 1976.
49. **Radziuk, B., Thomassen, Y., Van Loon, J. C., and Chau, Y. K.,** Determination of alkyllead compounds in air by gas chromatography and atomic absorption spectrometry, *Anal. Chim. Acta,* 105, 255, 1979.
50. **Reamer, D. C., Zoller, W. H., and O'Haver, T. C.,** Gas chromatography-microwave plasma detector for the determination of tetraalkyllead species in the atmosphere, *Anal. Chem.,* 50, 1449, 1978.
51. **Robinson, J. W., Kiesel, E. L., Goodbread, J. P., Bliss, R., and Marshall, R.,** The development of a gas chromatography-furnace atomic absorption combination for the determination of organic lead compounds, atomization processes in furnace atomizers, *Anal. Chim. Acta,* 92, 321, 1977.
52. **Zarneger, P. and Mushak, P.,** Quantitative measurements of inorganic mercury and organomercurials in water and biological media by gas-liquid chromatography, *Anal. Chim. Acta,* 69, 389, 1974.
53. **Filippelli, M.,** Determination of trace amounts of organic and inorganic mercury in biological materials by graphite furnace atomic absorption spectrometry and organic mercury speciation by gas chromatography, *Anal. Chem.,* 59, 116, 1987.
54. **Crappon, C. J. and Smith, J. C.,** Gas-chromatographic determination of inorganic mercury and organomercurials in biological materials, *Anal. Chem.,* 49, 365, 1977.

55. **Longbottom, J. E.,** Inexpensive mercury-specific gas chromatographic detector, *Anal. Chem.*, 44, 1111, 1972.
56. **Dressman, R. C.,** A new method for the gas chromatographic separation and detection of dialkylmercury compounds — application to river water analysis, *J. Chromatogr. Sci.*, 10, 472, 1972.
57. **Talmi, Y.,** The rapid sub-picogram determination of volatile organomercury compounds by gas chromatography with a microwave emission detector system, *Anal. Chem. Acta,* 74, 107, 1975.
58. **Talmi, Y. and Norvell, V. E.,** A rapid method for the determination of methylmercury chloride in water samples by gas chromatography with a microwave emission detector, *Anal. Chim. Acta,* 85, 203, 1976.
59. **Chiba, K., Yoshida, K., Tanabe, K., Haraguchi, H., and Fuwa, K.,** Determinaton of alkylmercury in seawater at the nanogram per liter level by gas chromatography/atomspheric pressure helium microwave induced plasma emission spectrometry, *Anal Chem.*, 55, 450, 1983.
60. **Panaro, K. W., Erickson, D., and Krull, I. S.,** Determination of methylmercury in fish by gas chromatography-direct current plasma atomic emission spectrometry, *Analyst,* 112, 1097, 1987.
61. **Measures, C. I. and Burton, J. D.,** Gas chromatographic method for the determination of selenite and total selenium in sea water, *Anal. Chim. Acta,* 120, 177, 1980.
62. **Uchida, H., Shimoishi, Y., and Toei, K.,** Gas chromatographic determination of selenium (-II,0), -(IV), and -(VI) in natural waters, *Environ. Sci. Technol.*, 14, 541, 1980.
63. **Chau, Y. K., Wong, P. T. S., and Goulden, P. D.,** Gas chromatography-atomic absorption method for the determination of dimethyl selenide and dimethyl diselenide, *Anal. Chem.*, 47, 2279, 1975.
64. **Radziuk, B. and Van Loon, J.,** Atomic absorption spectroscopy as a detector for the gas chromatographic study of volatile selenium alkanes from *Astragalus racemosus, Sci. Total Environ.*, 6, 251, 1976.
65. **Jiang, S., De Jonghe, W., and Adams, F.,** Determination of alkylselenide compounds in air by gas chromatography atomic absorption spectrometry, *Anal. Chim. Acta,* 136, 183, 1982.
66. **Daughtroy, R. H., Fitchett, A. W., and Massak, P.,** Quantitative measurements of inorganic and methylarsenides by gas-liquid chromatography, *Anal. Chim. Acta,* 79, 199, 1975.
67. **Schwett, G. and Ruessel, H. A.,** Gas chromatographic determination of arsenic as triphenylarsine, *Chromatographia,* 242, 1972.
68. **Talmi, Y. and Bostick, D. T.,** Determination of alkylarsenic acids in pesticides and environmental samples by gas chromatography with a microwave emission spectrometric detection system, *Anal. Chem.*, 47, 2145, 1975.
69. **Shaikh, A. U. and Tallman, D. E.,** Species-specific analysis for nanogram quantities of arsenic in natural waters by arsine generation followed by graphite furnace atomic absorption spectrometry, *Anal. Chim. Acta,* 98, 251, 1978.
70. **Braman, R. S. and Tompkins, M. A.,** Atomic emission spectrometric determination of antimony, germanium, and methylgermanium compounds in the environment, *Anal. Chem.*, 50, 1088, 1978.
71. **Andreae, M. O. and Froelich, P. N.,** Determination of germanium in natural waters by graphite furnace atomic absorption spectrometry with hydride generation, *Anal. Chem.*, 53, 287, 1981.
72. **Hambrick, G. A., Froelich, P. N., Andreae, M. O., and Lewis, B. L.,** Determination of methylgermanium species in natural waters by graphite furnace atomic absorption spectrometry with hydride generation, *Anal. Chem.*, 56, 421, 1984.
73. **Andreae, M. O., Asmode, J. F., Foster, P., and Van't dack, L.,** Determination of antimony(III), antimony(V), and methylantimony species in natural waters by atomic absorption spectrometry with hydride generation, *Anal. Chem.*, 53, 1766, 1981.
74. **Westoo, G.,** Determination of methylmercury in foodstuffs, I. Methylmercury compounds in fish, identification and determination, *Acta Chem. Scand.*, 20, 2131, 1966.
75. **Newsome, W. H.,** Determination of methylmercury in fish and in cereal grain products, *J. Agric. Ford Chem.*, 19, 567, 1971.
76. **Tattan, J. O'G. and Wagstaffe, P. G.,** Identification and determination of organomercurial fungicide residues by thin-layer and gas chromatography, *J. Chromatogr.*, 44, 284, 1965.
77. **Aue, W. A. and Kapila, S.,** The electron capture detector — controversies, comments and chromatograms, *J. Chromatogr. Sci.*, 11, 255, 1973.
78. **Kolb, B., Kemmner, G., Schleser, F. H., and Wiedeking, E.,** Element-spezfische getrennter, in metallverbindungen mittels atom-absorptions-spectroskopie (AAS), *Fresenius Z. Anal. Chem.*, 221, 166, 1966.
79. **Hahn, M. H., Mulligan, K. J., Jackson, M. E., and Caruso, J. A.,** The sequential determination of arsenic, selenium, germanium and tin as their hydrides by gas-solid chromatography with an atomic absorption detector, *Anal. Chim. Acta,* 118, 115, 1980.
80. **Chu, R. C., Barron, G. P., and Baumgarner, P. A. P.,** Arsenic determination at subnanogram levels by arsine evolution and flameless atomic absorption spectrophotometric technique, *Anal. Chem.*, 44, 1476, 1972.
81. **Chau, Y. K. and Wong, P. T. S.,** An element- and speciation-specific technique for the determination of organometallic compounds, in *Environmental Analysis,* Ewing, G. W., Ed., Academic Press, New York, 1977, 215.

82. **Forsyth, D. S. and Marshall, W. D.,** Performance of an automatic gas chromatograph-silica furnace-atomic absorption spectrometer for the determination of alkyllead compounds, *Anal. Chem.,* 57, 1299, 1985.
83. **Ebdon, L., Ward, R. W., and Leathard, D. A.,** Development and optimisation of atomic cells for sensitive gas chromatography-flame atomic-absorption spectrometry, *Analyst,* 107, 129, 1982.
84. **Segar, D. A.,** Flameless atomic absorption gas chromatography, *Anal. Lett.,* 7, 89, 1974.
85. **Parris, G. E., Blair, W. R., and Brinckman, F. E.,** Chemical and physical considerations in the use of atomic absorption detectors coupled with a gas chromatograph for determination of trace organometallic gases, *Anal. Chem.,* 49, 378, 1977.
86. **Grant, D. W.,** Emissivity detector for gas chromatography, in *Gas Chromatography,* Desty, D. H., Ed., Academic Press, New York, 1958, 153.
87. **Aue, W. A. and Hill, H. H.,** Selective determination of hetero-organics by a dual-channel detector based on flame conductivity and emission, *Anal. Chem.,* 45, 729, 1973.
88. **Aue, W. A. and Flinn, C. J.,** A photometric tin detector for gas chromatography, *J. Chromatogr.,* 142, 145, 1977.
89. **Krull, I. S. and Jordan, S.,** Interfacing GC and HPLC with plasma emission spectroscopy, *Am. Lab.,* 12, 21, 1980.
90. **McCormack, A. J., Tong, S. C., and Cooke, W. D.,** Sensitive selective gas chromatography detector based on emission spectrometry of organic compounds, *Anal. Chem.,* 12, 1470, 1965.
91. **Bache, C. A. and Lisk, D. J.,** Selective emission spectrometric determination of nanogram quantities of organic bromine, chlorine, iodine, phosphorus, and sulfur compounds in a helium plasma, *Anal. Chem.,* 39, 786, 1967.
92. **Moye, H. A.,** An improved microwave emission gas chromatography detector for pesticide analysis, *Anal. Chem.,* 39, 1441, 1967.
93. **Dagnall, R. M., Pratt, S. J., West, T. S., and Deans, D. R.,** The micro-wave excited emission detector in gas phase chromatography. I. Some studies with sulphur compounds, *Talanta,* 16, 797, 1969.
94. **Dagnall, R. M., West, T. S., and Whitehead, P.,** The determination of volatile metal chelates by using a microwave excited emission detector, *Analyst,* 98, 647, 1973.
95. **Kawaguchi, H., Sakamoto, T., and Mizuike, A.,** Emission spectrometric detection of metal chelates separated by gas chromatography, *Talanta,* 20, 321, 1973.
96. **Sakamoto, T., Kawaguchi, H., and Mizuike, A.,** Determination of traces of copper and aluminium in zinc by gas chromatography with the microwave plasma detector, *J. Chromatogr.,* 121, 383, 1976.
97. **Robbins, W. B. and Caruso, J. A.,** Development of hydride generation methods for atomic spectroscopic analysis, *Anal. Chem.,* 51, 889A, 1979.
98. **Beenakker, C. I. M.,** A cavity for microwave-induced plasma operated in helium and argon at atmospheric pressure, *Spectrochim. Acta,* 32B, 483, 1976.
99. **Beenakker, C. I. M.,** Evaluation of a microwave-induced plasma in helium at atmospheric pressure as an element-selective detector for gas chromatography, *Spectrochim. Acta Part B,* 31, 173, 1977.
100. **Tanabe, K., Chiba, K., Haraguchi, H., and Fuwa, K.,** Determination of mercury at the ultratrace level by atmospheric pressure helium microwave-induced plasma emission spectrometry, *Anal. Chem.,* 53, 1450, 1981.
101. **Lloyd, R. J., Barnes, R. M., Uden, P. C., and Elliott, W. G.,** Direct current atmospheric pressure argon plasma Echelle spectrometer as a specific metal gas chromatographic detector, *Anal. Chem.,* 50, 2025, 1978.
102. **Uden, P. C., Barnes, R. M., and DiSanzo, F. P.,** Determination of methyl-cyclopentadienylmanganesetricarbonyl (MMT) in gasoline by gas chromatography with interfaced direct current argon plasma emission detector, *Anal. Chem.,* 50, 852, 1978.
103. **Windsor, D. L. and Denton, M. B.,** Elemental analysis of gas chromatographic effluents with an inductively coupled plasma, *J. Chromatogr. Sci.,* 17, 492, 1979.
104. **Duebelbeis, D. O., Kapila, S., Yates, D. E., and Manahan, S. E.,** Gas chromatographic-inductively coupled plasma emission spectrometric determination of volatile organometallic species, *J. Chromatogr.,* 351, 465, 1986.
105. **Radziuk, B., Thomassen, Y., Butler, L. R. P., Van Loon, J. C., and Chau, Y. K.,** A study of atomic absorption and atomic fluorescence atomizer systems as detectors in the gas chromatographic determination of lead, *Anal. Chim. Acta,* 108, 31, 1979.
106. **Nakahara, T.,** Applications of hydride generation techniques in atomic absorption, atomic fluorescence and plasma atomic emission spectroscopy, *Prog. Anal. At. Spectrosc.,* 6, 163, 1983.
107. **Wong, P. T. S., Chau, Y. K., Luxon, L., and Bengert, G. A.,** Methylation of arsenic in the aquatic environment, in *Trace Substances in Environmental Health XI,* Hemphill, D. D., Ed., University of Missouri Press, Columbia, MO, 1977, 100.

108. **Ricci, G. R., Shepard, L. S., Colovos, G., and Hester, N. E.,** Ion chromatography with atomic absorption spectrometric detection for determination of organic and inorganic arsenic species, *Anal. Chem.*, 53, 610, 1981.
109. **Schwedt, G. and Russel, H. A.,** Gas-chromatographie von Tl, Se, Te, Hg, As, Sb, Bi, Sn als phenylverbindungen, *Z. Anal. Chem.*, 264, 301, 1973.

Chapter 8

SPECIATION OF TRACE ELEMENTS IN SEDIMENTS

Michael Kersten and Ulrich Förstner

TABLE OF CONTENTS

I. INTRODUCTION

During the last decade, the emphasis of sediment analysis techniques for the evaluation of the environmental impact of polluted sediments has changed from a simple determination of total concentrations towards a more sophisticated fractionation of the sediment compounds. This change in focus results from a recognition that biogeochemical and especially the ecotoxicological significance of a given pollutant input is determined by its specific binding form and coupled reactivity rather than by its accumulation rate in sediments. Effects of an environmental process on trace element behavior and fate can be understood only in terms of its species distribution. The total concentration of a trace element can be used to assess its environmental impact only in cases where it is present in the environment as a single, well-known species. This condition will be met only very rarely in water, and is even less likely in solid materials. Thus, the behavior and fate of an element in the environment (e.g., its bioavailability, toxicity, and distribution), cannot be reliably predicted on the basis of its total concentration.

The mobility, transport, and partitioning of trace metallic and metalloid elements in a natural aquatic and terrestrial system is a function of the chemical form of the element which, in turn, is controlled by the physicochemical and biological characteristics of that system. Major variations of these characteristics are found in time and space due to the dissipation and flux of energy and materials involved in the biogeochemical processes which drive the speciation reactions. The tendency of an element to be accumulated by organisms depends in particular upon the capacity of a sediment/water system to resupply dissolved trace elements removed from solution by biotic and abiotic processes. Solid components in sediments govern the dissolved levels of these elements via sorption/desorption and dissolution/precipitation reactions. Thus, particular trace metal species identification tends to be far more instructive than any total elemental concentrations.

In order to assess the environmental impact of a given pollutant, the following points must be addressed in contaminated sediment/water systems:

1. What is the reactivity of the metals introduced with solid materials from anthropogenic activities (hazardous waste, sewage sludge, atmospheric deposits, etc.) by comparison with the natural components?
2. Are the interactions of critical metals between solution and solid phases comparable for natural and contaminated systems?
3. Are the factors and processes of remobilization effective when either the solid inputs or the solid/solution interactions lead to weaker bonding of certain metal species in contaminated compared to natural systems?

The term "species" generally refers to the molecular forms of an element or a cluster of atoms of different elements in a given matrix (in this case, solid matter).[1] The term "form" is also used to indicate uncertainty or lack of knowledge about the exact nature of the species one expects to find in an environmental sample. Since the methods applied to date rarely yield information at the molecular level in the solid phases, the more general term, "form", is used, rather than "species", when referring to metals in soils and sediments.

The term "speciation" of trace metallic and metalloid elements in a mineral/water system encompasses three aspects:[1]

1. The actual distribution among molecular level entities in a given matrix.
2. The processes responsible for an observable distribution (species distribution).
3. The analytical methods used (species analysis).

This chapter will focus on the potential limitations of analytical methods used for speciation of trace elements in solid samples.

The measurement of speciation is a more complex task than the determination of total element contents. Speciation of particulate elements can theoretically be determined by either thermodynamic calculations or by experimental techniques. Thermodynamic models so far available only give suggestions as to the expected trace element species distributions, but may yield results far removed from the real solid speciation because of the important role of kinetically controlled processes in biogeochemistry. This approach, however, will not be considered here, but has been comprehensively reviewed by Sposito.[2] Experimental procedures include direct instrumental techniques and selective extraction procedures. The former have failed to provide information at the molecular level for all but organometallic species, while the latter suffer from being operationally defined, yielding data on metals "associated with" rather than "bound to" specific mineral phases or chemical components. The rationale behind the indirect chemical analysis is that metal associations present in a given sample are transformed selectively into a single species for which an analytical instrument is sensitive. In the past, such methods have required extensive empirical calibration studies, which will be discussed in Section II.

During sampling, storage, and analysis of an environmental sample, typically characterized by a more or less pronounced disequilibrium, species transformation may occur. In polluted ("stressed") systems, entropy increases and there is a concomitant increase in instability in both the physical and biological context.[3] The greater the stress in the environment, the more difficult is sample handling and storage prior to analysis. Many analytical techniques presented here are handicapped by disruptive preparation techniques which alter the chemical speciation of inorganic components or lead to loss of analyte before analysis (e.g., freezing, lyophilization, evaporation, oxidation, changes in pH, light-catalyzed reactions, reactions with the sample container, time delays before analysis with biologically active samples, etc.). It is just the stressed system where action is immediately needed and where, for an assessment or prognosis of possible adverse effects, the species and their transformations have to be evaluated. Thus, appropriate action must be taken to assure that the species to be determined does not change during the interval between sampling and analysis. Reliable and generally applicable methods for the preservation of the original distribution of the species in natural sediment and soil samples do not yet exist. Thus, it is important to be aware of the kinds and degrees of changes taking place in the soils being studied while they are being studied.

Once the impact of toxic compounds on natural systems has been measured or predicted using speciation results, a management plan can be formulated which usually includes costly engineering and political decisions. The economic "impact" of such studies can be significant. There are, however, a number of problems inherent in the interpretation of the acquired data. For this reason the value of the speciation of solids, at least in operational terms, for the study of pollution potential and sources will be considered at the end of this chapter, together with its significance for the management of metal-contaminated waste materials.

II. INSTRUMENTAL METHODS

A. Physical Separation Techniques

One of the oldest methods for sediment fractionation is to separate mineral grains, either manually with the aid of a microscope, or instrumentally, e.g., by magnetic separators or heavy mineral flotations, prior to chemical analysis. In practice, however, such techniques are extremely time-consuming. Total metal concentrations in soils and sediments tend to vary with particle size, higher concentrations (micrograms per gram) being found in the smaller particles (< 63 μm). Metal concentrations generally decrease in the silt and fine

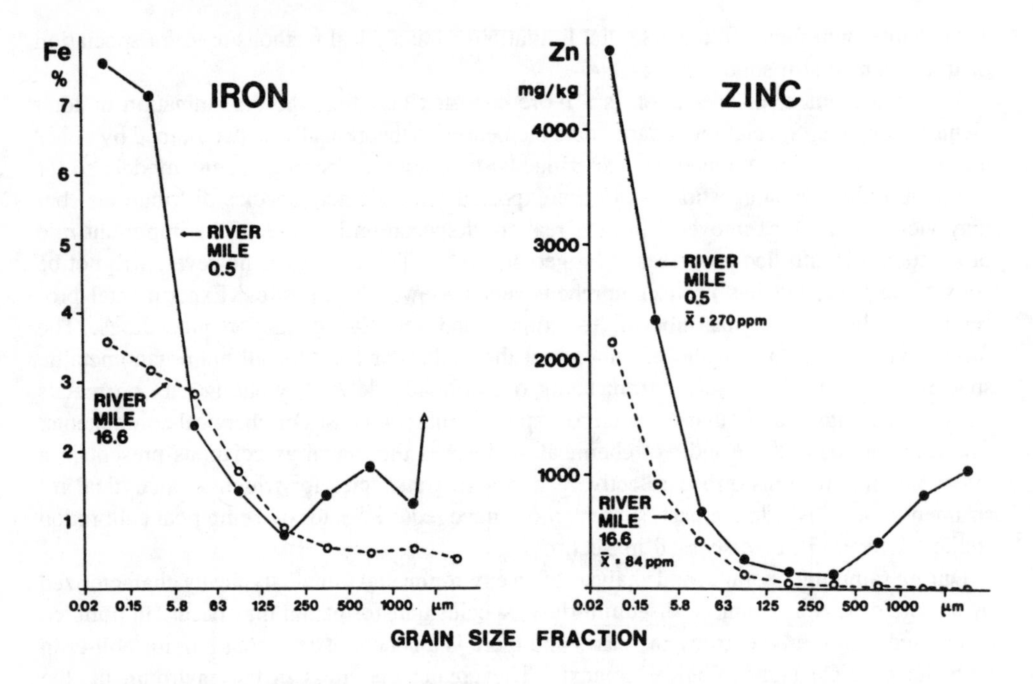

FIGURE 1. Grain size distribution of iron and zinc in two sediment samples from Saddle River, upstream (16.6 mi) and downstream (0.5 mi) from an urbanized area of Lodi, N.J. (From Förstner, U., *Hydrobiologia,* 91, 269, 1982. With permission.)

sand fractions, as they are dominated by quartz components. Coarser fractions contain heavy minerals which cause the metal content to again increase. A general decrease in metal concentrations with increasing particle size is shown in Figure 1, where a single sample was taken from a point upstream and one downstream of an urbanized area at Lodi, New Jersey.[4] A characteristic increase of metal concentrations in the medium and coarse sand fractions, particularly significant for Pb, Cu, Ni, Cr, Cd, Zn (Figure 1A) and also for Fe (Figure 1B), is probably due to the input of coarse waste particles. The decrease of metal concentrations in the range of 10 to 1000 μm grain size would probably be more pronounced if mechanical fractionation had not effectively separated individual particles according to their grain size.

Physical separation procedures of different size fractions usually involves sieving, gravity sedimentation, and differential centrifugation. Separation of the individual components in the finer size ranges is very difficult. The separation techniques assume that the metal-bearing soil or sediment components do not aggregate with each other. The fine-grained particles most commonly occur as complex aggregates, containing interlayered and intermixed iron and manganese oxide coatings and both viable and nonviable organic matter, often surrounding a clay particle. Norrish et al.,[5] e.g., showed that very fine precipitated ZnS may coat and impregnate any porous sediment matter, resulting in its presence in different physical fractions such as organic fragments and clay-carbonate aggregates. Disaggregation may result in a redistribution of the different types of surface sites available for metal binding and significant artifacts in the subsequent speciation measurements.

Density separation can be used for the study of accumulated metal phases. The rationale for this method is that components (such as organic matter, clays, and heavy minerals) differ sufficiently in specific gravity and will separate into distinct bands in mixtures of heavy liquids in a gradation of different densities. Simple float-sink types of density separation have been applied to solid associations of lead in soils,[7] roadside soils,[8] and urban dusts.[9] Recently, a scheme that involves particle size fractionation, density gradient fractionation and X-ray diffraction or scanning electron microscopy/energy dispersive X-ray analysis

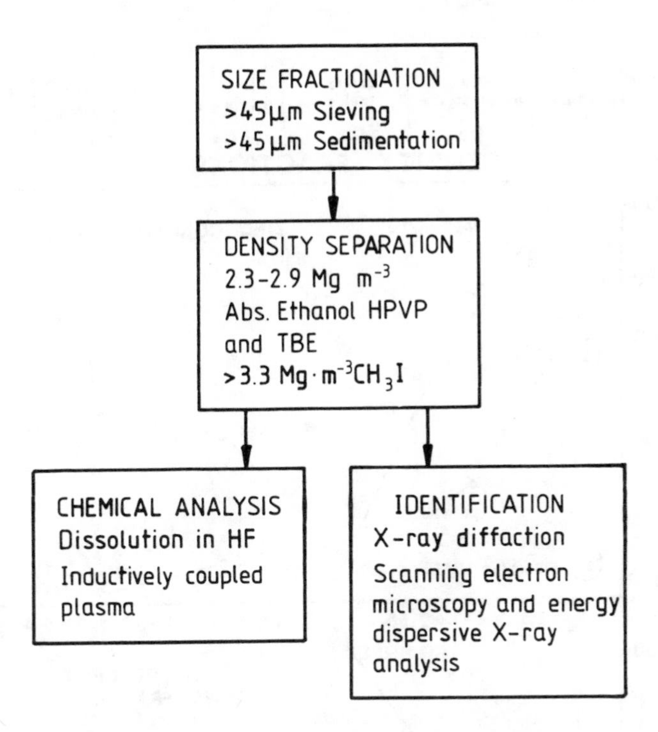

FIGURE 2. Analytical scheme for sediment partitioning by physical separation. (From Mattigot, S. V., *Scanning Electron Microsc.*, II, 611, 1982. With permission.)

(SEM/EDX) has been used to separate and directly identify solid-phase trace element species in such diverse material as fly ash,[10,11] mine waste-contaminated soil,[12] and sediments.[5,13,14] A flow chart of such an analytical scheme is shown in Figure 2.

Selected data from an extensive study on near-shore marine sediments collected from sites in Spencer Gulf, South Australia, offshore of a large lead smelter, are presented in Figure 3. These data are based on the use of a heavy liquid density gradient separation technique, developed by Warren and co-workers.[13,14] This scheme involves sediment separation into six distinct density subfractions, as listed in Table 1. Purified tetrabromoethane (TBE)/acetone mixtures were prepared to give a series of liquids of specific density from 2.2 to 2.75. The liquids, in order of decreasing density, were layered into centrifuge tubes. A small portion of a sized sediment fraction was placed on the surface of the topmost density layer. After centrifugation, distinct layers of particles of similar density formed at the interfaces between the liquid bands and were selectively removed. Each density subfraction was subjected to mineralogical analysis by X-ray diffraction, and finally the subfraction was chemically analyzed. The sediment components were separated into organic debris, conglomerates, quartz + calcite shells, magnesian calcite shells, aragonite shells, and heavy minerals. Three samples (C, D, and F) out of those presented by Dossis and Warren[14] in which the grain-size interval from 10 to 1000 µm was subdivided into the six density fractions (Table 1) are shown in Figure 3. Despite the varying absolute metal contents (e.g., zinc), the concentration curve of each sample, within a density spectrum, is relatively similar. The highest metal concentrations occur in the heavy mineral fraction (D6), having a density <2.95 g/cm^3. Another maximum occurs in the light density fraction (D1) and is markedly present in the organic debris of sample C. It is also relatively high in the agglomerates of sample D. Minimum concentrations occur entirely in the density grade (D3) between 2.55 and 2.66 g/cm^3, which is chiefly represented by quartz grains. In fraction D5 (shell fragments

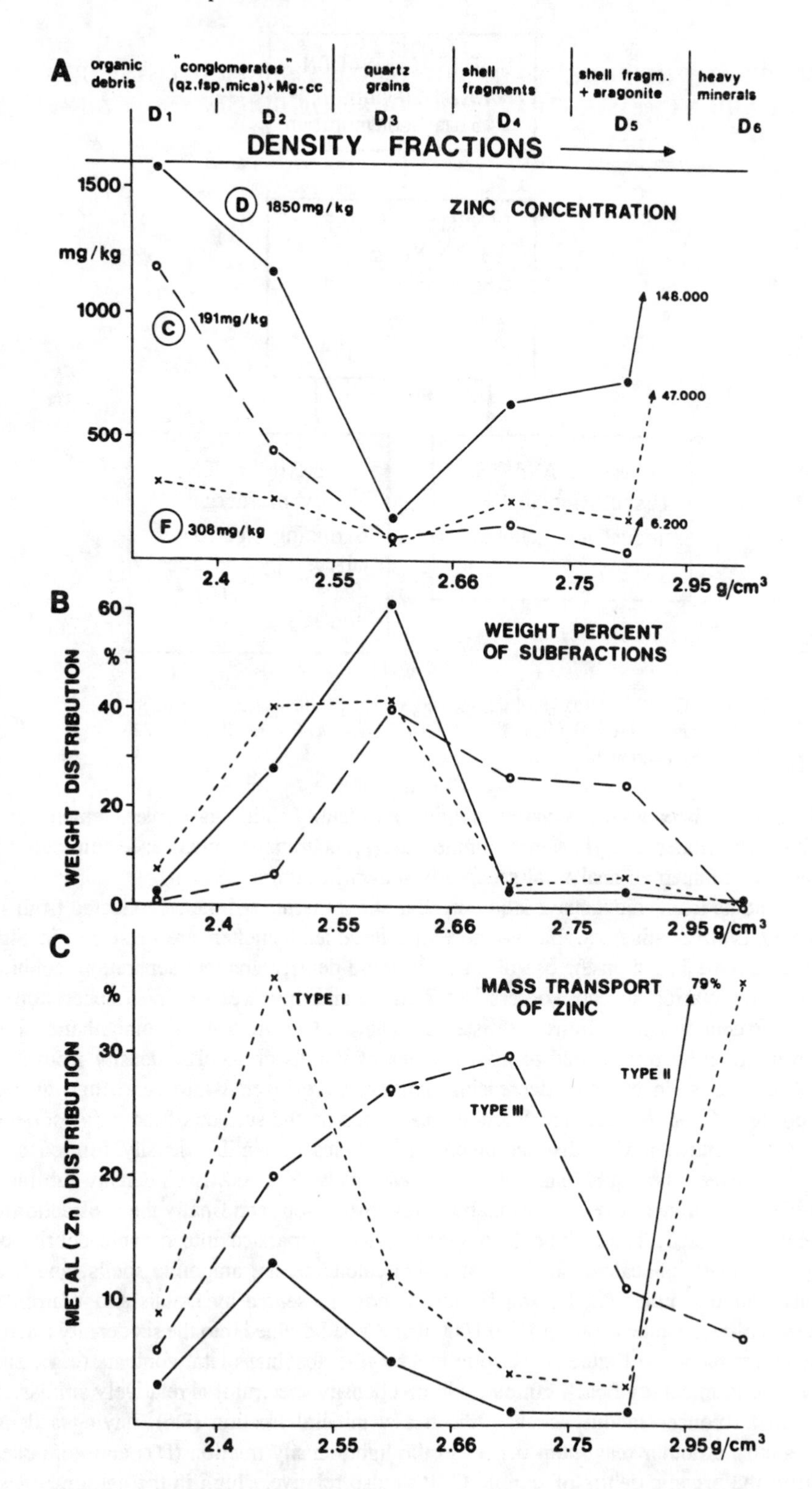

FIGURE 3. Density distribution of solid phases and zinc concentrations in three sediment samples from Spencer Gulf, South Australia. (From Förstner, U., *Hydrobiologia*, 91, 269, 1982. With permission.)

Table 1
MINERAL COMPONENTS OF THE DENSITY SUBFRACTIONS OF THE GRADIENT SEPARATION PROCEDURE OF DOSSIS AND WARREN

Density subfraction[a]	Mineral Components as determined by light microscope, XPD, and differential thermal analysis
<2.40	Organic debris; clay agglomerates[b] with magnesian calcite, quartz, and goethite
2.40—2.55	Agglomerates of magnesian calcite, quartz, mica, kaolin, traces of goethite and feldspar
2.55—2.66	Quartz and calcite shells
2.66—2.75	Magnesian calcite shells and shell fragments
2.75—2.95	Aragonite shells
>2.95	Heavy minerals (especially tourmaline, ilmenite, magnetite, and other opaque particles)

[a] Density separations in g/cm^3 using bromoform.
[b] The authors used the term ''conglomerates'' which may confuse with the geological term.

From Dossis, P. and Warren, L., in *Contaminants and Sediments,* Vol. 1, Baker, R., Ed., Ann Arbor Science, Ann Arbor, MI, 1980, 119. With permission.

and aragonite platelets), which shows a higher density, the zinc, lead, and cadmium contents are about three times higher in the calcite shells than in the aragonitic shells.

These investigations show that the heavy mineral fraction is the decisive component of trace metal concentrations in the polluted sediment samples, at least in the grain-size fraction 10 to 1000 μm. Thus, the environmental significance of lead in the mine waste-contaminated sample D was reduced, because it was present in insoluble heavy mineral particles. A different concentration-distribution relationship was observed in sample A, where the organic debris not only had the highest concentration of all three metals studied, but was also the major reservoir of the metals in the sediment.[14] Provided that the sediments contain only a small proportion of fine particles, accurate and reproducible data can be obtained by this technique.

A similar technique more suited to density separation of fine silt and clay particles is the linear density gradient method of Mattigod and Ervin.[15] Investigations performed by Loring[16,17] on sediments of the St. Lawrence estuary and the open gulf indicated that 61 to 98% of total Zn, Cu, Pb, Cr, Ni, and V concentrations were held detritally in various sulfide minerals (pyrite, chalcopyrite), oxide minerals (magnetite, ilmenite, chromite), and ferromagnesian silicates (amphiboles, chlorite, pyroxene, and garnet). Detrital host mineral concentrations were highest in the estuary, decreasing seaward due to the interior sources of terrigenous clastic material. In another study, in which sediments from the lower Elbe River (FRG) were split into three separate density fractions, Lichtfuss[18] showed an enrichment of Cr, Fe, Co, and Mn in the mineral fraction (>2.7 g/cm^3), of Mn, Fe, Co, and Pb in the organo-mineral aggregate fraction (2.2 to 2.7 g/cm^3), and of Hg, Zn, As, Cu, and Ni in the organic fraction (<2.2 g/cm^3). The metals bound to the organic fractions could be partially bioavailable.

A compilation of examples of autochtonous mineral formations and their effects on the enrichment or dilution of trace metals in recent lake and marine deposits, as obtained from physical separation experiments, is presented in Table 2. These data, however, should not be generalized to cover all cases, since the separation of mineral phases is often not completely satisfactory. In fact, Guy et al.[19] pointed to the problem that in real sediments, the components important to trace metal binding are not present as simple, discrete compounds, but as coatings on sand and clay particles, using a simple but impressive experiment to simulate real sediment. Clay (100 mg) and MnO_2 (20 mg) were suspended in distilled water for 24 h, centrifuged, dried, and ground. The resulting ''sediment'' when added to bromoform did not separate into two layers but formed a uniform dispersion of particles throughout; that

Table 2
EFFECTS OF AUTOCHTHONOUS MINERAL PHASES (Fe/Mn OXYHYDRATES NOT INCLUDED) ON THE ENRICHMENT (+,++) OR REDUCTION (−,−−) OF HEAVY METALS IN LAKE AND MARINE BASIN SEDIMENTS

	Type of source			
Mineral phase	**Endogenic[a]**	**Authigenic[b]**	**Lake and marine basin examples**	**Effect (examples)**
Carbonates				
Aragonite	×		Aci Crater Göl (Turkey)	−− all metals
Calcite	×	×	Bande-Amir (Afganistan)	− all metals
Mg calcite	×	×	Lake Balaton (Hungary)	− Fe, ± trace metals
Dolomite		×	L. Neusiedl (Austria/Hungary)	± Mn, Zn; − Ni, Cr
Rhodochrosite		×	L. Michigan (U.S.)	+ to ++ trace metals
Siderite		×	Black Sea, Birket Ram (Israel)	+ Mn, ± trace metals
Sulfates				
Gypsum	×		Australian Lakes	−− all metals
Glauberite	×		Curtain Springs (Australia)	−− all metals
Alunite	×		L. Brown (Australia)	± Pb, − other metals
Opaline silica	×		L. Malawi (East Africa)	− Cu, −− other metals
Sulfides				
Hydrotroilite	×	×	L. Constance (West Germany)	+ Ni, Cu, Cr
Pyrite		×	Black Sea, L. Kivu (East Africa)	± Pb, Zn, Mn ++ Cu, Ni, Co, Ni
Nontronite		×	L. Malawi	+ Mn, Co; ± other metals
Phosphates		×	L. Geneva (Switzerland)	+ Fe, Co, Cu, Zn, Pb(?)

[a] Endogenic fractions refer to minerals originating from processes occurring within the water column in response to effects such as precipitation, sorption, enrichment in organisms, and organometallic interactions. They exhibit a distinct temporal character, often as a result of the variation of the organic productivity, e.g., for carbonate precipitation.

[b] Authigenic (or diagenetic) fractions include those minerals which result from processes occurring within sediments once they have been deposited, mainly as a result of the decomposition of buried organisms.

Adapted from Förstner, U., in *Handbook of Strata-Bound and Stratiform Ore Deposits,* Wolfs, K. H., Ed., Elsevier, Amsterdam, 1981, 179. With permission.

is, no separation was achieved. Addition of a surfactant, however, may significantly improve separation of fine particles in TBE.

Amorphous Fe-oxide coatings appear to be the most significant in affecting both surface area and sediment-trace metal levels. This relationship has been recently demonstrated in a study of interrelations between specific surface area and trace element geochemistry in fluvial sediments.[20] The results supported the view that external surface area, as determined by the one-point BET method, is a function of both grain size and of composition (geochemical phase). Removal by a sequential extraction procedure (see below) of such geochemical adsorbents for trace elements as carbonates, oxyhydrate coatings, and organic matter, decreased the surface area. Results indicated that the same trace metal adsorbates (e.g., carbonate-extractable or acid-reducible coatings) may act as either a surface-area inhibitor (its removal produces an increase in surface area) or a contributor (its removal produces a decrease in surface area), depending upon the median grain size of the sample. Horowitz and Eldrick[20] stressed that although coatings may make a contribution to sample surface area in their own right, this contribution is only significant in the case of coarse-grained material which characteristically has a low surface area. For fine-grained material with a higher surface area, the effect is indirect in that the coatings cement fine grains together, forming water-stable aggregates. Removal of the "cement" breaks down the agglomerates to their original, smaller component particles which have larger surface areas than their coatings. Some of the long-standing disagreements in the field of sediment geochemistry are addressed in this work. The variable results in physical separation approaches may be due to differences in methods used to accomplish the separation. The more vigorous the separation, the less likely are aggregates to be retained. It is apparent that the break down of these aggregates by either physical separation or chemical leaching will increase the surface area of the sediment causing enhanced readsorption of either the major and/or the associated trace elements.[20]

The halogenated lower alkanes commonly used in heavy liquid separations, in particular TBE, are known to be carcinogenic. An acute TBE poisoning with serious liver damage to a chemist, who is exposed only 1 d (about 7 h) to a concentration of 1 to 2 ppm and during 10 min to a concentration of about 16 ppm, may occur, according to the latest edition of *Ullmann's Enzyklopädie der Technischen Chemie*. This problem, however, has been recently overcome by introduction of sodium polytungstate, $Na_6[H_2W_{12}O_{40}]$, as a new substance for the isopycnic density gradient centrifugation.[21] Among several other advantages, the neutral (pH 6) and chemically inert aqueous solution of the inorganic salt was shown to be nontoxic by ingestion, to have no obnoxious smell, and to permit a density range adjustable from 1 to 3.1 g/cm^3. The viscosity increase was insignificant up to a density of about 2.5 g/cm^3. Accordingly separations in the silt and clay size range are possible within shorter periods. Trace metal analysis can be performed providing the density fractions are cleaned with distilled water to remove any residue. In view of the safety and ease of handling, this material is a reasonable improvement over halogenated heavy liquids widely used in trace element solid speciation assessment by density fractionation.

B. Nondestructive Microanalysis Techniques

1. X-Ray Fluorescence Techniques

A large number of sophisticated instrumental techniques mainly based on the use of focused particle beams (photons, electrons, protons, ions, etc.) is available to directly determine element accumulations in sediments and soil components. Selected new developments of particular importance for geological analysis are presented in Table 3. Two instrumental approaches have been used to determine solid speciation. The first attempts to determine the associations of trace elements with major ones — the so-called *element-specific* approach. The second approach attempts to determine where trace elements are retained on or by solid matter — the so-called *phase-* or *site-specific* approach. Both types of analytical measurements may be performed on a single particle or on a field of particles, although individual particle analysis requires microprobe capabilities. Many techniques permit the user to visually select the analytical area, and perform both morphological and chemical analysis on the same site. This allows the determination of solid-phase association "chemical mapping" from the derived data.

Microanalysis techniques based on electron microprobes are the most widespread methods for the element-specific approach to solid speciation. However, several problems limit their utility to trace element speciation. The foremost is the co-excitation of a high bremsstrahlung background, which sets the detection limit usually higher than 100 μg/g. Norrish et al.,[5] e.g., reported detection limits for various elements using an electron microprobe equipped with wavelength-dispersive X-ray spectrometers based on counting statistics, as follows (in micrograms per gram): Zn 240, Pb 400, Cd 350, S 200, Mg 160, Ca 100, Si 140, Al 160,

Table 3
INSTRUMENTAL TECHNIQUES USED IN SOLID SPECIATION

Abbreviation	Technique	Ref.
AES	Auger electron spectroscopy	39,44
BA	Beta autoradiography	31
EPXMA	Electron probe X-ray microanalysis	5
EPR	Electron paramagnetic resonance (electron spin probe)	51,52,53
EXAFS	Extended X-ray absorption fine structure	34
HPLC-ICP/AAS	High-performance liquid chromatography, coupled with element-specific detectors	48 (see also Chapter 5)
LAMMA	Laser microprobe mass analyzer	49,50
PIXE	Proton-induced X-ray emission (proton probe)	28,29,30,31,32,46
SIMS	Secondary ion mass spectrometry (ion probe)	33,39,43
SEM/EDX	Scanning electron microscopy with energy-dispersive X-ray analysis	22,23,24,25,26,27
XPD	X-ray powder diffractometry	36,37,35
XPS(=ESCA)	X-ray photoelectron spectroscopy	39,44

Fe 350. This tends to limit the usefulness of these procedures to such cases as identification of specific trace metal or metalloid mineral particles or highly contaminated sediment samples. Where single grains rich in the element of concern can be located by electron imaging, quantitative elemental analyses can be carried out and the mineral phase can usually be identified. This technique is particularly useful for heavy mineral grains separated using density gradient centrifugation. Ramamoorthy and Massalski[22] succeeded in analyzing particles high in mercury in Ottawa River sediments by SEM/EDX. Suspended sediment samples were deposited on a polycarbonate membrane filter, dried, and then mounted on an aluminum stub for SEM/EDX analysis. Mercury was found to be associated with particles rich in sulfur and organic fibrous material of biologic origin. Jedwab[23] identified discrete particles in suspended matter from Atlantic and Pacific waters containing trace metals as major constituents by light microscopy and SEM/EDX. The structure of the particles extracted with a micromanipulator were examined with the aid of a microdiffraction X-ray camera. The combination of all these three techniques provided the detection and identification of individual particles of malachite, tenorite, paratenorite, laurionite, and sphalerite, as well as several intermediate compounds of Cu, Zn, Pb, Cl^-, $S^=$, OH^-, and $CO_3^=$.

Electron microprobe techniques in combination with X-ray diffraction are especially valuable in the identification of trace metals associated with sulfide minerals in recent marine sediments. Absence of significant oxidation of particulate sulfides upon drying, even when preparation is carried out in air, is an important prerequisite. Rapid oxidation of materials such as finely dispersed amorphous iron sulfide (hydrotroilite) in black mud may occur under aerobic conditions. In this case, the element-sensitive approach of microprobe analysis may provide more accurate results, since the elements remain in place. Ferris et al.[24] examined the origin and chemical nature of micron-sized sulfides associated with bacteria cells in sediments, and observed that nickle was precipitated in the form of millerite microcrystals. Luther et al.[25] provided micrographs of framboidal sulfide minerals which contained a cation other than iron as the major constituent in recent estuarine sediments. Nickel, the bulk concentration of which in these sediments was found to be rarely higher than 100 μg/g (dry weight), was enriched in pyrite crystals. Individual zinc sulfide framboids with a Zn/Fe ratio exceeding 3:1 have been found in sediments when the bulk zinc concentrations are 1000 μg/g or higher. In sediments with lower zinc concentrations, traces of zinc have been found in the various iron sulfide crystals. All crystals, whether separate or in aggregates, fall in a size range of 0.5 to 7 μm.

Norrish et al.[5] were the first to distinguish sulfide from sulfate sulfur, using a germanium analyzing crystal in their wavelength-dispersive X-ray spectrometer. The two SKα-lines, separated by 6′ (2θ), were partitioned and quantified using specific peak deconvolution equations. This represents a significant step forward in direct speciation, since no sophisticated separation is necessary for identifying sulfide phases by XRD or TEM structure analysis. Using this technique in combination with size and density gradient fractionation allowed the comparing and contrasting of elemental associations in a particular sediment component from sampling sites with different degrees of contamination. A significant shortcoming of this analytical technique due to the limited spatial resolution of the electron microprobe is that no clear-cut evidence could be provided as to whether the fine-grained precipitates are a single phase or mixtures of distinct sulfide phases.[5]

Microprobes also permit "chemical mapping" achieved by scanning an electron beam and image processing the signals derived from X-ray or electron energy loss spectrometers. High-resolution elemental maps can be obtained down to the low micrometer range. This technique is especially valuable in analyzing the trace metal associations in coatings of sediment and soil particles as well as element distributions in mixed-phase aggregates.[26] Use of chemical mapping avoids problems associated with the impurities in mineral separates.

Microanalysis has a severe disadvantage compared to conventional analysis of bulk material: representative measurements are difficult to obtain because of considerable variations in sample composition even within a set of apparently similar particles. In achieving the specificity associated with characterization of a single particle, one sacrifices statistical information unless a large number of particles is investigated serially, or a field of particles is studied in a single measurement. Instrumental solid speciation techniques have been recently advanced by the incorporation of expert systems in the routinely used electron probe X-ray microanalysis allowing "intelligent" particle selection, identification, and characterization. This technique was used by Bernard et al.[27] to automatically analyze suspended particulate matter for 11 major elements on a particle-by-particle basis using an image analysis system. The large acquired data set was interpreted using numerical multivariate analysis. This procedure provided 13 geochemically different particle types, the relative abundance of which in a study along the Ems estuary allowed the study of estuarine mixing. This represents the present stage in quantitative characterization and classification of fine particles by analytical microscopy for suspended matter and air pollutants.

In X-ray fluorescence microanalysis, to achieve detection limits in the nanogram- to microgram-per-gram level within an analytical domain in the square-micron range, which is the most important in geochemical sediment pollution studies, the signal to noise ratio has to be maximized. Relative to incident electrons, incident heavy charged particles, e.g., protons, produce orders of magnitude less of continuous radiation (bremsstrahlung) due to their much greater mass. The cross-sections for characteristic X-ray line emission are similar for 1- to 5-MeV protons and 10- to 50-KeV electrons. The X-ray signal to beam-induced background is consequently much higher for heavy charged particles compared with electrons.

Proton-induced X-ray emission (PIXE) microanalysis, or "proton microprobe", has been widely applied in geochemical and biological studies.[29-31] While most studies are performed on large-area samples, a few facilities have developed intense microbeams down to 2 μm by magnetic or electrostatic focusing.[30,32] Many old tandem (Van de Graaff) accelerators are presently available in physical institutes, which are with potential capability for proton microbeam X-ray fluorescence studies. For the 2-, 3-, and 4-MeV energies typically used

Table 4
DETECTION LIMITS OBTAINED FOR A SUSPENDED MATTER SAMPLE BY THE HAMBURG PIXE, USING A 2 × 2-μm PROBE

Elements	Kα[a] (ng/cm²)[b]	Lα[a] (ng/cm²)[b]	(μg/g)
Si	458		193
P	105		44
S	52		22
Cl	35		15
K	26		11
Ca	24		10
Sc	23		10
Ti	23		10
V	24		10
Cr	23		10
Mn	19		8
Fe,Co	20		8
Ni	21		9
Cu,Zn	24		10
Ga	29		12
Ge	37		16
As	44	634	18
Se	61	655	26
Br	84	684	35
Rb	125		53
Sr	143		60
Zr	220	290	93
Nb	289		122
Mo		97	41
Ag		49	21
Cd		47	20
In,Sn,Sb,Te,I		45	19
Cs,Ba		46	19
Hg		76	32
Tl		78	33
Pb,Bi		82	34
U		208	88

[a] X-rays emerging from the K- or L-shell, respectively.
[b] Element mass density on membrane filter loaded with 0.6 mg/cm² suspended matter (mainly inorganic).

From Grossmann, D., Kersten, M., Niecke, M., Puskeppel, A., and Voigt, R., *Mitt. Geol. Palaeontol. Inst. Univ. Hamburg,* 58, 619, 1985. With permission.

in PIXE analysis, the depths of X-ray production in a mineral are roughly 25, 60, and 100 μm.[31] Consequently, the depth and spatial resolution of a sample analyzed may be set by X-ray production, by X-ray absorption, or by the thickness of the sample. It is unnecessary to obtain beam sizes much smaller than 10 to 30 μm to acquire optimal spatial resolution when using standard petrographic thin sections or smear slides. Where better spatial resolution is required, ultrathin samples mounted on plastic films, as used in TEM analysis, may be used. This adaptation has recently been used to obtain spatial trace element analysis of suspended matter on Nuclepore® filter sheets.[32] The results and detection limits obtained in this study are summarized in Table 4. Major/minor element interferences typical for PIXE analysis of geological samples have been avoided using an Al absorber window on the Si(Li) detector, resulting in detection limits for trace metals in the lower microgram-per-gram level. Such amounts are undetectable in routine electron microprobe analysis. For trace element geochemistry, the proton microprobe technique offers significant advancements over conventional electron microprobes, and it is in this area that the greatest contributions for trace element association studies by X-ray fluorescence analysis are likely to be made.

2. Other Techniques

Elemental analysis based on X-ray fluorescence techniques are usually insensitive to chemical species, and trace element forms must be deduced from characteristic element associations. Direct structure analysis of the solid phases, using electron diffraction techniques, requires crystalline phases of a minimum size (ca. 20 to 100 nm) to obtain a diffraction pattern.[33] Trace metals and metalloids in contaminated recent sediments and soils are, however, often associated with amorphous or cryptocrystalline phases which additionally may be dispersed in major components. Information at least on coordination and valence states of elements in microdomains can be derived by detecting the chemical shifts in extended X-ray absorption fine structure (EXAFS). EXAFS differs from X-ray diffraction in as much as it is of shorter range and is thus capable of analyzing solid components which are amorphous in nature. This solid speciation method, though providing information at the molecular level, is to date restricted to high element concentrations and requires very sophisticated deconvulution techniques, but is very well suited to surface speciation studies.[34]

X-ray powder diffraction (XPD) analysis has been applied to the qualitative identification of trace metal phases collected from ambient aerosols, using dichotomous samplers to distinguish between the fine and the coarse particulate fractions. Direct XPD analysis of airborne particulate material collected on cellulose ester membrane filters in the St. Louis area has given results for both fine and coarse particle fractions; the coarse fraction samples being predominantly α-quarz and calcite, whereas the fine fractions are mainly combinations of sulfate and ammonium sulfate species.[35] The most frequently observed compounds were $(NH_4)_2SO_4$ and the double salts $(NH_4)_2SO_4 \cdot PbSO_4$ and $(NH_4)_2SO_4 \cdot H_2SO_4$. The lead sulfate salt was suggested to be associated with discrete source activity, although the formation of this compound by interaction of automotive PbBrCl with atmospheric $(NH_4)_2SO_4$ could not be ruled out. Conventional XPD is usually not used for quantitative analysis of crystalline compounds due to the relatively low sensitivity and severe matrix effects and a lack of appropriate standards which duplicate urban aerosols. A significant sensitivity improvement has recently been achieved by introducing a curved position-sensitive proportional counter for transmission XPD.[36] This method is suitable for quantitative X-ray analysis of multiphase mixtures and is based on transmission measurements of thin film samples (cf., e.g., Fukasawa et al.,[37] for preparation of appropriate airborne particulate samples). This method has the advantage that only a few (1 to 5) milligrams of substance are required when combined with adsorption measurements from defined circular parts of these samples. No limitation on the number of phases exist because matrix problems do not occur in the transmission mode. The automated method is obviously reliable, since the measured intensities and absorption factors can be checked for consistency. While no metal speciation studies are known by the authors to have been performed with this powerful tool, it should find wide acceptance in future work.

Since adsorption of pollutants on airborne and waterborne particles is a primary factor in determining the transport, deposition, reactivity, and potential toxicity of these components, analytical methods should be related to the chemistry of the surface of the particle and/or to the metal species highly enriched onto the surfaces. Keyser et al.[38] showed that a number of potentially toxic trace metal and organic species are highly enriched at the surfaces of many types of environmental particles. At present there are three techniques most frequently used in trace surface analysis: X-ray photoelectron spectroscopy (XPS or ESCA), Auger electron spectroscopy (AES, now also in high resolution scanning mode), and secondary ion-mass spectrometry (SIMS).

The XPS technique employs a soft X-ray source to eject core-level electrons from the sample surface. Analysis of the kinetic energy of the photoejected electrons provides chemical bonding information since the bonding energies of the core electron are sensitive to changes in the electronic structure ("chemical shifts") of the valence level. XPS spectra have been used to identify the oxidation state and quantitative chemical composition of surface species.[39] Dillard and co-workers have carried out extensive studies on the chemistry of specific ions adsorbed on the surface of a number of minerals, including the adsorption of chromium ions on clay and the adsorption of cobalt, lead, and nickel on ferromanganese nodules.[40-42] XPS measurements have shown that oxidation of Co(II) and Pb(II) occurs upon adsorption onto synthetic manganese oxide (disordered birnessite). In nodules containing predominantly σ-MnO_2, Co(III) was identified by examination of the photopeak structural features. In a nodule composed of todorokite, Cu(II) and Ni(II) were detected in oxide environments, but the bulk quantity of cobalt, though as much as 1820 μg/g, was insufficient to be studied in detail. Normally the detection limit of XPS for surface analysis lies in the range 0.1 to 1% by weight, which is insufficient to enable observation of trace constituents unless considerable surface enrichment is encountered. The utility of XPS for microprobe analysis is limited because of the difficulty of focusing X-rays to a diameter smaller than 1 mm, although X-ray microscopy with focused synchroton radiation has been developed recently.[33]

Rapid progress is presently being achieved with SIMS which has the highest detection power among the instrumental surface analysis techniques available. SIMS enables trace element surface speciation studies, even by microprobe analysis, in which the primary ion beam (most commonly negative oxygen ions) bombarding the sample surface can be focused to a diameter of about 3 to 5 μm.[38] Although only about 1 to 10% of the sputtered material is in the form of secondary ions that are mass analyzed by a double-focusing conventional mass spectrometer, it is possible to observe as little as 1 μg/g in the analytical volume. This technique has been used to determine surface predominance of Na, Li, K, S, C, Zn, Pb, Tl, Mn, Cr, and V on fly ash.[43] These findings were confirmed by other surface analytical techniques (AES, XPS) and solvent leaching of the fly ash particles.[44] Moreover, a combination of thermal treatment with acid dissolution procedures suggested the presence of carbonate species for Li, Ni, K, Rb, Mg, Sr, Ba, and Cu in fly ash samples.[44] Conditions for direct determination of metal speciation in solids are particularly favorable in oil fly ash, which may contain high amounts of vanadium ($\leq$10%) and nickel (2 to 3%). By means of X-ray diffraction, the presence of V_2O_5 and $VSO_4{\cdot}2H_2O$ was identified in total and water-soluble fractions of oil-fired fly ash samples.[45] More recently, VO_2^+ was identified as the principal vanadium species in fly ash samples by means of combined colorimetric ion chromatography (IC) and PIXE analysis.[46] These results are particularly remarkable, since the VO_2^+ cation, presumably as $(VO_2)MSO_4$ or $(VO_2)_2SO_4$ species in the acidic oil fly ash, is highly toxic.

For several reasons, speciation studies of metals on a molecular level are more promising for natural organic materials.[47] While the identification of the ligands as individual chemical compounds may be impossible, since the mechanisms of their formation lead to a continuum of structures which does not allow individual compounds to be isolated, the organic group-element bond is in most cases sufficiently stable over time and in different media to allow the separation of organometallic compounds from their matrix with concentration to levels required for detection and identification. This is important in analysis of organic matter extracted by sequential chemical extraction techniques. Instrumental fractionation and concentration techniques include the coupling of chromatographic separation with element-specific detection. Fractionation of organometallic species of Hg, As, Se, Pb, Sb, Sn, Ge, and Co with the aid of high-performance liquid chromatography (HPLC) in combination with conventional AAS and ICP techniques as sensitive element-specific detectors has been widely used[48] and is discussed in Chapter 6.

The interaction of organometallic forms with soil and sediment organic matter can be examined *in situ* with the laser microprobe mass analyzer (LAMMA). The principle of LAMMA is based on the excitation of a microvolume of the sample to an ionized state by a focused laser beam. The analytical information is derived from mass spectrometry of the molecular fragments. The nominal diameter of the laser focus on the sample can be varied by conventional optics down to 1 μm. The recorded LAMMA spectra show some similarity with the corresponding mass spectra obtained by SIMS. The disadvantage is that, in principle, ions generated both by laser-induced desorption and secondary formation in the microplasma cannot be discriminated. This technique is therefore more valuable as a complement to SIMS and ESCA. Although the usual thickness of the specimen for LAMMA analysis lies between 0.1 to 1 μm, Henstra et al.[49] demonstrated that microchemical analysis of thin sections of soils with a thickness of more than 15 μm could be achieved by ''laser milling''. Muller et al.[50] used this technique to identify vanadium complexes in natural asphaltenic fractions both *in situ* and in DMF-extractions.

Another technique worth noting for studying trace metal complexes in untreated and especially extracted fractions of soil organic matter is electron paramagnetic resonance (EPR) spectroscopy. Many attempts have been made to relate the hyperfine structures of EPR spectra to the forms of metals present in the organic solid matter. Theoretically EPR spectra can be used to estimate molecular weights of complexes formed and approximate metal-to-metal distances in the organic complexes.[51] EPR was applied to extracted organic fractions from an arable soil and showed that Cu was present partly as a copper porphyrin-type complex in the humic acid, but not in the fulvic acid fraction.[52] Senesi et al.[53] studied Cu^{2+} binding to an fulvic acid (FA) sample isolated from soil and showed that EPR spectra varied considerably with Cu/FA molar ratios. At high Cu^{2+} loading the signal was unresolved because of excess aqueous Cu^{2+}, but at low loading the resolved signal indicated the donor atoms, e.g., four oxygen or two oxygen plus two nitrogen, bound to the fulvic acid. Removal of excess Cu^{2+} with a strong cation-exchange resin left the copper complexes nearly unchanged, while resolving the spectra.

III. SELECTIVE EXTRACTION METHODS

A. General Considerations

Wet chemical extractions provide a convenient means to determine the major accumulative phases for metals in sedimentary deposits and mechanisms of their diagenetic transformation. A general goal of all studies involving selective chemical extractions is the accurate determination of partitioning of elements of environmental concern among discrete phases of a sample. Mineralogical constituents of sediment considered important in controlling metal concentrations in sediment are hydrous oxides of iron and manganese, organic matter, and clays. Fractionation is usually performed by a sequence of ''selective'' chemical extraction techniques which include the successive removal of these minerals and their associated metals. The concept of chemical leaching is based on the idea that a particular chemical solvent is either phase or mechanistic specific (e.g., buffered acetic acid will attack and dissolve only carbonates, neutral magnesium chloride will only displace adsorbates).[54] Interest in selective sequential extractions in geochemistry has grown since Le Riche and Weir[55] recognized various chemically bound forms of trace elements in soils. This has led to a variety of solvent sequences, developed for specific tasks.

Despite the clear advantages of a differentiated analysis over investigations of the bulk chemistry of sediments, verification studies conducted in recent years indicate that there are many problems associated with the "operational" speciation by partial dissolution techniques. It is common for studies in wet chemical extraction to point out that the various used extractants are not as selective as expected. Van Valin and Morse[56] concluded that, "the concept of an operationally defined element reactivity is generally used rather than attempts to individually characterize each phase." Some reviewers came to conclusions as pessimistic as, "these techniques represent nothing but an operational tool and complementary approach until physical techniques be available with the new generation of microprobes and other sophisticated instruments."[57] On the other hand, the results of the comparative study performed by Lion et al.[58] suggest that, "the role of sediment components may be evaluated from the perspective of competitive adsorption phenomena and that sediment adsorption characteristics and extractant-determined component-metal associations yield consistent information on the binding of metals." In light of the ease of performing trace element speciation with extraction techniques relative to any of the sophisticated instrumental and computer-aided modeling approaches, considerable frustrations may, however, arise, since, "a careless usage of these techniques without an appreciation of their pitfalls and limitations must lead to further generation of erroneous or misleading data."[59]

There is no general agreement in the literature on the solutions preferred for the various sediment and soil components to be extracted, due mostly to the "matrix effects"[57] involved in the heterogeneous chemical processes. The most appropriate extractants are determined by the aim of the study, by the type of solid material (sediment or soil, sewage sludge, fly ash, dredged harbor mud, street dusts, manganese nodules, and so on), and by the elements of interest. There exists a vast literature on specific research areas, in which appropriate extractant formulations may be found for a selected problem. More recent reviews exist in such fields as geochemical prospecting,[60] trace metal speciation in soils in general[61] and more specifically in sewage sludge-amended soils,[62] bioavailability of trace metals in estuarine oxic sediments,[63] deep sea geochemistry,[64] and fly ash characterization.[65] The original papers usually give many experimental details such as solid to solution ratios, treatment times, analytical technique problems, and similar useful suggestions. Some examples are presented in the last section on case studies.

Ideally, partial dissolution techniques should incorporate reagents which are sensitive to only one of the various soil or sediment components significant in trace metal binding. In sequential multiple extraction techniques, chemical extractants of various types are applied successively to the sample of soil or sediment, each follow-up treatment being more drastic in chemical action or of a different nature than the previous one. We will follow the hierarchical ranking of binding strength of metals to the various soil components with the most labile exchangeable fraction considered first, and the more stable carbonate, reducible, oxidizable, and residual fractions discussed subsequently. Although "selectivity" for a specific phase or binding form, in a strictly thermodynamic sense, cannot be expected for these procedures, there are also differences in the specificity between the various extractants and methods used. The compilation given by Pickering[59] (Figure 4) implies that any of the more common extractants to be discussed only crudely differentiate between the different forms of selected elements in sedimentary material. In particular, a single extraction method is not selective by itself. While the first "exchangeable metal binding form" is selectively displaceable by weak extractants, the reagents used for the components considered subsequently are all "nonselective" in that they are coextracting more or less extensively the more readily soluble components. The extractants discussed here, however, usually constitute one step in a scheme of sequential multiple extraction rather than being used as single extractants. Thus, our evaluation of the selectivity and effectiveness of reagents for partial dissolution techniques will be facilitated since we have only to take into account reagents which are able to differentiate between the component to be considered and the next less soluble component, but not the last extracted, more mobile component. A careful combination

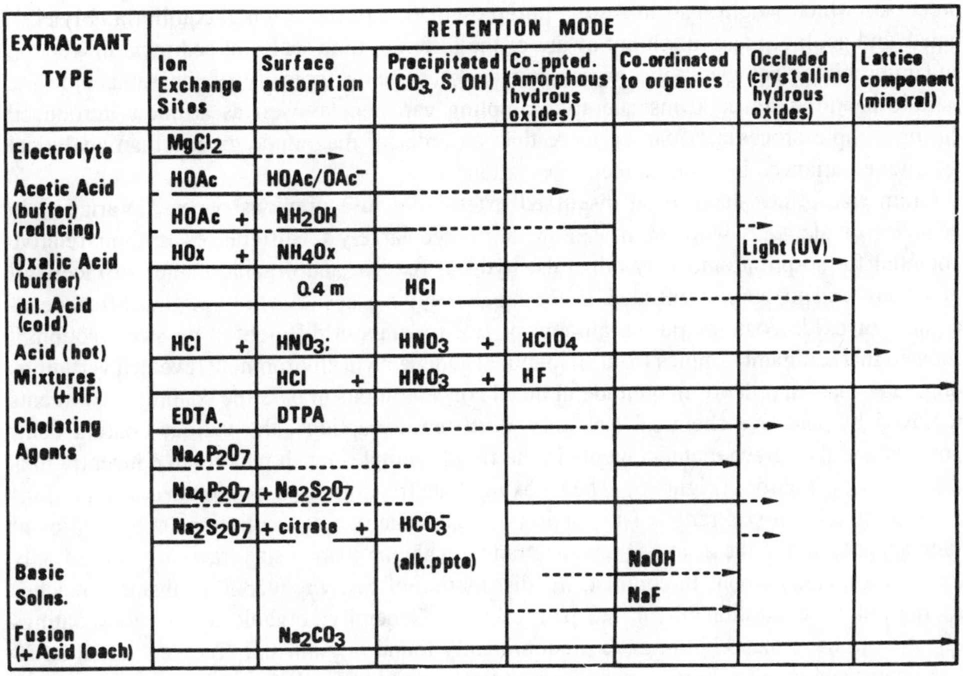

EXTRACTANT TYPE	RETENTION MODE: Ion Exchange Sites	Surface adsorption	Precipitated (CO_3, S, OH)	Co-ppted. (amorphous hydrous oxides)	Co-ordinated to organics	Occluded (crystalline hydrous oxides)	Lattice component (mineral)
Electrolyte	$MgCl_2$						
Acetic Acid (buffer)	HOAc	$HOAc/OAc^-$					
(reducing)	HOAc +	NH_2OH					
Oxalic Acid (buffer)	HOx +	NH_4Ox				Light (UV)	
dil. Acid (cold)		0.4 m	HCl				
Acid (hot)	HCl +	HNO_3;	HNO_3 +	$HClO_4$			
Mixtures (+HF)		HCl +	HNO_3 +	HF			
Chelating Agents	EDTA,	DTPA					
	$Na_4P_2O_7$						
	$Na_4P_2O_7$	+ $Na_2S_2O_7$					
	$Na_2S_2O_7$	+ citrate +	HCO_3^-				
Basic Solns.			(alk.ppte)		NaOH		
					NaF		
Fusion (+Acid leach)		Na_2CO_3					

FIGURE 4. Schematic representation of the ability of different extractants to release metals retained in different modes or associated with specific sediment fractions. Dashed segments indicate areas of uncertainty. (From Pickering, W. F., *CRC Crit. Rev. Anal. Chem.*, 12, 233, 1981.)

of such nonselective extractants in a sequence may then turn these "overlaps" in extraction efficiency to good use in an efficient scheme of selective extraction steps. Rather than discussing the vast variety of extractants so far published, we will concentrate on those extractants which are most commonly used in sequential schemes, with emphasis on artifacts potentially introduced while applying them.

In practice, three major factors may influence the success in selective extraction of sediment components: (1) the chemical properties of an extractant chosen, (2) its extraction efficiency, and (3) experimental parameter effects. Applying sequential extraction schemes, another three factors may involve (4) the sequence of the individual steps, (5) specific "matrix effects" such as cross-contamination and readsorption, and (6) heterogeneity as well as physical associations (e.g., coatings) of the various solid fractions. In evaluating the suitability of an extractant choosen for a specific investigation, all these factors have to be critically considered. Just choosing the most widely applied procedures could yield data of doubtful validity. The validity of selective extraction results will, however, be primarily dependent on the ways in which the samples are collected and preserved prior to analysis.

B. Sediment and Soil Sampling and Handling for Selective Extraction

1. Sampling

Because of the heterogeneity and complexity of solid matter in aqueous systems, care is required during sampling and analysis to minimize changes in speciation due to changes in the environmental conditions of the system. Effective interpretation of analytical results require that sample strategy and preparation be compatible with the questions to be resolved. For example, sampling for pollution mapping has to consider the heterogeneity of the soil matrix in the area of interest by methods such as particle size analysis and geochemical normalization. Sediment sampling must avoid alteration of natural-system biogeochemical processes which would lead to results unrepresentative of the original equilibria. Mineralogists and geologists are often not aware of the different time scales of sediment diagenesis and microbiological metabolic processes, the latter being typically of the same order as sample handling times. Consequently, sampling variance, as well as artifacts introduced during sample processing, can be more than an order of magnitude greater than analytical technique variances in trace element speciation.

Grain size effects have been discussed extensively in a previous paper.[66] Variation in behavior of elements with grain size are attributed largely to differences in their relative potential for sorption onto clay minerals, hydrous oxides, and organic matter surfaces, all of which tend to be concentrated in the smaller grain sizes. A study of the influence of organic or oxide coatings on the amount of fixed metals in different grain size fractions[67] showed that sediment samples from an identical geochemical environment revealed variations of greater than an order of magnitude in metal concentrations in possible coating constituents removed by selective chemical leaching. With few exceptions, the various coating components and their trace metal contents in the fluvial samples reach maximum concentrations in the silt size fractions (Figure 5). The Fe oxyhydrate fractions have maximum concentrations in the 2- to 20-μm-size range. The maximum organic carbon concentrations in the sediment samples occurred in the 2- to 6.3-μm-size range, whereas coarser size fractions showed only traces of organic carbon. In contrast, the distribution of easily reducible manganese reaches its highest concentration in the <2-μm fraction. Generally, organic and oxidic coatings effective in trace metal binding are predominantly found in grain size fractions less than 20 μm. Improved comparability among oxic sediment samples collected at different times and places from a given aquatic system and between different systems can best be obtained by analyzing the fine-grained fraction of the sediment. The <20-μm fraction is used with separation and preconcentration techniques using ultrasonic probes.[68,69] Physical manipulations, however, should be aimed at preserving sediment samples as closely as possible in their original condition. The more recommendable procedure calls for nondestructive grain size separation procedures rather than ultrasonic wet microsieving.[70]

Suspended sediment sampling is most frequently performed by filtration. However, several problems limit its utility in solid speciation studies. The foremost is the quantitative removal of the sediment from the membrane. Both ultrasonic scrubbing[68] and filter dissolution in a solvent extraction procedure[71] have been suggested. In the latter approach, the filter is completely dissolved in an organic phase, while the suspended matter remains in the aqueous phase enabling an effective separation. Removal of sediment from membranes before trace element extraction may not be necessary if the quantity of sediment being extracted is known. Trace elements extractable from the sediment are generally at high concentrations relative to the blank contributed by the membrane.

In recent years, suspended sediment recovery by continuous-flow centrifugation commonly has been used to obtain sufficient sample for speciation (at least some grams, to carry out all the analyses: particle size distribution, mineralogy, total and sequential extractions). Etcheber and Jouanneau[72] provided a comparative study of suspended matter separation by filtration, continuous-flow centrifugation, and shallow-water sediment traps. Though particles are separated by density rather than by size, the continuous-flow centrifugation technique was preferred due to its speed and high recovery rate. There is also less potential for postsampling alteration of suspended sediment since sample stabilization is accomplished more rapidly than with filtration or sediment traps. Flow rates up to 10 l/min are achievable with recoveries better than 99% in the >0.1-μm mineral particle size fractions. Collection of suspended matter by centrifugation results in material different from that obtained by filtration,[73-75] significantly in the small, low-density organic particle fractions, whereas the deviations in mineral composition of particles with grain sizes in the order of microns seem to be limited to 10 to 30%.[76] Recovery of colloidals is poor by either technique.[77] Filtration

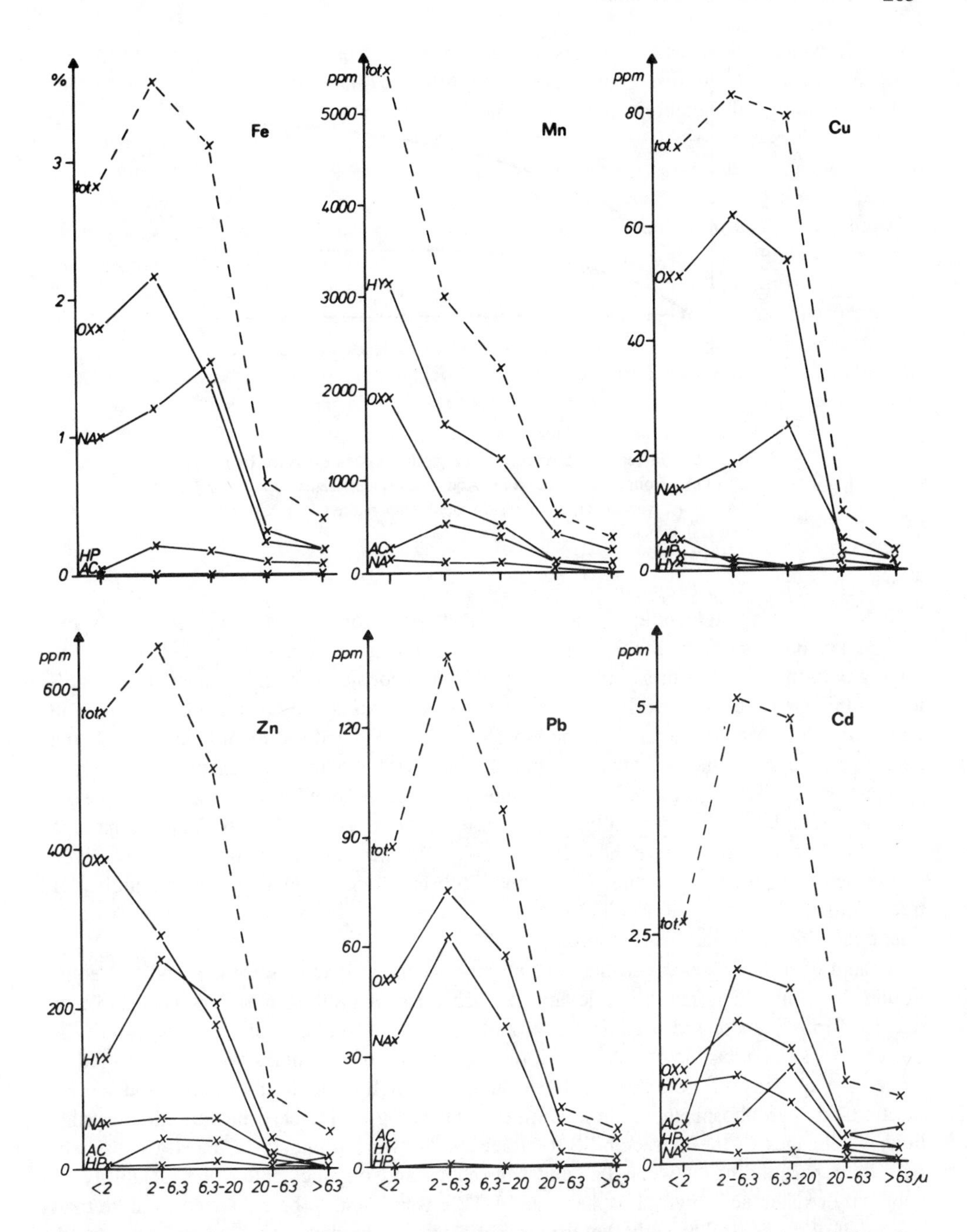

FIGURE 5. Grain size distribution of different metals and their chemical forms in a selected sample from the fluviatile Elbe river (From Schoer, J., *Environ. Technol. Lett.*, 6, 189, 1985. With permission.)

and batch centrifugation procedures were recently compared for their suitability in large-scale sediment sampling and trace metal reconnaissance programs.[78] Direct separation approach provided by centrifugation was found to be comparable with in-line filtration and can be carried out both precisely and accurately. The continuous-flow separation procedure is simplier to carry out, especially when suspended sediment concentrations (<150 mg/l) or trace-metal loads are low. These operationally defined separation procedures, however, still remain subject to a tremendous amount of interpretation and discussion, as no one has yet come up with a universally accepted method in this field.

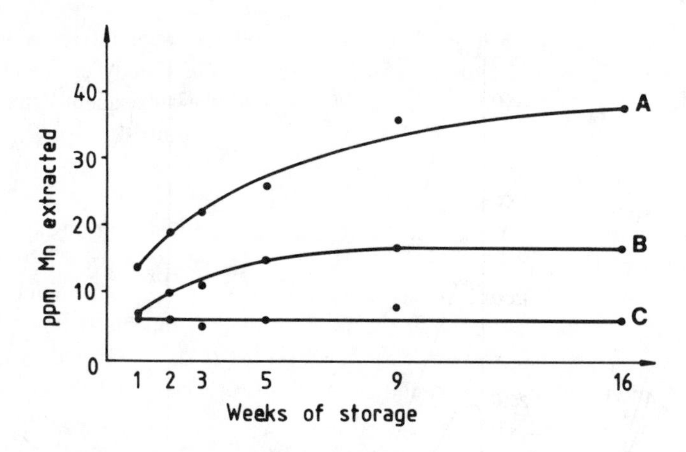

FIGURE 6. Increase with weeks of storage time in Mn extracted by pH 4.8 NH_4OAc from loamy fine sand (Aquic Udorthent) samples kept air dry (A), moist frozen (B), and moist at room temperature (C). (Data from Bartlett and James.[79])

2. Sample Storage and Preparation

Sample preservation techniques in trace element speciation studies of oxic sediments and sludges are different from those used for anoxic samples. Care should be taken to avoid mixing both together during sampling. For pollution reconaissance and bioavailability studies, sampling of the oxic overlying layer of sediments provides more relevant information on the remobilization of metals into the biosphere. The oxidized sediment layer controls the exchange of trace elements between sediment and overlying water in many aquatic environments and provides the pool of pollutants to which the benthic ecosphere is exposed. Sediment samples should generally be processed as soon as possible, because no storage techniques are able to preserve the delicately-poised *in situ* equilibrium particular to a given soil or sediment found in nature. Occasionally observations on changing extractability of trace elements by drying of soil and sediment samples have been reported. However, their importance has too often been ignored.

Air and oven drying cause instant and major changes in overall sediment and soil equilibrium by converting fractions relevant to trace element binding into highly unstable or reactive forms.[79] Increased organic matter and manganese solubility and exchangeability is one of the most noticeable effects of soil drying. Drying of sediments was also reported to reduce the quantity of Fe extracted by techniques which remove amorphous iron oxides (acetic acid, pyrophosphate, hydroxylamine), suggesting an increase in the oxide crystallinity.[80] Changes in the extractability of trace metals were found to be mostly consistent with their partitioning between Fe- and Mn-oxides and organic matter. Extractability of copper by oxalate acid, pyrophosphate, and DTPA was found to be enhanced to more than twice that of the control by sediment drying, reflecting the predominant binding of this metal by organic matter.[80] A remoistened sample may require a fairly long incubation time before it approaches the *in situ* chemical characteristics definable by chemical extraction of this sample immediately following field sampling.[79]

In practice, it is usually impossible to retroactively correct data that have been obtained from dried soils and sediments to those which exist originally in field. Such data may even be of limited value for comparing the bioavailable concentrations of trace metals in samples collected within the same environment. The effect of storage time on manganese extractability by pH 4.8 NH_4OAc, shown in Figure 6, indicates that characteristics of a dried soil sample also continue to change during dry storage.[79] Sieving and mixing in order to obtain a representative sample for bioavailability analysis may lead to precise but inaccurate results.

These effects make the preparation of stable soil and sediment reference materials for comparative speciation studies extremely difficult. A thorough study on the effects of different methods of oxidized sediment samples storage on the subsequent extraction of Cu, Zn, Fe, and Mn revealed, however, that drying and freezing of oxic samples are adequate methods of storage if the sediments are to be extracted only with diluted mineral acids.[80] These preparation procedures are adequate for pollution reconnaissance studies, since "non-residual" concentrations of trace metals are usually determined using a rapid HCl ($\leq$0.5 *M*) leaching technique, with sediment fractions usually sized <60-μm by dry sieving.

Wet storage of oxidized sediments and soils is inadequate due to a microbially induced shift from oxidizing to reducing conditions in the stored sediments. Extractability of the metal with the most insoluble sulfide (Cu) was reported to decline rapidly with this treatment.[80] Refrigeration should delay or inhibit the effects of wet storage, although Cu and Fe extractability by DTPA was found to be decreased and enhanced within 15 d of storage to nearly twice that of the immediately extracted subsamples.[80] Storage of aerobic soils at low but above-freezing temperatures (commonly 4°C) is an acceptable storage method, providing the samples are kept close to field capacity moisture, or slightly below, inside thin polyethylene bags permeable to both CO_2 and O_2.[79] There will, however, be a flush of bioactivity if the samples are brought back to room temperatures. Sterilization by gamma irradiation (a physical technique that should not change the geochemical equilibria in the samples) would probably minimize microbe-induced changes in the trace element partitioning.[79] Freezing is usually a suitable method to minimize microbial activity. However, it may lyse cells and thereby free organic excudates and any associated trace metals. Freezing was found to enhance water solubility of metals in the order Mn (8 to 17%) > Cu (7 to 15%) > Zn (6 to 12%) > Fe (3 to 7%). Storage significantly affected extractability of these metals by weak agents (ammonium acetate, DTPA, hydroxylamine) subsequent to freezing.[80]

In examining trace metal speciation in contaminated, usually anoxic dredged sediments by selective extraction procedures, Engler et al.[81] stressed that, "the anaerobic integrity of the samples must be maintained throughout manipulation and extraction." The effects of various preservation techniques (wet storage, freezing, freeze and oven drying) on metal speciation in anoxic samples has been recently reported.[82,83] In the absence of precautions to prevent exposure to atmospheric oxygen, several significant changes in trace metal concentrations were observed in all but the residual fractions of the five- to six-step sequential extraction procedures used. Among the different metals studied, iron and cadmium were particularly sensitive to sample pretreatment. The partitioning of both these metals in anoxic harbor basin mud is shown in Figure 7 with examples of the effect of agitating with oxygenated native water (elutriate test to estimate potential metal release with hydraulic dredging and open water disposal) and freeze and oven drying.

Chemical modeling with the pore water constituents of the sediment samples support the hypothesis that labile Fe (II)-carbonates and phosphates were present in the original anoxic freshwater sediments.[84] Cadmium, in contrast, is nearly quantitatively found in a H_2O_2-oxidizable fraction, supporting the hypothesis that it is predominantly bound to sulfides in anoxic sediments. Exposure of the samples to air caused the cadmium to be shifted in the extraction scheme to reducible and exchangeable fractions. After oven drying of the originally anoxic sediments, substantial Cd was extracted using NH_4OAc. One should be aware, however, that the high concentration in dissolved organic substances found in the first extraction steps of fresh anoxic sediments tends to suppress Cd peaks in atomic absorption spectroscopy analysis, which is not found with dried samples. Standard addition corrections are necessary to avoid such artifacts in studying anoxic samples.

Storage of anoxic sediments by freezing was found to cause the least change in the fractionation pattern of the various metals studied; only copper in the exchangeable fractions was increased, probably due to viable cell lysis in the sediments.[83] For anoxic sediments

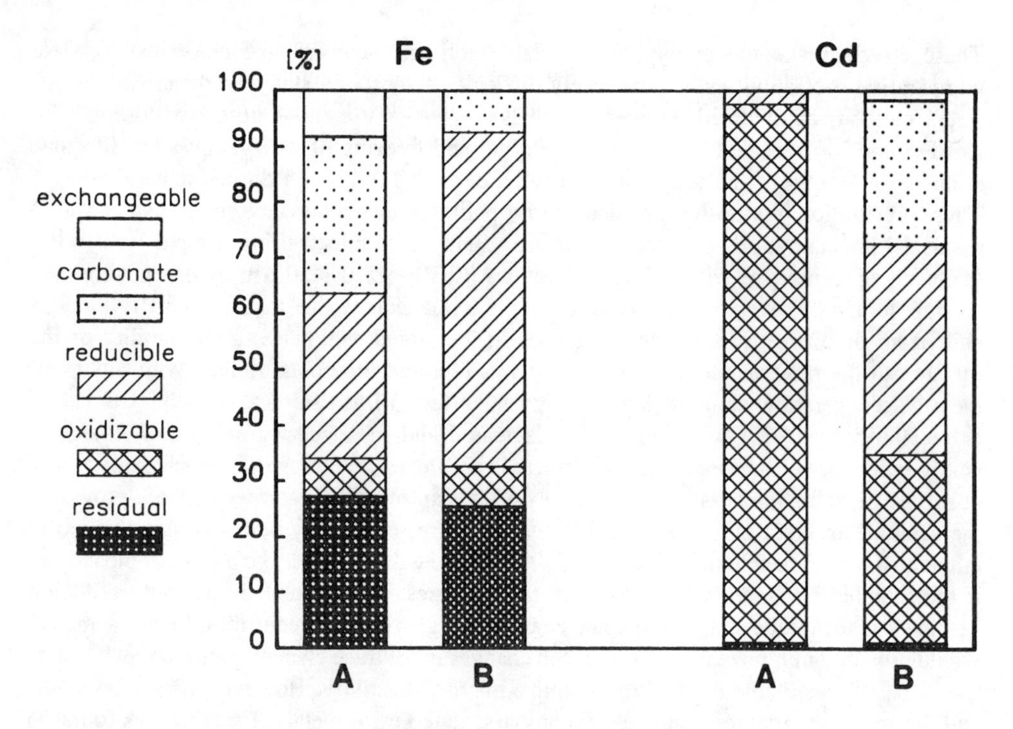

FIGURE 7. Sample drying as an important methodological error of the sequential extraction of anoxic sediment samples; (A) fresh samples extracted in glove box/argon atmosphere, (B) after freeze drying. (Data from Kersten and Förstner.[82])

and soils, the maintenance of oxygen-free conditions during sampling and extraction is of critical importance. It is clearly impossible to totally avoid changes in the *in situ* chemical speciation of trace elements, definable by extractants, unless the sediment and soil samples are extracted immediately upon collection. In this way, "it is important to be aware of the kinds and degrees of changes taking place in the soils we are studying while we are studying them."[79]

C. Limitations in the Selectivity of Extractants

1. Cation-Exchanging Extractants

Soil and sediment particles usually contain a wide range of individual components, each of which may bind trace elements in several ways. Exchangeable trace metals normally constitute only a small proportion of the total metals in soils and sediments. Partitioning and residence time of trace metals in natural systems, however, are both controlled by the chemical process occurring between the surfaces of these particles and the adjacent water phase. Findings from both sequential extraction studies and successful adsorption models suggest that natural organic films and hydrous oxide coatings of iron and manganese are the most important surface-active phases in environments ranging from seawater to soils. The overall adsorption characteristics of a natural sediment sample for a particular trace metal have been shown to vary with the relative abundance of the metal-adsorbing phases or the relative proportioning of surface sites among the components[58,77] Leckie[85] reviewed the processes occurring at the interface between the aqueous solution and the solid surface. He showed that the particle surface sites contain functional groups, wherein the acid-base and other coordinating properties of which are similar to those of their counterparts in soluble species. The reactions between the metal ions in solution and the binding sites at the solid surface are analogous to a multiligand solution system and thus follow all the concepts and mathematical formalism of coordination chemistry. The important difference, however, is the development of an electrical double layer at the surface/solution interface, which depends on pH (''pH adsorption edge'') and ionic strength, greatly affecting cation adsorption.

For a given adsorbent and pH, the fraction of an ion adsorbed varies with differing elements (at equivalent total element/adsorbent concentration ratios). This is attributed to differences in the surface-site binding energies for each element. The partitioning of a metal ion among the sediment or soil components and solution is thus influenced at equilibrium by

1. The density of available surface binding sites of each component, especially of high-energy sites.
2. The binding intensity of the metal ion to each component.
3. The abundance of each component in the sediment/soil, which may in part depend on the degree of interfacing of the components through aggregate formation.
4. The chemical characteristics of the water phase such as pH and dissolved ligand concentrations, which may influence the speciation of dissolved metal ions.
5. The concentration of other major and metal ions, which may compete with trace metals for the available binding sites.[86]

Determination of adsorbed metal concentrations can be accomplished by reversal of the adsorption equilibria either using chemical displacement of the metal ion from the surface binding sites, or by complexation by much less adsorbing ligands and, hence, reducing of adsorbing ion levels in solution. The third alternative of dissolving the trace element adsorbing phase will be discussed in subsequent sections. Performing the latter without inclusion of the two other alternate approaches would lead at most to only a redistribution of the released elements among the residual adsorbents.

Neutral salts such as $MgCl_2$, $CaCl_2$, and $NaNO_3$, are commonly used as ion-displacing extractants to promote displacement of metal ions physically bound by electrostatic attraction to negative sites on particle surfaces. The affinity of group I and II cations for most surface sites is, however, usually orders of magnitude less than the affinity of the trace metals of interest. In quantitative extraction studies, this problem must be offset by increasing the concentration of the competing ions. In practice, 1 *M* salt solutions have been widely used, although more dilute solutions (e.g., 0.05 *M* $CaCl_2$) have been preferred in some cases, resembling real natural conditions. For example, if the binding capacity of a sediment substrate for trace metals is 10 μmol/g,[86] then, on displacement from 1 g into 100 ml solution, the competing cation concentration will be four orders of magnitude higher than any trace metal concentrations to be exchanged. In many studies the cation displacement efficiency for oxic sediment and soil samples has been shown to decrease in the order $Cd > Zn > Cu \approx Pb$, despite the wide variety of sample compositions studied, which clearly represents the overall metal-site binding affinities. This is consistent with the decreasing pH values of the adsorption edges (pH_{50}: the point where dissolved element concentration decreases to half of the initial value, due to sorption) of these metals. The ion-exchangeable fraction of trace metals in anoxic sediments, on the other hand, was found to decrease in the order $Fe > Mn > Zn > Ni > Cr > Pb > Cd \geqslant Cu$, due to competing redox reactions.[87]

Only slight changes in the solubility of the more soluble sediment and soil components have been found upon contact with any neutral electrolyte solution, mostly due to activity coefficient values decreasing with increasing ionic strength. The magnitude of this effect tends to be smaller than the uncertainty associated with the analytical results. Pickering,[61] in his review on solid speciation, reported that $MgCl_2$ sediment leachates contain only low levels of Al, Si, and organic carbon, which implies that attack on clay minerals, sulfides, or organic matter is minimal. Fe concentrations were also found to be very low which was consistent with the low solubility of Fe/Mn oxides at neutral pH values. Only slight dissolution of carbonates (2 to 3%) indicates some effect from changing activity coefficient.

Robbins et al.,[64] in their review of the difficulties involved with exchangeable cation extraction, listed a number of studies in which NH_4Cl and NH_4OAc were found to dissolve considerable amounts of $CaCO_3$, $MgCO_3$, $BaCO_3$, $MgSO_4$, and $CaSO_4$. While these compounds may not be important in contributing to cation exchange capacity in soils and sediments, neutral NH_4OAc can additionally cause some dissolution of Mn-oxyhydrates and metal oxide coatings.[20,88] For this reason, Chester et al.[89] recently suggested the extraction of both carbonates and Mn-oxides by a hydroxylamine hydrochloride solution buffered by NaOAc. However, colloidal Fe released upon Mn reduction was found to be carried over with supernate as indicated by a typical red color, even after prolonged centrifugation. Dissolution of substrates is further enhanced in solutions containing acetic acid. Acetic acid alone is less effective in displacing cations from surface binding sites because of the lower level of competing ion present in solution ($[H^+] \approx 10^{-3} M$), but with its salts it forms a buffer system with high capacity and intensity which tends to eliminate any deleterious pH effects arising from differences in soil/extractant pH.

Nitrate salts have found wide acceptance in ion-exchangeable extractions. Nitrate ions have the advantage of not forming stable complexes with transition metals; hence, displacement reactions with nitrates are attributable solely to cationic competition for surface binding sites. In their thorough evaluation of different extractants for trace elements in soils, Sauerbeck and Rietz[90] accordingly showed that 0.1 and 1 *N* $Ca(NO_3)_2$ were more effective than the corresponding $NaNO_3$ solutions, which in turn have been suggested as promising extractants in predicting biological effects of accumulated metals in soils.[91] A series of standardized pot experiments conducted at five different European research institutes showed that 0.1 *M* $NaNO_3$-exchangeable Cd contents are a sensitive and accurate index of potential plant uptake from contaminated soils. As a consequence, the Swiss "Ordinance on Pollutants in Soils (VSBO/SR814.12)", effective from September 1, 1986, included a maximum permissible $NaNO_3$-exchangeable value for each metal in contaminated soils to be used for nourishment purposes. This represents an important success of the fractionation approach.

The extraction efficiencies of the various neutral salt solutions generally varies with the nature of the cations due to their various affinities for the surface binding sites described above, which tend to be in the order $H^+ \gg Ca > Mg > Na = NH_4^+$. Thus, 1 *M* H^+ may desorb more metal ions than 1 *M* Mg^{2+} which, in turn, may displace more than 1 *M* Na^+. In order to achieve an even higher displacement efficiency, copper salts have been used in some extraction studies (e.g., 0.1 *M* Cu-acetate).[92] Pickering[61] stressed that relative efficiency of exchanging extractants becomes more noticeable due to selective displacement for natural samples having sites of different bonding energies. In this study, 0.05 *M* $CaCl_2$ was found to displace more metal ions (10 to 20%) than 1 *M* $MgCl_2$ or 0.5 *M* NaCl, indicating that differences in affinity for the sorption sites was more important than chloro-complex formation. The presence of chloride ions can still enhance extraction efficiency significantly over what would be expected using neutral nitrate salt as the displacing agent, especially for cadmium. The bond strength of chloro-complexes of transition metals are much weaker than organic complexing agents, such as EDTA, and slightly less stable than those formed with acetate ions (pK_1-values, e.g., for Zn, Cd, Pb, and Cu for the OAc^- ion lie between 1.6 and 2.2.; and for the Cl^- ion between 0.7 and 2). Comparison of acetate and chloride salt extraction efficiences in single-component studies showed that recoveries of the former are 10% or even higher, representing the most efficient exchanging agents besides the acids.[61] Even more pronounced differences in extraction efficiencies of natural estuarine sediments are shown in Table 5.[93] In spite of their disadvantageous dissolving and complexing properties, extractions with both neutral 1 *M* NH_4OAc and $MgCl_2$ are yet the most widely used procedures for evaluation of exchangeable trace metal concentrations. These agents are also part of the more widely accepted sequential extraction schemes of Tessier et al.[94] and Salomons and Förstner.[95] Other complexing agents such as DTPA, EDTA, EGTA, and

Table 5
RELATIVE IMPORTANCE OF THE EXCHANGEABLE METAL FRACTION IN ESTUARINE SEDIMENT SAMPLES, BASED ON 100 UNITS OF METAL EXTRACTED BY AMMONIUM ACETATE

Reagents	Zn	Cu	Mn
1.0 *M* NH_4OAc[a]	100	100	100
1.0 *M* $MgCl_2$	11	63	84
1.0 *M* NH_4Cl	16	96	65
0.5 *M* MgOAc	25	124	122
1.0 *M* NH_4Cl + 1.0 *M* NH_4OAc	44	85	84
1.0 *M* NH_4OAc + 0.5 *M* MgOAc	55	109	134

[a] OAc = acetate.

From Badri, M. A. and Aston, S. R., in *Proc. Int. Conf. Heavy Metals in the Environment,* Ernst, W. H. O., Ed., CEP Consultants, Edinburgh, 1981, 709. With permission.

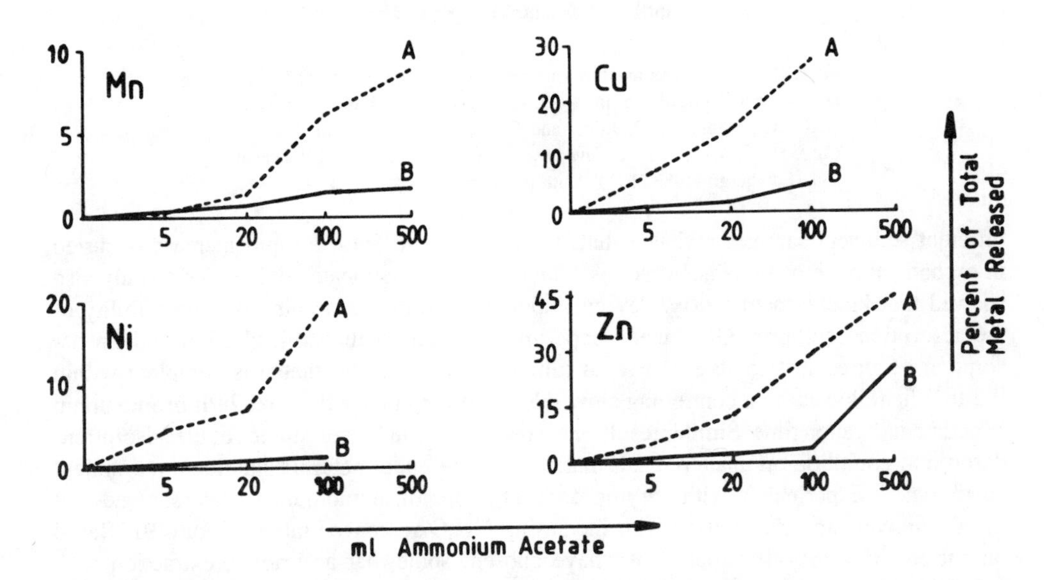

FIGURE 8. Metal removal by varying solid to solution ratio of 1 *M* ammonium acetate. Contact time 2 h. (A) Neckar River sediment with 17% $CaCO_3$, (B) deep-sea sediment with 2.5% $CaCO_3$.

HEDTA will not be considered here, since application of these extractants is not useful in sequential extraction studies. Their application in soil science has been comprehensively reviewed by Pickering.[61] The effect of pH and redox potential on many trace metal-chelate equilibria have been illustrated by Sommers and Lindsay.[96]

In displacement reactions, the electrostatic conditions imposed by the competitive major ions at the solid/solution interface are not a function of the concentration (moles of major cations $\gg$ moles of surface sites) of the solid. Since the narrow pH range of the adsorption edge of a trace ion is a function of the concentration of the solid, the particular adsorption-desorption behavior of trace ions in salt solutions will in fact depend on the concentration of the solid in the system.[97] The important influence of the solid-to-solution ratio on the exchange efficiency of NH_4OAc is shown in Figure 8. The results obtained from three

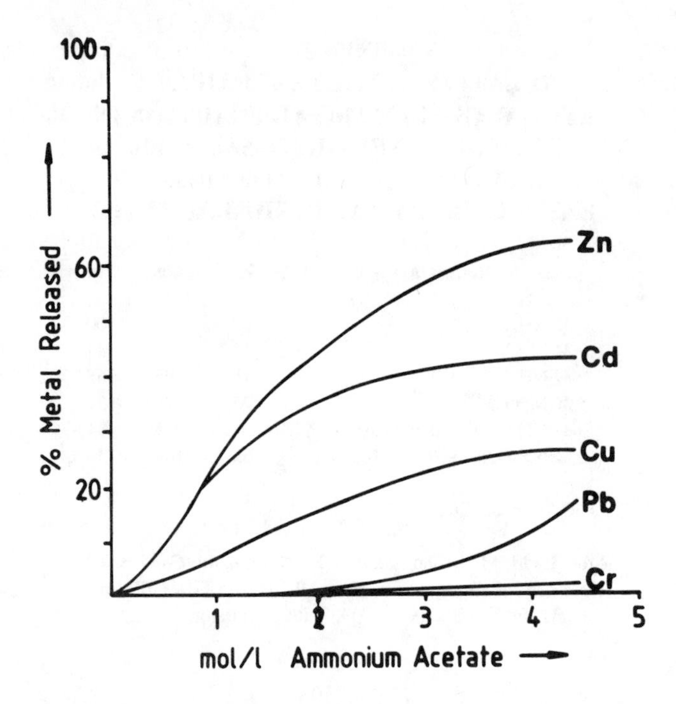

FIGURE 9. Metal removal by varying concentrations of ammonium acetate. Contact time 20 h in a 23.6% suspension. (From Van Hoek, G. L. M., Gommers, P. J. F., and Overwater, J. A. S., in *Proc. Int. Conf. Heavy Metals in the Environment,* Müller, G., Ed., CEP Consultants, Edinburgh, 1983, 1074. With permission.)

different sediment samples revealed that, for ratios above 1:40, a rapid increase occurred in exchanged trace metal concentrations, independent of carbonate content. This study also showed that kinetic factors may play an important role in determining the reversibility of the adsorption equilibria. Our kinetic experiments indicated that Mn, Cu, Ni, and Zn desorption occurred in two stages: a rapid initial release of metal that was complete within 0.5 to 2 h, followed by a continuing slow release of metal over the next 24 h of maximum experiment duration time. Similar results have been shown in kinetic studies of the adsorption-desorption equilibria on unaltered salt marsh sediment under natural estuarine environment conditions.[58] Experiments with varying concentrations of ammonium acetate showed that metal removal can be increased with increasing NH_4OAc concentration (Figure 9). Based on current information available, we have chosen, somewhat arbitrarily, extraction with 1 *M* NH_4OAc solutions at a solid/solution ratio of 1:10 to 1:520 after a 2-h time period to reflect a pseudoequilibrium value representative of the exchangeable trace metal portions of the sediment samples.

2. Carbonate-Dissolving Extractants

The preferred attack of carbonate minerals by acidified acetate buffer extractants has been incorporated in nearly all sequential schemes. Several studies on the efficiency of 1 *M* HOAc-NaOAc, pH 5, found that dissolution of carbonate and dolomite minerals with their enclosed trace metals was nearly quantitative without significant dissolution of organic matter, ferromanganese oxyhydrate, or aluminosilicate minerals.[94,99-102] The HOAc-NaOAc treatment is quite complete, extracting (>99.9%) the total carbonate-bound metals in sediments within 5 h.[94] In a study of a variety of sediments, chosen to reflect several normally encountered types of marine environments, 2.8% Fe and 14% Mn was extracted by HOAc-NaOAc.[56] Even higher extractable iron and manganese portions were obtained in fresh anoxic estuarine sediment samples (up to 36 and 80%, respectively), due to the solubility of both Fe(II)- and Mn(II)-carbonates and phosphates in acidic NaOAc solution.[2] The effects of buffered acetate solutions on Fe(II)- and Mn(II)-phase extraction substantially depends on the preservation of the redox conditions during the analytical procedure, due to the lability of the Fe/Mn(II) phases. As discussed earlier, substantial amounts of trace metals may also be extracted from sulfide phases oxidized during sediment drying.[82,97] Deaeration of the extractant by, e.g., bubbling N_2 gas through the solutions is essential. When oxygen is rigorously excluded during the handling and extraction of anoxic sediments, selectivity of the acetate extraction procedure with respect to amorphous sulfide phases was maintained, with acid volatile sulfide levels preserved near 87 to 100%.[83]

Displacement by strong-acid cation-exchange resins has been suggested as an alternative carbonate dissolution technique.[103] The acidic cation exchanger (4 to 5 g in H^+ form >100 mesh) is added to a 20-ml suspension of 0.5 g of fine-grained sample material and agitated in an open reaction vessel. The affinity of the exchanger resin for Ca^{2+} and Mg^{2+} was found to be sufficiently high to complete the reaction within 3 h. No significant remobilization of Fe and Mn was found in oxic sediments, probably due to the lack of free hydrogen ions available in the solution to attack other mineral species. The pH may change considerably due to CO_2 generation and formation of carbonic acid. Any clay particles adsorbing to the resin spherules may be easily scrubbed by a short ultrasonication and wet sieving. This approach has not become common practice, probably due to its inconvenience compared with the buffered acetate extraction.

Experimental parameters exert a significant influence on the rate and extent of trace metal exchange. Such heterogeneous reactions involving transfer through the sediment-water boundary are sensitive to contact time. Vigorous mechanical shaking is necessary to maintain the solid in uniform suspension. The equilibration time required for total dissolution of the carbonate phases by acidified acetate solutions depends on factors such as particle size, solid to solution ratio, type and crystallinity of the carbonate phase, and percentage of carbonate present.[104] Robbins et al.[64] have shown that the buffering capacity of the NaOAc-HOAc extractant solution is sufficient to dissolve all the $CaCO_3$ in a carbonate-rich sediment sample (68% $CaCO_3$), resulting in a final pH of 5.5 at an initial solid to solution ratio of 1:40.

The buffer extraction technique is considered to be promising, particularly where the aim is to differentiate between various weak-binding forms of metals and metal uptake/release processes in soils and sediments, as a function of pH. To characterize metals associated with carbonate formations, Span and Gaillard[101] undertook a sediment carbonate titration by a set of NaOAc solutions buffered to different pH values. These solutions, after equilibration with atmospheric pressure of CO_2, reached initial pH values ranging from 5.0 to 8.3. They could separate trace elements into two groups according to their correlation with calcium in their titration curves as shown in Figures 10 and 11. Trace elements such as cadmium and manganese follow the pattern of calcium, whereas the other metals studied departed from the distribution, indicating that these metals are, at best, only weakly associated with carbonate phases. The increase of Fe dissolution with decreasing pH can result from the dissolution of Fe-oxides due to the increased acetic acid concentrations, liberating Cu, Pb, and Zn adsorbed onto its surfaces. Span and Gaillard[101] stressed that it is impossible to isolate such desorption reactions from dissolution when performing sediment titrations, although the parallel evolution of Ca, Mg, Cd, and Mn strongly indicates that the metals are associated mainly with carbonates. A useful test of whether the latter reaction is fulfilled is to see if the molar fractions of the dissolved compounds, Me/Ca, remain constant during titration. Span and Gaillard[101] found essentially constant Cd/Ca and Mn/Ca ratios of $9.44 \times 10^{-6} \pm 0.8 \times 10^{-6}$ and $1.02 \times 10^{-3} \pm 0.12 \times 10^{-3}$. Lyle et al.,[100] however, found no such relationships with their deep-sea sediment extraction experiments. They examined the correlations between the concentration of a potentially incorporated metal and the fraction

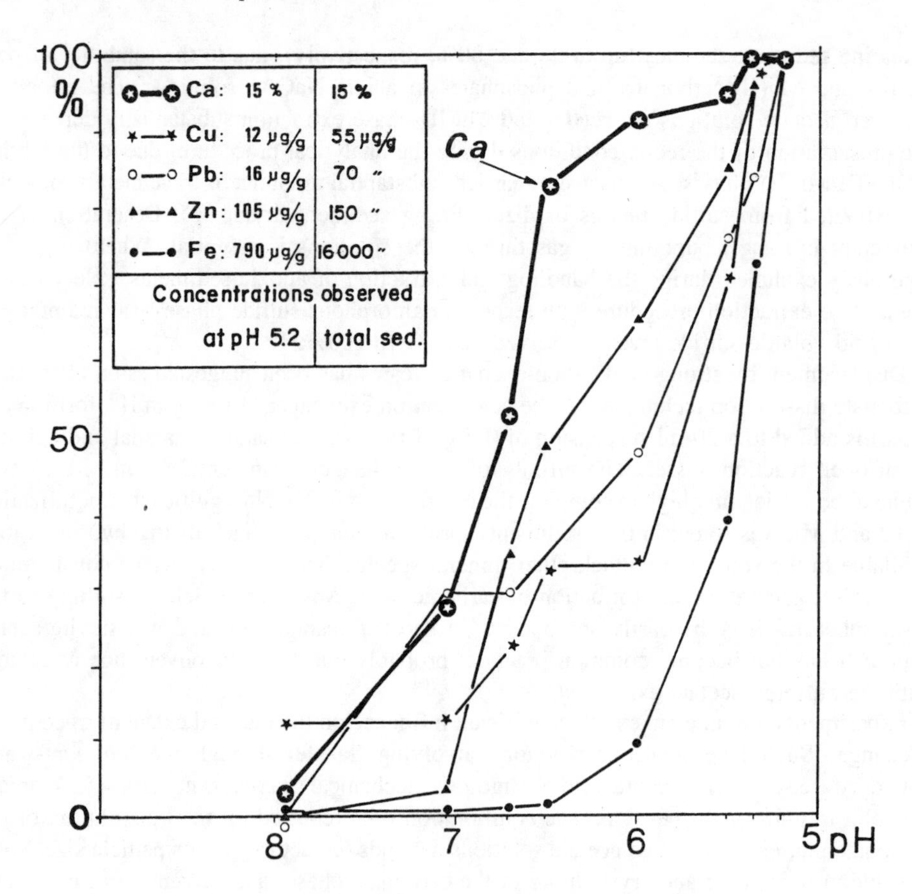

FIGURE 10. Ca, Cu, Pb, Zn, and Fe variations as a function of pH during sediment titration with 1 *M* NaOAc (100% represents the metal removal found at pH 5.2). (From Span, D. and Gaillard, J.-F., *Chem. Geol.*, 56, 135, 1986. With permission.)

of the noncarbonate portion of the sediment, i.e., the residue of the NaOAc-extractable. The intercept at zero noncarbonate thereby corresponds to the concentration of the incorporated element in calcite, whereas the intercept at 100% noncarbonate reflects the concentration of the element in other binding forms. Such a good negative correlation with a positive intercept was found only with strontium, suggesting that sedimentary Sr is primarily in calcium carbonate with a Sr/Ca molar ratio of 2×10^{-3}. Large negative intercepts for Fe and Mn, according to Lyle et al.,[100] imply that other processes control the distribution of these two metals.

3. *Acid-Reducing Extractants*

Fe(III)- and Mn(IV)-oxides, occurring as coatings on detrital particles, as cement between particles, and as pure concretions (the so-called ferromanganese nodules), are important components of both soils and sediments. Mineralogical structure of these oxides ranges from highly crystalline, through cryptocrystalline, to amorphous.[105] Their tendency to adsorb and control the solid/solution partitioning of trace metals is well documented. Associations of trace metals with manganese and iron oxides ranges from exchangeable forms (i.e., formation of surface complexes by reactions between interfacing solutes and functional groups attached to oxide surfaces), through moderately fixed (i.e., coprecipitated with amorphous oxides), to relatively strongly bound (e.g., occluded in goethite and other oxide mineral structures). Crystalline oxides are more resistant to attack than the noncrystalline oxyhydrates of Fe and

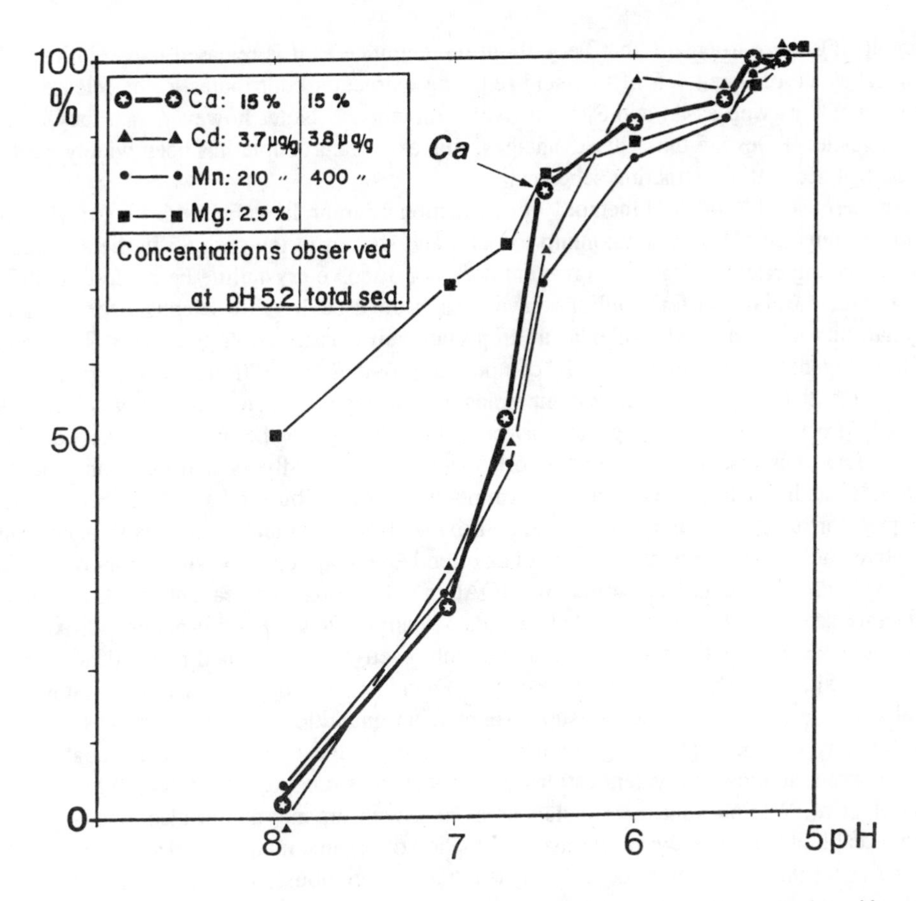

FIGURE 11. Ca, Mg, Cd, and Mn variations as a function of pH during sediment titration with 1 *M* NaOAc (100% represents the metal removal found at pH 5.2). (From Span, D. and Gaillard, J.-F., *Chem. Geol.*, 56, 135, 1986. With permission.)

Mn. Amorphous ferromanganese oxyhydrates with their associated elements can be dissolved under the effect of redox gradients in sediments, leading to a marked increase in trace metals in interstitial water. To be useful for environmental studies, a selective dissolution technique should be able to differentiate between amorphous and crystalline Fe oxides. Extraction experiments with emphasis on *easily reducible phases* best reflect the processes involved in diagenetic remobilization of various elements from these oxides. Thus, the methods applied in the differentiation of reducible forms of metals in sediment have been used to study pelagic deposits since the early days of manganese nodule research. Methods for Fe-oxides also commonly used in fractionation schemes originated from the early attempts to differentiate "horizons" for classifications in soil science.

Chemical leaching of oxide minerals with EDTA, dilute hydrochloric, and acetic acid has been used to determine the distribution of elements in both the authigenic phases and in detrital igneous minerals.[106] An extractant selective for oxyhydrates should, however, attack neither major silicate minerals nor organic matter. Due to the susceptibility of such sediment and soil components to attack by both HCl and EDTA solutions, these extractants have not found wide application. Studies of pelagic sediment extraction led Chester and Hughes[107] to introduce a combined acid-reducing agent of 1 *M* hydroxylamine hydrochloride and 25% (v/v) acetic acid for the separation of ferromanganese minerals, carbonate minerals, and adsorbed trace elements from oxidic deposits. Tessier et al.[94] reported that treatment with acetic acid/hydroxylamine hydrochloride reagent did not reduce the organic carbon content of sediments. A good extractant for Fe should have both a reducing agent and an acid or complexing agent to ensure that Fe will not reprecipitate as it is released into solution. The low pH value of about 1.5 of this acid-reducing extractant contributes to the release of Si from opal[107] as well as Al and Si from river sediments.[94] (Note, however, that they used a hot digestion!) Despite these disadvantages, this extractant couple has been widely applied as part of sequential extraction schemes.

Another useful "standard method" for extraction of amorphous Fe-oxides is 0.1 *M* oxalic acid, buffered to pH 3 by ammonium oxalate, and shaken in the dark.[105] In the absence of the catalyzing effect of light, this reagent does not dissolve crystalline Fe oxides. Acidified ammonium oxalate buffer solutions have long been used by soil scientists for assessing crystallinity of Fe and Al oxides as an important soil classification parameter. The ability of this reagent to attack amorphous Al compounds provides an additional reason for including it in sequential schemes, as these compounds may strongly influence the pH of acidic soils. Freshly precipitated Al gel (e.g., as may be formed during weathering [hydrolysis] of Na^+-form clays) has also been shown to sorb metal ions as avidly as hydrous iron oxides.[61] Strongly acidified hydroxylamine and ammonium oxalate buffer have both been widely adopted for removing amorphous ferric hydroxides from soils and sediments in many comparative studies.[108] The oxalate reagent has often being claimed to be slightly more specific for oxyhydroxides than hydroxylamine-HOAc.[109,110] The oxalate reagent attack on silicate minerals seems to be minimal.[111] Thus, its selectivity towards amorphous Fe oxides is sufficient, as long as the samples do not contain reactive, fine-grained magnetites[112] or Fe-rich layer silicates.[113] Wide occurrence of such Fe-smectites in deep-sea sediments, however, limits its use in marine geochemistry. Appreciable quantities of metals sorbed to organic matter may be coextracted with amorphous oxides,[55] due to the fact that oxalate forms complexes with most polyvalent cations, with pK_2 values ranging from 5.5 (Cd) to 8.5 (Cu). Slavek et al.[114] found that most metals associated with organic matter also tend to be ion-displaceable. Consequently, preliminary extraction of organic matter has been recommended (e.g., using the NaOCl treatment[115,116]), leaving any previously coated amorphous oxyhydroxides open to attack by reducing agents. In more recent studies, use of 0.25 *M* hydroxylamine hydrochloride in 0.25 *M* HCl,[20,108] a reducing reagent couple with selectivity similar to the acidified ammonium oxalate reagent, but with less complexing properties, has been preferred.

The efficiency of these reagents is determined by their reduction potential and ability to attack the various amorphous to crystalline forms of oxyhydroxides. For the reagents NH_2OH^+, $H_2C_2O_4$, and $S_2O_4^{2-}$, quoted standard reduction potentials, respectively, are -0.05, -0.49, and -1.87 V. Pfeiffer et al.[117] tested a sequential method comprised of all these three reagents to evaluate their potential in extraction of metals from reducible components of deep sea sediments and ferromanganese nodules. A weakly acidified reducing reagent, 0.1 *M* hydroxylamine hydrochloride in 0.01 *M* nitric acid,[118] was inserted before the oxalate reagent. It should perform well if set in sequence before an amorphous Fe-oxide extractant, since it is expected to solubilize the Mn and very little Fe ($<1-8\%$).[110,118] The nonsilicate, poorly-reducible iron oxides (mainly goethite) were extracted using a strongly reducing, citrate-buffered dithionite solution, according to Holmgren.[119] The time-dependent behavior of the extractions, the optimal solid/solution ratio, and the sequential application of the three extraction steps for transition metals in selected pelagic samples were investigated. Since the behavior of a specific extractant in repeated treatment steps is a very important factor for its selective applicability,[109] metal percentages released in three extractions with each reagent, in relation to the total contents after a hydrofluoric/perchloric acid digestion, were determined (Figure 12). The oxalate extractant has the best reproducibility for most metals in both deep-sea sediments and ferromanganese nodules (except for manganese in nodules). With the given solid to solution ratios and treatment periods, extraction with hydroxylamine hydrochloride and the dithionite buffer apparently does not occur in equilibrium for some elements and materials; thus, repetition of the extraction step results in further releases.

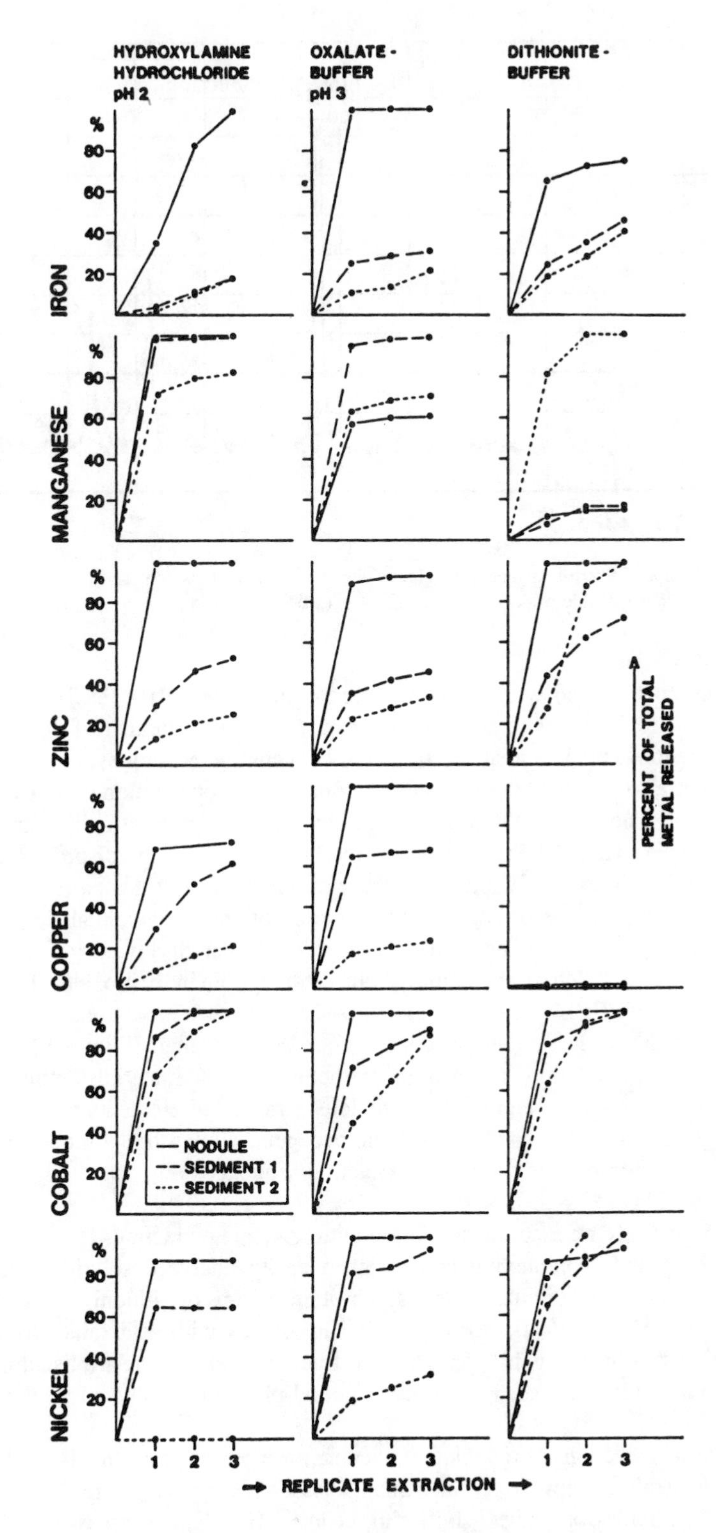

FIGURE 12. Mean recovery of metals by successive extractions with different acid-reducing reagents (From Pfeiffer, G., Förstner, U., and Stoffers, P., *Senckenbergiana Marit.*, 14, 23, 1982. With permission.)

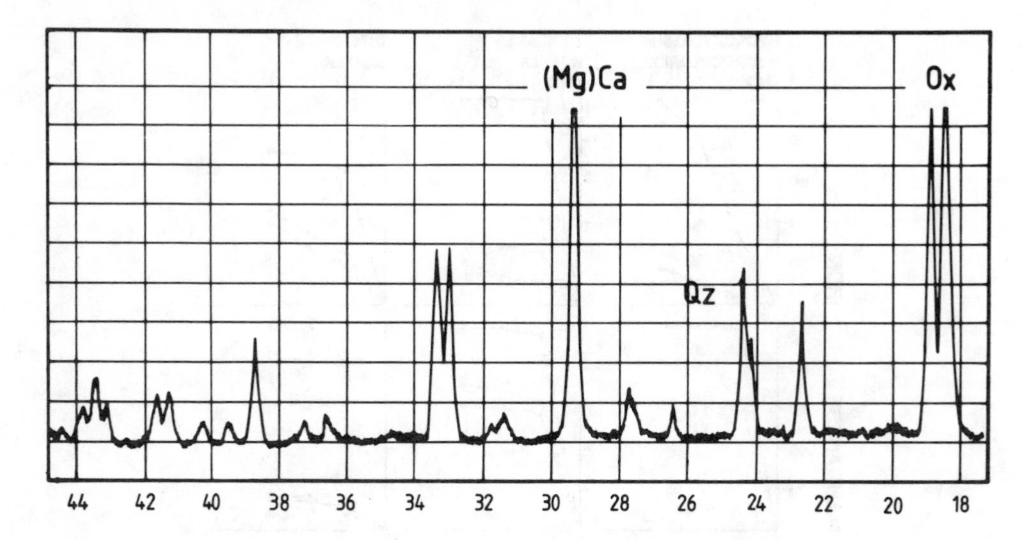

FIGURE 13. XPD plot of a ferromanganese nodule residue after oxalate extraction. Fe/Mn-oxalate precipitation is evidenced by the typical double-peak between 18 and 19° (2θ). Ox: oxalate, Qz: quartz, (Mg)Ca: magnesian calcite.

Problems are encountered with use of the citrate-dithionite buffer (DCB) extraction, as has been discussed by Tessier et al.[94] The reagent is highly contaminated with metals and its purification requires inconvenient use of cation-exchanging techniques. Frequent clogging of the burner was also observed during analysis of the extraction solution by AAS. Rozenson and Heller-Kallai[120] noted that DCB attacks Fe-rich layer silicates at similar rates as with well-crystallized Fe-oxides.[121] Sulfides are formed as a result of the disproportionation of dithionite in acidic solutions.[94,110] Based on these problems, use of this particular extractant cannot be recommended. The problem of precipitation of sparingly soluble oxalates of Ca, Ba, Fe, Mn, and Pb has often been mentioned as a similar disadvantage of the oxalate reagent. Chao and Liyi[108] found as much as about 1750 mg/l Fe in their oxalate leach without quoting problems with Fe(II) oxalate . In our Fe/Mn nodule we have found oxalate precipitatious when the dissolved Fe concentrations exceeded 1000 mg/l (Figure 13). Such high Fe levels, however, are rarely reached in extracting natural materials without severely overloading the extraction system.[117] The previous removal of elements which precipitate with oxalate by HOAc-NaOAc and hydroxylamine reagents allows the subsequent application of acidified ammonium oxalate within an extraction sequence.
quence.

Shuman[110] introduced an ascorbic acid-ammonium oxalate extraction. He pointed out that this acid-reducing and -complexing couple is an easily analyzed solution, well-defined chemically, without the contamination and solubility problems of dithionite, or the difficulty of standardizing UV light to use for crystalline Fe oxides with an oxalate solution. The extractants suggested by Shuman[110] for an oxide fractionation scheme are listed in Table 6, but requires more widespread testing on pure mineral phases to evaluate its selectivity (see Section III.C.6).

In all the various reactions of reducing extractants, protons are consumed and the solution becomes more alkaline. However, since manganese and iron is precipitated in alkaline solutions (as hydroxides or oxides), buffering at low pH values is necessary. Because of the importance of this fact, the chemistry of reducing hydroxylamine reactions should be considered carefully.[117] Hydroxylammonium chloride is weakly hydrolyzed in water ($pK_a = 5.8$), and in acidic solution is able to reduce MnO_2. Its reducing power, as defined by the Nernst equation, is pH dependent; thus, its reaction with both iron and manganese will

Table 6
REAGENTS USED FOR EXTRACTING SEDIMENT Fe- AND Mn-OXIDE FRACTIONS

Extractant	Solution	Time	Conditions	Ref.
Hydroxylamine hydrochloride[a]	0.1 *M* $NH_2OH{\cdot}HCl$, pH 2 /0.01 *M* HNO_3	30 min	Reciprocating	110,118
Oxalate buffer[a]	0.2 *M* $(NH_4)_2C_2O_4$, pH 3 /0.2 *M* $H_2C_2O_4$	4 h	Reciprocating, dark	105,110
Ascorbic chloride/oxalate[a]	Oxalate above + 0.1 *M* ascorbic acid	30 min	97°C, stir intermittently	110
Stannous acid/oxalate	Oxalate above + 0.1 g $SnCl_2$	30 min	97°C, stir intermittently	110
Hydroxylamine/sodium acetate	0.25 *M* $NH_2OH{\cdot}HCl$ + 1 *M* NaOAc, pH 5			89
Hydroxylamine/citrate buffer	1.0 *M* $NH_2OH{\cdot}HCl$ + 0.175 *M* Na citrate, pH 5	10 min	Vortex mix	64
Hydroxylamine/acetic acid	1.0 *M* $NH_2OH{\cdot}HCl$ in 25% (v/v) HOAc	30 min	Reciprocating	107
Hydroxylamine/hydrochloric acid	0.25 *M* $NH_2OH{\cdot}HCl$ in 0.25 *M* HCl	30 min	50°C, stir intermittently	108
Dithionite/citrate buffer (DCB)	20 g citrate + 2 g dithionite per 100 ml	Overnight	Reciprocating	119

[a] Procedures preferred by Shuman[110] in the order of his sequential extraction of reducible sediment components.

be enhanced by lowering the pH of the reagent from 3 to 1.[118] Reaction can be completed in a few minutes, depending on the physical structure of the sample such as surface area and the number of reactive sites. Extraction times of 30 to 60 min may be caused by an attempt to dissolve coarse fragments of Mn-nodules for complete dissolution, but multiple extractions may not be required. The concommitant reduction in the hydroxylamine concentration results in an logarithmic decrease in reaction rate.

The buffering capacity of the pure hydroxylamine hydrochloride solution is negligible, which is illustrated by the titration curves in Figure 14. The "buffer capacity" of hydroxylamine hydrochloride acidified by 0.01 *M* nitric acid is readily exceeded using concentrations as low as 2 to 3 mg/ml pure ferromanganese nodule material (which is at the lower end of the range used by Chao[118] in the original development of his method). Use of 5 mg/ml nodule material clearly affects the final pH value in the extraction experiment, as shown in Figure 15. The time dependency of pH shows strong increase, with reactions resulting in an "overloading" of this weak buffer system. Considerable precipitation of initially dissolved Fe and readsorption of both Pb and Cu (but not Mn, Zn, Co, and Ni) occurred during the extraction experiment (Figure 16).

Similar artifacts have recently been reported by Tipping et al.,[122] who found that treatment of a naturally occurring mixture of Mn and Fe oxides with acidified hydroxylamine (4 mg solids per milliliter of solution) solubilized almost all the Mn, Ca, Zn, and Ba present, together with about one third of the original Pb contents. Treatment of the residue with oxalate released all Fe and the remaining two thirds of Pb. Parallel examination of the solids by electron microscopy, coupled with electron probe microanalysis, showed, however, that the dissolution behavior of Ca and Pb did not reflect their distributions in the original mixture, where Ca had been associated with both the Mn and Fe oxides, but Pb largely with the Mn oxide alone. These discrepancies were attributed to overloading of the extractant in the course of the experiment. The transfer of a large amount of Pb from manganese to iron oxide occurred because during hydroxylamine treatment the pH of the suspension increased

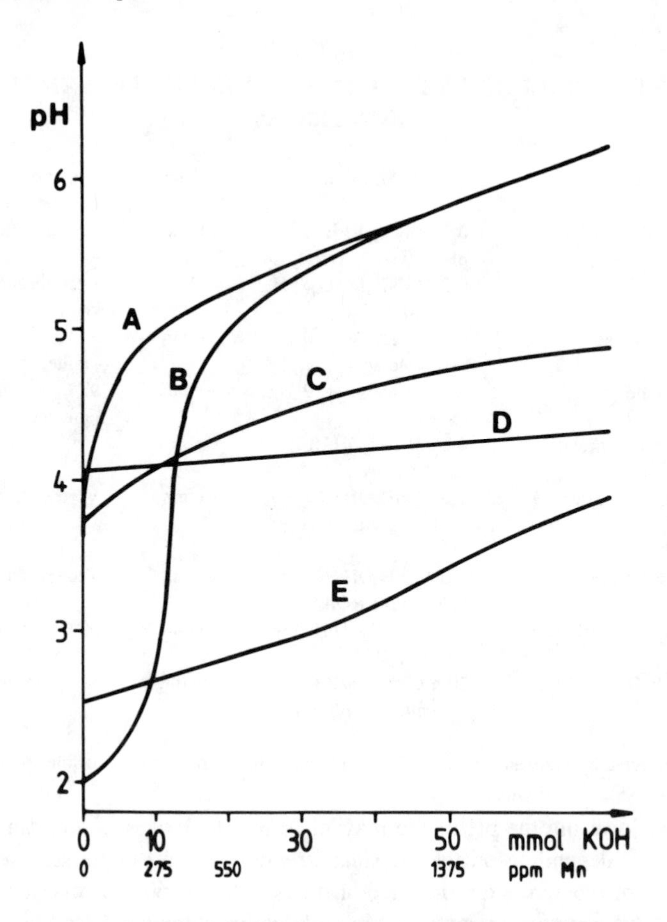

FIGURE 14. Titration curves of unbuffered 0.1-*M* hydroxylamine hydrochloride solutions (A) and those buffered with 0.01 *M* HNO_3 (B), 0.1 *M* sodium citrate (C), 0.5 *M* sodium acetate (D), and 0.1 *M* potassium phthalate (E), by KOH representing increased proton loss during reducing of Mn(IV) oxides.

dramatically, reaching a final value of 5.2.[122] These results clearly illustrate the readsorption problems predicted by Rendell et al.[123] on the basis of experiments with spiked sediment samples, and by Horowitz and Eldrick[20] from results obtained in studies of the interrelations between surface area and geochemical sediment phases (see discussion above). This phenomenon could result in an underestimation of the metal releases from solids, depending on their buffer capacity and oxide content, and could contribute to severe errors (''matrix effects'')[124] when a succession of extraction steps is carried out.

Robbins et al.[64] recommended buffering of the hydroxylamine solution by adding a sodium citrate buffer, pH 5, in order to obtain reproducible results. With this buffer pH during extraction remained relatively constant. Following extraction of nodule material containing 44% Mn and MnO_2 at 1:40 solid to solution ratio, a pH increase from 5.0 to only 5.2 was produced. Extraction with a solution containing ammonium citrate and hydroxylamine at pH 8.5 was first devised by Bloom[126] as a field chemical test in mineral prospecting work for the determination of cold-extractable metals in soils and sediments. Citrate serves as a chelating agent in the extracting solution to prevent readsorption of released metal ions at the prevailing high pH. While the acidity is low (pH 5 to 8.5), both the citrate and NaOAc buffers tend to enhance recovery of colloidal, amorphous, and poorly crystalline Fe compounds. In view of the failure to selectively extract Mn-oxides using buffered hydroxylamine,

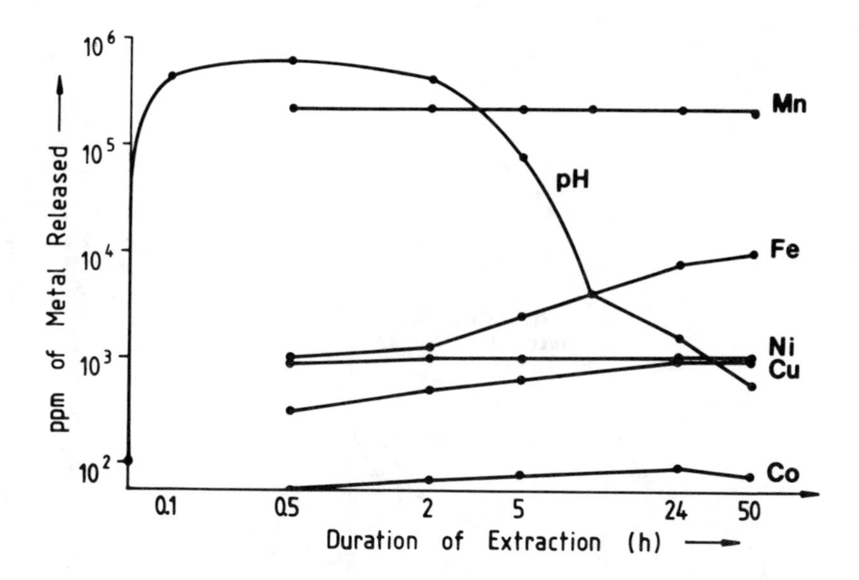

FIGURE 15. Changes of pH and metal removal during extraction of a ferromanganese nodule sample with 0.05 *M* hydroxylamine hydrochloride. Solid to solution ratio 1:200. pH scale = log (metal concentration scale).

we recommend, at least for anoxic sediments, to lump the two fractions together by an appropriate extractant.

The time dependency of pH change for unbuffered hydroxylamine is not easy to explain (Figure 15). After 2 h, the pH of the suspension decreases again, reaching nearly the initial values after several hours. Consequently, substantial iron and copper are released again into solution. As the redox reactions are completed after less than 0.1 h, the shift in pH is probably caused by decay of the extractant according to the reaction:[125]

$$4\ HONH_3^+ \rightarrow N_2O + 2\ NH_4^+ + 3\ H_2O + 2\ H^+$$

Thus, the stability of hydroxylamine depends on the final pH reached during the experiment. The extractant is sufficiently stable below pH 3. Mills[127] found in her kinetic experiments, using photometric analysis of the residual hydroxylamine concentrations, that these decrease after 10^2 to 10^3 min, depending on the type of sediment and the concentration of hydroxylamine used. Such release on lowering the pH is thus difficult to predict, and the best means of overcoming this problem is, again, strong buffering or avoidance of the hydroxylamine extractant.

4. Extractants Releasing Organic- and Sulfide-Bound Metals

A substantial proportion of trace metals in sediments are bound to viable or detrital organic matter, which may be considered as comprising both humified or "true" humic substances, and nonhumified materials. Organic-rich materials such as sewage sludge are also known to strongly accumulate metals. The mobility and translocation of trace metals such as Cu and Ni is strongly associated with the fate of organic matter in soils.[128] Organic-metal interactions are important in the distribution of trace metals both through competition or enhanced adsorption of organic-metal species and through films of (viable) organic matter on sedimentary particles. Natural organic matter such as humic and fulvic acids tends to have high complexation capacities for trace metals,[129-131] e.g., the total binding capacity of marine humic acid for metal ions is 200 to 600 μmol/g.[132] Hence, many extraction procedures contain steps designed to remove the organic fraction.

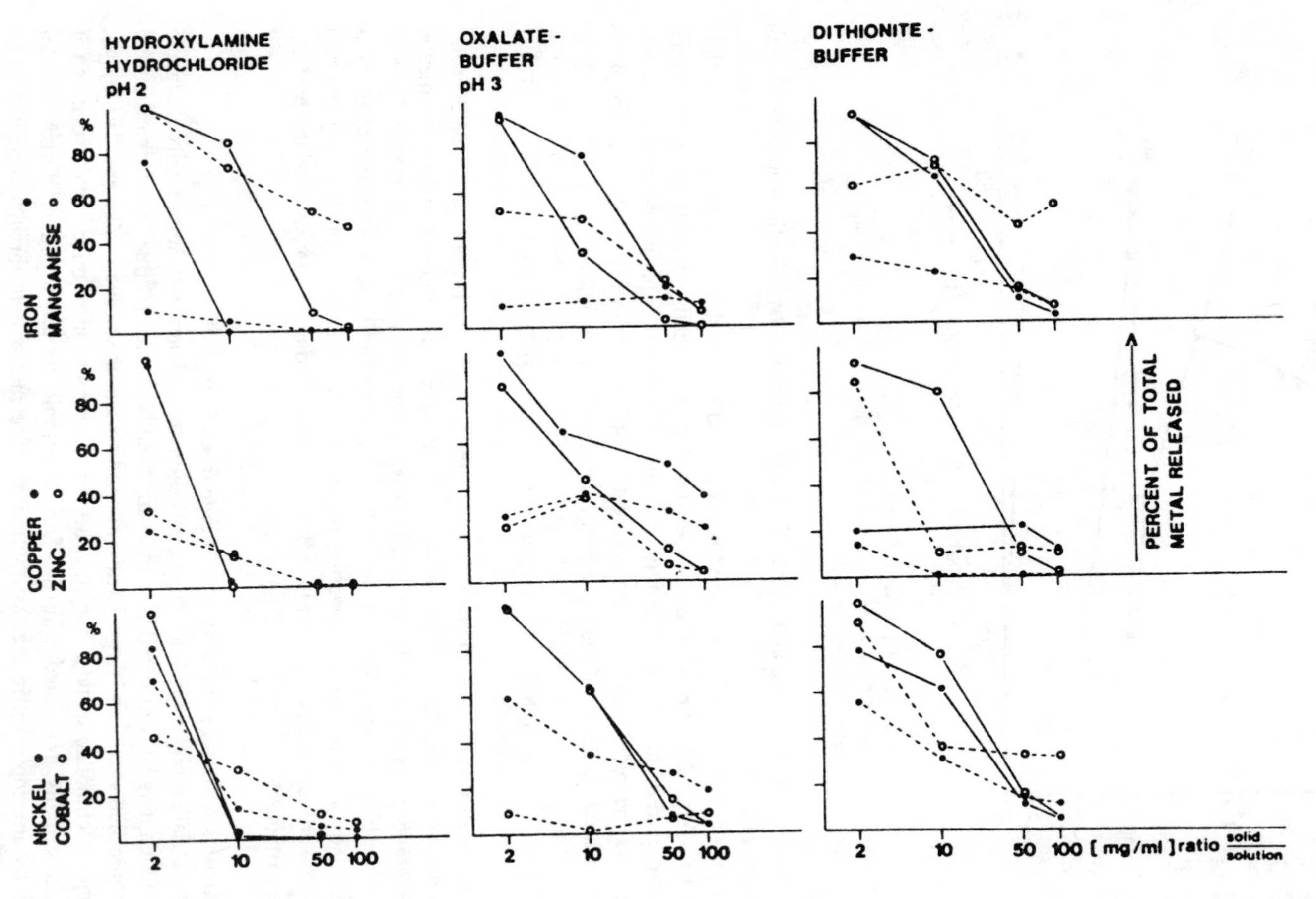

FIGURE 16. Effects of the solid to solution ratio on the release of metals from metalliferous sediment and nodule material. All investigated metals (Mn, Fe, Zn, Cu, Co, Ni) show a fall in the extraction rate with increasing concentration of solid substances due to an "overloading" of the buffer solution capacity. This decrease is generally more distinct for the ferromanganese nodule material (solid line) than for the metalliferous deep sea sediment samples (dashed line). (From Pfeiffer, G., Förstner, U., and Stoffers, P., *Senckenbergiana Marit.*, 14, 23, 1982. With permission.)

Although ''organically bound metal'' is frequently categorized as one distinct phase, it covers a broad spectrum of binding mechanisms including adsorption, complexation, and chelation. Approximately one third of the total binding capacity of marine humic acids is due to cation exchange with the remainder due to complexing sites.[132] There is inevitably an overlap in stability of these binding forms with others in any extraction scheme, and the basic conceptual question is raised as to the significance of metal portions entrapped in organic matter, when some organically sorbed metals are readily displaced by previous extraction steps. Slavek et al.[114] found that most metals associated with humic matter were ion displaceable. This contradicts the impression created by the usual order in most of the common sequential extraction schemes (i.e., organic fraction directly preceding the residual fraction for reasons discussed subsequently) that trace elements associated with the organic reaction are unavailable for short term chemical interactions. It is necessary to differentiate the organic fraction and isolate the ''labile organic fraction'' at an earlier stage of a sequence study, as recently proposed by Robbins et al.[64]

If removal of the organic fraction together with its trace metal load is considered important, it can be achieved either by oxidative destruction or dissolution in alkaline solutions. Both approaches, however, have substantial disadvantages. Oxidation of organic matter may result in the co-oxidation of sulfides, which may bind a substantial portion of the metals, especially in anoxic soils or sediments. Strong bases such as NaOH cause unacceptable dissolution of clay minerals. In alkaline solutions metals tend to hydrolyze and readsorb onto reactive surface sites formed by removal of their organic coatings. Since there is no generally accepted alternative, we will consider both approaches together with the appropriate measures to avoid their principal disadvantages, and also arguments for rejection of at least oxidative extractions.

The earliest reported reagent applied for extracting humic substances is sodium hydroxide, and to this day it remains that suggested for a standardized extraction procedure by the International Humic Substances Society. However, for reasons discussed above, this reagent is not suitable for trace metal recovery. Basic solutions of sodium-pyrophosphate ($Na_4P_2O_7$, with or without added NaOH) have historically been used for extracting metal-organic complexes from soils.[133] These are reasonably efficient in extracting the fulvic and humic acid fractions of organic matter, but not the refractory organic matter in samples. Thomson et al.[80] found that optical absorbance, as a measure of alkali-soluble organic matter from estuarine sediments, varied with extractant in the order NaOH $<$ DTPA $<$ pyrophosphate. The effectiveness of pyrophosphate solutions in removing the easily extractable organic matter (EOM) has been attributed to its ability to chelate with Ca and trivalent metal ions. In effect, the removal of cations bound to flocculated humates results in the subsequent conversion of the polyanion to its soluble sodium salts. Figure 17 demonstrates the relative removal efficiencies of these extractants for organic matter and aluminium from deep-sea sediments. The most important factor is the pH of the solution, which is commonly between 7 and 10. A lower pH tends to decrease the extraction of organic matter, but also tends to increase the extraction of amorphous Fe oxyhydrates. Gascho and Stevenson[134] indicated that increasing pH and salt concentration decreased the ratio of humic to fulvic acids, while an acidic mixture of dithionite and pyrophosphate was shown to effectively remove Fe from both oxides and organic matter in soils.[135] At pH 10, 0.1 M sodium pyrophosphate does not dissolve Fe oxyhydrates.[135] Shuman[110] demonstrated that $Na_4P_2O_7$ possibly extracts more Fe than just the organic fraction, since it solubilized more Fe than did the acidified ammonium oxalate agent. Its application may therefore be justified only if the amorphous oxides are extracted before the organic fraction by another extractant. The Fe portions found in pyrophosphate extracts are attributed in part also to cross-contamination arising from peptized sols.[136,137] The efficiency of basic pyrophosphate extractions also increases with repetitive treatment. Figure 18 shows that almost all the TPOM is removed after the fourth extraction

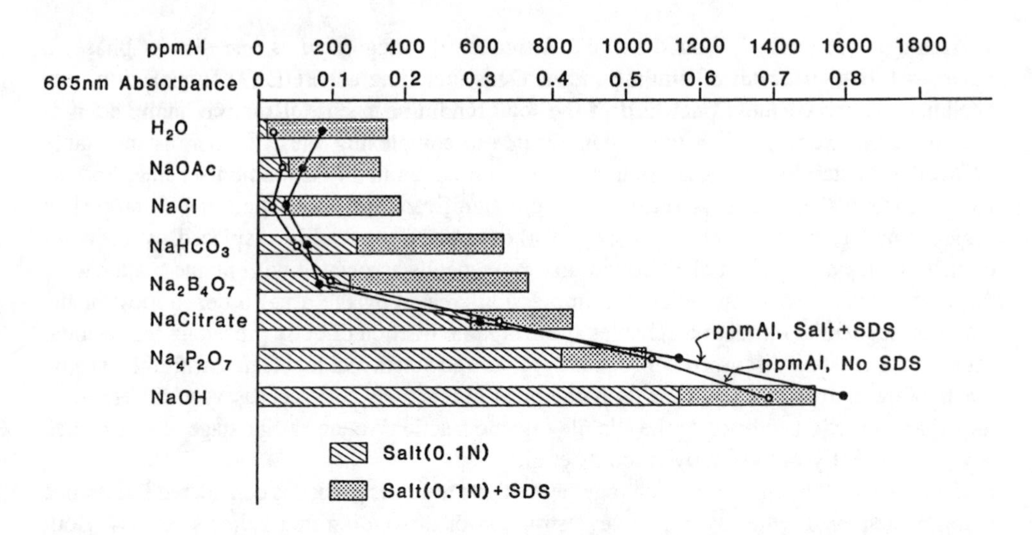

FIGURE 17. Relative extraction efficiency of organic matter and aluminum from a deep-sea organic-rich sediment by 16 different extractant solutions. In this experiment, 0.2 g material were extracted by a single 2-h, 100°C extraction in 30 ml of solution. In each case, addition of the surfactant (SDS) to the salt alone substantially improved organic matter dissolution, as shown by the increase in absorbance at 665 nm of the extraction solution. (From Robbins, J. M., Lyle, M., and Heath, G. R., Rep. 84-3, College of Oceanography, Oregon State University, Corvallis, 1984. With permission.)

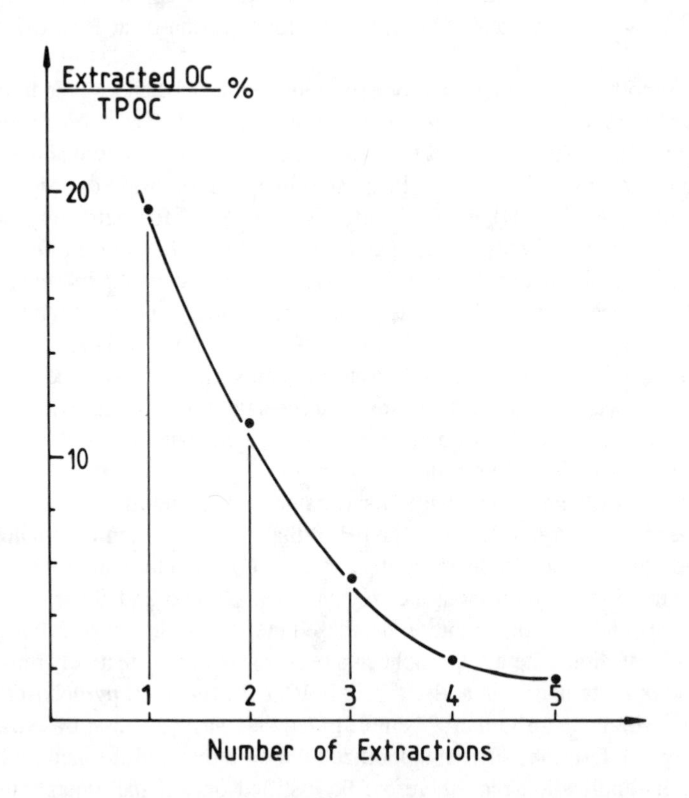

FIGURE 18. Organic matter extracted from suspended particles of the Gironde Estuary vs. number of extractions. The organic matter extracted is expressed by the ratio between the organic carbon removed by each step and the total particulate organic carbon (TPOC) in the initial sample. (From Berger, P., Etcheber, H., Ewald, M., Lavaux, G., and Belin, C., *Chem. Geol.*, 45, 1, 1984. With permission.)

Table 7
CHARACTERISTICS OF THE OXIDANTS AND THE OXIDANT SOLUTIONS SELECTED FOR THE EXTRACTION OF ORGANIC MATTER IN SEDIMENTS

Oxidant	Half-cell reaction	Solubility in water (mol/dm³)	Concentration (mol/dm³)	pH	E_h (V)
O_2	$O_2 + 4H^+ + 4e^- \Rightarrow 2H_2O$	1.3×10^{-3} [a]	2.7×10^{-4} [b]	2	1.06
NaOCl	$OCl^- + H_2O + 2e^- \Rightarrow Cl^- + 2OH^-$	≈5.0	0.3	8.5	1.40 [c]
H_2O_2	$H_2O_2 + 2H^+ + 2e^- \Rightarrow 2H_2O$	42.4	12.7	2	1.67

[a] $P_{O_2} = 1.01 \times 10^5$ Pa.
[b] $P_{O_2} = 2.10 \times 10^4$ Pa.
[c] $[Cl^-] = 1 \times 10^{-6}$ *M*.

in suspended matter samples.[138] Although pyrophosphate partially attacks clays and other aluminosilicates (Figure 17), it has been preferred to other strongly complexing reagents, such as citrate and EDTA, due to its lesser effects on oxides.

Solvent extraction of trace metals bound to sedimentary organic matter has rarely been reported. Soil chemists have advocated organic solvent pretreatments of soils to selectively remove lipids and waxes, but these procedures seem poorly suited to routine application. Hayes[139] has provided a comprehensive review of the work done to date on extraction with solvents such as EDTA, diaminoethane, *N,N*-dimethylformamide (DMF), and dimethylsulfoxide, either alone or in mixtures containing alkali, acid, or sodium pyrophosphate. Pyrophosphate was found to be a more suitable reagent than aqueous acetylacetone for the selective extraction of iron-organic complexes from soils.[140] One of the few studies of trace metals was that of Cooper and Harris,[141] who isolated the nonhumic acid compounds present in the organic fractions of soils and sediments (i.e., proteins, carbohydrates, peptides, fats, waxes, and resins) by various organic solvents. Summerhayes et al.[142] used a 1:1 methanol-benzene mixture to extract bitumen particles from New Bedford Harbor sediments.

There are more promising new developments with the application of surfactants, which can enhance the amount of organic matter extracted by basic solutions (Figure 17). Surfactants are especially effective dispersants where a substantial fraction of the component of interest is located at the phase boundary. Since humic and fulvic acids have high surface areas, surfactants should adsorb to these organics and disperse them. After extensive preinvestigation, Robbins et al.[64] decided to use hot 1% (w/v) sodium dodecyl sulfate (SDS, gel electrophoresis grade) buffered with a 0.2 *M* $NaHCO_3$ solution (pH 8.3) for the easily-extractable organic fraction. The higher ionic strength of the buffered solution favors flocculation of clays and oxyhydroxides also dispersed by the surfactant and thus reduces contamination of the organic fraction by these fine particles. This technique could be useful for separating metals associated with labile organic matter from those bound to labile sulfides in fresh anoxic materials. It would be valuable, however, to know the extraction efficiency on the basis of organic carbon measurements, and to consider the possible effects of high pH (precipitation, readsorption) upon extracted trace metals.

Oxidation is one of the most widely used techniques for the decomposition of organic matter. Solubility values are given in Table 7 to show the upper limits of concentrations of common oxidants in water. The oxidants are generally used at levels well below their solubilities and, thus, under conditions that could limit their effects on organic matter (Table 7). The accumulation of reaction products (e.g., Cl^- ions in the NaOCl solutions) during the organic matter decomposition lower the redox potentials, while solution pH has a direct effect on the redox potential of all three oxidants. The magnitude of this pH effect can be particularly dramatic in the O_2 system and somewhat less pronounced in the NaOCl and H_2O_2 systems. In general, the effectiveness of the O_2 and H_2O_2 oxidants are increased in acidic solutions, while the NaOCl is more efficient in basic solutions.

Sodium hypochlorite has initially been used to remove organic coatings from clays prior to X-ray diffraction analysis without destruction of the sesquioxide coatings or carbonates.[143] It was later adopted by Gibbs[88] in the first application of a sequential extraction scheme in the study of trace metal transport in rivers. NaOCl generally removes more organic carbon from sediments than does the stronger oxidant H_2O_2, due to the solubilization of organic matter such as humic acids in the alkaline medium, with a minimum removal of Si, Mn, Fe, and Al.[144] The method generally employed for NaOCl uses a 5 to 6% NaOCl solution at pH 8.5 to 9.5 and a solid to solution ratio of 1:2. The solutions, in centrifuge tubes, are placed in a boiling water bath for 15 to 30 min. Lavkulich and Wiens[144] tested up to five successive extractions and concluded that three were sufficient to remove up to 98% of the oxidizable organic carbon in soil samples. Shuman[145] showed that after two treatments, insignificant amounts of metals were dissolved from soils containing less than 3% organic carbon. He concluded that NaOCl is a valuable extractant for the organic fraction if used in earlier steps of sequential extraction schemes, since both alternate agents, hydrogen peroxide and sodium pyrophosphate, dissolve considerable portions of manganese and iron oxides.

The most widely used oxidizing agent is acidified hydrogen peroxide. The oxidation process is promoted by heating, either at 85°C for several hours, and/or by evaporating to dryness. Use of an acidic solution (usually pH 2) prevents scavenging of metal ions by any Fe(III)-hydroxide precipitated at higher pH values. Although a strong oxidant for organic matter, it is incomplete, and sulfide minerals are also affected to a large extent. According to Jackson,[146] the remaining organic matter (15 to 20% in sediment samples)[94] should consist of paraffin-like material and resistant (nonhumified) organic residues. Stronger oxidizing solutions usually rely upon the use of strong acids, which may, however, seriously attack major silicate minerals. As noted earlier, the principal mechanism of binding metals in humic acids is via adjacent functional groups. If the humic acid molecule could be degraded sufficiently to destroy these groups, then metal will be released into solution. The removal of bound metals should be complete after the H_2O_2 oxidation without affecting silicate minerals.[19,20,94] The choice of boiling 30% H_2O_2 represents a compromise between complete oxidation and alteration of silicate material. Since significant adsorption of added metal to residual clay particles was found to occur during an acidified hydrogen peroxide digestion procedure,[19,123] it is important to extract the residue immediately after heating. Gupta and Chen,[87] in sequestering trace metals from the organic and sulfide extractions, found that a larger fraction of the metals was leached when the sediment was oxidized with hydrogen peroxide and subsequently extracted with 1 *M* ammonium acetate in 6% nitric acid at about pH 2, in comparison to an extraction with 0.01 *M* pure nitric acid. Lead was substantially different in the two extractants due to the greater solubility of precipitated $PbSO_4$ in acidified ammonium acetate. Adsorption by active sites of trace metal ions released from organic coatings may be the reason for a diminished extraction even by 0.1 *M* nitric acid solution, whereas with an acidified ammonium acetate extraction, the adsorbed cations may be more effectively exchanged with ammonium ions. From these and other careful investigations, the hydrogen peroxide-acidified ammonium acetate couple has been widely adopted for the organic and sulfide extractions in both oxic and anoxic materials (see Table 8).

Among other problems with the peroxide reagent is its catalytic decomposition by MnO_2, which lowers its effectiveness, with the solution becoming sufficiently acidic, in the case of sulfide-rich anoxic materials, to attack clay minerals. Conditions promoting an expansion of clays during H_2O_2 treatment seem to enhance the ease and extent of structural cation loss.[149] Possible formation of oxalate as a result of the hydrogen peroxide treatment of soils

Table 8
SOME SEQUENTIAL EXTRACTION SCHEMES USED BETWEEN 1973 AND 1986

	Exchangeable[a]	Specifically sorbed, carbonate bound	Easily reducible substrates	Easily extractable organics	Moderately reducible oxides	Oxidizable oxides and sulfides	Crystalline Fe-oxides	Residual minerals	Ref.
A	$CaCl_2$	HOAc		$K_4P_2O_7$	NH_4Ox/HOx			HF	169
B	NH_4OAc	HOAc			$NH_2OH{\cdot}HCl$/ HOAc[4b]	H_2O_2[3]	(DCB[4])	$HNO_3/HF/HClO_4$	87
C	$MgCl_2$			NaOCl/DCB[3]			DCB[2]	Fusion	165
D	NH_4OAc		$NH_2OH{\cdot}HCl$ pH 2			H_2O_2/NH_4OAc	DCB	HF/HNO_3	81
E	NH_4OAc		$NH_2OH{\cdot}HCl$/ NH_4OAc			H_2O_2/NH_4OAc	$NH_2NH_2{\cdot}HCl$	HF	170
F		HOAc	$NH_2OH{\cdot}HCl$ pH 2		$NH_2OH{\cdot}HCl$/ HOAc[4]	H_2O_2/NH_4OAc		$HF/HNO_3/HClO_4$	147
G			$NH_2OH{\cdot}HCl$ pH 2[2]	NaOCl[1]	NH_4Ox/HOx		DCB	$HClO_4/HNO_3$	115
H	$MgCl_2$	NaOAc pH 5			$NH_4OH{\cdot}HCl$/ HOAc	H_2O_2/NH_4OAc		$HF/HClO_4$	94
I	NH_4OAc		$NH_2OH{\cdot}HCl$ pH 2			H_2O_2/NH_4OAc		$HF/HClO_4$	139
J	NH_4OAc/MgOAc		$NH_2OH{\cdot}HCl$ pH 2[3]			H_2O_2/NH_4OAc[2]		—	93
K	NH_4OAc[1]/ CuOAc[2]/NaOAc[4]			$Na_4P_2O_7$[3]			HNO_3	—	92
L			$NH_2OH{\cdot}HCl$ pH 2[2]	NaOCl[1]	$NH_4OH{\cdot}HCl$/ HOAc			HNO_3	116
M	$BaCl_2$	NaOAc pH 5[3]			$NH_2OH{\cdot}HCl$/ HOAc[4]	H_2O_2/NH_4OAc[2]		$HF/HClO_4/HNO_3$	171
N	$MgNO_3$		$NH_2OH{\cdot}HCl$ pH 2[3]	NaOCl[2]	NH_4Ox/HOx			$HF/HNO_3/HCl$	145
O	$MgCl_2$	NaOAc pH 5			$NH_2OH{\cdot}HCl$/ HOAc	H_2O_2/NH_4OAc		$HF/HClO_4/HCl$	26
P	NH_4citrate		$NH_2OH{\cdot}HCl$ pH 2	NaOCl/HCl			$NH_2NH_2{\cdot}HCl$	—	172

Table 8 (continued)
SOME SEQUENTIAL EXTRACTION SCHEMES USED BETWEEN 1973 AND 1986

Exchangeable[a]	Specifically sorbed, carbonate bound	Easily reducible substrates	Easily extractable organics	Moderately reducible oxides	Oxidizable oxides and sulfides	Crystalline Fe-oxides	Residual minerals	Ref.
Q	NaOAc pH 5		SDS/$NaHCO_3$	$NH_2OH{\cdot}HCl$/ ci-trate			HF/HNO_3	64
R $MgCl_2$	NaOAc pH 5	$NH_2OH{\cdot}HCl$ pH 2		$NH_2OH{\cdot}HCl$/ HOAc[5]	H_2O_2/NH_4OAc		—	174
S NH_4OAc	NaOAc pH 5	$NH_2OH{\cdot}HCl$ pH 2		NH_4Ox/HOx	H_2O_2/NH_4OAc		HNO_3	95
T NH_4OAc	NaOAc pH 5	$NH_2OH{\cdot}HCl$ pH 2		$NH_2OH{\cdot}HCl$/ HOAc	H_2O_2		HNO_3/HCl/HF	175
U $MgCl_2$	$NH_2OH{\cdot}HCl$/NaOAc pH 5		$Na_4P_2O_7$	$NH_2OH{\cdot}HCl$/ HOAc	H_2O_2/NH_4OAc		—	89

[a] Some schemes extract the interstitial and water-soluble fraction before the exchangeable, which is not indicated here.
[b] Order of attack noted by superscript numbers, where it differs from left to right.

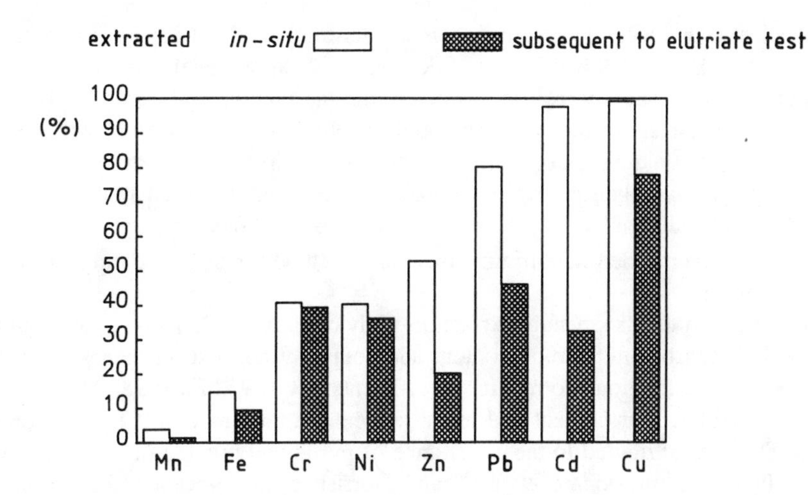

FIGURE 19. Average changes of oxidizable metal fractions in anoxic sediment samples from the Hamburg Harbor upon aeration of the suspension during the elutriate test. (Data from Kersten and Förstner.[82])

has been reported, and this may dissolve oxides of iron and manganese.[150] Thus, this extractant should preferably be applied after oxide reducing reagents in a sequential extraction scheme, even through most of the extractants previously used for reduction of oxides will most probably extract minor portions of trace metals from the organic fraction. These intimate interactions have particularly been observed for estuarine particulate matter, where oxide coatings protect the organic matter from oxidation by H_2O_2 and vice versa, organic matter seems to inhibit reduction of oxidic components.[151]

Kitano et al.[152] suggested pretreatment of anoxic samples with 0.1 *M* HCl solution with air bubbling for 12 h at room temperature to extract the labile sulfide fraction before applying the peroxide digestion for organic matter. The authors found that pyrite in the sediment was not completely dissolved by the 0.1-*M* HCl treatment, but that the subsequent 30% H_2O_2 treatment dissolved pyrite completely. We have shown recently that weak oxidation by bubbling air through the suspension has in fact ambiguous effects on the oxidizable forms of the various trace metals.[82] Figure 19 suggests that the average percentages of metals recovered in the H_2O_2-fraction as the fifth step of a six-step sequential extraction of anoxic freshwater samples increase in the order Mn $<$ Fe $<$ Cr $\leqslant$ Ni $<$ Zn $<$ Pb $<$ Cd $<$ Cu, which corresponds in part to the respective metal sulfide solubility. A common pattern, except for Mn, Fe, Cr, and Ni, is a significant reduction of the H_2O_2-extractable metal fraction after aeration. The relative effectiveness of this slight aeration step on the oxidizable metal forms was thereby found to decrease in the order Cd $>$ Zn $>$ Pb $>$ Cu, which indicates that these predominantly anthropogenically derived metals probably form the most labile sulfide forms in the heavily contaminated samples of Hamburg Harbor studied.[82] Gupta and Chen[87] similarly found a strong relationship between the percentages of nonresidual Cd, Cu, Pb, and Zn in the sulfidic/organic fraction with sulfide concentrations, which indicated that they are less mobilized under prevailing anoxic conditions. Such results, gained from sequential extraction studies, are important in studies of dredged material disposal.[153] The majority of the sulfide content, however, is obviously not extractable by this easily oxidizing treatment. Chao and Sanzalone,[154] in a comparative study of the chemical dissolution of nine common crystalline sulfide minerals (galena, chalcopyrite, cinnabar, molybdenite, pyrite, sphalerite, and tetrahydrite), found that up to 43 to 66% of the total sulfur content was removed by a combined hydrogen peroxide/ascorbic acid leach, while stibnite and orpiment were only slightly attacked.

More efficient techniques for selective chemical dissolution of sulfides have recently been suggested by Klock et al.[155] Artificial sulfide phases of known elemental content, mixed with well-analyzed silicates, were used to determine the relative and absolute efficiencies of six different oxidants: bromine water/carbon tetrachloride, nitric/hydrochloric acid, hydrogen peroxide/ammonium citrate, bromine/methanol, and hydrogen peroxide/ascorbic acid couples. Except for hydrogen peroxide/ammonium citrate, these reagents all dissolved the sulfide phases, but additionally attacked the silicates in proportions ranging from 3 to 100%. Klock et al.[155] recommended the bromine-methanol method for differentiating sulfide from silicate metal forms.

Kitano et al.,[152] in their sequential extraction study of anoxic sediments, found significant correlations between organic carbon content and copper, zinc, and manganese contents in the peroxide-soluble fraction (correlation coefficients: Cu 0.97, Zn 0.98, Mn 0.81), and also between pyritic sulfur content and metal contents in this fraction (Cu 0.95, Zn 0.96, Mn 1.00). This was attributed to the covariance of pyrite and organic matter in the anoxic sediments. In fact, both Donard et al.[156] and Norrish et al.,[5] recently demonstrated by microprobe analysis that pyrite can precipitate within organic particles and, thus, both phases can build up intimate aggregates. These and similar studies indicate that it is almost impossible in selective chemical extractions to obtain a clear-cut discrimination of trace metals bound to organic matter, and in sulfide or oxide forms, no matter which sequence is chosen.

5. Strong Acid Extractants

After removal of the above chemical components from the sediment sample, the residue consists of silicates and other resistant minerals. Although these "residual" sediment components are not environmentally significant, determination of this fraction has been considered useful at least as a control for comparing the sum of all individual leachate steps to the result for the bulk analysis of the respective element. Such mass balances should be established to ascertain that all metal forms have been detected.

The determination of total element contents of a sediment or soil sample is not a trivial task. Problems of accurately and precisely determining the weights of the residual fraction of the sediment have been reported, e.g., by Badri and Aston,[93] who found significant deviations between the residual metal contents and the mathematical difference between the total element contents and the sum of the nonresidual fractions. These deviations are attributed to problems of choosing an acceptable chemical method for the acid digestion of this fraction. Few sophisticated analytical methods intrinsically measure total element contents (e.g., instrumental neutron activation analysis, or X-ray fluorescence analysis). Most of the more common methods only determine concentrations of certain species or forms of elements (e.g., metals in solution by atomic absorption or emission spectroscopy).

Fortunately, the efficiency and "selectivity" of a total element "extractant" may be easily checked visually, as at best no residue has to be left after treatment. Procedures for direct total digestion and dissolution of the most resistant silicate materials are based on strong-acid treatments at elevated temperatures.[157] The residual sediment components can be decomposed by hot digestion with hydrofluoric acid in combination with other oxidizing mineral acids such as nitric, perchloric, or hydrochloric acid. Acid mixtures such as HF-HNO_3-$HClO_4$,[13] HF-HNO_3-HCl,[158] or more simply, mixtures of HF and one of the other three acids, are preferred. At elevated temperatures, the use of sealed Teflon® or polypropylene bottles,[159] or the so-called Teflon® bombs[160] prevents the loss of volatile fluorides of Cr, Ge, Se(IV), Si, Sn, Te, Tl, Ti, and Zr. In the case of Si, no volatilization of SiF_4 occurs at temperatures $<100°C$.[161]

Acid digestion procedures are effective for all but a few inherently resistant minerals, which require alkaline fusions before acid dissolution.[157] Digestion of sediment samples by acid formulations, however, rarely results in clear solutions. Anoxic sulfide-rich samples yield mixtures of sulfate and elemental S, which may lead to white precipitates.[157] Using $HClO_4$ for sediment digestion may also yield precipitates of $KClO_4$,[13] while hydrofluoric acid may precipitate secondary fluorides such as $MgAlF_5 \cdot H_2O$, $NaAlF_4 \cdot H_2O$, and CaF_2.[163] Coprecipitation of some Fe, Al, Zn, and Ti can occur with CaF_2. Addition of boric acid has been proposed to dissolve any precipitated fluoride formed during HF-HNO_3 digestion.[158,159] In more concentrated solutions (>500 mg/100 ml), a gelatinous precipitate of borosilicates may form after a few days due to the high concentration of silicon. This has no adverse effects on the trace element determinations for which only the clear supernatant is used. Digestion of organic-rich material may yield black organic residues, but any metals initially associated with the organic components are presumed to be oxidized and then dissolved as fluoride complexes. Digestion of Mn-rich metalliferous sediments and nodules with HF/HNO_3, without previous selective extraction of the Mn, results in the formation of a Mn(IV)-oxide precipitate, which needs reduction to Mn(II) (e.g., by adding $NH_2OH \cdot HCl$).[64] The result of the combined digestion and dissolution procedure is a clear solution for analysis by flame atomic absorption spectrophotometry. Lower trace metal contents may require application of the graphite furnace. However, this can rapidly be damaged if the solutions contain excess amounts of HF or $HClO_4$.

With respect to their source, the metals present in an environmental sediment sample can be classified as either lithogenic or anthropogenic. Anthropogenic metal pollution is the nonlattice-held or, in sequential extraction terms, the "total nonresidual (TNR)" fraction often differentiated by a single extraction with dilute mineral acid or a strong chelating reagent (e.g., EDTA).

A comparison of the binding strength in lower Rhône River sediments of stable metal nuclides and their unstable counterparts, derived from radioactive emissions from nuclear power and processing plants, indicated similar differences in element extractability.[163] The extratabilities of both cesium (stable Cs and ^{137}Cs) and manganese (stable Mn and ^{54}Mn) isotopes were studied with a five-step leaching sequence. The anthropogenically derived radionuclides are nearly quantitatively extractable in the nonresidual fractions, whereas substantial proportions of the stable Cs and Mn are bound by residual mineral phases (Figure 20). It is clear that the difference in the leachability of manganese compounds mainly occurs within the reducible fractions, whereas the major difference between the stable Cs and ^{137}Cs can be seen within the residual fraction of the extraction sequence used.

Dilute HCl extractions, used to evaluate pollution effects or bioavailable metal concentrations, are claimed to ensure release of all unstable secondary species of interest from sediment components (organic matter, clays, hydrous oxides, carbonates) without causing significant attack of any mineral fragments present in the samples. Fiszman et al.[164] submitted their sediment samples to both single-acid extraction techniques and a four-step sequential extraction procedure. Results achieved with both techniques were compared to find out which single-acid extraction method worked best without attacking the geological matrix. Among the individual acid extractants choosen to leach the sediments were HNO_3 and HCl in different concentrations (0.1 to 5 *M*). It was found that only the weakest hydrochloric acid solution gave metal extraction efficiencies which were not significantly different from the sum of nonresidual steps applied in an extraction procedure.

Another possibility is to minimize primary mineral attack by grain size exclusion. In a sequential extraction study on the transport phases of trace elements in the Amazon and Yukon rivers, Gibbs[165] found characteristic relationships between different metal associations with particulates and their grain size spectra (Figure 21). It is evident that exclusion of the coarser grain size affects mainly the crystalline residual fraction, but not the fractions adsorbed to primary minerals and coatings. Serial extraction with 0.3 *M* HCl conducted by Malo[166] indicates that the dissolution of trace metals more closely follows the dissolution of Fe and Mn than the structural components Si and Al. The efficiency of the single

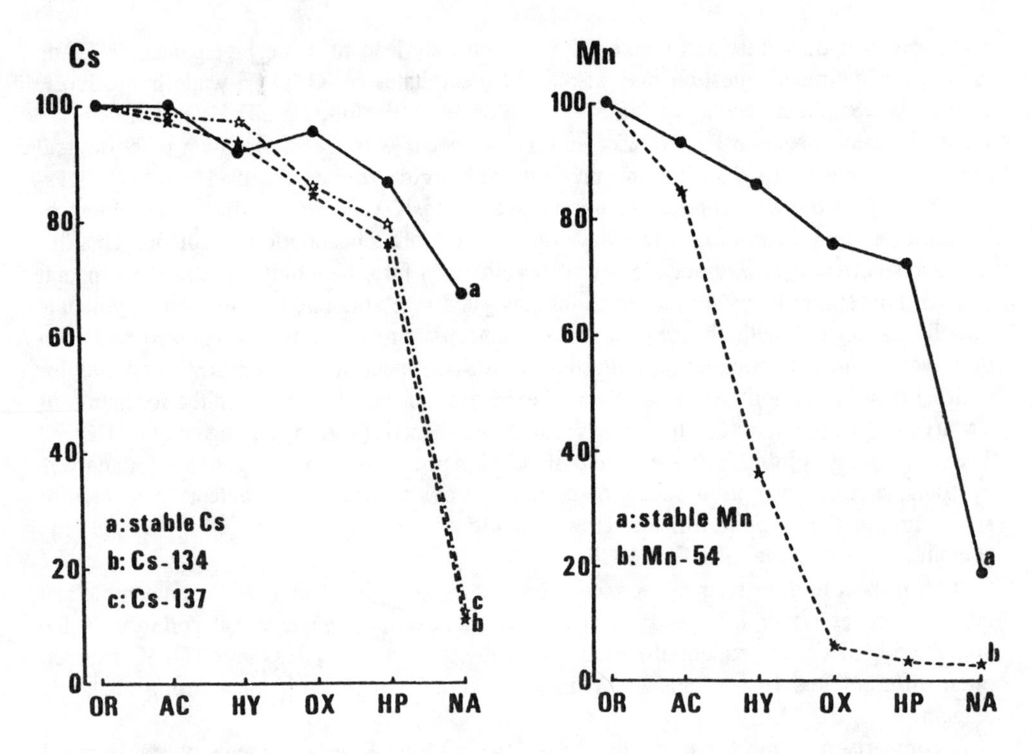

FIGURE 20. Extraction curves for stable cesium, ^{134}Cs and ^{137}Cs, stable manganese, and ^{54}Mn in a sediment sample from the lower Rhône River. (OR) Original sediment, (AC) element percentages residual after acetate, (HY) hydroxylamine, (OX) oxalate, (HP) hydrogen peroxide, and (NA) nitric acid extraction. (From Förstner, U. and Schoer, J., *Environ. Technol. Lett.*, 7, 295, 1984. With permission.)

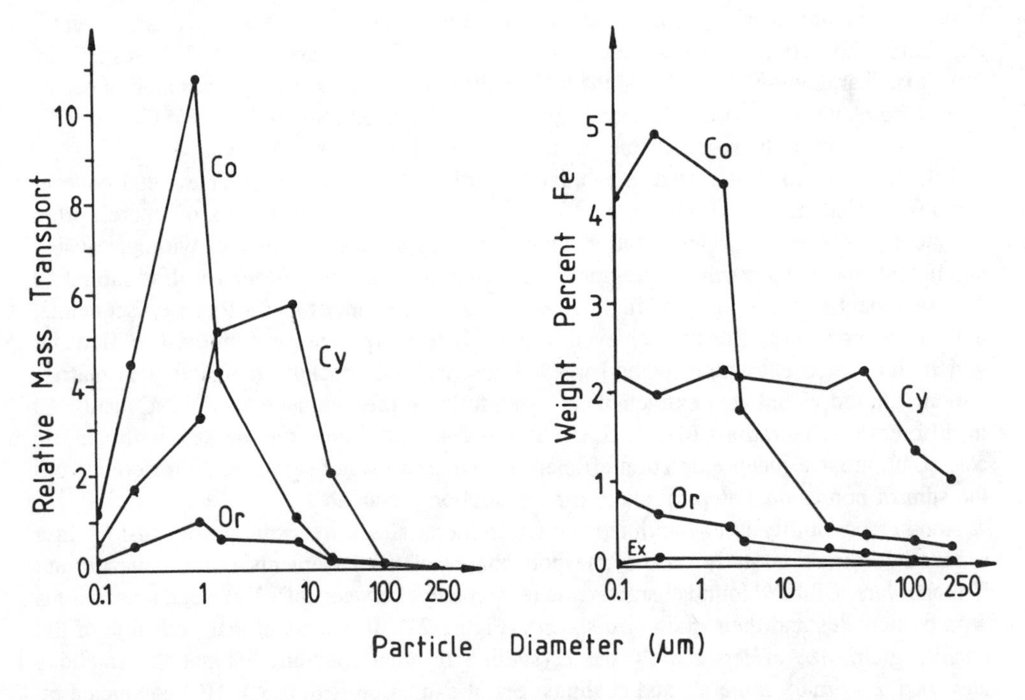

FIGURE 21. Mass transport and concentration in Amazon River of Fe in various particle sizes and for various transporting phases: (Co) Fe in coatings, (Cy) in crystalline matrix, (Or) in organic matrix, and (Ex) on exchangeable sites. (From Gibbs, R. J., *Geol. Soc. Am. Bull.*, 88, 829, 1977. With permission.)

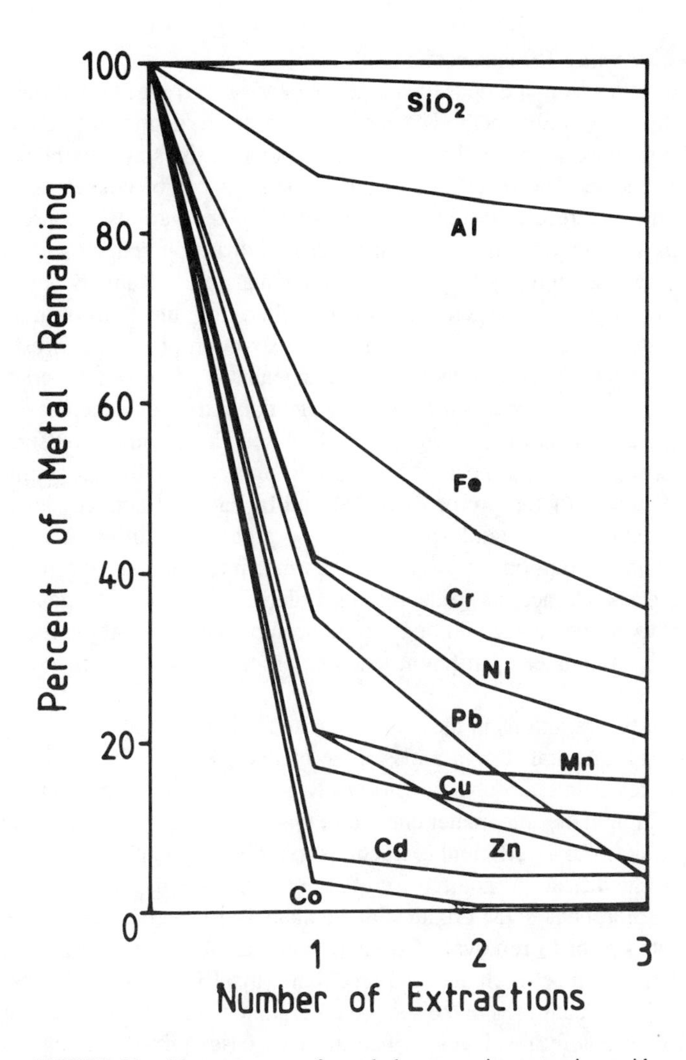

FIGURE 22. Mean recovery of metals by successive extractions with 0.3 *M* HCl solution. Solid to solution ratios 1:20 to 1:40. (From Malo, B. A., *Environ. Sci. Technol.*, 11, 277, 1977. With permission.)

0.3-*M* HCl leach indicates that replicate treatments dissolved only slightly more metals, except in the case of Pb (Figure 22). The efficiency of hydrochloric acid in extracting metals, as expected, depends on the concentration of the acid, the solid to solution ratio, and the temperature of the extraction.[166] Hydrochloric acid can dissolve even iron oxides of varying degrees of crystallinity, depending on the concentration of the acid and the temperature of reaction.[167] At elevated temperatures, however, hydrochloric acid can attack residual minerals such as clays. Thus, hydrochloric acid is commonly used as a "cold extractant" for pollution reconnaissance surveys. The percentage of TNR metal contents released by either dilute EDTA or HCl was found to vary between elements, with the lithogenous (residual fraction) character of the elements decreasing in the order Si > Al > Cu > Mn > Fe > Co > Cr > Ni > Pb > Zn > Cd,[87,88] or Si > Al > Fe > Cr > Ni > Pb > Mn > Cu > Zn > Cd > Co.[166] Chester et al.[168] have shown recently that a rapid 0.5-*M* HCl leaching technique can successfully be used as a first stage in a reconnaissance survey designed to identify trace metal pollution in stream sediments. The TNR metal concentrations are shown to be extremely variable in the sediments, and a number of approaches were assessed in order to interpret the data in an environmentally useful manner.

6. Sequential Extraction Procedures

If the single extraction methods discussed so far are arranged in proper sequence, a sequential multiple extraction method can be formulated. The degree of fractionation required depends on the purpose of the study, as does the choice of the single extractant in each step in a sequential scheme. The specificity of the single steps can be considerably improved by incorporation into a carefully designed combination of the various "nonselective" single extraction steps in a sequential extraction scheme. Extraction schemes involving three to eight different fractions have been proposed as summarized in Table 8.

While the least aggressive reagents should be applied first, there has been little uniformity with respect to the reagents used, or the order of extraction (Table 8). Most of the recommended schemes seek to first displace the exchangeable fraction as a separate entity, using $MgCl_2$ or NH_4OAc (pH 7) treatments. Most sequential extraction schemes call for removal of carbonates present as the next step (using HOAc, with or without buffering by NaOAc to pH 5). In subsequent steps, post-1977 proposals show more distinct similarities because they are modifications of the protocols introduced by Engler[81] and Tessier et al.,[94] which introduced separation of reducible phases. These procedures differ in minor operational details (e.g., solid to solution ratios, treatment time, interstep washing procedures). Most schemes seek to use extractants in decreasing order of pH values. In general, pH considerations appear to be secondary, although it has been pointed out that the metal fractionation pattern yielded by a number of different leaching procedures can be related to the pH value distribution used.[56]

While most of the variations in the schemes arise from the initial problem definition, i.e., differences in experimental design and sample characteristics, other variations reflect opposing logical viewpoints as to the position of organic matter extraction. Those favoring early decomposition of organic matter claim that this step releases entrapped mineral particles and thereby promotes more efficient extractions in the subsequent fractionation steps. Other teams argue that acidification required for efficient oxidation displaces metal ions sorbed to carbonates and amorphous Fe/Mn-oxides, and that acid-displacement and reducible fractions should be isolated prior to removal of organic material. Another alternative proposal is to insert the oxidation step between the easily reducing (by $NH_2OH{\cdot}HCl$, pH 2) and moderately reducing steps.[81,147] As outlined in the section on organic matter destruction, these alternative viewpoints are not readily resolvable, since in many cases mixed coatings are present and one could not expect to encounter clearly-defined entities in natural particulate material. Some authors tried to overcome this by introducing nonoxidizing extractants for organic matter removal (e.g., Robbins et al.[64]); and most recently, Chester et al.[168] proposed application of a nonoxidizing extractant ($Na_4P_2O_7$) for humic-associated trace metals before oxalate-extraction of amorphous iron oxides and an oxidizing reagent (H_2O_2) subsequently for residual organics and sulfides.

The degree of interaction between solid phases and extractant solutions can be altered by changes in experimental parameters such as reagent concentration, final suspension pH, solid to solution ratio, temperature, and contact time and intensity. The absence of standardized conditions makes it difficult to compare experimental data derived from studies in which such parameters are significantly different or even not listed, although during recent years investigators have tended to use similar extraction sequences. Possible approaches to evaluating extraction sequences are summarized in Table 9.

As one of the first groups, Guy et al.[19] used well-defined model sediments to evaluate three extractants for fractionation of metals in sediment: 1 *M* NH_4Cl (ion-exchangeable), H_2O_2 (organic), and dilute HNO_3 (total nonresidual metal), both single and in sequence. They have outlined that, "one problem with the use of real sediments to test the extraction procedures is the difficulty of evaluating the efficiency of extraction and the possibility that intercomponent interactions may be producing erroneous results." To overcome these dif-

Table 9
POSSIBLE APPROACHES FOR THE EVALUATION OF EXTRACTION SELECTIVITY

Use of pure phases (with and without spiking of trace elements)
- Alone
- In model sediments
- Spiked into natural sediments

Analysis of extractants and/or residual sediment for various "complementary" parameters
- Major elements such as Al, Si, Ca, Mg, K, P, S, etc.
- Organic C
- Inorganic C
- Acid-volatile sulfide

Successive extractions with the same reagent

Comparison with results obtained from other techniques such as direct instrumental analysis or chemical modeling

ficulties, they used model sediments consisting of a mixture of bentonite, MnO_2, and humic matter (Aldrich Chemical Co.) spiked with copper and lead. The release of metal from the bentonite or humic matter by the ammonium chloride or hydrogen peroxide treatment in the presence of the other substrates was found to yield low results, which were attributed to the released metal being absorbed by the other residual components.

In their study of extraction selectivity, Rapin and Förstner[99] applied their five-step extraction scheme to a number of well-defined solid phases, notably several pure carbonates (Ca, Cd, Mn, Pb). In each case they determined the fate during extraction of both the major metal and any metals present as impurities. Selectivity proved to be sufficient for the carbonate phases (>85% extraction in the appropriate fraction), acceptable for most metals associated with the amorphous oxides (>80% in the appropriate fraction, except for Pb, Cd, Cr, and Zn which were found in part in the exchangeable and residual fractions, probably due to various impurities such as clays), but poor for those metals present as impurities in the amorphous and crystalline sulfides (Figure 23). The poor selectivity of the sequential extraction procedure for the sulfide samples is probably caused by partial oxidation, since the experiments were not conducted under oxygen-free conditions. The clay sample (chlorite) was, as expected, attacked by the last "residual" step; however, the acid-reducing step appears to solubilize about 15% of the iron. Chester and Hughes[107] also reported that chlorite is attacked by this step. Part of the Cd, Pb, and Cu was leached during the first two steps, indicating that they are absorbed onto the particle surfaces. Fe(II)-phosphates, such as vivianite, were attacked in the reducing and residual steps to an equal extent. Since this phase is also attacked selectively during the second HOAc-NaOAc step, these effects are phase is also sensitive to aeration in aqueous suspension, these effects are presumably attributable to artifacts such as impurities and oxidation reactions during the experiments.

A similar approach has been applied by Nirel and co-workers for testing Tessier's scheme.[124] Both individual and mixed artificial sediment substrates were used to examine the specificity and reproducibility of extractants for various trace elements, especially the stable isotopes from the various fission and activation products found in the aquatic environment. The application of Tessier's scheme to the individual substrates gave the following results: (1) a high percentage (averaging 95%) of all the measurable elements was recovered from the carbonates in the appropriate fraction, (2) the recovery from the spiked humic acid (Fluka® AG) was also rather good (70 to 90% of As, Ce, Co, Cs, Cr, and Fe), but (3) in the iron and manganese oxides, only Ce, Co, Cr, and Cs were totally recovered, whereas 20 to 60% of the remaining elements measured (As, Fe, Hf, Sb, and Zr) were found adsorbed onto the vial walls, and (4) extraction of certain spiked elements (As, Cs, Hf, and Zr) from the

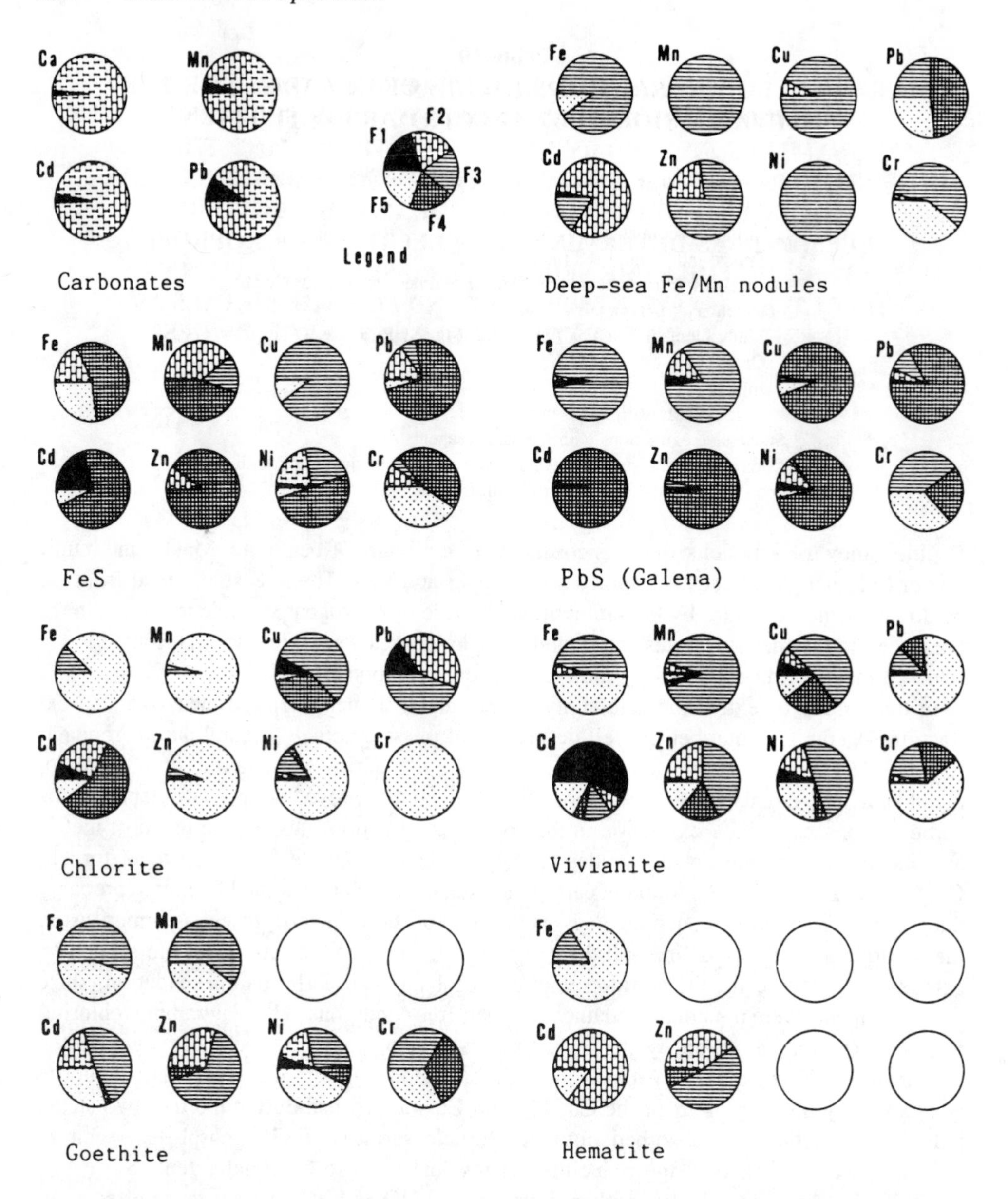

FIGURE 23. Fractionation patterns of metals in pure solids based on the slightly modified sequential extraction procedure of Tessier et al.[94] (From Rapin, F. and Förstner, U., in *Proc. Int. Conf. Heavy Metals in the Environment*, Müller, G., Ed., CEP Consultants, Edinburgh, 1983, 1074. With permission.)

montmorillonite clay sample by the exchanging $MgCl_2$ reagent was unexpectedly low. Moreover, the results obtained with individual substrates were poorly reproduced in the mixing experiment, showing that the sequential extraction procedure is significantly biased by matrix effects. A systematic deficiency was found in the first two fractions, which was counterbalanced by excesses in the last fractions. The underestimations found in the first steps were attributed to readsorption artifacts as evidenced by Rendell et al.[123] It was concluded that application of sequential extraction to natural sediments will give reproducible results only if the relative abundances of their organic and inorganic components do not show large variations, while a comparison of a given fraction in different sample matrices may be questionable.[124]

Table 10
SEQUENTIAL EXTRACTION RESULTS OF Cd AND Zn FOR THE DRIED BUT ORIGINALLY ANOXIC HARBOR SEDIMENT SAMPLES STUDIED BY LEE AND KITTRICK[26] AND THE PROBABILITY (WITH INTERVAL AT 95% CONFIDENCE LEVEL) OF BOTH METALS FOR ASSOCIATION WITH OTHER ELEMENTS AS DETERMINED BY ELECTRON MICROPROBE STUDY OF THE SAME SEDIMENTS (ASSUMING PARTICLES THAT GIVE >100 COUNTS PER SECOND FOR THE METALS ARE REPRESENTATIVE OF THE WHOLE), IN PERCENTAGES

	Extraction		Microprobe		
Fraction	**Cd**	**Zn**	**Zn**	**Cd**	**Element**
Exchangeable	34.0 ± 0.8	3.8 ± 0.1	10 (2—27)	11 (1—34)	Cl
Carbonate	36.2 ± 0.9	50.3 ± 1.9	0	6 (0—27)	Ca
Reducible	21.9 ± 1.6	38.6 ± 0.2	17 (6—35)	7 (1—22)	Fe,Mn
Sulfide/organic	0.5 ± 0.0	0.6 ± 0.1	83 (65—94)	89 (66—99)	S
Residual	7.4 ± 0.1	6.8 ± 0.3			

Composite samples, prepared to simulate hydrous coatings of Mn- and Fe-oxides on stream alluvium, were also used by Robinson[172] to compare the efficiency and specificity of three sequential extraction schemes, namely those proposed by Tessier et al.,[94] Filipek et al.,[147] and by himself. The specificity and efficiency of the extraction procedures were measured by comparing the mean partitioning, as determined by each procedure, with the calculated partitioning based on the total dissolution of constituents in composite samples. All three procedures were quite efficient for Mn and Zn extraction, but differed in the details of partitioning and their efficiency for Fe. None of the procedures were sufficiently selective or efficient to allow an interpretation of metal partitioning relationships without partial biasing by the different reagents used in the procedure. One would conclude from the data obtained with the Tessier procedure that very little Zn occurs in the composite samples as surface-bound ions or in the organic fraction, but data from the Filipek procedure indicate that both fractions are very important for Zn. In the Filipek procedure, however, acetic acid is used to displace exhchangeable metals instead of the weaker $MgCl_2$ reagent. In addition, organic matter in this procedure is extracted at an earlier stage (Table 8). The Fe-oxide portion of coatings was shown by Robinson[173] to contain the largest proportion of Cu and Zn, but in his study, the excess Zn was extracted from the residual mineral substrates by the hydrazine solution used for the reducible fraction. Robinson and co-workers stressed that a meaningful comparison of metal partitioning relationships from separate studies are very questionable because previous interpretations of metal partitioning in coatings and sediment are based on the tested sequential extraction procedures and appear to incorporate a reagent effect.

An evaluation of sequential extraction procedures for natural sediments was reported by Lee and Kittrick.[26] In a study of anoxic harbor sediment heavily contaminated with cadmium, the electron beam microprobe was used to obtain direct evidence of trace metal partitioning, showing cadmium to be most frequently associated with sulfur (in about 90% of the particles). Thermodynamic calculations also supported the existence of cadmium sulfides. In contrast, exchangeable (34%), carbonate (36%), and reducible oxide-bound(22%) cadmium represented the most important fractions from these samples (Table 10). It is quite probable, however, that this disagreement is due to improper sample handling because the authors dried their anoxic sediment samples prior to applying the extraction procedures. This example suggests that some arguments against solid speciation on polluted sediments are still based on wrong assumptions, i.e., on the discrepancies between thermodynamic data and results from extraction sequences with invalid sampling protocols. In our experiments on anoxic sludges, performed under oxygen-free conditions, nonresidual trace metals have indeed predominantly been found in the oxidizable (organic + sulfide) fraction, while the previous four extraction steps did not affect significant release of metals from these samples.[82] Proper sampling preparation and experimental conditions are the main prerequisite for attaining reliable data from chemical extraction as well as instrumental techniques.[173]

The forementioned evaluations generally suffer from use of artificial (nonequilibrium) samples or inadequate sample handling. The crystalline and aged substrates prepared or dried at elevated temperatures may not be fully representative of their natural equivalents, and the impact of metal transfer from a given fraction to another is also obviously dependent on specific experimental conditions. The limitations reported in these and other studies have led us, and others, to the conclusion that results given by sequential extraction experiments are representative of certain operationally defined groups of metal forms and associations than of the true speciation in sediments. These problems also highlight the limitations of some of the terminology often used (such as the "organically bound fraction" and "oxy-hydroxide fraction"). While there is a link between the reagent chosen and the soil component most strongly attacked, secondary reactions reduce the validity of these convenient labels. Of equal (and possibly better)[61] value would be groupings such as water soluble, acid soluble, salt displacement, acetate extractable, acid reducible or oxidizable, etc., i.e., the labels defined by the extractants chosen. If this practical obstacle is recognized, the application and the potential usefulness of this speciation approach in environmental studies can be logically discussed.

IV. CASE STUDIES

A. Mobility and Bioavailability of Trace Metals in Sediments

Despite the numerous limitations described, speciation assessment techniques are warranted as a differentiated approach in the study of interactive processes between water/biota and operationally defined solid phases. Since the estimation of trace element speciation by instrumental methods has not become an easily performable and common approach, the use of chemical extractants remains one of the few tools available for the examination of trace-metal chemical associations with natural particulate materials. In environmental pollution studies on solid materials, sequential extraction techniques have been performed for the following.

1. Identification of the Main Binding Sites of Trace Metals

Geochemists first of all attempted to provide a compilation of the most important mechanisms of metal accumulation in sedimentary particles, to evaluate pollution levels, and to understand the mechanisms of trace element transport in natural as well as polluted environments. The complexity of the overall problem is made apparent from recent reviews,[6,18,153] which have examined the types of binding of trace metals in sediments and sludges. Results obtained from extraction of fluvial suspended matter and marine sediments with a modification of Tessier's scheme are shown for 20 elements in Figure 24 and 25. Obviously the residual fraction of elements in the less polluted marine samples is the most important and variable. Both the carbonate and the organic/sulfide fractions represent less than 30% in these oxidic samples. Exchangeable metal fractions are more important in the suspended matter samples than in the deposited sediments. The second important fraction in both

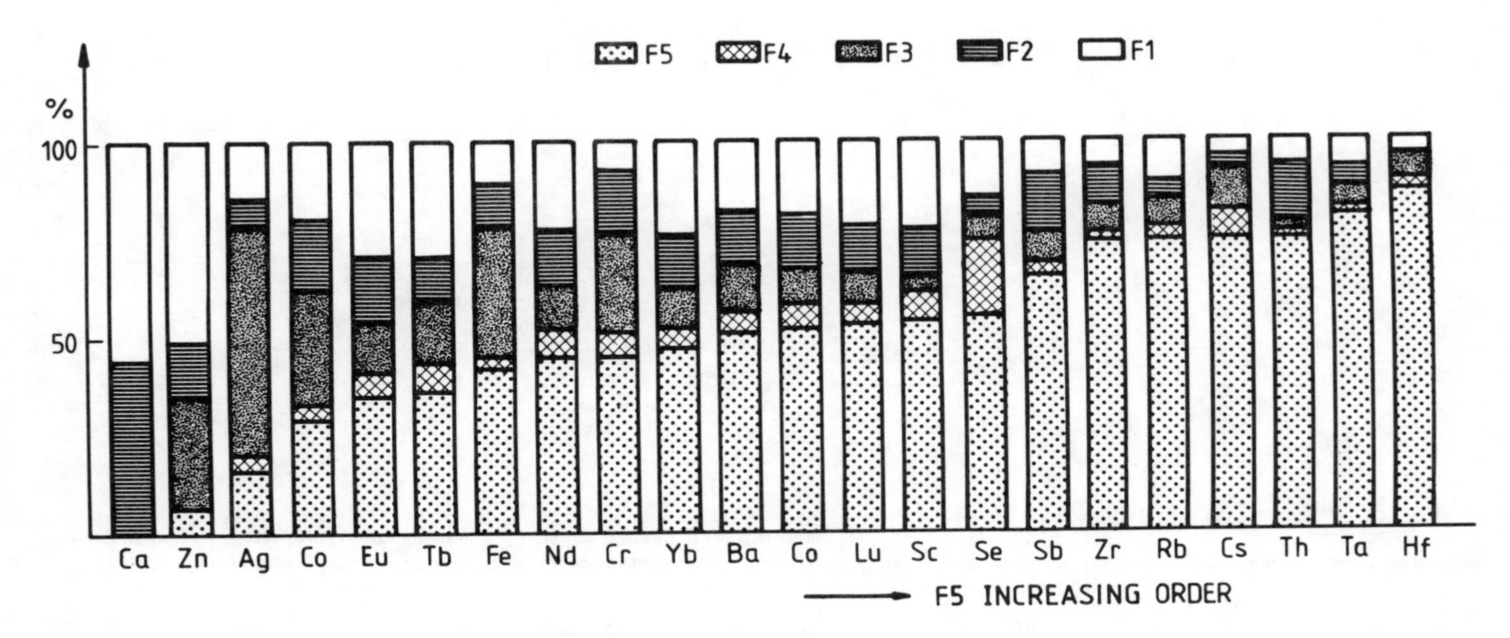

FIGURE 24. Fractionation patterns of elements in suspended matter based on extractions using the modified Tessier procedure. Means based on results from the Loire, Rhône, Gironde, and Huanghe Rivers. F1: exchangeable, F2: carbonate, F3: Fe,Mn oxides, F4: organic matter and sulfides, F5: residual. (From Martin, J.M., Nirel, P., and Thomas, A.J., *Mar. Chem.*, 22, 313, 1987. With permission.)

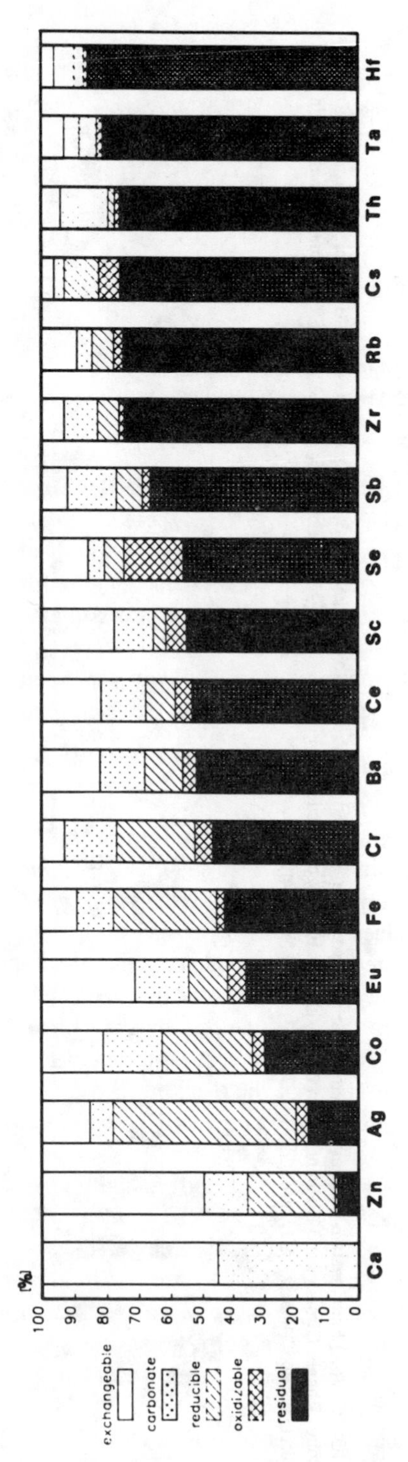

FIGURE 25. Fractionation patterns of elements in marine sediment from the Mediterranean Sea off Thraki, northern Greece, based on extractions using the modified Tessier procedure. Means based on results from 177 samples. (Data from Sakellariadou.[178])

Table 11
MECHANISMS AND SUBSTRATES FOR METAL SPECIATION IN SEDIMENTS

	Rock debris, solid waste material, incl. organics	Metal-hydroxide, -carbonate, -sulfide	Organic substance (reactive)	Hydrous Fe/Mn-oxide (mainly authigenic)
Inert bonding	×××			×
Precipitation		×××		××
Adsorption (phys.)	×	×	×	×
Coprecipitation		××		×××
Coating (surface)			×××	×××
Flocculation			×	××

From Förstner, U., in *Chemical Methods for Assessing Bio-Available Metals in Sludges and Soils*, Leschber, R., Davis, R. D., and L'Hermite, P., Eds., Elsevier, New York, 1985, 1. With permission.

environments is the acid reducible, which is in accord with the well-known role that oxyhydrate coatings play as trace metal carriers.[176] Statistical analyses within the specific subsets of data obtained by chemical extraction studies confirm that trace metals are partitioned among several substrates in most sediments. These substrates compete with one another for the metals, and the outcome of the competition is strongly influenced by the concentrations of the different substrates in the sediment.[177] Table 11 summarizes some of the most important alternative mechanisms, products, and substrates of metal enrichment and their estimated significance in both natural and man-affected aquatic and terrestrial systems as deduced from sequential extraction studies.

2. Assessment of Sources by Characterization of Typical Inputs

There are few studies in this field. One example is a regional survey performed on the distribution on thallium in soils in an area where two point sources were significant.[180] One was the mine tailing deposit of an abandoned lead-zinc mine, the other was the chimney of a cement plant which had used sulfidic roasting residues as additives to special cement products. Leaching experiments with ammonium acetate/EDTA on polluted soil samples were able to differentiate the pollution sources of the samples. In the mine area, the extractable proportions were in the range 0.5 to 5%, whereas from the soils affected by the cement plant emissions, up to 40% could be extracted. The lower absolute concentrations in the latter area were therefore more available. In fact, 75 to 100% of the thallium in flue gas dusts was found to be soluble in water, leading in some cases to increased toxicity and accumulation of thallium in plants.[181]

Studies of source differentiation by other speciation approaches were performed by Pilkington and Warren[13] (physical fractionation of sediments adjacent to a lead smelter) and Norrish et al.[5] (electron microprobe study of the forms of metals present in sediments adjacent to a lead-zinc smelter from Spencer Gulf, South Australia). For source discrimination of airborne emissions, major progress to date has been achieved using physical separation methods. A differentiated separation scheme was applied to characterize urban dusts according to their source, initially for microanalytical lead tracing,[9] then also for multielement analysis.[182] A flow chart of this separation scheme is shown in Figure 26.

Linton et al.[9] established identities of individual particles (automobile exhaust, paint chips) on the basis of characteristic morphology and chemical composition. It was shown that automobile exhaust, apart from its dominance in street dust samples, contributed substantially to dusts collected in the vicinity of buildings which have leadpainted trim. Surface-sensitive

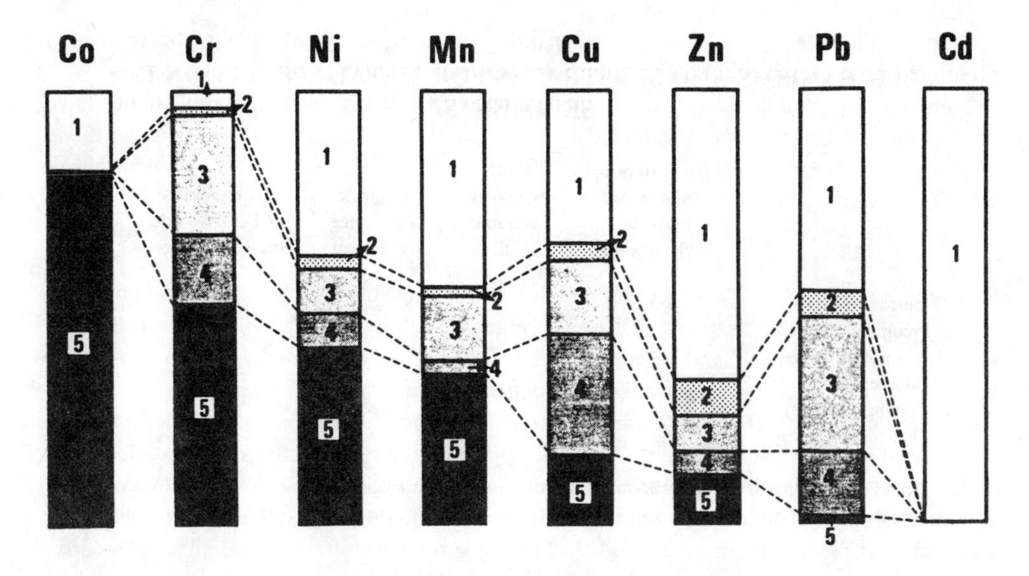

FIGURE 26. Fractionation patterns of trace metals in Standard Reference Material NBS 1648 (''urban particulate matter''). (From Salomons, W. and Förstner, U., *Metals in the Hydrocycle*, Springer-Verlag, Berlin, 1984. With permission.)

analytical techniques indicated that automobile exhaust particles contain the elements Br, Cr, Cu, Mn, Ni, P, Pb, and Tl on their outer surface.[9] In a similar study, element concentrations from 30 physically fractionated subsamples (each having unique particle size, density, and ferromagnetic characteristics) obtained from the intersection of two moderately traveled streets in Urbana, IL, were subjected to target transformation-factor analysis.[182] The distribution profiles showed three distinct sources of zinc. Comparison with the normalized elemental concentration for several specific sources (input test vectors) suggested that these sources were tire wear, soil, and cement.

3. Estimation of Biological Availability of Metal Pollutants

The availability of trace metals to organisms may be influenced by factors such as

1. The physiological and ecological characteristics of the organisms.
2. The form of dissolved trace elements, as well as the chemical and physical characteristics of the water.
3. The form of trace elements in solids.

Much solid speciation work has been done on the relation between chemical leachability and uptake of metals by plants.[81,183] Early applications in soil science have been expanded to encompass the estimation and prediction of trace element bioavailability. This original idea contributed to the development of the single-leaching methods, but generally has failed. Many papers have shown that with respect to bioavailability, as distinct from geochemical mobility, greater caution must be exercised in the interpretation of chemical extractant data for trace metals in soils. The extractable metal content at which similar and uniform phyto- or zootoxic concentrations in plants or activities of microorganisms are induced, are not consistent when soils have varying physical, chemical, and biological characteristics. This handicap is primarily due to a lack of information about the specific mechanism by which organisms actively translocate trace element species.[185] Thus, an ideal reagent emulating the biological system cannot normally be established, and a more empirical approach has been adopted, with confusing results. Attempts to simulate biological species behavior is also limited by the different time scales involved. For example, plant uptakes occur over a prolonged time period, while extraction tests are usually short-lived.

Some encouraging developments in the use of extractants for bioavailability studies have been obtained. Experiments conducted by Häni and Gupta[186] demonstrated that the metal concentration in the solution phase of soil can be successfully simulated through neutral salt extraction, estimating biological effects of accumulated metals irrespective of soil characteristics and environmental factors such as temperature and degree of aeration. Consequently, the new Swiss "Ordinance on the Tolerable Heavy Metal Content in Soils" include two maximum permissible values for each metal in soil: (1) HNO_3-soluble (total metal content), and (2) $NaNO_3$-soluble (bioavailable metal content).[186] Another interesting approach is the combination of chemical and biological tests. In such an experiment, Diks and Allen[174] found a high correlation in the uptake of copper present in the easily reducible fraction of their sequential extraction scheme and suggested that the redox potential and pH in the gut of the studied macrobenthic species was such that manganese coatings and associated trace metal loads were dissolved. Another integrated approach correlating the bioassay data of a two-chambered exchange system to the results from a six-step chemical extraction sequence system was designed in our laboratory.[187] This system allows effects of salinity, pH, redox, and added complexing agents on both accumulation and toxicity of trace elements to be studied, while separating the algal population from suspended solids by a filter membrane.

Sequential extraction experiments have emphasized the protective role of iron oxyhydrates in prevention of biotoxicity of trace metals. Luoma and Bryan[188] showed that the ratio Pb/Fe in 1 *M* HCl extracts reflects availability of Pb to a bivalve species. Similar results have been found for uptake of Cu, Zn, and Pb by freshwater pelecypods.[189] Iron oxyhydrates in sediments appear to be a competitor to biogenic mechanisms for metal uptake as predicted for organic chelators in the water column ("equilibrium" or "speciation" hypothesis).[190] Another promising approach is the application of biological complexing agents for studying the effect of leaching of contaminants from airborne particles in the lungs of mammals. It has been suggested that biological chelators such as cysteine and other serum proteins are important natural leaching agents in removal of trace metals from fly ash particles *in vivo*.[191]

4. Estimation of Waste Disposal Conditions Affecting Metal Mobility

Speciation can be used to estimate the remobilization of metals under changing environmental conditions. The mobility of elements in sediments is strongly affected by pH, redox potential, and complexing ligands (both organic and inorganic). While there is a predominance of simple mineral-solution equilibria for the major elements in the aquatic environment, the behavior of many trace elements is more complex. Trace element equilibrium is determined by such mechanisms as coprecipitation, surface effects, and interactions with organic phases.[192] Many interactions, however, can be estimated from the simple scheme given in Table 12, which includes data from several studies on soil and dredged material. Changes from reducing to oxidizing conditions can be expected to strongly increase the mobility of metals such as Mo, V, U, and Se, to a lesser degree of Hg, Cu, Cd, and probably also of Zn, Co, Ni, and Pb. Mobility of Mn and Fe is, however, characteristically lowered under oxidizing conditions. Geochemical mobility, in response to environmental acidification, will significantly increase, if Al, Mn, Zn, and, to a lesser degree, Cd, Co, and Ni are also present in the nonresidual fraction of the sediment. The elements forming anionic species, such as S, As, Se, Cr, and Mo, are appreciably solubilized, e.g., from fly ash sluicing/ponding systems, at neutral to alkaline pH conditions.[193] Lowering of pH will affect the exchangeable, then the carbonate and parts of the acid-reducible fraction of sediments, the latter consisting of Fe-oxyhydrates in less crystallized forms.

5. Evaluation of Sedimentary Diagenetic Effects on Metal Mobility

The importance of sediments to the biogeochemical cycling of contaminants such as trace

Table 12
RELATIVE MOBILITIES OF ELEMENTS IN SOILS AND SEDIMENTS

	Electron activity		Proton activity	
Relative mobility	**Reducing**	**Oxidizing**	**Neutral-alkaline**	**Acidic**
Very low mobility	Al, Cr, Mo V, S, Se, Hg, Cu, U, Cd, Pb, B	Al, Cr, Fe, Mn	Al, Cr, Cu, Hg, Ni	Si
Low mobility	Si, K, P, Fe, Ni,	Si, K, P, Pb	Si, K, P, Pb, Fe, Zn, Cd	K, Fe(III)
Medium mobility	Mn	Ni, Cu, Hg, Zn, Cd	Mn	Al, V, Cu, Cr, Pb
High mobility	Ca, Na, Mg, Sr	Ca, Na, Mg, Sr, Mo, V, Se, U	Ca, Na, Mg, Sr	Ca, Na, Mg, Sr, Co, Ni, Zn, Cd, Hg
Very high mobility	Cl, I, Br	Cl, I, Br, S, B	Cl, I, Br, Mo, V, U, Se, S, B	Cl, I, Br, S, B

From Förstner, U., in *Chemical Methods for Assessing Bio-Available Metals in Sludges and Soils,* Leschber, R., Davis, R. D., and L'Hermite, P., Eds., Elsevier, New York, 1985, 1. With permission.

metals is well known.[95] Knowledge of the chemical diagenesis of sediments is therefore essential to an understanding of contaminant mobility in marine and freshwater environments. Once dissolved metal pollutants are accumulated by transport and sedimentation, significant secondary release of them from sediments can occur as a result of the following processes:

1. Postdepositional remobilization by oxidation and decomposition of organic detritus, mediated by microbiological activities.
2. Control of the solid/solution partitioning by early diagenetic effects such as changing the surface chemistry of oxyhydroxide mineral coatings.
3. Authigenic production/dissolution of metal precipitates such as sulfides, with the reduced forms generally more insoluble than the oxidized forms.
4. Desorption from clay minerals and other substrates due to formation of soluble organic and inorganic complexes.

In the estuarine environment the exchangeable fraction is particularly affected. However, changes of pH and redox potential could also influence other easily extractable phases. The high proportion of Cd in exchangeable fractions of polluted oxic sediments is reflected in the typical mobilization effects of this element in the estuarine environment.[194] Under postoxic conditions, i.e., during the reduction of nitrate and manganese oxides, redox changes affect the easily reducible forms of metals. In strongly reducing environments of highly polluted sediments, the moderately reducible fraction is affected too, especially when it is comprised of oxyhydrate coatings. It is the kind of particle association and the underlying elemental chemistry of contaminants which controls their behavior and ultimate fate. Transformations of metal forms during early diagenesis have been successfully studied by sequential chemical extraction techniques in some recent studies. These studies and their implications have been adequately reviewed elsewhere.[195-197]

B. Trace Element Associations in Atmospheric Deposits

The imposed stress on the biosphere by the distribution of particulates in the atmosphere has increased. The dramatic impact of atmospheric lead emission is representative of this development. The concentrations of lead in air in prehistoric times has been estimated as

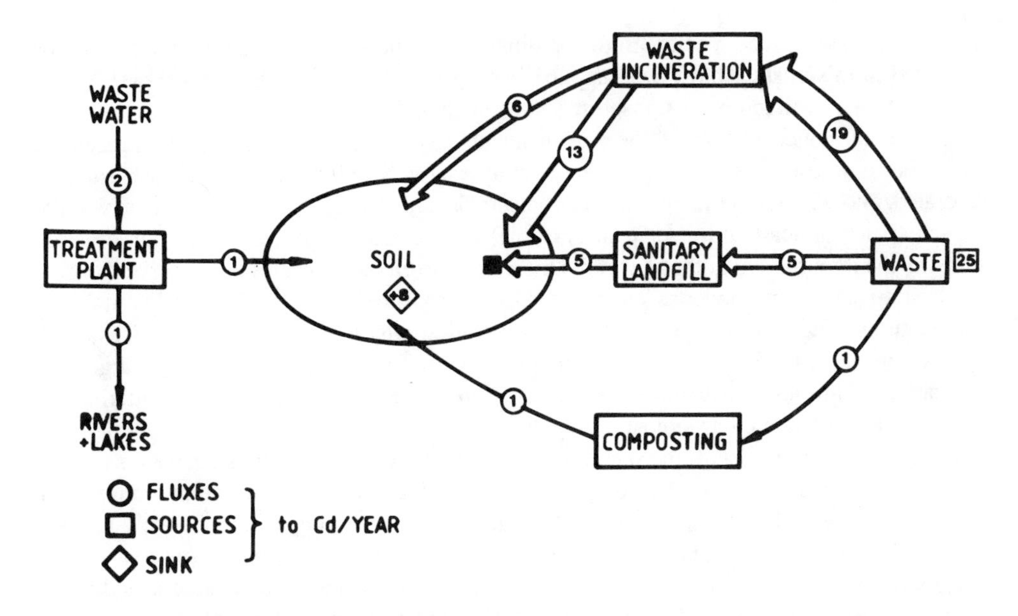

FIGURE 27. Cadmium input into soil from different sources, based on the Swiss example. (From Keller, L. and Brunner, P. H., *Ecotoxicol. Environ. Safety,* 7, 141, 1983. With permission.)

approximately 0.4 ng/m^3. Today, even in remote regions of North America, it is 10 ng/m^3, and in urban areas, concentrations range between 500 and 10,000 ng/m^3.[198] As a consequence, present-day body burdens of lead in American and European people are on the average two to three orders of magnitude higher than those of early man.[199] While there is already a marked drop in the overall lead levels from the decreased use of leaded gasoline in many industrialized countries, the developing countries like Venezuela, with a growing number of motor vehicles and continuing use of leaded gasoline, cause still serious pollution of their urban environments.[200]

Accumulation of cadmium in agricultural soils from atmospheric deposits and its increased uptake by plants is also of world-wide concern compared to relatively few critical situations in aquatic systems. This aggrevating situation is demonstrated from an example of Switzerland (Figure 27), which shows that the atmospheric inputs of cadmium have been increased some 100-fold by anthropogenic activities, including dispersion of combustion residues (especially fly ash), smelting of ores, urban and industrial waste incineration, etc.[201]

Particulate loads of toxic chemicals play a significant degrading role in strongly stressed atmospheric and terrestrial systems. Mobility of a trace metal in the terrestrial and aquatic environment is reflected by the partitioning between dissolved and solid forms. This ratio is primarily influenced by the respective inputs and secondarily by the interactions taking place within the different environmental compartments. For atmospheric input the percentage of dry deposition (by which aerosols or gaseous compounds are deposited on surfaces such as soil particles and plant leaves) has been observed between 10 and 90% of bulk Cd deposition, depending on the emission sources, climatic conditions, and, in particular, on pH.[202] A 90% wet deposition is described from the situation in West Germany, where the deposited rain water showed pH values between 3.9 and 4.4.[203]

The speciation of solid surfaces plays an important role in mediating the behavior and fate of atmospherically derived trace elements. In relating this to environmental problems, Farmer and Linton[204] stressed that, "accessibility to the environment (via washout, rainout, groundwater leaching, lung fluids, etc.) is governed by both metal surface accessibility (extent of surface enrichment) and metal surface solubility (surface speciation)." Methods for determining the reactivity of particle-associated atmospheric pollutants are therefore mainly related to surface species. Both instrumental techniques and solvent leaching methods discussed in this chapter have been applied in the last 10 years to atmospheric particulates to estimate the availability of transported trace elements.

Preferential concentration of toxic trace metals in smaller particles is very apparent in atmospheric aerosols. X-ray fluorescence analyses for 29 elements in a fly ash sample, separated into 17 well-defined size fractions, indicated that the concentrations of volatile elements such as lead and arsenic increase as particle size decreases.[205] This observation supports the important hypothesis that the more volatile element species are vaporized during combustion and then condense on the surfaces of coentrained fly ash particles at lower temperatures. These results prove critical justification why emission control devices, such as mechanical dust collectors, electrostatic precipitators, scrubbers, etc., remove larger fly-ash particles, but are inefficient in preventing vapors and finer particles from entering the terrestrial and aquatic environment.

In the human environment, the smallest particles (<1 μm), which deposit in the pulmonary region of the respiratory tract, are of greatest concern. Toxic species which predominate in submicrometer-sized particles come into intimate contact with body fluids, due to their high surface area, and thereby enter the bloodstream.[206] Leaching experiments on coal fly ash using physiological solutions of biological chelates, such as canine blood serum, indicate a significant leaching of Zn, V, Cu, and even Fe. The observed time-dependent drop in leaching rates probably reflects the depletion of surface accumulations of soluble metal components such as sulfates, halides, and phosphates in the fly ash.

Combustion residues are still a major source of atmospherically deposited pollutants. Recent studies show that the volatile and other trace metals and metalloids in fly ash are concentrated on the surface of the particles.[65] A combined approach has been proposed to quantify the relative concentration of elements in the alumosilicate matrix and in nonmatrix or surface material of coal fly ash in order to study the trace metal partitioning in the fly ash particles.[207] Particle size (as derived from the specific surface area) and acid extractability (with 0.029 to 0.57 *M* hydrofluoric acid) of the elements were normalized to the percentage of nonmatrix fraction and the two sets of data were compared. Major differences between the nonmatrix concentrations were found for Ca, Ni, Sr, Sc, La, and the rare earths (too low size-dependence factor) and for Pb, Ba, and W (too low solubility dependence). The latter anomalies are probably caused by insoluble phases either present in the ash (WO_3) or formed during the extraction. The results indicate that more than 70% of the Ti, Na, K, Mg, Hf, Th, and Fe is associated with the alumosilicate matrix; more than 70% of the As, Se, Mo, Zn, Cd, V, U, and Sb is associated with surface material on the ash particles; and that the elements Mn, Be, Cr, Cu, and Co are intermediate in behavior, i.e., are distributed about equally between matrix and surface material.[207] Recently, Hansen et al.[208] examined the leaching behavior of fly ash samples with various reagents after differentiation into size classes. The existence of combinations of major anion species other than silicate, i.e., fluoride, phosphate, and sulfate, in the surface material were related to typical associations with metal cations of Zn, Cd, Co, Cu, and Cr.

In an extensive investigation, Wu et al.[209] examined the effects of different combustion conditions, coal types, and fly ash collection devices on the chemical nature of the ash. They used an extraction scheme which differentiated water-soluble, acetic acid-leachable, easily reducible, H_2O_2-oxidizable, moderately reducible, and residual fractions. The ''leachable fraction'', comprising metals soluble in water or being ion-exchangeable, together with those associated with carbonates and amorphous oxides, was generally highest for B, Cd, V, Se, and As; followed by Cr, Cu, Ni, Pb, and Zn. The matrix-elements iron and aluminium were leached to a much lesser extent by water, 1 *M* acetic acid, and a solution of 0.1 *M* hydroxylamine hydrochloride in 0.01 *M* nitric acid (easily reducible fraction). Up to 30 to 50% of the nonresidual concentrations of Cu, Zn, V, and Ni were found in the oxidizable

Table 13
SUMMARY OF THE RESULTS OF CHEMICAL FLY ASH PARTICLE SURFACE CHARACTERIZATION

Group	Elements	Surface concentration	Effect of leaching in the immediate surface region	Leachability
A	Al, Ti, Si	No	Insignificant	Low
B	Fe, Na, Li, K, S	Yes	Insignificant	Moderate to high
C	Pb, Tl, Mn, Cr, V	Yes	Depletion after solvent leaching	Moderate to high
D	Mg, Ca	No	Insignificant	Moderate

After Natusch, D. F. S. and Taylor, D. R., in Chemical and Physical Characterization of Coal Fly Ash, Rep. 600/3-80-093, U.S. Environmental Protection Agency, Washington, D.C., 1980.

fraction. Incomplete combustion at larger coal particle sizes (stoker furnace) and lower combustion temperatures (lower heat content of coal, fluidized-bed combustion) may result in higher trace metals leachable from the ash. Increases were primarily found in the acetic acid-leachable and oxidizable phases, probably associated with unburned carbonate and coal residue in the ash.[209]

A summary of the combined data from solvent-leaching experiments and surface microanalysis by Natusch and Taylor[44] showed that typical elements can be categorized into four general groups (Table 13). The Group A elements (Si and Al) are major constituents of the glassy fly ash particle matrix, exhibiting low leachabilities. Due to its close association with the silicate impurities, Ti is incorporated as a minor component in the fused fly ash. The surface predominance observed for the Group B and C elements apparently is related to their typical behavior during volatilization and subsequent deposition. However, the higher leachability in the Group B elements coupled with the relatively minor changes in observed surface concentration could well represent an artifact due to surface migration of Na, Li, K, and S under ion bombardment.[44] The Group D elements (Ca, Mg) exhibit similar behavior to that of the Group A elements. The alkaline earths are often present as carbonates in coal and subsequently decompose to form oxides during combustion which may be transferred into soluble sulfates.

Major progress has also been achieved in evaluation of the environmental significance of airborne emissions at their sink, i.e., in street dirt and urban soils. To identify lead species in street dust, samples were analyzed by X-ray diffractometry after magnetic and density-gradient separation as well as by leaching experiments.[210] The most frequently observed crystalline compound was $PbSO_4$. Its presence is explained by weathering of $PbSO_4 \cdot (NH_4)_2SO_4$. Since the bulk of the compounds liberated to the environment by car exhausts are lead halides, reactions of airborne species such as PbBrCl with neutral and acid sulfates may also be involved. Since $PbSO_4$ has a low but significant water solubility, the transformation of lead species is suspected to be caused by reactions involving essentially insoluble organic material, with precipitation by carbonate and sorption by hydrous oxides apparently being of secondary importance. Biggins and Harrison,[210] however, concluded, "that crystalline compounds of lead can account at most for only a few percent of total lead in the samples examined, and that alternative approaches to speciation of the lead are required."

In studies of metals in street dust and urban soil, high proportions of Cd, Zn, Pb, and Ni were found in soluble and/or exchangeable forms, suggesting that these elements may have immediate impact on the quality of receiving waters or may be easily available for organisms in soil (Figure 28).[277] Respective data on the fractionation of solid Cd and Pb (Table 14) for street dust (A,B), roadside soil (C,D), and urban soil (E) revealed fairly distinct parti-

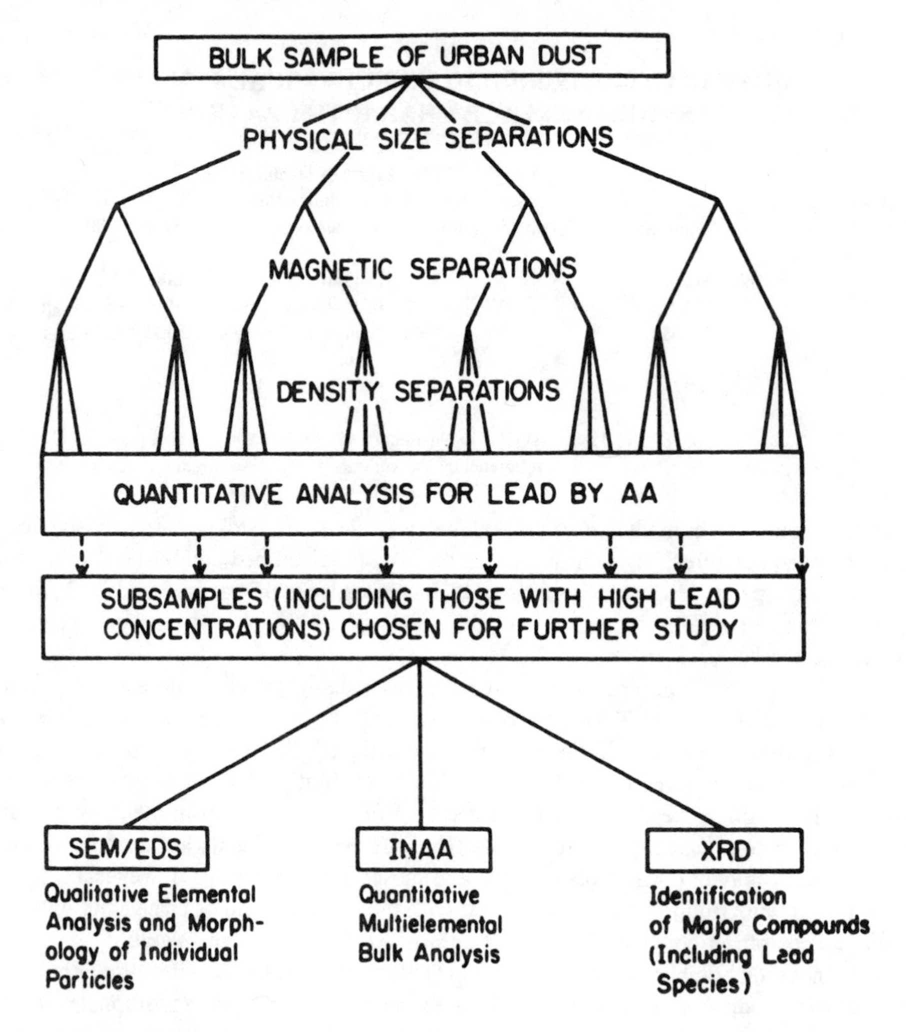

FIGURE 28. Flow chart for urban dust fractionation and lead solid speciation. (After Linton, R. W., Natusch, D. F. S., Solomon, R. L., and Evans, C. A., Jr., *Environ. Sci. Technol.*, 14, 159, 1980. With permission.)

tioning patterns for lead and cadmium, confirming that this type of sequential chemical extraction procedure provides valuable information on the identification of trends in the main chemical associations of toxic trace metals in the urban environment. The study by Harrison et al.[212] of six street dust (A in Table 14) and four roadside soils (C in Table 14) indicated that cadmium was present to an appreciable extent in the exchangeable fraction, while lead was predominantly associated with carbonate and Fe/Mn oxyhydrates. While total concentrations for these metals suggested an order of contamination of lead > cadmium, the detailed sequential extraction data indicated the reverse order for release or mobility. The high cation and chloride concentrations in the exchanging solution (1 M $MgCl_2$) of the sequential extraction scheme (that of Tessier et al.[94]) may reflect conditions in soils contaminated with deicing salt as well as transformations that might occur on mixing with seawater. In fact, Chester et al.[213] found a fairly significant relation between seawater solubility of trace elements in atmospheric particulates and the percentage in loosely-bound fractions (Figure 29).

Compared to the street dust leaching data, experiments on roadside soils (examples C,D) as the major sink of the polluted aerosols indicated significantly lower extractions in the

Table 14
MEAN CHEMICAL PARTITIONING (PERCENT OF TOTAL) OF TRACE METALS IN LANCASTER (A)[212] AND GLASGOW (B)[175] STREET DUST, CARACAS (C)[200] AND LANCASTER (D)[212] ROADSIDE SOIL, GLASGOW URBAN SOIL (E),[175] AND INCINERATED SEWAGE SLUDGE FROM HAMILTON MUNICIPAL INCINERATOR[214]

	Lead (%)						Cadmium (%)					
Fraction	A[a]	B	C[a,b]	D[a]	E	F[a]	A[a]	B	C	D[a]	E	F[a]
Exchangeable	2	13	0	1	2	0	20	27		26	19	0
Carbonate	43	28	22	26	11	0	38	19		24	13	0
Easily reducible		5			0			4			4	
	38		1	44		1	28			25		1
Moderately reducible		27			51			12			15	
Organic/sulfide	7	14	39	12	19	1	8	22	8	7	10	
Residual	10	13	38	17	17	98	6	16		18	42	89

[a] The reducible fraction was not differentiated in these studies.
[b] Order of extraction steps differs: organic/sulfide before reducible fraction.

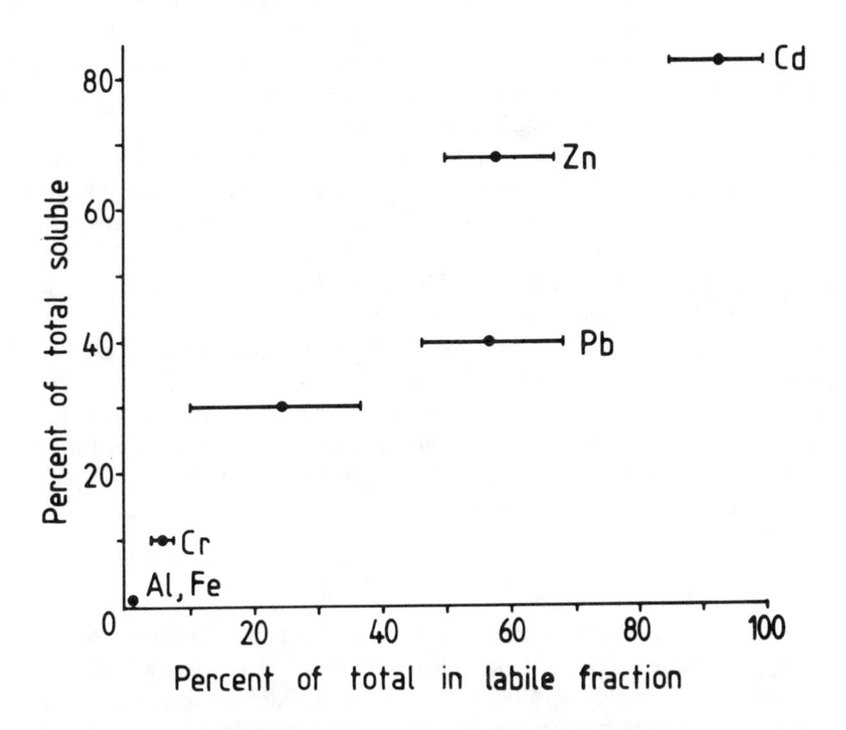

FIGURE 29. Relation between seawater solubility of trace metals from polluted atmospheric aerosols and the percentage associated with loosely bound fractions. (From Chester, R., Murphy, K. J. T., Towner, J., and Thomas, A., *Chem. Geol.*, 54, 1, 1986. With permission.)

more labile phases and an increase of the residual metal percentages. This is connected with the lower total metal contents in the roadside samples compared to dust samples. The finding that Pb is mainly associated with other than ion-exchangeable forms has practical implications, because Pb would be highly immobile and difficult to leach down the soil profile. It was expected to be taken up by plant roots and soil biota only to a minor extent.[200] Differences in metal partitioning between soil and street dust are also influenced by variations in physical characteristics (e.g., grain sizes). They can be increased by the number and abundance of exchange sites in the street dust, leading to substantial increases in the relative importance of the exchangeable fraction.[175] In urban soils (D,E), a major difference was the enhanced significance of the carbonate fraction for lead and cadmium in sample D (Lancaster urban soil). This was probably a consequence of its more alkaline nature (pH 7.2 to 8.0) compared with the more acidic Glasgow soils (samples E). Harrison et al.[212] noted that the carbonate fraction was also an important component in their street dust because of its generally higher calcite content. The general assumption that the more enriched elements are also the more reactive ones (see above) does not fit all atmospheric waste materials, as exemplified by the data from incinerated sludge ash (sample F), where the potentially toxic elements are highly enriched but are rather immobile.[214]

Not so surprising is the finding that there is a pronounced difference in significance of the organic fraction of lead found in roadside samples C and D, considering the fact that different sequential extraction schemes were used. Garcia-Miragaya[200] (sample C) applied the procedure of McLaren and Crawford[169] (A in Table 8) which extracts the organic fraction before the oxide fraction, whereas Harrison et al.[214] (sample D) used that of Tessier et al.[94] (H in Table 8). The problems encountered in the order of organic and oxide soil substrate dissolution have been discussed above. Harrison and co-workers discussed three earlier studies in which the association of lead in soils with the organic fraction may have been overestimated due to experimental artifacts.

Gibson and Farmer[175] pointed out that the chemical partitioning studies may help to clarify the controversy surrounding the relative sources of contributions to the intake of lead by young children from a range of potential sources (petrol, paint, lead water pipes) and exposure routes (ingestion of street dust, soil, food and water, inhalation of atmospheric fine particulates and gases). They used 0.07 *M* hydrochloric acid as a chemical extractant to simulate human stomach acidity, thus estimating the daily ingestion of 100 mg of street dust at an uptake of 26 μg lead per day, a value close to the daily dietary uptake of 25 μg for a 2-year-old child in an inner city environment. It was also demonstrated that the release of lead from street dust by 0.07 *M* hydrochloric acid was broadly related to the size of the combined exchangeable, carbonate, and easily reducible fractions. If this relationship holds for urban soil, then an average 13% of soil (sample E) lead would be made "available" for adsorption following ingestion, using the geometric mean of 160 mg/kg for Glasgow soil (E), 21 μg lead per gram of soil swallowed.[175] For cadmium, however, contributions from soil and street dust ingestion were shown to be negligible relative to dietary sources, despite the high amounts of this element found in labile fractions.

V. CONCLUSION

The present state of knowledge of solid matter speciation of trace elements is still somewhat unsatisfactory because the appropriate techniques are only operational tools with associated conceptual and practical problems. With respect to estimating bioavailable element concentrations, one such conceptual problem is the effect of competition between binding sites on the solid substrate and selective mechanisms of metal translocation by the different organisms involved, a situation which cannot as yet be improved by more sophisticated speciation approaches. On the other hand, the usefulness of a differentiated approach to the interactive processes between water/biota and, even only operationally defined, solid phases has been clearly evidenced. The possible environmental implications, e.g., during dredging operations, after land disposal of waste material, from acid precipitation, for redox changes in the subsoil, and for ingestion of polluted urban dust, can be qualitatively estimated, particularly when the physicochemical conditions of the interacting environmental milieu are taken into consideration. The applications of the speciation approaches, as quoted in this chapter, are therefore convincing and increasing especially in the fields of civil engineering and geosciences.

Most of the instrumental techniques available to date are too sophisticated to be routinely included in trace element speciation studies, though rapid progress is presently being achieved with analytical microscopy, PIXE, and the surface analysis techniques AES and SIMS. These are particuarly valuable, providing collaborative information on the partitioning of trace elements in fine particles, and should be used in proving effectiveness of the sequential chemical extraction techniques. The method of sequential chemical extraction is the least sophisticated and most convenient technique available for a speciation assessment. While geoscientists involved in environmental pollution studies have discussed this method extensively, it has also been shown that, to again quote Pickering,[59] "a careless usage of this technique without an appreciation of its pitfalls and limitations must lead to further generation of erroneous or misleading data." We must be certain that we fully understand what is happening during extraction, minimize the possibility of producing artifacts, and choose standard procedures to ensure we are generating comparable results. Situations where extractive techniques lack specificity and where application can cause erroneous results have been defined in this chapter. The primary importance of proper sampling protocol has been emphasized, since the artifacts introduced by improper sample handling can override any limitations of the various reagents and extraction approaches. The number of fractionation steps required depends on the purposes of the study. Among the listed procedures, the scheme of Tessier et al.[94] and its modifications has been most widely applied. It has proved adequate for the specific tasks being undertaken, ranging from estuarine sediment to waste material, and even to street dust. This scheme is often too sophisticated for many purposes, leading to greatly enhanced time involvement and overall cost. A more sophisticated sequential extraction scheme such as Tessier's, however, significantly improves the specificity and efficiency of extraction, by a carefully designed combination of various single, non-selective extractants, and the quality of information and comparability with other studies made in this field.

ACKNOWLEDGMENTS

The authors wish to express their appreciation to the many colleagues who provided both conceptual and practical improvements to the speciation approaches used in our laboratory, especially to Drs. Wolfgang Calmano, Rudolf Deurer, Sambasiva R. Patchineelam, Georg Pfeiffer, François Rapin, and Jürgen Schoer. The discussions and research findings made available by Dr. James M. Robbins (Oregon State University College of Oceanography) provided valuable information to this chapter. Last, not least, we are indebted to Dr. Graeme Batley for his well-honed efforts in editing this review.

REFERENCES

1. **Bernhard, M., Brinckman, F. E., and Irgolic, K. J.,** Why speciation?, in *The Importance of Chemical Speciation in Environmental Processes,* Bernhard, M., Brinckman, F. E., and Sadler, P. J., Eds., Springer-Verlag, Berlin, 1986, 7.
2. **Sposito, G.,** Chemical models of inorganic pollutants in soils, *CRC Crit. Rev. Anal. Chem.*, 15, 1, 1984.
3. **Wood, J. M., Chakrabarty, A. M., Craig, P. J., Förstner, U., Fowler, B. A., Herms, U., Krull, I. S., Mackay, D., Olson, G. J., Russell, D. H., Salomons, W., and Silver, S.,** Group report on the speciation in systems under stress, in *The Importance of Chemical Speciation in Environmental Processes,* Bernhard, M., Brinckman, F. E., and Sadler, P. J., Eds., Springer-Verlag, Berlin, 1986, 425.

4. **Wilber, W. G. and Hunter, J. V.,** The impact of urbanization on distribution of heavy metals in bottom sediments of the Saddle River, *Water Resour. Bull.*, 15, 790, 1979.
5. **Norrish, K., Rosser, H., and Warren, L. J.,** A geochemical study of the forms of metals present in sediments from Spencer Gulf, South Australia, *Appl. Geochem.*, 1, 117, 1986.
6. **Förstner, U.,** Accumulative phases for heavy metals in limnic sediments, *Hydrobiologia,* 91, 269, 1982.
7. **Norrish, K.,** Some phosphate minerals in soils, in *Trans. 9th Int. Cong. Soil Sci.*, 2, 713, 1968.
8. **Olson, K. W. and Scogerboe, R. K.,** Identification of soil lead compounds from automotive sources, *Environ. Sci. Technol.*, 9, 227, 1975.
9. **Linton, R. W., Natusch, D. F. S., Solomon, R. L., and Evans, C. A., Jr.,** Physicochemical approach to lead tracing, *Environ. Sci. Technol.*, 14, 159, 1980.
10. **Mattigot, S. V.,** Characterization of fly ash particles, *Scanning Electron Microsc.*, II, 611, 1982.
11. **Mattigot, S. V. and Ervin, J.,** Scheme for density separation and identification of compound forms in size fractionated fly ash, *Fuel,* 62, 927, 1983.
12. **Mattigot, S. V., Page, A. L., and Thornton, I.,** Identification of some trace metal minerals in a mine-waste contaminated soil, *Soil Sci. Soc. Am. J.*, 50, 254, 1986.
13. **Pilkington, E. S. and Warren, L. J.,** Determination of heavy-metal distribution in marine sediments, *Environ. Sci. Technol.*, 13, 295, 1979.
14. **Dossis, P. and Warren, L.,** Distribution of heavy metals between the minerals and organic debris in a contaminated marine sediment, in *Contaminants and Sediments,* Vol. 1, Baker, R. A., Ed., Ann Arbor Science, Ann Arbor, MI, 1980, 119.
15. **Loring, D. H.,** Geochemistry of zinc, copper and lead in the sediments of the estuary and Gulf of St. Lawrence, *Can. J. Earth Sci.*, 15, 757, 1978.
16. **Loring, D. H.,** Geochemistry of cobalt, nickel, chromium, and vanadium in the sediments of the estuary and open Gulf of St. Lawrence, *Can. J. Earth Sci.*, 16, 1196, 1979.
17. **Lichtfuss, R.,** Schwermetalle in den Sedimenten schleswigholsteinischer Flieβgewässer — Untersuchungen zu Gesamtgehalten und Bindungsformen, Ph.D. thesis, University of Kiel, W. Germany, 1977.
18. **Förstner, U.,** Recent heavy metal accumulations in limnic sediments, in *Handbook of Strata-Bound and Stratiform Ore Deposits,* Wolfs, K. H., Ed., Elsevier, Amsterdam, 1981, 179.
19. **Guy, R. D., Chakrabarti, C. L., and McBain, D. C.,** An evaluation of extraction techniques for the fractionation of copper and lead in model sediments, *Water Res.*, 12, 21, 1978.
20. **Horowitz, A. J. and Eldrick, K. A.,** Surface area and its interrelation with grain size, geochemical phase, and sediment-trace element chemistry, *Appl. Geochem.*, 2, 437, 1987.
21. **Plewinsky, B. and Kamps, R.,** Sodium metatungstate, a new medium for binary and ternary density gradient centrifugation, *Makromol. Chem.*, 185, 1429, 1984.
22. **Ramamoorthy, S. and Massalski, A.,** Analysis of structure-localized mercury in Ottawa river sediments by scanning electron microscopy/energy-dispersive X-ray microanalysis technique, *Environ. Geol.*, 2, 351, 1979.
23. **Jedwab, J.,** Cooper, zinc, and lead minerals suspended in ocean waters, *Geochim. Cosmochim. Acta,* 43, 101, 1979.
24. **Ferris, F. G., Fyfe, W. S., and Beveridge, T. J.,** Bacteria as nucleation sites for authigenic minerals in a metal-contaminated lake sediment, *Chem. Geol.*, 63, 225, 1987.
25. **Luther, G. W., III, Meyerson, A. L., Krajewski, J. J., and Hires, R.,** Metal sulfides in estuarine sediments, *J. Sed. Petrol.*, 50, 1117, 1980.
26. **Lee, F. Y. and Kittrick, J. A.,** Elements associated with the cadmium phase in a harbor sediment as determined with the electron beam microprobe, *J. Environ. Qual.*, 13, 337, 1984.
27. **Bernard, P. C., Van Grieken, R. E., and Eisma, D.,** Classification of estuarine particles using automated electron microprobe analysis and multivariate techniques, *Environ. Sci. Technol.*, 20, 467, 1986.
28. **Folkmann, F.,** Analytical use of ion-induced X-rays, *J. Phys. E,* 8, 429, 1975.
29. **Johansson, S. A. E. and Johansson, T. B.,** Analytical application of particle induced X-ray emission, *Nucl. Instrum. Methods,* 137, 473, 1976.
30. **Cahill, T. A.,** Proton microprobes and particle-induced X-ray analytical systems, *Annu. Rev. Nucl. Part. Sci.*, 30, 211, 1980.
31. **Burnett, D. S. and Woolum, D. S.,** In situ trace element microanalysis, *Annu. Rev. Earth Planet. Sci.*, 11, 329, 1983.
32. **Grossmann, D., Kersten, M., Niecke, M., Puskeppel, A., and Voigt, R.** Determination of trace elements in membrane filter samples of suspended matter by the Hamburg proton microprobe, in "Transport of Carbon and Minerals in Major World Rivers," Part 3, Degens, E. T., Kempe, S., and Herrera, R., Eds., *Mitt. Geol. Palaeontol. Inst. Univ. Hamburg,* 58, 619, 1985.
33. **Grasserbauer, M.,** In situ microanalysis — a survey of some major new developments, *Trends Anal. Chem.*, 5, VII, 1986.
34. **Hayes, K. F., Roe, A. L., Brown, G. E., Jr., Hodgson, K. O., Leckie, J. O., and Parks, G. A.,** In situ X-ray absorption study of surface complexes: selenium oxyanions on α-FeOOH, *Science,* 238, 783, 1987.

35. **O'Connor, B. H. and Jaklevic, J. M.,** Characterization of ambient aerosol particulate samples from the St. Louis area by X-ray powder diffractometry, *Atmos. Environ.*, 15, 1681, 1981.
36. **Wölfel, E. R.,** A new method for quantitative X-ray analysis of multiphase mixtures, *J. Appl. Crystallogr.*, 14, 291, 1981.
37. **Fukasawa, T., Iwatsuki, M., and Tillekeratne, S. P.,** X-ray diffraction analysis of airborne particulates collected by an Andersen sampler. Compound distribution vs. particle size, *Environ. Sci. Technol.*, 17, 596, 1983.
38. **Keyser, T. R., Natusch, D. F. S., Evans, C. A., Jr., Linton, R. W.,** Characterizing the surfaces of environmental particles, *Environ. Sci. Technol.*, 12, 768, 1978.
39. **Linton, R. W., Harvey, D. T., and Cabaniss, G. E.,** Environmental applications of surface analysis techniques: PAS, XPS, Auger electron spectroscopy, SIMS, in *Analytical Aspects of Environmental Chemistry,* Natusch, D. F. S. and Hopke, P. K., Eds., John Wiley & Sons, New York, 1983, 137.
40. **Murray, J. W. and Dillard, J. G.,** The oxidation of cobalt(II) adsorbed on manganese dioxide, *Geochim. Cosmochim. Acta,* 43, 781, 1979.
41. **Koppelman, M. H. and Dillard, J. G.,** Adsorption of $Cr(NH_3)_6^{3+}$ and $Cr(en)^{3+}$ on clay minerals and the characterization of chromium by X-ray photoelectron spectroscopy, *Clays Clay Miner.*, 28, 211, 1980.
42. **Dillard, J. G., Crowther, D. L., and Calvert, S. E.,** X-ray photoelectron spectroscopic study of ferromanganese nodules: chemical speciation for selected transition metals, *Geochim. Cosmochim. Acta,* 48, 1565, 1984.
43. **Linton, R. W., Loh, A., Natusch, D. F. S., Evans, C. A., Jr., and Williams, P.,** Surface predominance of trace elements in airborne particles, *Science,* 191, 852, 1976.
44. **Natusch, D. F. S. and Taylor, D. R.,** Environmental effects of western coal combustion. IV, in Chemical and Physical Characterization of Coal Fly Ash, Rep. 600/3-80-093, U.S. Environmental Protection Agency, Washington, D.C., 1980.
45. **Henry, W. M. and Knapp, K. T.,** Compound forms of fossil fuel fly ash emissions, *Environ. Sci. Technol.*, 14, 450, 1980.
46. **Eatough, D. J., Eatough, N. L., Hill, M. W., Mangelson, N. F., and Hansen, L. D.,** Identification of VO_2^+ in particles from the flue lines of oil-fired power plants, *Environ. Sci. Technol.*, 18, 124, 1984.
47. **Andreae, M. O.,** Chemical species in seawater and marine particulates, in *The Importance of Chemical Speciation in Environmental Processes,* Life Sciences Res. Rep. 33, Bernhard, M., Brinckman, F. E., and Sadler, P. J., Eds., Springer-Verlag, Berlin, 1986, 301.
48. **Irgolic, K. J. and Brinckman, F. E.,** Liquid chromatography element-specific detection systems for analysis of molecular species, in *The Importance of Chemical Speciation in Environmental Processes,* Life Sciences Res. Rep. 33, Bernhard, M., Brinckman, F. E., and Sadler, P. J., Eds., Springer-Verlag, Berlin, 1986, 667.
49. **Henstra, S., Bisdom, E. B. A., Jongerius, A., Heinen, H. J., and Meier, S.,** Microchemical analysis on thin sections of soils with the laser microprobe mass analyser (LAMMA), *Fresenius Z. Anal. Chem.*, 308, 280, 1981.
50. **Muller, J. F., Berthé, C., and Magar, J. M.,** LAMMA analysis of organo-metallic compounds, *Fresenius Z. Anal. Chem.*, 308, 312, 1981.
51. **Tempelton, G. D., III and Chasteen, N. D.,** Vanadium-fulvic acid chemistry: conformational and binding studies by electron spin probe techniques, *Geochim. Cosmochim. Acta,* 44, 741, 1980.
52. **Cheshire, M. V., Berrow, M. L., Goodman, B. A., and Mundie, C. M.,** Metal distribution and nature of some Cu, Mn and V complexes in humic and fulvic acid fractions of soil organic matter, *Geochim. Cosmochim. Acta,* 41, 1131, 1977.
53. **Senesi, N., Bocian, D. F., and Sposito, G.,** Electron spin resonance investigation of copper(II) complexation by soil fulvic acid, *Soil. Sci. Soc. Am. J.*, 49, 114, 1985.
54. **Horowitz, A. J.,** A Primer on Trace Metal-Sediment Chemistry, U.S. Geological Survey Water-Supply Paper 84-2277, Doraville, GA, 1985.
55. **LeRiche, H. H. and Weir, A. H.,** A method of studying trace elements in soil fractions, *J. Soil Sci.*, 14, 225, 1963.
56. **Van Valin, R. and Morse, J. W.,** An investigation of methods commonly used for the selective removal and characterization of trace metals in sediments, *Mar. Chem.*, 11, 535, 1982.
57. **Martin, J. M., Nirel, P., and Thomas, A. J.,** Sequential extraction techniques: promises and problems, *Mar. Chem.*, 22, 313, 1987.
58. **Lion, L. W., Altmann, R. S., and Leckie, J. O.,** Trace-metal adsorption characteristics of estuarine particulate matter: evaluation of contributions of Fe/Mn oxide and organic surface coatings, *Environ. Sci. Technol.*, 16, 660, 1982.
59. **Pickering, W. F.,** Selective chemical extraction of soil components and bound metal species, *CRC Crit. Rev. Anal. Chem.*, 12, 233, 1981.
60. **Chao, T. T.,** Use of partial dissolution techniques in geochemical exploration, *J. Geochem. Explor.*, 20, 101, 1984.

61. **Pickering, W. F.,** Metal ion speciation — soils and sediments (a review), *Ore Geol. Rev.,* 1, 83, 1986.
62. **Lake, D. L., Kirk, P. W. W., and Lester, J. N.,** Fractionation, characterization, and speciation of heavy metals in sewage sludge and sludge-amended soils: a review, *J. Environ. Qual.,* 13, 175, 1984.
63. **Campbell, P. G. C., Lewis, A. G., Chapman, A. A., Crowder, A. A., Fletche, W. K., Imber, B., Luoma, S. N., Stokes, P. M., and Winfrey, M.,** *Biologically Available Metals in Sediments,* NRCC Report No. 27694, National Research Council of Canada, Ottawa, 1988.
64. **Robbins, J. M., Lyle, M., and Heath, G. R.,** A Sequential Extraction Procedure for Partitioning Elements among Coexisting Phases in Marine Sediments, Rep. 84-3, College of Oceanography, Oregon State University, Corvallis, 1984.
65. **Förstner, U.,** Chemical forms and environmental effects of critical elements in solid-waste materials — combustion residues, in *The Importance of Chemical Speciation in Environmental Processes,* Bernhard, M., Brinckman, F. E., and Sadler, P. J., Eds., Springer-Verlag, Berlin, 1986, 465.
66. **Förstner, U., Calmano, W., and Schoer, J.,** Heavy metals in bottom sediments and suspended material from the Elbe, Weser and Ems estuaries and from the German Bight (south eastern North Sea), *Thalassia Jugosl.,* 18, 97, 1982.
67. **Schoer, J.,** Iron-oxo-hydroxides and their significance to the behaviour of heavy metals in estuaries, *Environ. Technol. Lett.,* 6, 189, 1985.
68. **Jenne, E. A., Kennedy, V. C., Burchard, J. M., and Ball, J. W.,** Sediment collection and processing for selective extraction and for total trace element analyses, in *Contaminants and Sediments,* Vol. 2, Baker, R. A., Ed., Ann Arbor Science, Ann Arbor, MI, 1980, 169.
69. **Ackermann, F., Bergmann, H., and Schleichert, U.,** Monitoring of heavy metals in coastal and estuarine sediments — a question of grain size: <20 μm versus <60 μm, *Environ. Technol. Lett.,* 4, 317, 1983.
70. **Daeschner, H. W.,** Wet sieving with precision electroformed sieves, *Powder Technol.,* 2, 349, 1969.
71. **Tillekeratne, S., Hiraide, M., and Mizuike, A.,** Selective leaching of trace heavy metals associated with suspended matter in fresh waters, *Microchim. Acta,* III, 69, 1984.
72. **Etcheber, H. and Jouanneau, J. M.,** Comparison of the different methods for the recovery of suspended matter from estuarine waters: deposition, filtration and centrifugation; consequences for the determination of some heavy metals, *Estuarine Coastal Mar. Sci.,* 11, 701, 1980.
73. **Duinker, J. C., Nolting, R. F., and Van Der Sloot, H. A.,** The determination of suspended metals in coastal waters by different sampling and processing methods (filtration, centrifugation), *Neth. J. Sea Res.,* 13, 282, 1979.
74. **Ongley, E. D. and Blachford, D. P.,** Application of continuous flow centrifugation to contaminant analysis of suspended sediment in fluvial systems, *Environ. Technol. Lett.,* 3, 219, 1982.
75. **De Mora, S. J. and Harrison, R. M.,** The use of physical separation techniques in trace metal speciation studies, *Water Res.,* 17, 723, 1983.
76. **Salomons, W. and Mook, W. G.,** Biogeochemical processes affecting metal concentrations in lake sediments (Ijsselmeer, The Netherlands), *Sci. Total Environ.,* 16, 217, 1980.
77. **Stumm, W. and Morgan, J. J.,** *Aquatic Chemistry,* John Wiley & Sons, New York, 1981.
78. **Horowitz, A. J.,** Comparison of methods for the concentration of suspended sediment in river water for subsequent chemical analysis, *Environ. Sci. Technol.,* 20, 155, 1986.
79. **Bartlett, R. and James, B.,** Studying dried, stored soil samples — some pitfalls, *Soil. Sci. Soc. Am. J.,* 44, 721, 1980.
80. **Thomson, E. A., Luoma, S. N., Cain, D. J., and Johansson, C.,** The effect of sample storage on the extraction of Cu, Zn, Fe, Mn and organic materials from oxidized estuarine sediments, *Water Air Soil Pollut.,* 215, 1980.
81. **Engler, R. M., Brannon, J. M., and Rose, J.,** A practical selective extraction procedure for sediment characterization, in *Chemistry of Marine Sediments,* Yen, T. F., Ed., Ann Arbor Science, Ann Arbor, MI, 1977, 163.
82. **Kersten, M., and Förstner, U.,** Chemical fractionation of heavy metals in anoxic estuarine and coastal sediments, *Water Sci. Technol.,* 18, 121, 1986.
83. **Rapin, F., Tessier, A., Campbell, P. G. C., and Carignan, R.,** Potential artifacts in the determination of metal partitioning in sediments by a sequential extraction procedure, *Environ. Sci. Technol.,* 20, 836, 1986.
84. **De Rooij, N. M., Bril, J., Kerdijk, H. N., and Salomons, W.,** Geochemical processes in a large scale disposal of contaminated sludge, in *Proc. Int. Conf. Heavy Metals in the Environment,* Vol. 2, Lekkas, T. D., Ed., CEP Consultants, Edinburgh, 1985, 225.
85. **Leckie, J. O.,** Adsorption and transformation of trace element species at sediment/water interfaces, in *The Importance of Chemical Speciation in Environmental Processes,* Bernhard, M., Brinckman, F. E., and Sadler, P. J., Eds., Springer-Verlag, Berlin, 1986, 237.

86. **Luoma, S. N. and Davis, J. A.,** Requirements for modeling trace metals partitioning in oxidized estuarine sediments, *Mar. Chem.,* 12, 159, 1983.
87. **Gupta, S. K. and Chen, K. Y.,** Partitioning of trace metals in selective chemical fractions of nearshore sediments, *Environ. Lett.,* 10, 129, 1975.
88. **Gibbs, R. J.,** Mechanisms of trace metal transport in rivers, *Science,* 180, 71, 1973.
89. **Towner, J. V.,** Studies of Chemical Extraction Techniques Used for Elucidating the Partitioning of Trace Metals in Sediments, Ph.D. thesis, University of Liverpool, England, 1984.
90. **Sauerbeck, D. R. and Rietz, E.,** Soil-Chemical Evaluation of Different Extractants for Heavy Metal in Soils, Rep. EUR 8022(CA 99:193726), Commission of the European Communities, Brussels, 1983.
91. **Häni, H. and Gupta, S.,** Reasons to use neutral salt solutions to assess the metal impact on plants and soils, in *Chemical Methods for Assessing Bio-Available Metals in Sludges and Soils,* Leschber, R., Davis, R. D., and L'Hermite, P., Eds., Elsevier, New York, 1985, 42.
92. **Boust, D. and Saas, A.,** A selective chemical extraction procedure applied to trace metals; comparison between several reagents on two types of sediments (Seine and Gironde estuaries), in *Proc. Int. Conf. Heavy Metals in the Environment,* Ernst, W. H. O., Ed., CEP Consultants, Edinburgh, 1981, 709.
93. **Badri, M. A. and Aston, S. R.,** A comparative study of sequential extraction procedures in the geochemical fractionation of heavy meatls in estuarine sediments, in *Proc. Int. Conf. Heavy Metals in the Environment,* Ernst, W. H. O., Ed., CEP Consultants, Edinburgh, 1981, 705.
94. **Tessier, A., Campbell, P. G. C., and Bisson, M.,** Sequential extraction procedure for the speciation of particulate trace metals, *Anal. Chem.,* 51, 844, 1979.
95. **Salomons, W. and Förstner, U.,** *Metals in the Hydrocycle,* Springer-Verlag, Berlin, 1984.
96. **Sommers, L. E. and Lindsay, W. L.,** Effect of pH and redox on predicted heavy metal-chelate equilibria in soils, *Soil Sci. Soc. Am. J.,* 43, 39, 1979.
97. **Balistrieri, L. S. and Murray, J. W.,** The adsorption of Cu, Pb, Zn, and Cd on goethite from major ion seawater, *Geochim. Cosmochim. Acta,* 46, 1253, 1982.
98. **Van Hoek, G. L. M., Gommers, P. J. F., and Overwater, J. A. S.,** Removal of heavy metals from polluted sediments — ion-exchange resins, a possible solution, in *Proc. Int. Conf. Heavy Metals in the Environment,* Vol. 2, Müller, G., Ed., CEP Consultants, Edinburgh, 1983, 856.
99. **Rapin, F. and Förstner, U.,** Sequential leaching technique for particulate metal speciation: the selectivity of various extractants, in *Proc. Int. Conf. Heavy Metals in the Environment,* Müller, G., Ed., CEP Consultants, Edinburgh, 1983, 1074.
100. **Lyle, M., Heath, G. R., and Robbins, J. M.,** Transport and release of transition elements during early diagenesis: sequential leaching of sediments from MANOP Sites M and H. I. pH 5 acetic acid leach, *Geochim. Cosmochim. Acta,* 48, 1705, 1984.
101. **Span, D. and Gaillard, J.-F.,** An investigation of a procedure for determining carbonate-bound trace metals, *Chem. Geol.,* 56, 135, 1986.
102. **Kersten, M.,** Geochemische und Numerische Untersuchungen zum Mechanismus der Schwermetallfre-isetzung aus Elbschlick, Ph.D. thesis, Technological University of Hamburg-Harburg, 1988.
103. **Deurer, R., Förstner, U., and Schmoll, G.,** Selective chemical extraction of carbonate-associated trace metals in recent lacustrine sediments, *Geochim. Cosmochim. Acta,* 42, 425, 1978.
104. **Kunze, G. W.,** Pretreatment for mineralogical analysis, in *Methods of Soil Analysis,* Black, C. A., Ed., American Society of Agronomy, Madison, WI, 1965, 568.
105. **Schwertmann, U.,** Die fractionierte Extraction der freien Eisenoxide im Boden, ihre mineralogische Formen und ihre Enstehungsweisen, *Z. Pflanzenernaehr. Dueng. Bodenkd.,* 84, 194, 1959.
106. **Goldberg, E. D. and Arrhenius, G. O.,** Chemistry of Pacific pelagic sediments, *Geochim. Cosmochim. Acta,* 13, 153, 1958.
107. **Chester, R. and Hughes, M. J.,** A chemical technique for the separation of ferro-manganese minerals, carbonate and adsorbed trace elements from pelagic sediments, *Chem. Geol.,* 2, 249, 1967.
108. **Chao, T. T. and Liyi, Z.,** Extraction techniques for selective dissolution of amorphous iron oxides from soils and sediments, *Soil Sci. Soc. Am. J.,* 47, 225, 1983.

soils and sediments, *Soil Sci. Soc. Am. J.,* 47, 225, 1983.

109. **Heath, G. R. and Dymond, J.,** Genesis and transformation of metalliferous sediments from the East Pacific Rise, Bauer Deep and central region northwest Nazca plate, *Geol. Soc. Am. Bull.,* 88, 723, 1977.
110. **Shuman, L. M.,** Separting soil iron- and manganese oxide fractions for microelement analysis, *Soil Sci. Soc. Am. J.,* 46, 1099, 1982.
111. **Landa, D. R. and Gast, R. G.,** Evaluation of crystallinity in hydrated ferric oxides, *Clays Clay Miner.,* 21, 121, 1973.
112. **Rhoton, F. E., Bigham, J. M., Norton, L. D., and Smeck, N. E.,** Contribution of magnetite to oxalate-extractable iron in soils and sediments from the Maumee river basin of Ohio, *Soil Sci. Soc. Am. J.,* 45, 645, 1981.

113. **Arshad, M. A., Arnaud, R. J. St., and Huang, P. M.,** Dissolution of trioctahedral layer silicates by ammonium oxalate, sodium dithionite-citrate-bicarbonate, and potassium pyrophosphate, *Can. J. Soil. Sci.,* 52, 19, 1972.
114. **Slavek, J., Wold, J., and Pickering, W. F.,** Extraction of metal ions associated with humic acids, *Talanta,* 29, 743, 1982.
115. **Hoffman, S. J. and Fletcher, W. K.,** Extraction of Cu, Zn, Mo, Fe and Mn from soils and sediments using a sequential procedure, in *Geochemical Exploration,* Watterson, J. R. and Theobald, P. K., Eds., Association of Exploration Geochemists, Rexdale, Ontario, 1978, 289.
116. **Bogle, E. W. and Nichol, I.,** Metal transfer, partition and fixation in drainage waters and sediments in carbonate terrain in southeastern Ontario, *J. Geochem. Explor.,* 15, 405, 1981.
117. **Pfeiffer, G., Förstner, U., and Stoffers, P.,** Speciation of reducible metal components in pelagic sediments by chemical extraction, *Senckenbergiana Marit.,* 14, 23, 1982.
118. **Chao, T. T.,** Selective dissolution of manganese oxides from soils and sediments with acidified hydroxylamine, *Soil Sci. Soc. Am. Proc.,* 36, 764, 1972.
119. **Holmgren, G. S.,** A rapid citrate-dithionite extractable iron procedure, *Soil Sci. Soc. Am. Proc.,* 31, 210, 1967.
120. **Rozenson, I. and Heller-Kallai, L.,** Reduction and oxidation of Fe^{3+} in dioctahedral smectites. III. Oxidation of octahedral Fe in montmorillonite, *Clays Clay Miner.,* 26, 88, 1978.
121. **Pawluk, S.,** Measurement of crystalline and amorphous iron removal in soils, *Can. J. Soil Sci.,* 52, 119, 1972.
122. **Tipping, E., Hetherington, N. B., Hilton, J., Thompson, D. W., Bowles, E., and Hamilton-Taylor, J.,** Artifacts in the use of selective chemical extraction to determine distributions of metals between oxides of manganese and iron, *Anal. Chem.,* 57, 1944, 1985.
123. **Rendell, P. S., Batley, G. E., and Cameron, A. J.,** Adsorption as a control of metal concentrations in sediment extracts, *Environ. Sci. Technol.,* 14, 314, 1980.
124. **Nirel, P., Thomas, A. J., and Martin, J. M.,** A critical evaluation of sequential extraction techniques, in *Speciation of Fission and Activation Products in the Environment,* Bulman, R. H. and Cooper, J. R., Elsevier, New York, 1985, 19.
125. **Jones, K.,** Nitrogen, in *Comprehensive Inorganic Chemistry,* Bailar, J. C., Jr., Ed., Pergamon Press, Oxford, 1973, 147.
126. **Bloom, H.,** A field method for the determination of ammonium citrate-soluble heavy metals in soils and alluvium, *Econ. Geol.,* 50, 535, 1955.
127. **Mills, B. A.,** Selective Kinetic Extraction of Some Northeastern Equatorial Pacific Pelagic Sediments, M.S. thesis, University of Wisconsin, Madison, 1978.
128. **Sterrit, R. M. and Lester, J. N.,** The value of sewage sludge to agriculture and effects of the agricultural use of sludges contaminated with toxic elements: a review, *Sci. Total Environ.,* 16, 55, 1980.
129. **Buffle, J.,** Natural organic matter and metal-organic interactions in aquatic systems, in *Metal Ions in Biological Systems,* Sigel, H., Ed., Marcel Dekker, New York, 1984, chap. 6.
130. **Kerndorff, H. and Schnitzer, M.,** Sorption of metals on humic acid, *Geochim. Cosmochim. Acta,* 44, 1701, 1980.
131. **Sposito, G.,** Sorption of trace metals by humic materials in soils and natural waters, *CRC Crit. Rev. Environ. Control,* 16, 193, 1986.
132. **Rashid, M. A.,** *Geochemistry of Marine Humic Substances,* Springer-Verlag, New York, 1985, chap. 4 and 7.
133. **Bremner, J. M. and Lees, H.,** Studies in soil organic matter. II. The extraction of soil organic matter by neutral reagents, *J. Agric. Sci.,* 39, 274, 1949.
134. **Gascho, G. J. and Stevenson, F. J.,** An improved method for extracting organic matter from soil, *Soil. Sci. Soc. Am. Proc.,* 32, 117, 1968.
135. **McKeague, J. A. and Day, J. H.,** Dithionite and oxalate extractable Fe and Al as aids in differentiating various classes of soils, *Can. J. Soil Sci.,* 46, 13, 1966.
136. **Jeanroy, E. and Guillet, B.,** The occurrence of suspended ferruginous particles in pyrophosphate extracts of some soil horizons, *Geoderma,* 26, 95, 1981.
137. **Kassim, J. K., Gafoor, S. N., and Adams, W. A.,** Ferrihydrite in pyrophosphate extracts of podzol B horizons, *Clay Miner.,* 19, 99, 1984.
138. **Berger, P., Etcheber, H., Ewald, M., Lavaux, G., and Belin, C.,** Variation of organic matter extracted from particles along the Gironde Estuary (France), *Chem. Geol.,* 45, 1, 1984.
139. **Hayes, M. H. B.,** Extraction of humic substances from soils, in *Humic Substances in Soil, Sediment, and Water,* Aiken, G. R., McKnight, D. M., Wershaw, R. L., and MacCarthy, P., Eds., Wiley-Interscience, New York, 1985.
140. **Bascomb, C. L. and Thanigasalam, K.,** Comparison of aqueous acetylacetone and potassium pyrophosphate solutions for selective extraction of organic-bound Fe from soils, *J. Soil Sci.,* 29, 382, 1978.

141. **Cooper, B. S. and Harris, R. C.,** Heavy metals in organic phases of river and estuarine sediment, *Mar. Pollut. Bull.*, 5, 24, 1974.
142. **Summerhayes, C., Ellis, J., Stoffers, P., Briggs, S., and Fitzgerald, M.,** Fine-grained Sediment and Industrial Waste Distribution in New Bedford Harbor and Western Buzzards Bay, Massachusetts, Rep. WHOI-76-115, Woods Hole Oceanographic Institution, Woods Hole, MA, 1976.
143. **Anderson, J. U.,** An improved treatment for mineralogical analysis of samples containing organic matter, *Clays Clay Miner.*, 10, 380, 1963.
144. **Lavkulich, L. M. and Wiens, J. H.,** Comparison of organic matter destruction by hydrogen peroxide and sodium hypochloride and its effects on selected mineral constituents, *Soil Sci. Soc. Am. Proc.*, 34, 755, 1970.
145. **Shuman, L. M.,** Sodium hypochlorite methods for extracting microelements associated with soil organic matter, *Soil Sci. Soc. Am. J.*, 47, 656, 1983.
146. **Jackson, M. L.,** *Soil Chemical Analysis,* Prentice-Hall, Englewood Cliffs, NJ, 1958.
147. **Filipek, L. H. and Owen, R. M.,** Analysis of heavy metal distributions among different mineralogical states in sediments, *Can. J. Spectrosc.*, 23, 31, 1978.
148. **Salomons, W. and Förstner, U.,** Trace metal analysis on polluted sediments. II. Evaluation of environmental impact, *Environ. Technol. Lett.*, 1, 506, 1980.
149. **Farmer, V. C., Russel, J. D., McHardy, W. J., Newman, A. C. D., Ahlrichs, J. L., and Rimasaite, J. Y. H.,** Evidence for loss of protons and octahedral iron from oxidized biotites and vermiculites, *Mineral. Mag.*, 38, 121, 1971.
150. **Farmer, V. C. and Mitchell, B.,** Occurrence of oxalates in soil clays following hydrogen peroxide treatment, *Soil Sci.*, 96, 221, 1963.
151. **Etcheber, H., Bourg, A. C. M., and Donard, O.,** Critical aspects of selective extractions of trace metals from estuarine suspended matter. Fe and Mn hydroxides and organic matter interactions, in *Proc. Int. Conf. Heavy Metals in the Environment,* Vol. 2, Müller, G., Ed., CEP Consultants, Edinburgh, 1983, 1200.
152. **Kitano, Y., Sakata, M., and Matsumoto, E.,** Partitioning of heavy metals into mineral and organic fractions in a sediment core from Tokyo Bay, *Geochim. Cosmochim. Acta,* 44, 1279, 1980.
153. **Kersten, M.,** Geochemistry of priority pollutants in anoxic organics, in *Management of Mine Tailings and Dredged Materials,* Salomons, W. and Förstner, U., Eds., Springer-Verlag, Berlin, 1988, 170.
154. **Chao, T. T. and Sanzolone, R. F.,** Chemical dissolution of sulfide minerals, *U. S. Geol. Surv. J. Res.*, 5, 409, 1977.
155. **Klock, P. R., Czamanske, G. K., Foose, M., and Pesek, J.,** Selective chemical dissolution of sulfides: an evaluation of six methods applicable to assaying sulfide-bound nickel, *Chem. Geol.*, 54, 157, 1986.
156. **Donard, O., Bourg, A. C. M., Etchebar, H., and Le Ribault, L.,** Direct evidence of coatings on particulate matter and implications for heavy metal distributions in an estuarine environment, in *Proc. Int. Conf. Heavy Metals in the Environment,* Vol. 2, Müller, G., Ed., CEP Consultants, Edinburgh, 1983, 1009.
157. **Sulcek, Z., Povondra, P., and Dolezal, J.,** Decomposition procedures in inorganic analysis, *CRC Crit. Rev. Anal. Chem.*, 6, 255, 1977.
158. **Rantalla, R. T. T. and Loring, D. H.,** Cadmium in marine sediments: determinationby graphite furnace atomic absorption spectroscopy, *ICES Techniques in Marine Environmental Sciences,* Vol. 3, ICES Copenhage, 1987.
159. **Farmer, J. G. and Gibson, M. J.,** Direct determination of cadmium, chromium, copper and lead in siliceous standard reference materials from a fluoroboric acid matrix by graphite furnace atomic absorption spectrometry, *At. Spectrosc.*, 2, 176, 1981.
160. **Kotz, L., Kaiser, G., Tschöpel, P., and Tölg, G.,** Aufschluβ biologischer Matrices für die Bestimmung sehr niedriger Spurenelementgehalte bei begrenzter Einwaage mit Salpetersäure unter Druck in einem Teflongefäβ, *Fresenius Z. Anal. Chem.*, 260, 207, 1972.
161. **Langmyhr, F. J. and Graff, P. R.,** Studies in the spectrophotometric determination of silicon in materials decomposed by hydrofluoric acid. I. Loss of silicon by decomposition with hydrofluoric acid, *Anal. Chim. Acta,* 21, 334, 1959.
162. **Langmyhr, F. J. and Kringstad, K.,** An investigation of the composition of the precipitates formed by the decomposition of silicate rocks in 38—40% hydrofluoric acid, *Anal. Chim. Acta,* 35, 131, 1966.
163. **Förstner, U. and Schoer, J.,** Diagenesis of chemical associations of ^{136}Cs and other artificial radionuclides in river sediments, *Environ. Technol. Lett.*, 7, 295, 1984.
164. **Fiszman, M., Pfeiffer, W. C., and Drude de Lacerda, L.,** Comparison of methods used for extraction and geochemical distribution of heavy metals in bottom sediments from Sepetiba Bay, R. J., *Environ. Technol. Lett.*, 5, 567, 1984.
165. **Gibbs, R. J.,** Transport phases of transition metals in the Amazon and Yukon rivers, *Geol. Soc. Am. Bull.*, 88, 829, 1977.
166. **Malo, B. A.,** Partial extraction of metals from aquatic sediments, *Environ. Sci. Technol.*, 11, 277, 1977.

167. **Sorensen, R. C., Oelsligle, D. D., and Knudsen, D.,** Extraction of Zn, Fe and Mn from soils with 0.1*N* hydrochloric acid as affected by soil properties, solution:soil ratio and length of extraction period, *Soil Sci.*, 111, 352, 1971.
168. **Chester, R., Kudoja, W. M., Thomas, A., and Towner, J.,** Pollution reconnaissance in stream sediments using non-residual trace metals, *Environ. Pollut. B,* 10, 213, 1985.
169. **McLaren, R. G. and Crawford, D. V.,** Studies of soil copper. I. The fractionation of copper in soils, *J. Soil Sci.*, 24, 172, 1973.
170. **Gatehouse, S., Russell, D. W., and Van Moort, J. C.,** Sequential soil analysis in exploration geochemistry, *J. Geochem. Explor.*, 8, 483, 1977.
171. **Meguellati, N., Robbe, D., Marchandise, P., and Astruc, M.,** A new chemical extraction procedure in the fractionation of heavy metals in sediments — interpretation, in *Proc. Int. Conf. Heavy Metals in the Environment,* Vol. 2, Müller, G., Ed., CEP Consultants, Edinburgh, 1983, 1090.
172. **Robinson, G. D.,** Sequential chemical extractions and metal partitioning in hydrous Mn-Fe-oxide coatings: reagent choice and substrate composition affect results, *Chem. Geol.*, 47, 97, 1984.
173. **König, I., Knauth, H.-D., Koopmann, Chr., Wagner, F. E., and Wagner, U.,** Mössbauer studies of sediments and suspended matter from the river Elbe, *Hyper. Int.*, 41, 811, 1988.
174. **Diks, D. M. and Allen, H. E.,** Correlation of copper distribution in a freshwater-sediment system to bioavailability, *Bull. Environ. Contam. Toxicol.*, 30, 37, 1983.
175. **Gibson, M. J. and Farmer, J. G.,** Multi-step sequential chemical extraction of heavy metals from urban soils, *Environ. Pollut. B,* 11, 117, 1986.
176. **Jenne, E. A.,** Trace element sorption by sediments and soils — sites and processes, in *Symposium on Molybdenum,* Vol. 2, Chappell, W. and Petersen, K., Eds., Marcel Dekker, New York, 1977, 425.
177. **Luoma, S. N. and Bryan, G. W.,** A statistical assessment of the form of trace metals in oxidized estuarine sediments employing chemical extractants, *Sci. Total Environ.*, 17, 165, 1981.
178. **Sakellariadou, F.,** The geochemistry and the partitioning of various metals in selective chemical fractions of sediments collected from the Samothraki Plateau (N. Greece), in *Proc. Int. Conf. Chemicals in the Environment,* Lester, J. N., Perry, R., and Sterrit, R. M., Eds., Selper, London, 1986, 578.
179. **Förstner, U.,** Chemical forms and reactivities of metals in sediments, in *Chemical Methods for Assessing Bio-Available Metals in Sludges and Soils,* Leschber, R., Davis, R. D. and L'Hermite, P., Eds., Elsevier, New York, 1985, 1.
180. **Schoer, J.,** Thallium, in *The Handbook of Environmental Chemistry,* Vol. 3/C, Hutzinger, O., Ed., Springer-Verlag, Berlin, 1984, 143.
181. **Lehn, H. and Schoer, J.,** Thallium transfer from soils to plants: relations between chemical forms and plant uptake, in *Proc. Int. Conf. Heavy Metals in the Environment,* Vol. 2, Lekkas, T. D., Ed., CEP Consultants, Edinburgh, 1985, 286.
182. **Hopke, P. K., Lamb, R. E., and Natusch, D. F. S.,** Multi-elemental characterization of urban roadway dust, *Environ. Sci. Technol.*, 14, 164, 1980.
183. **Singh, B. R. and Narwal, R. P.,** Plant availability of heavy metals in a sludge-treated soil. II. Metal extractability compared with plant metal uptake, *J. Environ. Qual.*, 13, 344, 1984.
184. **Luoma, S. N.,** Bioavailability of trace metals to aquatic organisms: a review, *Sci. Total Environ.*, 28, 1, 1983.
185. **Cataldo, D. A., and Wildung, R. E.,** Soil and plant factors influencing the accumulation of heavy metals by plants, *Environ. Health Perspect.*, 27, 149, 1978.
186. **Gupta, S. K. and Häni, H.,** Methode zur Bestimmung der Biorelevanten Schwermetallgehalte im Boden und Klärschlamm als Zusatzkriterium zur Festlegung von Schwermetallrichtwerten im Boden, Rep. Swiss Federal Research Station for Agricultural Chemistry and Hygiene of Environment, CH-3097, Liebefeld-Bern, Switzerland, 1985.
187. **Ahlf, W.,** Verhalten sedimentgebundener Schwermetalle in einem Algentestsystem, characterisiert durch Bioaccumulation und Toxizität, *Vom Wasser,* 65, 183, 1985.
188. **Luoma, S. N. and Bryan, G. W.,** Factors controlling the availability of sediment-bound lead to the estuarine bivalve Scrobicularia plana, *J. Mar. Biol. Assoc. U.K.*, 58, 793, 1978.
189. **Tessier, A., Campbell, P. G. C., and Auclair, J. C.,** Relationships between trace metal partitioning in sediments and their bioaccumulation in freshwater pelecypods, in *Proc. Int. Conf. Heavy Metals in the Environment,* Vol. 2, Müller, G., Ed., CEP Consultants, Edinburgh, 1983, 1086.
190. **Allen, H. E., Hall, R. H., and Brisbin, T. D.,** Metal speciation: effects on aquatic toxicity, *Environ. Sci. Technol.*, 14, 441, 1980.
191. **Harris, W. R. and Silberman, D.,** Time-dependent leaching of coal fly ash by chelating agents, *Environ. Sci. Technol.*, 17, 139, 1983.
192. **Plant, J. A. and Raiswell, R.,** Principles of environmental geochemistry, in *Applied Environmental Geochemistry,* Thornton, I., Ed., Academic Press, London, 1983, chap. 1.
193. **Dreesen, D. R., Gladney, E. S., Owens, J. W., Perkins, B. L., Wienke, C. L., and Wangen, L. E.,** Comparison of levels of trace elements extracted from fly ash and levels found in effluent waters from a coal-fired power plant, *Environ. Sci. Technol.*, 11, 1017, 1977.

194. **Ahlf, W.,** The river Elbe: behaviour of Cd and Zn during estuarine mixing, *Environ. Technol. Lett.*, 4, 405, 1983.
195. **Kersten, M. and Förstner, U.,** Effect of sample pretreatment on the reliability of solid speciation data of heavy metals — implications for the study of early diagenetic processes, *Mar. Chem.*, 22, 299, 1987.
196. **Kersten, M. and Förstner, U.,** Cadmium associations in freshwater and marine sediment, in *Cadmium in the Aquatic Environment,* Vol. 3, Nriagu, J. O. and Sprague, J. B., Eds., John Wiley & Sons, New York, 1987, 51.
197. **Kersten, M.,** Geobiological effects on the mobility of contaminants in marine sediments, in *Pollution of the North Sea: An Assessment,* Salomons, W., Bayne, B., Duursma, E., and Förstner, U., Eds., Springer-Verlag, Berlin, 1988, 36.
198. **Förstner, U.,** Changes in metal mobilities in aquatic and terrestrial cycles, in *Speciation, Separation and Recovery of Metals,* Patterson, J. W., Ed., Lewis, Chelsea, MI, 1987, 3.
199. **Boeckx, R. L.,** Lead poisoning in children, *Anal. Chem.*, 58, 274A, 1986.
200. **Garcia-Miragaya, J.,** Levels, chemical fractionation, and solubility of lead in roadside soils of Caracas, Venezuela, *Soil Sci.*, 138, 147, 1984.
201. **Keller, L. and Brunner, P. H.,** Waste-related cadmium cycle in Switzerland, *Ecotoxicol. Environ. Safety,* 7, 141, 1983.
202. **Nriagu, J. O., Ed.,** Cadmium in the atmosphere and in precipitation, in *Cadmium in the Environment,* Vol. 1, John Wiley & Sons, New York, 1980, 71.
203. **Nürnberg, H. W., Valenta, P., and Nguyen, V. D.,** The wet deposition of heavy metals from the atmosphere in the Federal Republic of Germany, in *Proc. Int. Conf. Heavy Metal in the Environment,* Vol. 1, Müller, G., Ed., CEP Consultants, Edinburgh, 1983, 115.
204. **Farmer, M. E. and Linton, R. W.,** Correlative surface analysis studies of environmental particles, *Environ. Sci. Technol.*, 18, 319, 1984.
205. **Smith, R. D., Campbell, J. A., and Nielson, K. K.,** Concentration dependence upon particle size of volatilized elements in fly ash, *Environ. Sci. Technol.*, 13, 553, 1979.
206. **Natusch, D. F. S., Wallace, J. R., and Evans, C. A., Jr.,** Toxic trace elements: preferential concentration in respirable particles, *Science,* 183, 202, 1974.
207. **Hansen, L. D. and Fisher, G. L.,** Elemental distribution in coal fly ash particles, *Environ. Sci. Technol.*, 14, 1111, 1980.
208. **Hansen, L. D., Silberman, D., Fisher, G. L., and Eatough, D. J.,** Chemical speciation of elements in stack-collected, respirable-size, coal fly ash, *Environ. Sci. Technol.*, 18, 181, 1984.
209. **Wu, E. J., Choi, W. W., and Chen, K. Y.,** Chemical affiliation of trace metals in coal ash, *AIChE Symp. Ser.*, 210, 177, 1980.
210. **Biggins, P. D. E. and Harrison, R. M.,** Chemical speciation of lead compounds in street dusts, *Environ. Sci. Technol.*, 14, 336, 1980.
211. **Lum, K. R., Betteridge, J. S., and MacDonald, R. R.,** The potential availability of P, Al, Cd, Co, Cr, Cu, Fe, Mn, Ni, Pb and Zn in urban particulate matter, *Environ. Technol. Lett.*, 3, 57, 1982.
212. **Harrison, R. M., Laxen, D. P. H., and Wilson, S. J.,** Chemical associations of lead, cadmium, copper, and zinc in street dust and roadside soils, *Environ. Sci. Technol.*, 15, 1378, 1981.
213. **Chester, R., Murphy, K. J. T., Towner, J., and Thomas, A.,** The partitioning of elements in crust-dominated marine aerosols, *Chem. Geol.*, 54, 1, 1986.
214. **Fraser, J. L. and Lum, K. R.,** Availability of elements of environmental importance in incinerated sludge ash, *Environ. Sci. Technol.*, 17, 52, 1982.
215. **Gibson, M. J. and Farmer, J. G.,** Chemical partitioning of trace metal contaminants in urban street dirt, *Sci. Total Environ.*, 33, 49, 1984.

Chapter 9

TRACE ELEMENT SPECIATION IN BIOLOGICAL SYSTEMS

T. Mark Florence

TABLE OF CONTENTS

I. INTRODUCTION

Trace element speciation is a life-controlling factor in all living organisms, from unicellular algae to the human species. Almost half of the several hundred known enzymes are metalloenzymes, with the metal at the catalytic (active) center, or performing structural or ion-directing roles. Metals are also associated with a wide range of proteins in blood serum, and the distribution of a metal in serum between amino acids, proteins, and enzymes, i.e., its speciation, often has important pathological implications.

A total of 30 elements are now believed to be essential to life (Figure 1).[1] They can be divided into the 6 structural elements, 5 macrominerals, and 19 trace elements. Cotzias[2] considered that a trace element is essential if it meets the following criteria: (1) it is present in all healthy tissues of all living things, (2) its concentration from one animal to the next is fairly constant, (3) its withdrawal from the body induces, reproducibly, the same physiological and structural abnormalities regardless of the species studied, (4) its addition either reverses or prevents these abnormalities, (5) the abnormalities induced by deficiency are always accompanied by pertinent, significant biochemical changes, and (6) these biochemical changes can be prevented or cured when the deficiency is corrected.

Liebscher and Smith[3] proposed that the shape of the distribution curve for a trace element in a particular tissue can be used as a test for essentiality. Essential elements have a normal distribution, whereas elements present as environmental contaminants have a skewed distribution pattern (Figure 2). Certain elements, although not satisfying all the requirements for essentiality, sometimes have a stimulating effect on growth or survival. These elements include Li, Ti, Ga, Ge, Rb, Zr, Sb, Ba, Au, Hg, and some of the lanthanides.[4]

Every essential element is toxic if taken in excess, and there is a safe window of essential dose between deficiency and toxicity (Figure 3). For some elements (e.g., Ca and Mg) this window is wide, whereas for others (e.g., Se and F), the window is narrow, and an excess will rapidly lead to toxicity and death.[5] Trace element antagonism is well established in both plant and animal species.[6] Figure 4 shows a trace element interaction chart for plants; similar stimulation and antagonism effects exist for man.[6]

A high degree of chemical speciation can exist for any trace element in biological systems.[7] For example, not all forms of iron are effective in the treatment of iron deficiency, not all forms of chromium can be utilized as glucose tolerance factor, and not all selenium compounds are protective against selenium-responsive diseases. The valency of an element also greatly affects its biological action. Chromium is effective only in the (III) state, Cr(VI) being very toxic. Similarly, arsenic (III) is much more toxic than arsenic (V), and mercury alkyl complexes are extremely toxic to aquatic organisms. Cobalt in vitamin B_{12} is a classic example of trace element biospecificity, this vitamin being the only known form of cobalt in living organisms.[8] Most other elements, however, have a wide range of highly specialized forms. Iron, for example, has storage proteins and transport proteins; it exists as porphyrins in hemoglobin, myoglobin, cytochromes, peroxidases and catalases, is associated with sulfur in ferredoxins, and in invertebrates is present as the oxygen-carrying pigment, hemerythrin (Section IV.D).

The great diversity of trace element speciation in living systems presents a formidable challenge to the bioinorganic chemist. Speciation analysis of a biological sample with its complex organic matrix is much more difficult than analysis of a water sample. More sophisticated techniques are required, and the results are often difficult to interpret. Nevertheless, it is essential that research in this area be increased if we are to understand in more detail the role of trace elements in health and disease. The study of human nutrition also requires much more information about trace element speciation in food. Just as total metal analysis of a water sample provides little information about the potential toxicity of the water to an aquatic organism,[9] so total analysis of an essential element in a food sample gives no

Bulk Elements

Essential Trace Elements

Essential Trace Elements (Proposed)

H																	He
Li	Be											B	C	N	O	F	Ne
Na	Mg											Al	Si	P	S	Cl	Ar
K	Ca	Sc	Ti	V	Cr	Mn	Fe	Co	Ni	Cu	Zn	Ga	Ge	As	Se	Br	Kr
Rb	Sr	Y	Zr	Nb	Mo	Tc	Ru	Rh	Pd	Ag	Cd	In	Sn	Sb	Te	I	Xe
Cs	Ba		Hf	Ta	W	Re	Os	Ir	Pt	Au	Hg	Tl	Pb	Bi	Po	At	Rn
Fr	Ra																

Lanthanides

Actinides

FIGURE 1. Essential elements and the Periodic Table. (Adapted from Frieden, E., *J. Chem. Educ.*, 62, 917, 1985. With permission.)

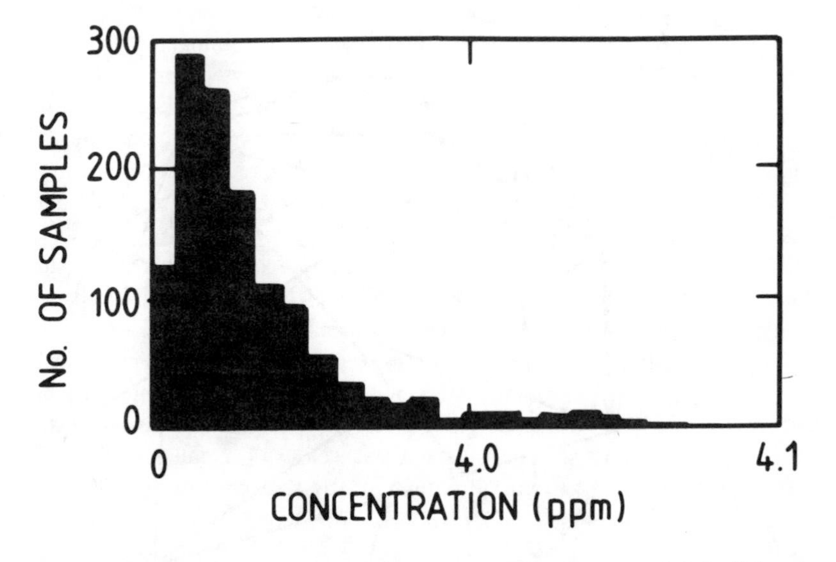

FIGURE 2. Distribution of arsenic (a nonessential element) in human scalp hair. (Adapted from Liebscher, K. and Smith, H., *Arch. Environ. Health*, 17, 881, 1968. With permission.)

indication of the uptake (bioavailability) of that element by an animal that eats the food. Future legislation may require that labile, as well as total, metal be measured in water samples, and that the bioavailable fractions of the essential elements in a food be shown on the package.

II. TECHNIQUES FOR SPECIATION ANALYSIS OF BIOLOGICAL SAMPLES

Before speciation analysis of biological samples is attempted, the analytical chemist should demonstrate his ability to accurately and precisely measure *total* metal in the samples. Even total metal analysis is often a formidable task, and the accepted concentrations of some metals in biological samples have consistently fallen (as they have in seawater!)[10] over the years as superior analytical techniques become available and, particularly, as more care has

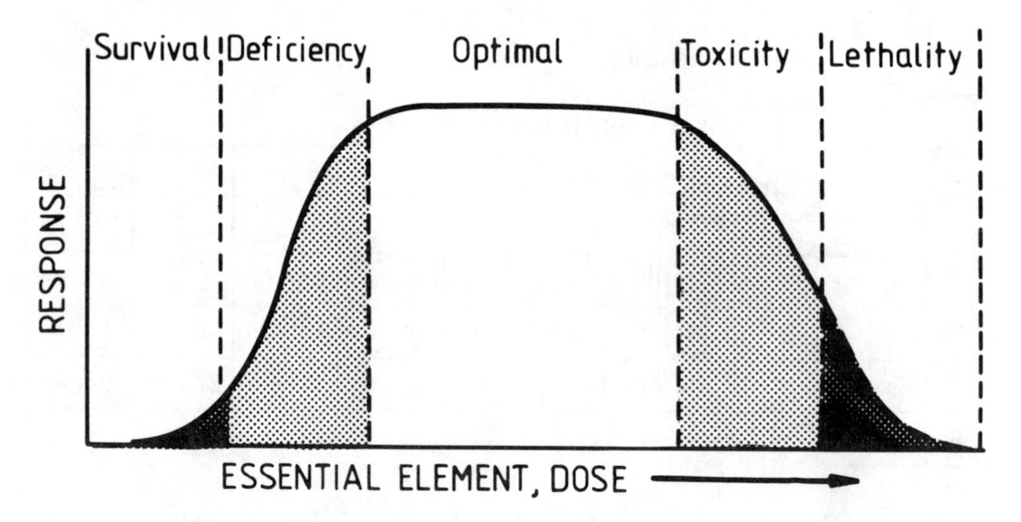

FIGURE 3. Dose-response curve for an essential element. (Adapted from Frieden, E., in *Biochemistry of the Essential Ultratrace Elements,* Frieden, E., Ed., Plenum Press, New York, 1984, 1. With permission.)

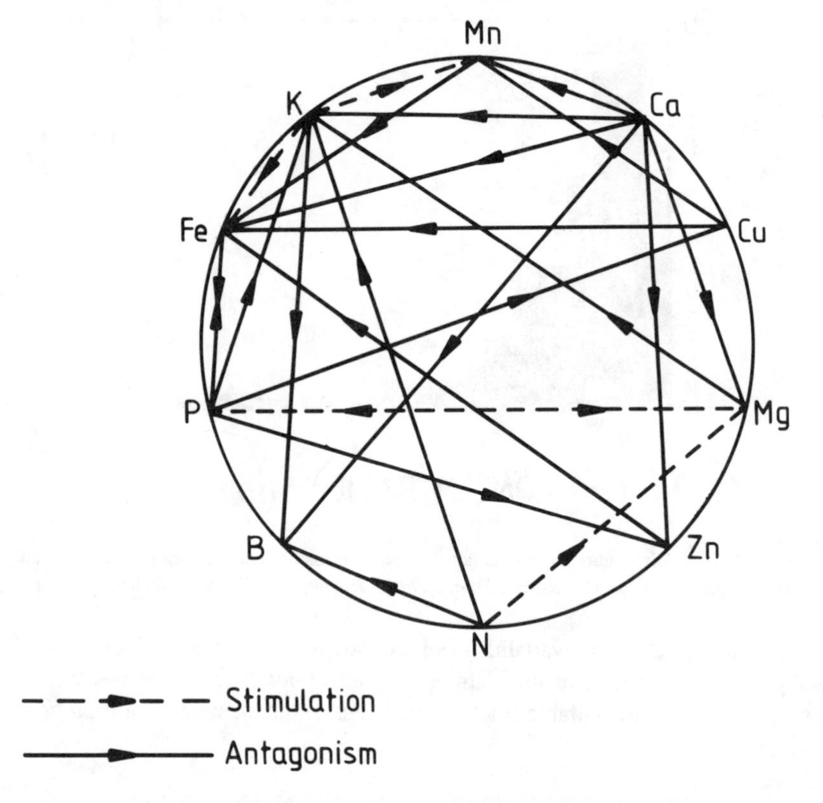

FIGURE 4. Trace element interaction chart for plants. (Adapted from Fiabane, A. M. and Williams, D. R., *The Principles of Bio-Inorganic Chemistry,* The Chemical Society, London, 1977, 68. With permission.)

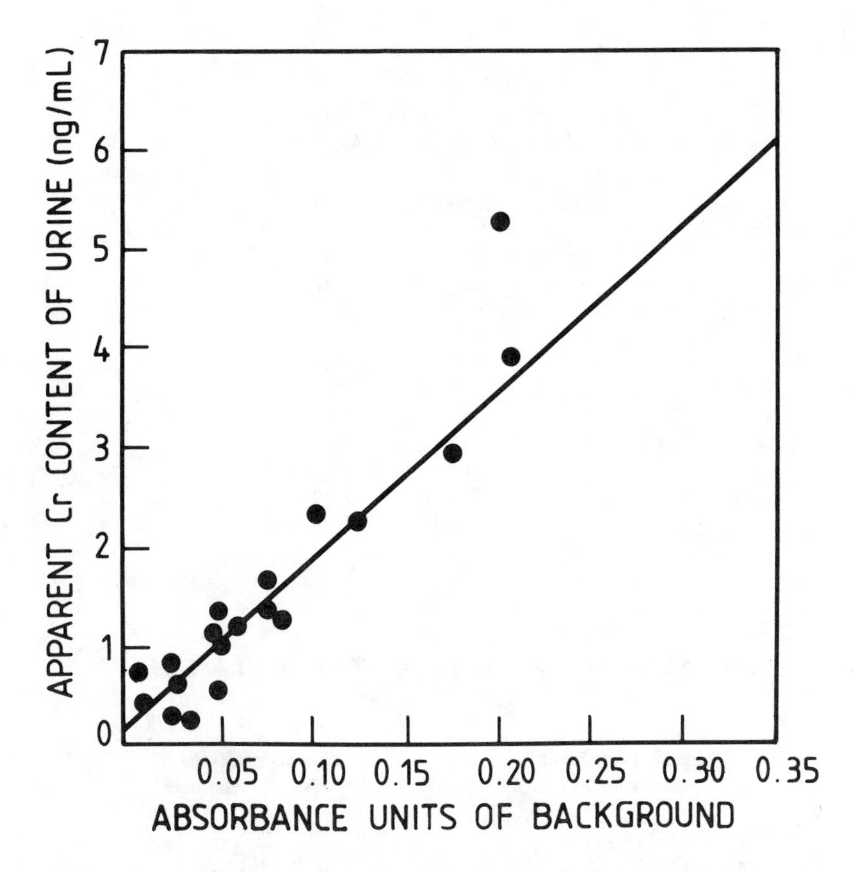

FIGURE 5. Correlation between atomic absorption background absorbance and apparent chromium concentration in urine. (Adapted from Veillon, C., *Anal. Chem.*, 58, 851A, 1986. With permission.)

been taken to avoid contamination during sampling and analysis. An excellent example of this trend is the measurement of chromium in blood and urine.[11] Between 1964 and 1970, values of 18 to 1500 μg Cr per day were reported for 24-h urine samples. By 1978 the values were down to 3 to 10 μg Cr per day, and today most urine measurements yield 0.5 to 2 μg Cr per day. Prior to 1978, human plasma was believed to contain about 20 μg Cr per liter; today the accepted value is closer to 0.2 μg Cr per liter.[11] The errors arose from failure to adequately correct for background adsorption in atomic absorption spectrometry (AAS) (Figure 5) and failure to recognize sources of gross contamination, such as stainless steel hypodermic needles and chromium blanks in collection vials.[11]

The safest way to ensure that one has contamination during analysis under control, and that one can make accurate and precise measurements, is to use a range of standard reference materials. Veillon[11] has listed the biological trace element standard reference materials available at present.

A. Theoretical Considerations

Ahrland et al.[12] proposed that all metal ions could be separated into three categories, class A, class B, and borderline, based on equilibrium constants for their metal-ligand complexes. Pearson[13] refers to class A metal ions as "hard" acids, and Class B metal ions as "soft" acids. Class A metals are those which, based on the magnitude of the equilibrium constant, have the following preference sequence for ligands:[14]

$$F^- > Cl^- > Br^- > I^-$$

Table 1
CLASSIFICATION OF METAL IONS INTO REACTIVITY CLASSES

Class A	Borderline	Class B
Cs^+	Pb^{2+}	Au^+
K^+	Sn^{2+}	Ag^+
Na^+	Cu^{2+}	Tl^+
Li^+	Cd^{2+}	Cu^+
Ba^{2+}	Fe^{2+}	Hg^{2+}
Ca^{2+}	Co^{2+}	Pt^{2+}
Sr^{2+}	Ni^{2+}	Bi^{3+}
Mg^{2+}	Zn^{2+}	Tl^{3+}
La^{3+}	Mn^{2+}	Pb(IV)
Lu^{3+}	Sb(III)	$Hg(CH_3)_2^{2}$
Sc^{3+}	Fe(III)	$Pb(CH_3)_3^+$
Be^{2+}	As(III)	
Al^{3+}	Sn(IV)	

Table 2
LIGANDS PREFERRED BY DIFFERENT CLASSES OF METAL IONS[a]

Ligands preferred by class A metal ions	Ligands preferred by class B metal ions
F^-, O^{2-}, OH^-, H_2O, CO_3^{2-}, SO_4^{2-}, $ROSO_3^-$, NO_3^-, HPO_4^{2-}, PO_4^{3-}, ROH, $RCOO^-$, $-\overset{|}{C}=O$, ROR	I^-, R^-, CN^-, S^{2-}, RS^-, R_2S, R_3As, CO, H^-, Cl^-, Br^-, N_3^-, NO_2^-, SO_3^{2-}, NH_3, RNH_2, R_3N

[a] Borderline metal ions can react with all ligands shown, but may exhibit preferences.

and for metal-binding donor atoms in ligands:

$$O > N > S \approx Se$$

Class B metal ions exhibit completely the opposite order of preference for both ligands and donor atoms. The borderline metal ions form an intermediate group, in which there is no distinct preference sequence. A listing of metal ions according to this classification is given in Table 1.[14] Note that the metals are organized on thermodynamic considerations; in some cases kinetic effects may change the order of preference. Table 2 lists class A and B metal ion preferences for ligands commonly found in biological systems. In general, class A metals seek out oxygen donors, and class B, nitrogen and sulfur donor ligands. Borderline metal ions react with all categories of ligands. Table 3 shows functional groups commonly sought by class A and class B metal ions in biological molecules.[14]

Ochiai[15] has divided metal toxicity into three categories: (1) blocking the essential biological functional groups of molecules, (2) displacing the essential metal ion in biomolecules, and (3) modifying the active conformation of biomolecules. Class B metal ions are more toxic than borderline ions, which are more toxic than class A ions, although there are some exceptions according to the test organism. For example, Be^{2+} (class A ion) is extremely toxic in mammals because it can displace Mg^{2+} in magnesium-dependent enzymes, causing

Table 3
BINDING SITES IN PROTEINS ACCORDING TO METAL ION CLASS

Functional groups sought by class A metal ions	Functional groups sought by class B metal ions
Carboxylate	Sulfhydryl
Carbonyl	Disulfide
Alcohol	Thioether
Phosphate	Amino
Phosphodiester	Heterocyclic nitrogen

faults in DNA replication which may lead to cancer.[14,15] Amongst the various macromolecular components of the cell, the electron donor atoms of the nucleic acids provide multiple sites for cation binding. If binding occurs predominantly at the phosphate groups (class A metal ions), neutralization of the negative charges can stabilize the conformaton of DNA. However, with Cd^{2+}, Cu^{2+}, and Pb^{2+} (borderline metal ions), binding occurs predominantly at the heterocyclic bases, which can cause bond breakage and destabilization of the molecule.[14] Class B metal ions, e.g., Hg^{2+} and Cu^{+}, may cross-link adjacent DNA strands, again causing severe destabilization.

B. Computer Modeling Techniques

Models of trace element biochemistry are based upon the following concepts:[16] (1) transition metals exist in biological fluids in a situation close to equilibrium and are mobile between free hydrated ion (extremely low concentration), low-molecular-weight complexes, and labile exchangeable protein, (2) metalloproteins are also present, in which the metal is tightly bound and unavailable for exchange, (3) the biological processes involve competition between a range of ligands and a range of metal ions. The importance of low-molecular-weight complexes arises because it is well established that intestinal absorption, blood plasma reactions, passage through the blood-brain barrier, and renal and biliary excretion, all involve low-molecular-weight transition metal ion complexes, usually ternary amino acid anionic complexes.[16]

Table 4 summarizes some of the chemical modeling results by Williams[16] for copper, lead, cadmium, and zinc in the low-molecular-weight fraction of blood plasma, and Table 5 gives the results for zinc in human milk.[16] Charged complexes are capable of being excreted through the kidneys, whereas neutral complexes can penetrate cell membranes into tissues or be excreted through the bile duct.

Computer modeling can also be used to predict the effect of drug administration. Table 6 shows the calculated effect of some drugs on the distribution of copper in the low-molecular-weight fraction of blood plasma (cf. Table 4),[17] while Figure 6 displays the calculated distribution of salicylic acid species in blood plasma (pH 7.4) and the stomach.[18] At stomach pH of 1.6, most of the salicylic acid (weak acid) is in the unionized form, and so will be rapidly absorbed from the stomach and enter the blood stream.

C. Nuclear Magnetic Resonance

Nuclear magnetic resonance (NMR) analysis relies on an inherent property of certain nuclei, i.e., spin angular momentum. Isotopes of hydrogen (^{1}H), carbon(^{13}C), nitrogen (^{15}N), and phosphorus (^{31}P) all have spin-$^{1}/_{2}$ nuclei and so give rise to well-defined NMR signals. The relative sensitivities of the four biologically important nuclei are shown in Table 7.[19] A great advantage of NMR for biological speciation measurements is that the analysis can be made in living organisms. Cellular pH can be determined with high precision by ^{1}H

Table 4
COMPUTER MODELING RESULTS FOR COPPER, LEAD, CADMIUM, AND ZINC IN LOW-MOLECULAR-WEIGHT FRACTION OF HUMAN BLOOD PLASMA[a]

Species[b]	% of total	Species[b]	% of total
Cu		Pb	
Cu.his.ser.^{0}	8	Pb.cis.H^+	5
Cu.his.thr^0	13	Pb.cys.cta^{3-}	7
Cu.his.thr.OH^-	13	Pb.cys^0	80
Cu.his$_2$^{0}	15		
Zn		Cd	
Zn.his$^+$	4	Cd.cys.H^+	8
Zn.cys.his$^-$	17	Cd.cys.OH^-	14
Zn.cys^0	38	Cd.cis^0	27
		Cd.cys^0	47

[a] Nonprotein-bound metal.
[b] his = histidine; ser = serine; thr = threonine; cys = cysteine; cis = cystine; cta = citrate.

Table 5
COMPUTER MODELING RESULTS FOR SPECIATION OF NONPROTEIN ZINC IN HUMAN MILK

Zinc species[a]	% of total zinc[b]
Zn.gly$^+$	6
Zn.cta.OH^{2-}	13
Zn.cta$^-$	46

[a] gly = glycine; cta = citrate
[b] Assuming 3.7×10^{-6} *M* picolinic acid. If picolinic acid = 3.1×10^{-4} *M*, Zn.pic$_2^0$ = 50% of total nonprotein zinc.

Table 6
CALCULATED EFFECT OF SOME DRUGS (1×10^{-6} *M*) ON COPPER SPECIATION IN HUMAN BLOOD PLASMA

Copper species[a]	% of total copper in low-molecular-weight fraction
Cu.tetraethylenepentamine^{3-}	99
Cu.EDTA^{2-} (1×10^{-3} *M* EDTA)	84
Cu.triethylenetetramine^{2+}	99
Cu.penicillamine.histidine[b]	54
Cu. *N,N'*-bis(2-aminoethyl)-1,3-propanediamine	100

[a] Speciation in low-molecular-weight fraction (nonceruloplasmin, non-albumin).
[b] Penicillamine (1×10^{-3} *M*) in oxidized form.

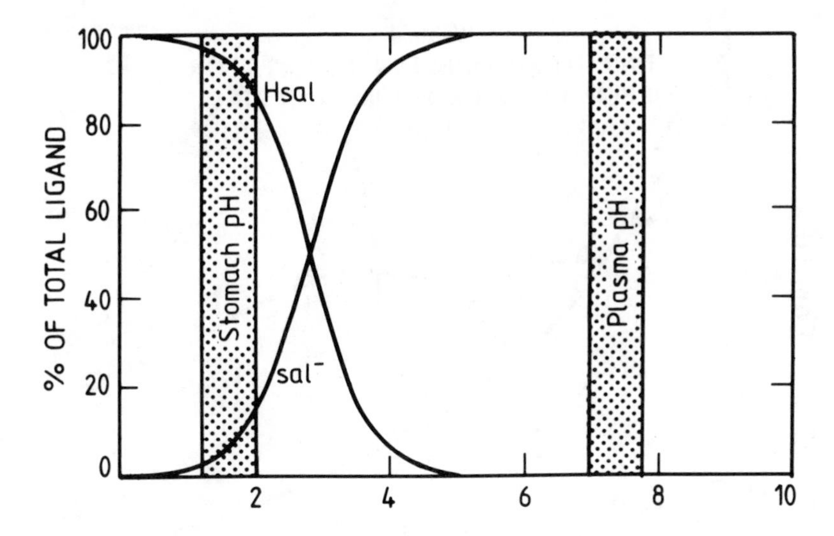

FIGURE 6. The effect of pH on the speciation of salicylic acid in biological systems. (Adapted from Fiabane, A. M. and Williams, D. R., *The Principles of Bio-Inorganic Chemistry,* The Chemical Society, London, 1977, 68. With permission.)

Table 7
NMR SENSITIVITIES OF THE FOUR BIOLOGICALLY IMPORTANT NUCLEI

Nucleus	Natural abundance, %	Relative sensitivity of the nuclei
Hydrogen(^{1}H)	99.98	1.00
Carbon(^{13}C)	1.11	1.59×10^{-2}
Nitrogen (^{15}N)	0.37	1.04×10^{-4}
Phosphorus (^{31}P)	100	6.63×10^{-2}

NMR since, where a molecule has a group with a suitable pKa value, the nuclei associated with the relevant ionizable group exhibit pH-dependent chemical shifts within the pH range of the pKa value. The internal pH of the human erythrocyte was found[19] to be 6.88 ± 0.05 by this method.

Binding of metals to human cellular glutathione (GSH) is very simply studied by ^{1}H spin echo NMR.[19] GSH is present in human erythrocytes at a concentration of 2×10^{-3} *M*, and plays an important role in metal detoxification. It is a major binding site for many soft (class B) ions or species such as Zn^{2+}, methylmercury, and triethyllead. Binding of species by glutathione can cause two different effects in NMR.[19] If binding is solely through the thiol residue (e.g., methylmercury), changes occur in the cysteinyl residues of GSH. However, when the metal binds both to the thiol residue and to hemoglobin protein (e.g., Zn^{2+}), a loss of the GSH signal occurs. This is because GSH then becomes part of the macrostructure of the red cell, and thus changes its environment.

In a recent study of the effect of arsenic on whole cells, Reglinski and Smith[19] observed a considerable change in the intracellular thiol status. There was a large decrease in GSH, and some loss of ergothioneine (Figure 7). Many similar studies of biopsy tissue samples, biological fluids, and even whole animals can be made nondestructively by NMR to elucidate the role of heavy metals in health and disease. Table 8 shows the NMR classification of some spin-$^1/_2$ nuclei and some quadrupolar metal nuclei.[20] Unfortunately, all quadrupolar nuclei in asymmetrical environments suffer line-broadening problems, which may make detection impossible.

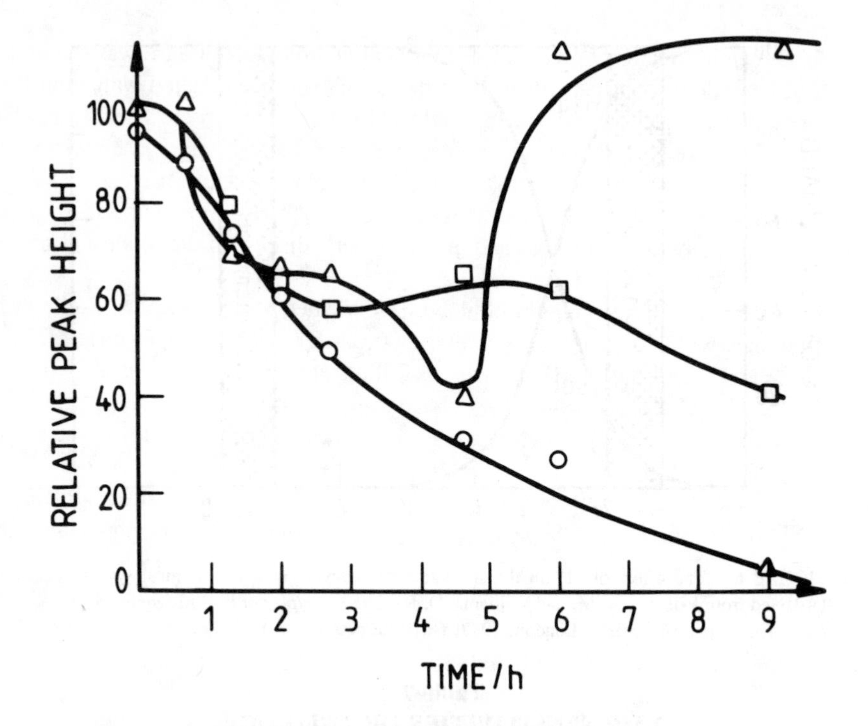

FIGURE 7. Changes in the ^{1}H spin echo NMR spectrum of human erythrocytes incubated with dimethylarsonic acid. ○ dimethylarsenic acid, □ glutathione, △ ergothioneine. (Adapted from Reglinski, J. and Smith, W. E., *Trends Anal. Chem.*, 5, 190, 1986. With permission.)

Table 8
NMR CLASSIFICATION OF SOME SPIN-$^{1}/_{2}$ NUCLEI

Magnetic strength	Natural abundance		
	High (>90%)	Medium	Low(<10%)
Strong	^{1}H, ^{19}F, ^{31}P	^{205}Tl	^{3}H, ^{3}He
Medium	—	^{113}Cd, ^{195}Pt, ^{199}Hg, ^{207}Pb	^{13}C, ^{15}N, ^{77}Se, ^{119}Sn, ^{125}Te
Weak	^{89}Y, ^{103}Rh	^{109}Ag, ^{183}W	^{57}Fe, ^{187}Os

The oxidation state of platinum in biological samples can be readily determined by ^{195}Pt NMR spectroscopy, since signals for Pt(II) occur at lower frequencies (higher field) than those of Pt(IV).[20] The different halide complexes of platinum are also distinguishable. Although similar discrimination cannot be achieved for cadmium because it is a labile metal ion, Cd-metallothionein can be distinguished from simple cadmium complexes.[20]

D. Dialysis and Ultrafiltration

Both dialysis and ultrafiltration use membranes with very small pore size to separate molecules on the basis of their molecular size, which is related to molecular weight. The membrane materials, however, may be quite different, and dialysis membranes usually have a smaller pore size (2 to 5 nm) compared to ultrafiltration membranes (5 to 50 nm). Ultrafiltration is often performed under inert gas pressure, while dialysis can be performed in a variety of configurations, ranging from bags to tubes and discs.

There has been considerable interest in the ultrafiltration speciation of aluminum in blood plasma and dialysis fluid, since dissolved aluminum was shown to be involved in the etiology of dialysis dementia (Section IV.G). This often fatal neurological disease was shown to be the result of hemodialysis of patients using water which had been treated with alum floc to remove organic matter. Low-molecular-weight (M_r) aluminum species could apparently diffuse through the dialysis tubing, cross the blood-brain barrier, and cause severe neurological damage (dialysis encephalopathy). Aluminum in blood has also been associated with dialysis osteodystrophy, a disease involving reduction in bone formation rate in patients receiving chronic parenteral nutrition, and in Alzheimer's disease of the senile and presenile type.[21]

Studies of ultrafilterable aluminum in blood have yielded variable results. Zudin et al.[22] using membranes with a M_r cut-off of 6000 to 8000 Dal, found that an average of 40% of plasma aluminum was ultrafilterable, whereas Elliott et al.[23] and Graf et al.[24] found an ultrafilterable fraction of 10 to 25%. This variation is not surprising, since clogging of the membrane pores by proteins or polymerized aluminum can significantly affect the nominal M_r cut-off value.[25]

The nature of the aluminum complex that crosses the blood-brain barrier remains unidentified, although it is likely to be an organo-aluminum complex, with compounds similar to maltol possibly being the ligand.[26]

Ultrafiltration under pressure was used to study nickel binding in serum from humans, rats, lobsters, dogs, and rabbits.[27] Equilibrium dialysis was employed by Glennan and Sarkar[28] to investigate the Ni(II) binding site of human albumin. The use of ^{63}Ni allowed the identification of a specific Ni-binding site on the protein.[28]

A problem with ultrafiltration and dialysis, common to most other separation techniques, is the difficulty of adjusting the experimental conditions to accurately represent the *in vivo* situation. The results should therefore be interpreted with caution.

E. Electrophoresis

Electrophoresis has been applied to the study of protein-binding characteristics of a variety of metals. In general, these methods involve electrophoretic separation of proteins on a gel, followed by determination of metal in the localized areas. For example, Conradi et al.[29] studied the serum protein binding characteristics of lead by incubation of the serum with ^{210}Pb, separation of the proteins by isoelectric focusing, and assay of small sections of the gel for the presence of lead. This technique showed that lead associated primarily with the orosomucoid (a glycoprotein) fraction, and weakly with α-antichymotrypsin and IgG. Bruns et al.[30] used gel electrophoresis and ^{45}Ca to show that a calcium-binding protein exists in rat intestine. An elegant technique for the study of metal-protein binding was described by Bradwell and Burnett.[31] The technique, two-dimensional immunoelectrophoresis, involves electrophoresis in one direction on an agrose gel, and immunoelectrophoresis in a perpendicular direction. This method was combined with autoradiography using ^{59}Fe, ^{65}Zn, ^{109}Cd, ^{63}Ni, and ^{45}Ca to study the protein-binding characteristics of these metals.[31]

F. Gel Filtration and Ion Chromatography

Besides the free amino acids, some of the proteins present in blood are responsible for the rapid distribution of trace elements through the body after uptake through the gastrointestinal tract and are at least partly responsible for the deposition of trace elements in specific organs.[32]

Trace elements in plasma are present in three main forms — bound to albumin, to low-molecular-weight substances (e.g., amino acids, metabolites), and incorporated in metalloproteins. Albumin is the usual carrier of the metabolically active trace elements which are loosely bound and are therefore easily circulated to binding sites in the various tissues. Albumin-bound trace elements exchange fairly rapidly with radiotracer metal and are easily removed by chelating agents such as EDTA.[32]

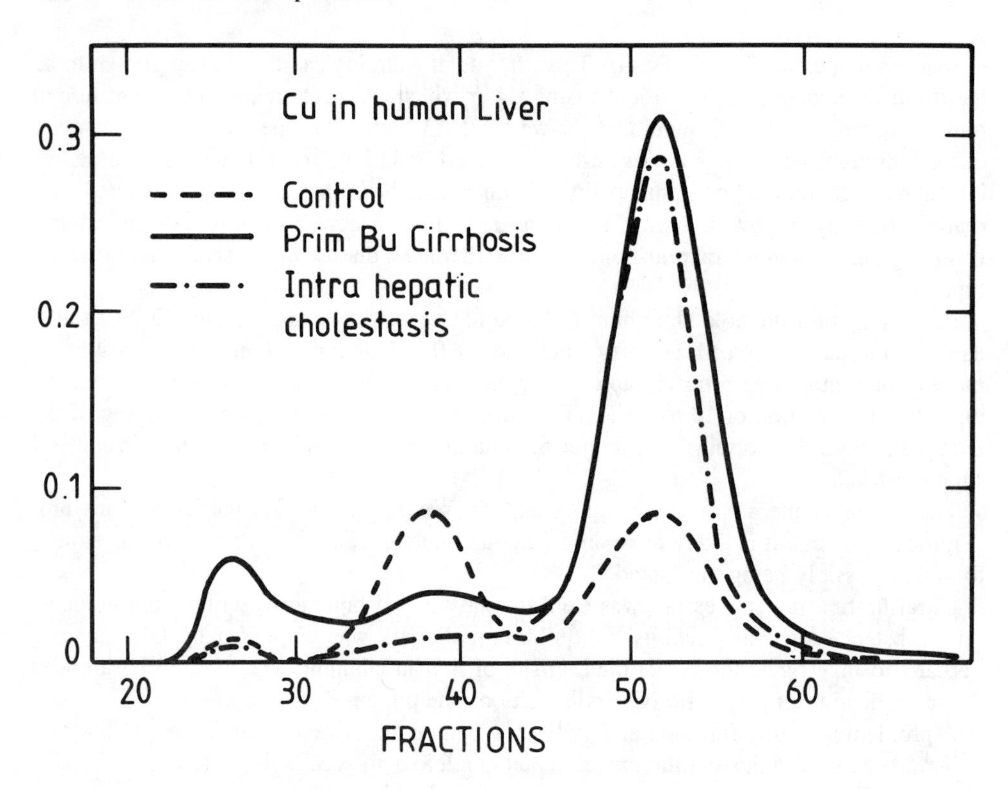

FIGURE 8. Distribution of copper over cytoplasmic liver proteins separated by gel chromatography. (Adapted from Van den Hamer, C. J. and Houtman, J. P., in *Trace Element Analytical Chemistry in Medicine and Biology*, Bratter, P. and Schramel, P., Eds., Walter de Gruyter, Berlin, 1980, 233. With permission.)

One of the most versatile and useful techniques for studying metal binding in biological samples is gel filtration chromatography, a technique which allows molecules to be separated on the basis of the size exclusion characteristics of the gel. Wilson's disease was probably the first human disorder to which gel chromatography was applied.[32] In this disease, liver copper is unusually high, and blood copper very low. Sternlieb et al.,[33] using ^{64}Cu tracer, showed that copper accumulates in the liver of affected people and is associated with a protein, probably a metallothionein, with a M_r of 10,000, whereas in control subjects copper appears in the chromatographic fraction corresponding to a M_r of 30,000. Similar liver copper distributions were found for some other diseases, such as primary biliariry cirrhosis and intrahepatic cholestasis, where a high percentage of cytoplasmic copper was also found in the 10,000-Dal fraction (Figure 8).[33]

G. Other Techniques

In all these studies of liver and other tissue samples, trace element distribution is normally carried out on the soluble, cytoplasmic components. The first step, therefore, is homogenization, followed by centrifugation, then chromatographic separation of the soluble proteins. In a few special cases, however, trace element speciation in solid tissue has been attempted. Sarx and Baechmann[34] used selective reduction with Ar/H_2 to sequentially volatilize arsenic species from biological matrices, with detection by atomic absorption spectrometry or microwave-induced plasma optical emission spectroscopy. Selenium speciation in tissue samples has been determined by a similar technique.[35] The nuclear technique of particle-induced X-ray emission (PIXE) has been used to map the spatial distribution of trace elements in different cells and in subcellular particles.[32]

Table 9
PROPERTIES OF METALLOTHIONEINS

1. Certain metals induce the synthesis of thionein
2. Low molecular weight (6,000 to 10,000)
3. High cysteine content (30%)
4. Absence of disulfide bonds and aromatic amino acids
5. Heat stability
6. Located principally in the cytoplasm
7. Absent in biological fluids

III. METALLOTHIONEINS

Metallothioneins (MTs) are ubiquitous low-molecular-weight proteins that are characterized by an unusually high cysteine content (30 mol %) and a selective capacity to bind heavy metal ions such as zinc, cadmium, and copper (Table 9).[36] There are two major groups of metallothioneins, which are referred to as MT-I and MT-II, based on electrophoretic properties.[37]

Injection of animals with cadmium or zinc salts causes MTs to accumulate in the liver and kidney. Durnam and Palmiter[38] have shown that this is due to transcriptional activation of MT genes by heavy metal ions that bind to the proteins. These and earlier findings led to the hypothesis that MTs function as a protective system against heavy metal toxicity.[37] Indeed, MTs do seem to play such a role in cultured mammalian cells, but it seems unlikely that protection against heavy metals would be their sole function. Firstly, these ions are not present at high concentrations in most biotopes. A single human liver may contain as much as 200 mg of MT, but most other animals have much lower concentrations. Secondly, if the role of MTs was purely protective, these proteins should be found only after exposure to toxic heavy metals; in fact, the basal level of MT is quite high.[37]

The MTs that are naturally present in an animal's liver and kidney serve as the major storage form for the essential trace elements, copper and zinc. The level of these elements is highly influenced by the dietary status of the animal; once the level of the metal reaches a certain threshold, the transcription rate of MT genes increases and more protein is synthesized to bind the excess metal. When copper or zinc is in short supply, MTs are rapidly degraded.[37]

Several metals have been shown to induce an MT response in animals (Table 10).[36,39-42] Recent cytological observations have raised the possibility that MTs are involved in the control of cellular growth.[43] High concentrations of MT are present in tissue undergoing rapid growth and development to supply zinc and possibly copper for nucleic acid metabolism, protein synthesis, and other metabolic processes. In fetal rat liver just before birth, the concentrations of zinc and copper rise dramatically. More than 50% of the zinc and 25% of the copper is associated with MT.[40] Increase in stress leads to a rapid rise in MT concentration, as does UV irradiation.[37,40] Metallothioneins are also extraordinarily efficient scavengers of the hydroxyl radical (but not superoxide radical).[37] Hydroxyl radical damage to MT occurs at the metal-thiolate clusters, which is readily repaired by GSH.[37] Possible biological roles for MTs are summarized in Table 11.

The structure of MTs has been studied by a variety of biophysical and biochemical techniques including UV, ESR, and NMR spectroscopy, X-ray diffraction, amino acid sequencing, and partial proteolysis.[39] Determination of MTs in tissue supernatants is usually carried out by ion exchange or gel chromatography, with UV detection and metal analysis of the eluted fractions, or voltammetrically using the sulfhydryl groups for quantitation.

Table 10
INDUCTION OF METALLOTHIONEINS IN VARIOUS ORGANS OF EXPERIMENTAL ANIMALS BY METALS

Metal	Organ[a] Liver	Kidney	Spleen
Bismuth	?	√	×
Cadmium	√	√	√
Copper	√	√	×
Gold	?	√	×
Lead	×	×	×
Mercury	×	√	×
Silver	√	√	?
Zinc	√	×	×

[a] √ = synthesis; × = no synthesis; ? = no evidence.

Table 11
POSSIBLE BIOLOGICAL ROLES OF METALLOTHIONEINS

1. Metal storage
2. Detoxification of metals
3. Metal transport
4. Participate in immune response
5. Metabolism of essential metals

Table 12
SOME IMPORTANT COPPER METALLOENZYMES

Enzyme	Source	Function
Cytochrome oxidase	Heart	Electron transport
Ceruloplasmin	Plasma	Iron transport
Uricase	Liver and kidney	Catabolism of uric acid
Dopamine hydroxylase	Adrenal	Hydroxylation of dopamine
Amine oxidase	Plasma	Oxidation of monoamines
Diamine oxidase	Kidney	Oxidation of diamines
Tyrosinase	Skin and liver	Oxidation of tyrosine

IV. SPECIATION OF SOME ELEMENTS IN BIOLOGICAL SAMPLES

A. Copper

Copper has long been known to be an essential component of many chemical compounds found in biological systems. Until recently, the major function of copper was thought to be related to hematopoiesis. Now many copper enzymes are known (Table 12).[46]

More than 60% of erythrocyte copper occurs as a nearly colorless copper protein, erythrocuprein, which contains about 0.34% Cu and has a M_r of 35,000. Erythrocuprein functions as a superoxide dismutase, and is also possibly a scavenger of singlet oxygen.[47]

Copper in plasma occurs in two main forms; 5 to 10% is loosely bound to albumin and histidine, while the remainder (90 to 95%) is strongly bound in the blue copper protein ceruloplasmin, which is a glycoprotein of M_r 132,000, containing six copper atoms per molecule. Ceruloplasmin is a true oxidase (ferroxidase) and is essential to the metabolism of iron, where it oxidizes iron(II) to (III), enabling iron to be taken up by the transport protein, transferrin. Ceruloplasmin is also the main copper transport protein[46,49] and the most important serum antioxidant.[50-53] Its concentration in serum increases in the later stages of pregnancy, in women taking oral contraceptives, and in certain diseases such as leukemia and hepatitis, where it is present as an acute-phase reactant. Copper in sweat increases with increasing serum ceruloplasmin, although copper in sweat is present as low-molecular-weight species.[54]

Table 13
CALCIUM-MODULATED ENZYMES

Phosphorylase b kinase
Phosphodiesterase
Adenylate cyclase
α-Glycerophosphate dehydrogenase
Myofibrillas ATPase
Isocitric dehydrogenase[a]
Pyruvate dehydrogenase phosphate phosphatase[a]
Pyruvate dehydrogenase kinase[a]
α-Ketoglutarate dehydrogenase[a]

[a] Mitochondrial enzyme.

B. Zinc

Zinc distribution in animal tissues is not uniform; high concentrations are found in the prostate, skin and appendages, choroid, liver, pancreas, bone, and blood. Zinc is also found in high concentrations in keratins; in animals covered with hair, fur, or wool, a large proportion of total body zinc is found in these materials, e.g., 38% of body zinc in rats is in the skin and hair, while 20% is in human skin.[46]

Zinc in human blood is distributed 75 to 88% in erythrocytes (mostly as carbonic anhydrase), 12 to 22% in plasma, and 3% in leukocytes. One third of zinc in plasma is loosely bound to serum albumins, the remainder being more firmly attached to α-globulins, with minor fractions complexed by histidine and cysteine.[16,46,55] Plasma zinc levels fall in late pregnancy and when taking oral contraceptives.[54]

Zinc is associated with many enzyme systems, both as metalloenzyme and enzyme activator, as well as filling a structural role. Zinc deficiency leads to impaired DNA synthesis, delayed wound healing, and decrease in collagen synthesis.[46] Zinc deficiency can occur, even though dietary intake is sufficient, if the diet contains too much phytic acid, fiber, or other substances that strongly adsorb zinc and interfere with its adsorption.[56]

C. Calcium

The realization that calcium, in addition to playing an essential role as a structural element in mineralized tissues, is uniquely important as a general biological messenger is rather recent.[57] It now appears evident that Ca^{2+} influences the generation of electrical signals in cells, carries the signals to the target in the cell interior, and, in the transfer of information from the plasma membrane to some intracellular activities, Ca^{2+} is associated with a soluble protein, which binds it with high specificity. This small acidic protein, calmodulin (M_r = 17,000), binds four calcium atoms, is widely distributed in the body, and performs a variety of functions.[58] From the standpoint of regulation of Ca^{2+} fluxes, of primary interest is perhaps the activation by calmodulin of the plasma membrane Ca^{2+}-pumping ATPase, and the Ca^{2+}-ATPase in sarcoplasmic reticulum. Calmodulin also reacts strongly with Pb^{2+}; lead binding of this protein has been suggested as a toxic mechanism for lead.[58]

The calcium-modulated enzymes (Table 13) are activated by very low Ca^{2+} concentrations and so must have a high affinity for the cation.[57] The concentration of ionized calcium inside the cell is at least 1000 times lower than in the extracellular environment. In the cytosol of most cells, Ca^{2+} is between 0.13 and 1.3 μM. Mitochondria from all tissues studied so far can transport Ca^{2+}. These organelles accumulate large amounts of calcium, although specific uptake and export mechanisms are present to avoid calcification of mitochondrial membranes.[57]

Gel filtration has been used to study calcium speciation in human blood serum.[21] Separate peaks for bound calcium were identified as due to protein-bound calcium and the calcium complexes of phosphate, citrate, sulfate, and lactate. More detailed chromatography showed that protein binding of calcium involved a range of proteins, including IgM, α-macroglobulin, albumin, transferrin, and orosomucoid.[21]

D. Iron

A normal adult human contains 4 to 5 g of iron, most of which is present in complex forms bound to protein either as porphyrin or heme compounds, especially hemoglobin and myglobin, or as nonheme protein compounds such as ferritin and transferrin which constitute about 25% of total body iron.[59] Blood hemoglobin represents about 65% of total iron in a human, with 3% associated in myoglobin. The most common and widespread disease of the world, anemia, is due partly to a dietary insufficiency of iron, although most of the iron in the body is continuously recycled. Very little is absorbed from the diet and very little is excreted.[59]

Hemoglobin (M_r = 65,000) is a complex of globin and four ferroprotoporphyrin, or heme, moieties. The iron content is about 0.35%.[60] Ferritin, the main iron storage protein (M_r = 900,000) is brown, water soluble, and contains a maximum of 4500 iron atoms per molecule, which occur in the protein as micelles of a hydrated iron oxide/phosphate complex of approximate composition $(FeOOH)_8(FeO{:}OPO_3H_2)$.[61] The role of ferritin is to sequester iron in a form which is nontoxic, soluble and readily available. Ferritin is stored mainly in the liver, spleen, and bone marrow.

Transferrin (M_r = 76,000) is a glycoprotein with two identical iron-binding sites each capable of binding one atom of iron.[62] Transferrin serves as the principal carrier of iron in the blood, but has another important function, that of participating in the defense mechanisms of the body against infection. In normal individuals, only 30 to 40% of the iron binding capacity of transferrin is used, the remainder being known as the latent iron-binding capacity. Free, ionic, or molecular iron is present in neglible quantities in most samples of blood and other biological fluids. However, it has been suggested that "decompartmentalization" of iron is involved in several diseases.[63] Release of "free" iron or iron present as low-molecular-weight complexes, such as Fe-ATP, could catalyze reactions, such as the Haber-Weiss reaction, that release highly damaging free radicals.[64]

$$H_2O_2 + O_2^{\overline{\cdot}} + H^+ \xrightarrow{\text{Fe-ATP}} O_2 + H_2O + OH\cdot \qquad (1)$$

Iron is probably a universal nutrient for all living cells; even in cells acquiring energy via the non-iron-utilizing processes of glycolysis and fermentation, the element is still required for synthesis of ribotide reductase, a DNA precursor.[65] A problem, however, for microorganisms to utilize iron is that most of the iron of the earth exists in the ferric form, which has an extremely small solubility product ($<10^{-38}$ M). All aerobic and facultative anaerobic organisms solve this problem by synthesising or utilizing a group of low-M_r, ferri-specific ligands, known collectively as siderophores, and having a matching membrane receptor for the iron-laden siderophore complex.[65] The siderophores (M_r 500 to 1000) are hexadentate, oxygen-bonding ligands, virtually specific for iron with which they form complexes with equilibrium constants in the range 10^{30} to 10^{50} M^{-1}. They penetrate to the cytoplasm where the ferric iron is reduced and released, and the siderophore is then exported unchanged from the cell for further use in iron transport.[65] An hydroxamate type of siderophore, desferrioxamine (or "desferal"), is produced by Ciba Pharmaceutical Company for therapeutic use to treat diseases involving iron overload.

E. Manganese

A normal adult contains a total of only 10 to 20 mg of manganese which is distributed widely throughout the tissues and fluids without notable concentration in any organ, although tissues rich in mitochondria tend to be higher. A deficiency of manganese, however, results in a wide variety of structural and physiological defects, including reduced growth rate, skeletal abnormalities, and impaired reproduction.[66]

Manganese absorbed from the gastrointestinal tract enters into the portal blood where most of it becomes bound to α_2 macroglobulin in a 1:1 ratio.[67] Free Mn^{2+} and α_2-macroglobulin-bound Mn^{2+} in the portal blood are rapidly taken up by the liver. A fraction is oxidized to Mn(III), possibly by ferroxidase I or ceruloplasmin, and is then bound to transferrin, in which from it is taken up by extrahepatic tissue. Within cells, manganese is found predominantly in the mitochondria, where its uptake and export may be via a calcium carrier.[67,68]

As with many of the essential transition elements, the interaction between manganese and enzymes results in two distinct types of associations: (1) metalloenzymes and (2) metal-enzyme complexes. The Mn-containing metalloenzymes are very few and include pyruvate carboxylase, avimanganin, and Mn-superoxide dismutase.[67] There are numerous (e.g., the glycosyltransferases) enzymes that are activated by manganese. Pyruvate carboxylase contains four atoms of Mn per molecule, and manganese is present as Mn(II).[68] However, Mn(III) is the dominant valency state of manganese in many biological tissues and in natural waters.[69] Superoxide dismustase, for example, contains manganese as Mn(III).[67]

Most manganese is paramagnetic, and the presence of the unpaired electron endows the metal ion with magnetic properties that can be utilized in NMR and electron paramagnetic resonance (EPR) studies. Free Mn^{2+} shows a sextet hyperfine spectrum in EPR, whereas bound Mn^{2+} does not.[67] Using EPR, Ash and Schram[70] found 35 nmol Mn^{2+} per milliliter of cell water in rat hepatocytes. The ratio of free to bound Mn^{2+} in hepatocytes was dependent on the physiological state; the free Mn^{2+} concentration in whole cells was 0.71 nmol/ml of cell water in fed rats, and 0.25 nmol/ml in fasted rats.[70]

Despite being an essential element, excessive manganese intake can induce severe neurological damage, with symptoms similar to those of Parkinson's disease. The best description of manganism emanates from the studies of manganese miners in South America, particularly in the mining villages of Chile.[71] Acute manganese intoxication is characterized in man by disorientation, memory impairment, acute anxiety, and hallucinations. In the chronic stage of the disease, its manifestations more closely resemble Parkinson's disease or Wilson's disease, with rigidity and hypokinesia.[71,72] The neurological effects of manganese are believed to be due to losses of the neurotransmitter, dopamine, as a result of manganese-catalyzed oxidation of this compound.[73]

$$\text{dopamine} + 2\,Mn^{3+} \longrightarrow \text{dopamine quinone} + 2\,Mn^{2+} + 2\,H^{+} \qquad (2)$$

The manganese (III) is produced by the reactions:

$$Mn^{2+} + O_2 \xrightarrow[\text{dopamine}]{\text{xanthine}} Mn^{3+} + O_2^{\bullet -} \qquad (3)$$

$$Mn^{2+} + O_2^{\bullet -} + 2\,H^{+} \rightarrow Mn^{2+} + O_2^{\bullet -} \qquad (4)$$

$$\text{net:} \quad 2\,Mn^{2+} + O_2 + 2\,H^{+} \rightarrow 2\,Mn^{3+} + H_2O_2 \qquad (5)$$

These reactions proceed only in the presence of certain ligands, such as phosphate, xanthine, uric acid, and epinephrine. Manganese catalysis of dopamine oxidation is effectively quenched *in vitro* by ascorbic acid, dehydroascorbic acid, or thiamine.[73]

F. Selenium

Selenium shares with sulfur an affinity for heavy metals such as cadmium, silver, mercury, and methylmercury.[74] Forms of selenium that occur in living organisms include the low-molecular-weight compounds, selenocysteine, selenohomocystine, selenomethionine, dimethylselenide, selenotaurine, and the enzymes gluthathione peroxidase, formate reductase, and glycine reductase. In some bacteria, a seleno-transfer RNA is present.[75]

The absorption of selenium from the gastrointestinal tract varies with the chemical form of the element and the amount of the element ingested. Cantor et al.[76] studied the availability of selenium in foodstuffs to chickens. In most plant foods, selenium was highly available, ranging from 71% for wheat, 86% for corn, 60% for soybean, and 210% for alfalfa meal, when sodium selenite was used as a 100% control. Selenium had much lower availability from animal foods, e.g., 22% from tuna and 15% from beef. Foods highest in selenium are fish and whole grains.[74]

After absorption, selenium is carried mainly in the plasma, associated with plasma proteins. They deliver selenium to all of the tissues, including the bones, hair, the erythrocytes, and the leucocytes.[77] Albumin appears to be the immediate receptor of selenium, from which it is released to be bound by α_2- and β_1-globulins. Selenium occurs in all the cells and tissues of the body. The kidney, and especially the kidney cortex, has the highest concentration, followed by the glandular tissues, especially the pancreas, the pituitary, and the liver. Cadmium, arsenic, and copper inhibit selenium uptake and, if fed in large amounts, can induce selenium deficiency in animals.[75,77]

Gluthathione peroxidase is the most extensively studied selenium enzyme because of its critical importance in preventing lipid peroxidation in cell membranes. It has a M_r of 85,000, with four identical subunits, each containing one atom of selenium. Gluthathione peroxidase catalyzes the reactions,[78]

$$2\ GSH + H_2O_2 \rightarrow GSSG + 2\ H_2O \tag{6}$$

$$2\ GSH + ROOH \rightarrow GSSG + ROH + H_2O \tag{7}$$

Thus, glutathione peroxidase removes dangerously reactive hydrogen peroxide and lipid peroxides and enables us to survive in a hostile aerobic environment. To be able to intercept H_2O_2 as it is formed during respiration, glutathione peroxidase is concentrated in the mitochondrial matrix space where it not only removes aberrant H_2O_2, but protects the inner mitochondrial membrane, which is high in unsaturated fatty acids and prone to oxidation.[79]

Selenium has been found to lower cancer rates in experimental animals, and is probably the most potent anitcancer agent so far tested. An international study showed an inverse relationship between cancer incidence of the country, and selenium concentration in the diet.[80]

G. Aluminum

In Western countries, the average daily intake of aluminum is 20 mg, although this can vary greatly, depending on diet. Tap water can be high in aluminum if alum was used in the treatment process, processed cheese contains up to 0.7% Al, baking powders, 5%, and aluminum salts are commonly used as buffers in drugs. The recommended dose of a typical antacid preparation contains 2 g of elemental aluminum.[81]

The gastrointestinal tract is an effective barrier to aluminum absorption, which is usually only about 0.5% of intake. Greatly increased absorption, however, is obtained in the presence of some complexing agents, e.g., citrate.[82] In human plasma, aluminum is partitioned between ultrafiltrable nonprotein-bound and protein-bound fractions. Studies using ultrafiltration, gel filtration, and atomic absorption spectrometery showed that at normal aluminum blood concentrations (40 ± 7 μg Al per liter), about half is protein bound to albumin and transferrin and the remainder is nonprotein bound.[21,81,83]

A unique property of the transferrins is the requirement of CO_2 for binding Fe^{3+}, and this property is maintained for the binding of Al^{3+}.[81]

Renal failure patients who received phosphate-binding aluminum gels for prolonged periods and who were dialyzed against alum-treated water, often developed a fatal encephalopathy which was characterized by high brain concentrations of aluminum.[84] At the pH of plasma, a high percentage of soluble aluminum is likely to be present as neutral hydrolyzed species which may easily pass through glomerular or capillary membranes.[85] Patients suffering from senile dementia of the Alzheimer type also often have elevated aluminum levels (specifically, aluminosilicates at the core of the senile plaques) in brain neocortical areas, which has led to the suggestion that the element is involved in the pathogenesis of this disease.[84]

H. Lead

The total body burden of lead in an adult man living in an industrialized city is in the range 100 to 300 mg, with a mean of 120 mg in the U.S. About 90% of this lead is in the skeleton, with most soft tissues containing 0.2 to 1.5 ppm Pb wet weight.[86,87]

Lead is readily absorbed through the gastrointestinal tract to the extent of 5 to 10% in man, the diet contributing an average of 10 μg Pb per day.[87] Lead is also rapidly absorbed through the skin by reaction with lactic acid and amino acids in sweat, with which it forms lipid-soluble complexes.[88] Lead metal powder, or solutions of inorganic lead salts (e.g., $Pb(NO_3)_2$ or $PbCl_2$) placed on the forearm, then covered with plastic wrap, is absorbed via the sweat glands and ditributed throughout the body in 1 to 2 h. Sweat samples taken on the opposite forearm show lead concentrations 5 to 20 times the normal value (10 μg Pb per liter) (Figure 9). Lead battery factory workers had lead sweat concentrations as high as 500 μg Pb per liter.[88]

The major contributor of lead in the environment is the combustion of leaded gasoline in motor vehicles. In 1968, 241,000 tons of lead in fuel additives was used in the U.S. alone.[87] The usage, however, of lead additives has decreased substantially in recent years. Lead in automobile exhausts consists principally of PbBrCl particles (<0.5 μm diameter), 90% of which are respirable. Alkyllead vapor is much more efficiently absorbed by lungs than particulate lead and is highly toxic to the central nervous system.[87]

In blood, 95% of the lead is in red blood cells (RBC) and 5% in the plasma. Part of the RBC lead is associated with the cell membrane, and the remainder with hemoglobin. It is not necessarily the total amount of lead in the body that determines toxicity, but rather the amount of diffusible lead that produces toxic effects. Lead may flow from the renal tubular capillary, through the tubular lining cell of the tubular lumen, or may be readsorbed from the lumen in a nondiffusible form bound to the RBC, and a diffusible form in the plasma, where it is bound to small organic ligands of peptides, probably cysteine and albumin.[16] Only the diffusible, or ligand-bound, lead can be transported across membranes or renal tubular lining cells. In lead poisoning, the increase in urinary lead is largely ligand-bound lead. The urinary excretion of lead after the administration of a chelating agent (EDTA or penicillamine) is considered to reflect the mobilizable pool of lead in blood, soft tissues, and perhaps soft bone.[87]

Lead interacts with other metals, notably calcium. Diets low in calcium result in increased lead tissue concentrations. Iron deficiency appears to enhance the intestinal absorption of

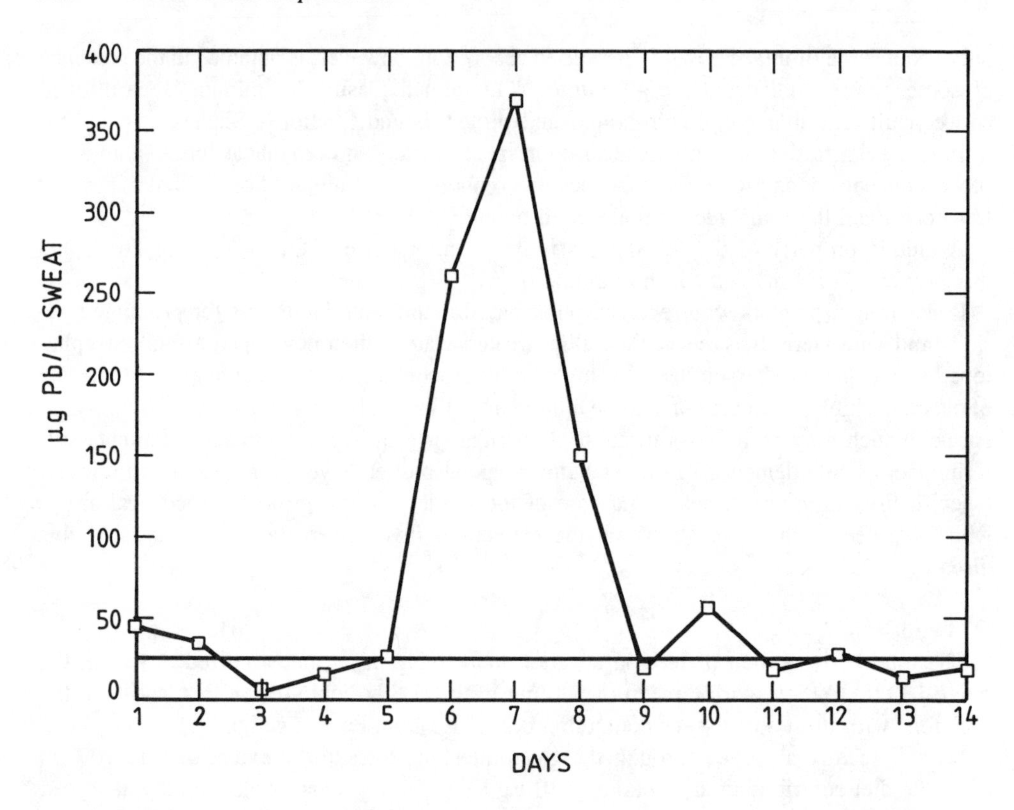

FIGURE 9. Absorption of inorganic lead compounds through the skin. The sweat samples analyzed were taken from the opposite arm to that exposed to the lead compound, in this case, lead nitrate in solution, on day 5. Average lead in sweat level indicated by horizontal line.

lead from 10% to 25%, and in children there is an inverse correlation between blood lead and serum iron levels. Lead interferes with heme synthesis by inhibiting the heme synthesizing enzyme, ferrochelatase. As a consequence, lead-induced anemia results in iron accumulation in the form of ferritin in the mitochondria of reticulocytes.[87,89]

Low dietary zinc also enhances lead absorption, and high lead intake causes a decrease in serum ceruloplasmin concentrations.[87]

REFERENCES

1. **Frieden, E.,** New perspectives on the essential trace elements, *J. Chem. Educ.*, 62, 917, 1985.
2. **Cotzias, G. C.,** Proc. First Annu. Conf. Trace Substances in Environmental Health, Hemphill, D. D., Ed., University of Missouri, Columbia, 1984, 5.
3. **Liebscher, K. and Smith, H.,** Essential and non-essential trace elements, *Arch. Environ. Health,* 17, 881, 1968.
4. **Frieden, E.,** A survey of the essential biochemical elements, in *Biochemistry of the Essential Ultratrace Elements,* Frieden, E., Ed., Plenum Press, New York, 1984, 1.
5. **Feinendegen, L. E. and Kasperek, K.,** Medical aspects of trace element research, in *Trace Element Analytical Chemistry in Medicine and Biology,* Bratter, P. and Schramel, P., Eds., Walter de Gruyter, Berlin, 1980, 1.
6. **Fiabane, A. M. and Williams, D. R.,** *The Principles of Bio-Inorganic Chemistry,* The Chemical Society, London, 1977, 68.

7. **Bernhard, M., Brinckman, F. E., and Irgolic, K. J.,** Why speciation?, in *The Importance of Chemical Speciation in Environmental Processes,* Bernhard, M., Brinckman, F. E., and Sadler, F., Eds., Springer-Verlag, Berlin, 1986, 7.
8. **Underwood, E. J.,** *Trace Elements in Human and Animal Nutrition,* Academic Press, New York, 1977, 146.
9. **Florence, T. M.,** Trace element speciation and aquatic toxicology, *Trends Anal. Chem.,* 2, 162, 1983.
10. **Florence, T. M.,** The speciation of trace elements in waters, *Talanta,* 29, 345, 1982.
11. **Veillon, C.,** Trace element analysis of biological samples, *Anal. Chem.,* 58, 851A, 1986.
12. **Ahrland, S., Chatt, J., and Davies, N. R.,** The relative affinities of ligand atoms for acceptor molecules and ions, *Q. Rev. Chem. Soc.,* 12, 265, 1958.
13. **Pearson, R. G.,** Hard and soft acids and bases, *J. Am. Chem. Soc.,* 85, 3533, 1963.
14. **Nieboer, E. and Richardson, D. H.,** The replacement of the non-descript term "heavy metals" by a biologically and chemically significant classification of metal ions, *Environ. Pollut. Ser. B,* 1, 3, 1980.
15. **Ochiai, E.,** *Bioinorganic Chemistry: An Introduction,* Allyn and Bacon, Boston, 1977, 468.
16. **Williams, D. R.,** Computer models of metal biochemistry and metabolism, in *Chemical Toxicology and Clinical Chemistry of Metals,* Brown, S. S. and Savory, J., Eds., Academic Press, New York, 1983, 167.
17. **May, P. M. and Williams, D. R.,** Computer simulation of chelation therapy, *FEBS Lett.,* 78, 134, 1977.
18. **Fiabane, A. M. and Williams, D. R.,** *The Principles of Bio-Inorganic Chemistry,* The Chemical Society, London, 1977, 88.
19. **Reglinski, J. and Smith, W. E.,** Nuclear magnetic resonance in living systems, *Trends Anal. Chem.,* 5, 190, 1986.
20. **Sadler, P. J.,** Multinuclear NMR methods for the in situ characterization of chemical species, in *The Importance of Chemical Speciation in Environmental Processes,* Bernhard, M., Brinckman, F. E., and Sadler, P. J., Eds., Springer-Verlag, Berlin, 1986, 563.
21. **Savory, J., Bertholf, R. L., and Wills, M. R.,** Speciation and fractionation of toxic metals in biological materials, in *Chemical Toxicology and Clinical Chemistry of Metals,* Brown, S. S. and Savory, J., Eds., Academic Press, New York, 1983, 183.
22. **Ludin, A. P., Curuso, C., Sass, M., and Berlyne, G. M.,** Ultrafiltrable aluminum in serum of normal man, *Clin. Res.,* 26, 636A, 1978.
23. **Elliott, H. L., Dryburgh, F., Fell, G. S., Sabet, S., and MacDougall, A. I.,** Aluminum toxicity during regular hemodialysis, *Br. Med. J.,* 1, 1101, 1978.
24. **Graf, H., Stummvoll, H. K., and Meisinger, V.,** Dialysate aluminium concentration and aluminium transfer during haemodialysis, *Lancet,* i, 46, 1982.
25. **Hydes, D. J. and Liss, P. S.,** The behaviour of dissolved aluminium in coastal and estuarine waters, *Estuarine Coastal Mar. Sci.,* 5, 755, 1977.
26. **Orvig, C. and Nelson, W. O.,** The chemistry of neurologically-active, neutral, and water-soluble aluminum complexes, Proc. 194th American Chemical Society Natl. Meet., Environmental Chemistry Div., New Orleans, August 1987, 462.
27. **Hendel, R. C. and Sunderman, F. W.,** Species variations in the proportions of ultrafiltrable and protein-bound serum nickel, *Res. Commun. Chem. Pathol. Pharmacol.,* 4, 141, 1972.
28. **Glennon, J. D. and Sarkar, B.,** Nickel(II) transport in human blood serum, *Biochem. J.,* 203, 15, 1982.
29. **Conradi, S., Ronnevi, L., and Stibler, H.,** Serum protein binding of lead in vitro in amyotrophic lateral sclerosis patients and controls, *J. Neurol. Sci.,* 37, 95, 1978.
30. **Bruns, M. E., Fleischer, E. B., and Avioli, L. V.,** Control of vitamin D-dependent calcium-binding protein in rat intestine by growth and fasting, *J. Biol. Chem.,* 252, 4145, 1977.
31. **Bradwell, A. R. and Burnett, D.,** Improved methodology and precision using a straight base line technique for the quantitation of proteins by two dimensional immunoelectrophoresis, *Clin. Chim. Acta,* 58, 283, 1975.
32. **Van den Hamer, C. J. and Houtman, J. P.,** Special forms of bound trace elements; their analysis and interest in medicine, in *Trace Element Analytical Chemistry in Medicine and Biology,* Bratter, P. and Schramel, P., Eds., Walter de Gruyter, Berlin, 1980, 233.
33. **Sternlieb, I., Van den Hamer, C. J., Morell, A. G., Alpert, S., Gregoriadis, G., and Scheinberg, I.,** Lysosomal defect of hepatic copper excretion in Wilson's disease, *Gastroenterology,* 64, 99, 1973.
34. **Sarx, B. and Baechmann, K.,** Speciation of arsenic compounds by volatilization from solid compounds, *Z. Anal. Chem.,* 316, 621, 1983.
35. **Hanamura, S., Smith, B. W., and Winefordner, J. D.,** Speciation of inorganic and organometallic compounds in solid biological samples by thermal vaporization and plasma emission spectrometry, *Anal. Chem.,* 55, 2026, 1983.
36. **Cherian, M. G. and Goyer, R. A.,** Metallothioneins and their role in the metabolism and toxicity of metals, *Life Sci.,* 23, 1, 1978.
37. **Karin, M.,** Metallothioneins: proteins in search of function, *Cell,* 41, 9, 1985.

38. **Durnam, D. M. and Palmiter, R. D.,** Transcriptional regulation of the mouse metallothionein-I gene by heavy metals, *J. Biol. Chem.*, 256, 5712, 1981.
39. **Hamer, D. H.,** Metallothionein, *Annu. Rev. Biochem.*, 55, 913, 1986.
40. **Brady, F. O.,** The physiological function of metallothionein, *TIBS*, 7, 143, 1982.
41. **Kojima, Y. and Kagi, J.H.,** Metallothionein, *TIBS*, 3, 90, 1978.
42. **Kagi, J. H., Kojima, Y., Berger, C., Kissling, M. M., Lerch, K., and Vasak, M.,** Metallothionein: structure and evolution, in *Metalloproteins*, Weser, U., Ed., Georg Thieme Verlag, New York, 1979, 194.
43. **Le Beau, M. M., Diaz, M.O., Karin, M., and Rowley, J. D.,** Metallothionein gene cluster is split by chromosome 16 rearrangements in myelomonocytic leukemia, *Nature (London)*, 313, 709, 1985.
44. **Roesijadi, G. and Hall, R. E.,** Characterization of mercury-binding proteins from the gills of marine mussels exposed to mercury, *Comp. Biochem. Physiol.*, 70C, 59, 1981.
45. **Olafson, R. W. and Sim, R. G.,** An electrochemical approach to quantitation and characterization of metallothioneins, *Anal. Biochem.*, 100, 343, 1979.
46. **Fisher, G. L.,** Function and homeostasis of copper and zinc in mammals, *Sci. Total Environ.*, 4, 373, 1975.
47. **Underwood, E. J.,** *Trace Elements in Human and Animal Nutrition*, Academic Press, New York, 1977, 62.
48. **Williams, D. R., Furnival, C., and May, P. M.,** Computer analysis of low molecular weight copper complexes in biofluids, in *Inflammatory Diseases and Copper*, Sorenson, J. R., Ed., Humana Press, Clifton, N.J. 1982, 45.
49. **Bremner, I.,** Absorption, transport and distribution of copper, in *Biological Roles of Copper*, Ciba Foundation Symp. 79 (New Series), Excerpta Medica, Amsterdam, 1980, 23.
50. **Cass, A. E. and Hill, H. A.,** Copper proteins and copper enzymes, in *Biological Roles of Copper*, Ciba Foundation Symp. 79 (New Series), Exerpta Medica, Amsterdam, 1980, 71.
51. **Evans, G. W.,** Copper homeostasis in the mammalian system, *Physiol. Rev.*, 53, 535, 1973.
52. **Lee, G. R., Williams, D. M., and Cartwright, G. E.,** Role of copper in iron metabolism and heme biosynthesis, in *Trace Elements in Human Health and Disease*, Vol. 1, Prasad, A. S., Ed., Academic Press, New York, 1976, 373.
53. **Frieden, E.,** Ceruloplasmin: a multi-functional cupro-protein of vertebrate plasma, in *Inflammatory Diseases and Copper*, Sorenson, J. R., Ed., Humana Press, Clifton, N.J., 1982, 159.
54. **Stauber, J. L. and Florence, T. M.,** The determination of trace metals in sweat by anodic stripping voltammetry, *Sci. Total Environ.*, 60, 263, 1987.
55. **Giroux, E. L., Durieux, M., and Schechter, P. J.,** A study of zinc distribution in human serum, *Bioinorg. Chem.*, 5, 211, 1976.
56. **Reinhold, J. R., Faradji, B., Abadi, P., and Ismail-Beigi, F.,** Binding of zinc to fibre and other solids of wholemeal bread, in *Trace Elements in Human Health and Disease*, Vol. 1, Prasad, A. S., Ed., Academic Press, New York, 1976, 163.
57. **Carafoli, E.,** The physiological role and the biological fitness of calcium, in *Disorders of Mineral Metabolism*, Vol. 2, Bronner, F. and Coburn, J.W., Eds., Academic Press, New York, 1982, 2.
58. **Habermann, E. and Richardt, G.,** Intracellular calcium binding proteins as targets for heavy metal ions, *TIBS*, 11, 298, 1986.
59. **Underwood, E. J.,** *Trace Elements in Human and Animal Nutrition*, Academic Press, New York, 1977, 13.
60. **Rifkind, J.M.,** Hemoglobin and myoglobin, in *Inorganic Biochemistry*, Vol. 2, Eichhorn, G. L., Ed., Elsevier, Amsterdam, 1973, 832.
61. **Harrison, P. M. and Hoy, T. G.,** Ferritin, in *Inorganic Biochemistry*, Vol. 1, Eichhorn, G. L., Ed., Elsevier, Amsterdam, 1973, 253.
62. **Aisen, P.,** The transferrins, in *Inorganic Biochemistry*, Vol. 1, Eichhorn, G. L., Ed., Elsevier, Amsterdam, 1973, 280.
63. **Willson, R. L.,** Free radicals and electron transfer in biology and medicine, *Chem. Ind.*, March, 183, 1977.
64. **Florence, T. M.,** The production of hydroxyl radical from hydrogen peroxide, *J. Inorg. Biochem.*, 22, 221, 1984.
65. **Neilands, J. B.,** Biomedical and environmental significance of siderophores, in *Trace Metals in Health and Disease*, Kharasch, N., Ed., Raven Press, New York, 1979, 27.
66. **Underwood, E. J.,** *Trace Elements in Human and Animal Nutrition*, Academic Press, New York, 1977, 170.
67. **Keen, C. L., Lonnerdal, B., and Hurley, L. S.,** Manganese, in *Biochemistry of the Essential Ultratrace Elements*, Frieden, E., Ed., Plenum Press, New York, 1984, 89.
68. **Leach, R. M. and Lilburn, M. S.,** Manganese metabolism and its function, *World Rev. Nutr. Diet*, 32, 123, 1978.

69. **Stauber, J. L. and Florence, T. M.,** Interactions of copper and manganese, *Aquat. Toxicol.,* 7, 241, 1985.
70. **Ash, D. E. and Schramm, V. L.,** Determination of free and bound manganese (II) in hepatocytes from fed and fasted rats, *J. Biol. Chem.,* 257, 9261, 1982.
71. **Donaldson, J. and Barbeau, A.,** Manganese neurotoxicity: possible clues to the etiology of human brain disorders, in *Metal Ions in Neurology and Psychiatry,* Gabay, S., Harris, J., and Ho, B. T., Alan R. Liss, New York, 1985, 259.
72. **Leach, R. M.,** Metabolism and function of manganese, in *Trace Elements in Human Health and Disease,* Vol. 2, Prasad, A. S. and Oberleas, D., Eds., Academic Press, New York, 1976, 235.
73. **Florence, T. M., Stauber, J. L., and Fardy, J. J.,** Ecological studies of manganese on Groote Eylandt, Proc. Conf. on Manganese and Metabolism, Darwin, Australia, June 1987, 23.
74. **Young, V. R., Nahapetian, A., and Janghorbani, M.,** Selenium bioavailability with reference to human nutrition, *Am. J. Clin. Nutr.,* 35, 1076, 1982.
75. **Shamberger, R. J.,** Selenium, in *Biochemistry of the Essential Ultratrace Elements,* Frieden, E., Ed., Plenum Press, New York, 1984, 201.
76. **Cantor, A. H., Scott, M. L., and Noguchi, T.,** Biological availability of selenium in feedstuffs and selenium compounds for restoring blood plasma glutathione peroxidase activity in selenium-depleted chicks, *J. Nutr.,* 105, 96, 1975.
77. **Underwood, E. J.,** *Trace Elements in Human and Animal Nutrition,* Academic Press, New York, 1977, 302.
78. **Burk, R. F.,** Selenium and glutathione peroxidase, in *Trace Elements in Human Health and Disease,* Vol. 2, Prasad, A. S. and Oberlease, D., Ed., Academic Press, New York, 1976, 165.
79. **Shamberger, R. J.,** *Biochemistry of Selenium,* Plenum Press, New York, 1983, 17.
80. **Schrauzer, G. N.,** Selenium and cancer: historical developments and perspectives, in *Selenium in Biology and Medicine,* Spallholz, J. E., Martin, L. L., and Ganther, H. E. Eds., Avi Publishing, Westport, CT, 1981, 98.
81. **McLachlan, D. R. and Farnell, B. J.,** Aluminum and neuronal degeneration, in *Metal Ions in Neurology and Psychiatry,* Gabay, S., Harris, J., and Ho, B. T., Alan R. Liss, New York, 1985, 69.
82. **Alfrey, A. C.,** Aluminum metabolism and toxicity, Proc. 194th American Chemical Society Natl. Meet., Environmental Chemistry Div., New Orleans, August 1987, 458.
83. **Underwood, E. J.,** *Trace Elements in Human and Animal Nutrition,* Academic Press, New York, 1977, 430.
84. **Ferry, G.,** Alumino silicates at the centre of dementia, *New Sci.,* February, 1986, 23.
85. **Helliwell, S., Batley, G. E., Florence, T. M., and Lumsden, B. G.,** Speciation and toxicity of aluminium in a model fresh water, *Environ. Technol. Lett.,* 4, 141, 1983.
86. **Underwood, E. J.,** *Trace Elements in Human and Animal Nutrition,* Academic Press, New York, 1977, 410.
87. **Goyer, R. A.,** Lead, in *Disorders of Mineral Metabolism,* Bronner, F. and Coburn, J. W., Academic Press, New York, 1981, 159.
88. **Stauber, J. L. and Florence, T. M.,** The determination of trace metals in sweat by anodic stripping voltammetry, Proc. 9th Australian Symp. on Analytical Chemistry, Royal Australian Chemical Institute, Sydney, 1987, 569.
89. **Webb, M.,** Metabolic targets of metal toxicity, in *Clinical Chemistry and Chemical Toxicology of Metals,* Brown, S. S., Ed., Elsevier, New York, 1977, 51.

INDEX

A

B

C

F

G

H

I

J

K

L

M

N

O

P

Q

R

S

T